TRUTH, POSSIBILITY AND PROBABILITY

New Logical Foundations of Probability
and Statistical Inference

NORTH-HOLLAND MATHEMATICS STUDIES 166
(Continuation of the Notas de Matemática)

Editor: Leopoldo NACHBIN

Centro Brasileiro de Pesquisas Físicas
Rio de Janeiro, Brazil
and
University of Rochester
New York, U.S.A.

NORTH-HOLLAND – AMSTERDAM • NEW YORK • OXFORD • TOKYO

TRUTH, POSSIBILITY AND PROBABILITY

New Logical Foundations of Probability and Statistical Inference

Rolando CHUAQUI

Facultad de Matemátics
Pontificia Universidad Católica de Chile
Casilla 6177, Santiago 22
Chile

1991

NORTH-HOLLAND – AMSTERDAM • NEW YORK • OXFORD • TOKYO

ELSEVIER SCIENCE PUBLISHERS B.V.
Sara Burgerhartstraat 25
P.O. Box 211, 1000 AE Amsterdam, The Netherlands

Distributors for the U.S.A. and Canada:

ELSEVIER SCIENCE PUBLISHING COMPANY, INC.
655 Avenue of the Americas
New York, N.Y. 10010, U.S.A.

ISBN: 0 444 88840 3

© ELSEVIER SCIENCE PUBLISHERS B.V., 1991

All rights reserved. No part of this publication may be reproduced, stored in a retrieval system, or transmitted, in any form or by any means, electronic, mechanical, photocopying, recording or otherwise, without the prior written permission of the publisher, Elsevier Science Publishers B.V. / Physical Sciences and Engineering Division, P.O. Box 103, 1000 AC Amsterdam, The Netherlands.

Special regulations for readers in the U.S.A. – This publication has been registered with the Copyright Clearance Center Inc. (CCC), Salem, Massachusetts. Information can be obtained from the CCC about conditions under which photocopies of parts of this publication may be made in the U.S.A. All other copyright questions, including photocopying outside of the U.S.A., should be referred to the publisher.

No responsibility is assumed by the publisher for any injury and/or damage to persons or property as a matter of products liability, negligence or otherwise, or from any use or operation of any methods, products, instructions or ideas contained in the material herein.

Printed in the Netherlands

*To my wife Kathy
and to the memory of our son John
who was a ray of light in our lives*

Preface

When one speaks about the foundations of probability, there are two subjects that come to mind: the axiomatic foundations of the calculus of probability, which is a well developed, independent mathematical discipline, and the study of the possible interpretations for the calculus, especially in connection with its applications. The axiomatic foundations were laid out by Kolmogorov in [56], and his axioms or similar ones are accepted by everyone. In the Preliminaries and Part 2, we discuss a simple version of Kolmogorov's axioms and some of their consequences.

On the other hand, as it is well-known, there are several conflicting interpretations of probability, such as considering probability as a measure of the degree of belief or as the limit of relative frequencies. In this book, I present a new interpretation of probability satisfying Kolmogorov's axioms, which I think captures the original intuitions on the subject. Thus, most of the content of the book deals with the second of the foundational subjects mentioned above, which is, in a large measure, philosophical in character. My new interpretation, however, leads to the construction of mathematical models, and a large part of the book deals with the mathematical development of these models.

The interpretation of probability presented in this book is new in the sense that it is different from the interpretations that are at this moment considered reasonable by those who work in the subject. I think, however, that it is rooted in the traditional view of probability dating back to the 17th and 18th centuries, when the calculus of probability originated.

Another purpose of this book is the study of the foundations of statistical inference and decision theory. The mathematical models by themselves cannot provide an interpretation of these subjects, in part, because they allow several interpretations. Hence, I need, besides the models, extra principles for this analysis. The same general philosophical position about probability that has led me to the models mentioned above, has also led me to the formulation of principles, which I believe are the basis for statistical inference and decision theory. I think that these principles, together with my probability models, provide a system for rationally deciding between the different statistical techniques that have been proposed.

I believe that the principles for statistical inference presented in this book are prevalent in usual statistical practice, although I think that they have never been explicitly stated in a coherent and complete fashion. It seems that their ultimate inspiration is in the justification of R. A. Fisher of significance tests (but not in his fiducial probability), although my results are mostly in accord (with many exceptions) with the techniques of Neyman and Pearson. For decision theory, however, I accept a Bayesian foundation.

Since as human beings we need to understand what we do, the study of the foundations of a science is justified in its own right. I believe that it is especially important for the practitioners of a science to understand its foundations; in particular, to understand the meaning of the terms they are using, to realize the significance of the methods of the science and their limitations, to visualize the formal structure of the science, etc. Probability is extensively used in many sciences, not only in statistics, but also in theoretical physics and other sciences. Through statistics, it is also applied to all the natural and social sciences. Hence, the foundations of probability should be of interest to all of these different types of scientists. In fact, anyone involved in the philosophy of science is naturally drawn into the study of the foundations of probability.

There are many instances of philosophical ideas influencing the development of a science. But statistics is the only case I know of where philosophical ideas are not just relevant for the development of the science, but also have immediate practical implications. It is well-known that the different interpretations of the concept of probability, which are based on competing philosophical ideas, lead to different statistical techniques. For instance, those who favor a subjectivist or logical interpretation are Bayesian; the believers in a frequency interpretation choose the methods of Neyman-Pearson or Fisher. Although sometimes the application of the different techniques may steer us to similar results, it is also not infrequent that they may yield mutually contradictory consequences.

A "practical" field where foundations of probability has had importance in the last years, is that of artificial intelligence, especially the construction of expert systems. This is due to the fact that for many such applications the concept of probability needs to be clear. Although I do not discuss extensively this subject, the considerations in Chapter V are relevant to artificial intelligence.

I believe that each conception of probability is rooted in a particular epistemological position. To argue for my interpretation of probability from my epistemological position, however, and then argue in favor of this epistemological position, would take us too far in this book, and is better left for other publications. I believe, however, that my position is in accordance with common sense ideas about probability. So I will argue here for my conception mainly by appealing to intuitions about probability. I will often argue from the consequences of the different definitions and interpretations, for instance by the construction of examples and counterexamples based on them. Once they are produced, I shall often take the intuitive content of these examples to be obvious.

A few words are in order about the title of the book. Although it is called

"Truth, possibility and probability", there is no exhaustive discussion of truth or possibility in it. Truth and possibility are only discussed with reference to probability, in fact, as the basis for the definition of probability advanced in this book. Truth, especially, is only briefly touched upon, and there is no treatment of the different theories of truth.

After a brief chapter with preliminaries, which contains a very short survey of the mathematical theory of probability, the book consists of four parts. Part 1 includes a philosophical discussion of the foundations of my interpretation of probability, a general outline of the models, and the main intuitive ideas that are basic to my approach, including what I accept as the basic principles of statistical inference and decision theory.

Part 2 contains a brief survey of the elementary mathematical theory of probability, including the notions of infinitesimal analysis that are needed for my development. The reason for this need of infinitesimal analysis, which is presented in its nonstandard version due essentially to A. Robinson, [78], is that I build only finite models, and, in order to take into account all probability spaces, I have to approximate infinite spaces by spaces with an infinite natural number of elements (what are called hyperfinite spaces). I believe that one of the main interests of the book is the building of probability theory upon a hyperfinite basis. This construction is partly inspired in Nelson's [71].

The study of nonstandard analysis in the literature has, for the most part, been devoted to the obtaining of new standard results and better proofs of old standard results. The point of view adopted in this book, which I think is also Nelson's, is the study of nonstandard analysis for its own sake. Thus, most results are left in their nonstandard form. For instance, the general representation theorem in Chapter XV, mentioned below, is an interesting nonstandard theorem, and I don't know whether it has an interesting standard version or not.

Part 3 includes a formal development of the models for my interpretation of probability. In Chapter XV, which is in this part, I prove a general representation theorem that states that all stochastic processes can be approximated, in a precise nonstandard sense, by processes defined over one of my models. This theorem also has an interpretation which is independent of my theory: it says that any stochastic process can be approximated by a process defined on a probability space based on equiprobable outcomes.

In Part 4, I give a short survey of statistical inference and decision theory, and I analyze the principles behind the practice of these disciplines, giving examples and applications. In particular, I shall develop a theory of provisional acceptance of hypotheses, which I think is consistent with current statistical practice. In fact, most of the Neyman-Pearson techniques for statistical inference are justified on the basis of principles that are different from those advanced by Neyman and Pearson themselves and other classical statiticians. On the other hand, for decision theory, which is sharply distinguished from statistical inference proper, a Bayesian approach is adopted. I believe that this division of labor between classical and Bayesian statisticians is reflected in the current practice: most statisticians that work in the testing of scientific hypotheses are classical, and most of those who work in areas where decision theory is important, such as economics, are Bayesian.

Also included are two appendices. Appendix A contains a complete set of axioms for the nonstandard analysis needed in the book, and a sketch of the proof that these axioms are consistent. Appendix B contains a brief sketch of the foundations of the theory of integration and measure form a nonstandard standpoint.

Part I is rather informal. The only prerequisites are some understanding of probability and statistical inference, although I believe that the Preliminaries provides a brief outline of the main ideas on probability. The rest of the book is heavily mathematical, but I believe that it can be followed by any person who has mathematical maturity. For Part 4, and also somewhat for Chapter IV in Part I, some acquaintance with statistical techniques is desirable, but not essential.

There are many people that have commented on successive drafts of the book. I am grateful to W. N. Reinhardt and N. C. A. da Costa for a careful reading of an earlier version of parts of the book. They both made many interesting comments and suggestions, some of which are noted in parts of the book.

Thanks are due to N. Bertoglio, C. Gutiérrez, and R. Lewin for comments and corrections. I especially thank A. Cofré, who read a large part of the next to last version. A. Cofré pointed out several errors and made many suggestions for improvement of the text. I thank J. C. Lira for some of the illustrations that LaTeX did not draw.

Although, I started working on foundations of probability in my Ph. D. dissertation (1965), done under Professor D. Blackwell, actual work on the book began in 1984, while I held a Guggenheim fellowship at the Institute for Mathematical Studies in the Social Sciences in Stanford University. The probability seminar of Professor P. Suppes, then and in the period 1986–1989, when I worked on a project there, was an inspiration for many of my ideas, and my conversations with him have helped me to clarify my views and state them in a more precise form.

Besides the Simon Guggenheim Foundation, whose support was important at a crucial moment, I have also to thank for partial financial support the Program for Scientific and Technological Development of the Organization of American States, 1978–1985, Project "Theory of models and applications", and several projects from FONDECYT, the Chilean government foundation for science.

I am very grateful to North-Holland Pub. Co., its editor, A. Sevenster, and especially, the editor of the series, L. Nachbin, without whose interest the book would probably never have been published.

Special thanks must go the Pontificia Universidad Católica de Chile, where I have worked most of the time of preparation of the book. The constant support of my colleagues and the authorities of the university, through several (DIUC) research projects and otherwise, have been essential for the success of this enterprise.

Santiago, Chile, 1991

Contents

Preface	v
List of Figures	xvii
Preliminaries	**1**
1. Notation	1
2. Probability	4
Conditional probability	6
Independent events	7
Random variables	8
Constant random variables	11
Indicator or Bernoulli random variables	11
Binomial random variables	11
Inequalities	12
Law of large numbers	12
Product measures and average measures	13

Part 1
The concept of probability

Chapter I. The problem of foundations	**17**
1. The uses of probability	17
2. The interpretations of probability	21
The classical definition of probability	23
Frequency theory	25
Propensity theory	29
Personalistic or subjectivistic theories	30
Logical theories	33
3. The Principle of Direct Inference	34
4. Foundations, an a priori inquiry	38

Chapter II. Truth, possibility, and probability — 41

1. Truth as related to probability — 41
2. Possibility — 44
3. Probability based on possibility — 49
4. The Principle of Direct Inference — 54

Chapter III. Probability models — 57

1. Outcomes — 57
2. The symmetry or first invariance principle — 59
   Bertrand's mixture paradox — 61
   Bertrand's random chord paradox — 62
   Principles of Symmetry — 63
3. The family of events — 67
4. The External Invariance Principle — 69
   Dependence and independence — 69
   Compound outcomes — 70
   Probability space — 70
   Second Invariance Principle — 73
   Models for inference and decision — 76
5. Probability models — 77
   Random variables as part of a language — 79
   Probability models and languages — 81

Chapter IV. Principles of inference and decision — 85

1. The basis for decisions — 86
   Principle of Direct Inference — 88
2. The basis for statistical inference — 89
   Kyburg's lottery paradox — 90
   Acceptance and rejection of hypotheses — 92
   The Principle of Inverse Inference — 97
   Examples — 103
3. Probability and degree of belief — 107
   Example of impossibility of assigning degrees — 108
   Comparison with subjectivists' views — 108
   Coherence of the two principles — 109
4. Probabilities and frequencies — 110
   Rules for the acceptance of frequencies as estimates of probabilities — 112

Chapter V. A medical example — 117

1. The disease model — 117

2.	The Bayesian model	119
3.	The classical statistical inference model	121

Part 2
Elements of Probability Theory

Chapter VI. Probability theory		129
1.	Probability	129
2.	Conditional probability	132
	Bayes formula	133
	Independent events	134
3.	Random variables	134
4.	Expectation, variance and covariance	136
	Constant random variables	138
	Indicator or Bernoulli random variables	138
	Binomial random variables	139
	Expectation of a function of a random variable	139
5.	Inequalities	140
6.	Algebras of random variables	141
7.	Algebras of events	145
8.	Conditional distributions and expectations	149
9.	Constructions of probability spaces	152
10.	Stochastic processes	155
Chapter VII. Disjunctive spaces		159
1.	Disjunctive algebras of sets	159
2.	Finite disjunctive probability spaces	160
3.	Probabilities invariant under groups	164
Chapter VIII. Elements of infinitesimal analysis		171
1.	Infinitesimals	173
2.	Standard, internal and external sets	175
3.	Internal analogues of convergence	183
4.	Series	185
5.	Convergence of functions	187
6.	Continuity	188
7.	Differentials and derivatives	189

Chapter IX. Integration — 197

1. Hiperfinite sums — 197
2. Integrals — 203
3. Riemann integrals in higher dimensions — 206
4. Transformations — 210
5. Curves and surfaces — 212
6. Change of variables — 216
7. Improper integrals — 222

Chapter X. Probability distributions — 227

1. Hyperfinite probability spaces — 227
2. Discrete distributions — 228
   Bernoulli and binomial distributions — 232
   Poisson distribution — 233
3. Continuous distributions — 235
   Uniform distributions — 240
4. Normal distributions — 243
   De Moivre-Laplace Theorem — 244
5. Samples from a normal population — 247

Chapter XI. Hyperfinite random processes — 253

1. Properties that hold almost everywhere — 253
2. The law of large numbers — 257
   Weak law of large numbers — 258
   Strong law of large numbers — 258
3. Brownian motion and Central Limit Theorem — 262

Part 3
Probability models

Chapter XII. Simple probability structures — 279

1. Relational systems — 280
2. The structure of chance — 282
3. The probability space — 285
4. Simple probability structures — 288

Chapter XIII. The structure of chance — 293

1. Binary relations and causal relations — 294

2.	Relational systems with causal structures	298
3.	Chance structures	303
4.	The group of internal invariance	308
5.	Classical structures	315

Chapter XIV. Equiprobability structures 319

1.	Equiprobability structures	320
2.	Measures on super classical structures	328
3.	Homomorphisms	334
4.	Induced measures	339
5.	The external invariance principle	340

Chapter XV. Probability structures 343

1.	Existence of classical structures	345
2.	Measure-determined sets	345
3.	Probability structures	347
4.	Representation theorem	348

Chapter XVI. Examples of probability structures 353

1.	Distributions of r balls into n cells	353
2.	Selection of a chord at random	356
3.	The theory of errors	357
4.	Brownian motion	359
5.	Models for inference and decision	361
	Models for inference	361
	Models for decision	363

Chapter XVII. Logical probability 365

1.	Languages for simple structures	365
2.	Languages for compound structures	369
	Examples	374

Part 4
Statistical inference and decision theory

Chapter XVIII. Classical statistical inference 379

1.	General framework	380
2.	Discriminating experiments	384

3.	Significance tests	390
4.	Hypotheses tests	394
5.	The Sufficiency Principle	400
6.	Point estimation	403
7.	Confidence region estimation	406

Chapter XIX. Problems in statistical inference — 413

1. The Neyman-Pearson theory — 413
   Outline of the Neyman-Pearson justification — 413
   Gillies' examples — 415
2. Alternative hypotheses — 417
3. Unsound techniques for estimation — 419
   Randomized tests — 419
   Prediction intervals — 420
4. Likelihood and conditionality — 421
   The Likelihood Principle — 422
   The Conditionality Principle — 422
5. Initial and final precision — 424
6. Gillies' falsification rule — 427
7. Bayesian estimation — 428

Chapter XX. Decision theory — 431

1. Decision Theory and Consequentialism — 431
2. Maximizing expected utility — 433
3. Decisions based on evidence — 435
4. Decisions in the medical example — 438
5. Decision Theory in the law — 439

Appendix A. Foundations of nonstandard analysis — 443

1. Axioms — 444
   Statement of the axioms — 445
2. Equivalence of axioms to transfer — 451
   Extension and Transfer Principles — 452
   Proof of the Extension and Transfer Principles from Axioms (1)–(7) — 452
   Proof of Axioms (1)–(7) from the Extension and Transfer Principles — 454
3. Construction of the nonstandard universe — 455
   Ultrafilters. — 455
   Ultrapowers — 455

Appendix B. Extensions of integrals and measures	461
1. Internal abstract theory of integration	461
2. Null sets	467
3. Real valued probability measures	468
Bibliography	469

List of Figures

I.1	Model of inference and decision	22
III.2	Carnap's model of logic	65
III.3	A probability structure with a disjunctive algebra of events that is not an algebra	68
III.4	Hosiasson's first game	72
III.5	Hosiasson's second game	72
III.6	Hosiasson's third game	72
IX.7	Intersection of curve with rectangles.	214
IX.8	Transformation of a square.	218
XI.9	A T-Wiener walk obtained from the process ξ	268
XIII.10	Causal relations	297
XIX.11	Gillies' first example	416
XIX.12	Gillies' second example	416

Preliminaries

1. Notation

Most of the notation introduced in this section will not be used in Part 1. The set theoretic notation that will be used is mostly standard. For instance, the set of objects that satisfy a certain condition $\varphi(x)$ is denoted by $\{x \mid \varphi(x)\}$, and the set of elements x of a certain set A that satisfy $\varphi(x)$, by $\{x \in A \mid \varphi(x)\}$. The *union* and *intersection* of two sets A and B are written $A \cup B$ and $A \cap B$, respectively. The intersection $A \cap B$ is also written AB, following the usage in Probability Theory. These notations are extended to the union and intersection of finitely many sets.

If $\mathcal{A}$ is a family of sets, then $\bigcup \mathcal{A}$ denotes the *union of all elements of* $\mathcal{A}$. This union is written sometimes $\bigcup_{x \in \mathcal{A}} x$. I shall also use $\bigcup \{x \mid \varphi(x)\}$, that is, the union of all sets x that satisfy $\varphi(x)$. Thus, for instance, $\bigcup \mathcal{A} = \bigcup \{x \mid x \in \mathcal{A}\}$. Similar notations will be used for intersections: $\bigcap \mathcal{A}$ denotes the intersection of all elements of $\mathcal{A}$, and $\bigcap \{x \mid \varphi(x)\}$ is the intersection of all sets x that satisfy $\varphi(x)$.

As is usual, $A \subseteq B$ means that A *is a subset of* B, and $A \subset B$, that A *is a proper subset of* B. We also write $B \supseteq A$ and $B \supset A$ with the obvious meanings. $A - B$ is the *set-theoretic difference between* A *and* B. When we have a fixed universe Ω, we write A^c for the *complement* of A, i.e., $A^c = \Omega - A$. If A is any set, we write $\mathcal{P}(A)$ for *the power set of* A, i.e., the family of all subsets of A.

The set of all *natural numbers is denoted by* $\mathbb{N}$ and the set of all *real numbers*, $\mathbb{R}$.

A finite sequence of n elements or n-*tuple*, will be written $\langle a_1, \ldots, a_n \rangle$. In particular, an ordered pair is identified with a 2-tuple and written $\langle a, b \rangle$. The set of all n-tuples of elements of A is written $^n A$. If $\tau(x_1, \ldots, x_n)$ is any term, we shall write

$$\{\tau(x_1, \ldots, x_n) \mid \varphi(x_1, \ldots, x_n)\},$$

for the set of $\tau(x_1, \ldots, x_n)$ that satisfy $\varphi(x_1, \ldots, x_n)$. For instance, *the Cartesian* *product of A and B is*

$$A \times B = \{\langle a,b\rangle \mid a \in A \text{ and } b \in B\}.$$

The set of n–tuples is, then

$$^n A = \{\langle a_1,\ldots,a_n\rangle \mid a_1,\ldots,a_n \in A\}.$$

An n–tuple of elements of a set A, is identified with a function from the set of numbers $\{1,2,\ldots,n\}$ into A. Thus, if we identify n with $\{1,2,\ldots,n\}$, $^n A$ is the set of functions from n into A.

We generalize this notation for other sets of functions. The set *of all functions with domain B and range included in A* is $^B A$. That is

$$^B A = \{f \mid f : B \to A\},$$

writing $f : B \to A$ for the expression 'f is a function whose domain is B and whose range is included in A'. Thus, the set of all infinite (countable) sequences of elements of A is denoted by $^\mathbb{N} A$.

A function f is *one-one* if $f(x) = f(y)$ implies that $x = y$, for every x and y in the domain of f. If f is one-one, f^{-1} is the *inverse function*. A function $f : A \to B$ is *onto* B, if for every $y \in B$ there is an $x \in A$ such that $f(x) = y$. A one-one function from A onto A is called a *permutation of A*.

The function f whose domain is B and such that for every $b \in B$, $f(b) = \tau(b)$, where τ is a term, denoting, for instance, an ordered pair, is written by

$$f = \langle \tau(b) \mid b \in B\rangle.$$

Similarly, when the domain consists of the objects which satisfy the condition $\varphi(x)$, we write

$$f = \langle \tau(b) \mid \varphi(b)\rangle.$$

For instance, the infinite sequence a_n for $n \in \mathbb{N}$, is written $\langle a_n \mid n \in \mathbb{N}\rangle$. A function $\langle R_i \mid i \in I\rangle$ shall also be called a *system with index set I*. For sequences with domain the natural numbers or a particular natural number, we also write

$$\{x_n\}_{n\in\mathbb{N}} = \langle x_n \mid n \in \mathbb{N}\rangle,$$

or simply $\{x_n\}$, when there is no danger of confusion.

If f and g are functions, then the *composition of f and g* is written $f \circ g$, i.e.

$$f \circ g = \langle f(g(x)) \mid x \text{ is in the domain of } g \text{ and } g(x) \text{ is in the domain of } f\rangle.$$

If f is a function, we write $f''A$ for the *image of A under f*, i.e

$$f''A = \{f(x) \mid x \in A\}.$$

The *inverse image of A under f* is written $f^{-1''}A$, i.e

$$f^{-1''}A = \{x \mid f(x) \in A\}.$$

The *restriction of f to the set B* is written $f \restriction B$, that is

$$f \restriction B = \langle f(x) \mid x \in B\rangle.$$

1. NOTATION

When there is no danger of confusion, we also use the more customary notation, and write the image, $f''A$, simply as $f(A)$, and the inverse image, $f^{-1}{''}A$, as $f^{-1}(A)$.

The set of 1-tuples of elements of A is ^1A. That is

$$^1A = \{\langle a \rangle \mid a \in A\}.$$

Abusing a little the notation, we identify $\langle a \rangle$ with a, and ^1A with A.

The set $^\emptyset A$, which we shall also write ^0A by identifying $\emptyset$ with the number 0, contains just the function with empty domain, which is the empty set, i.e.

$$^0A = \{\emptyset\}.$$

We shall also have the opportunity to use *the generalized Cartesian product*. If for each $i \in I$, we have that $\tau(i)$ is a set, then

$$\prod_{i \in I} \tau(i) = \prod \langle \tau(i) \mid i \in I \rangle$$
$$= \{f \mid f \text{ is a function on } I \text{ and, for each } i \in I, f(i) \in \tau(i)\}.$$

If the set I is defined by the condition $\varphi(i)$, we also write

$$\prod \langle \tau(i) \mid \varphi(i) \rangle.$$

For instance, if A_i, for $i \in I$, is an indexed family of sets (i.e., A is a function with domain I and values A_i), we write the generalized Cartesian product of this family as

$$\prod \langle A_i \mid i \in I \rangle.$$

An *n-ary relation with field A*, or *over A*, is any subset of ^nA. In particular, due to the identification of ^1A with A, a 1-ary relation with field A is just a subset of A. An *n-ary operation Q with field A*, or *over A*, is a function from ^nA into A. Thus, an n-ary operation over A is an $n+1$-ary relation over A. A 0-ary operation Q is a function from ^0A into A, i.e., $Q : \{\emptyset\} \to A$. The operation Q is, then, $\{\langle \emptyset, a \rangle\}$, for a certain element a of A. In certain situations, we may identify Q with this element a.

When we say 'relation', unless explicitly excepted, we mean 'binary relation'. For a binary relation R, we write '$a\,\text{R}\,b$' instead of '$\langle a, b \rangle \in \text{R}$'

A few special kinds of relations will be mentioned. A relation R is *reflexive on A* if for every $x \in A$, we have $x\text{R}x$. When we say just *reflexive*, we mean reflexive on its domain. The relation R is *symmetric*, if $x\text{R}y$ implies $y\text{R}x$, and R is *transitive* if $x\text{R}y$ and $y\text{R}z$ imply $x\text{R}z$. Finally, R is an *equivalence relation on A*, if R is reflexive on A, symmetric and transitive. It is well known, that if R is an equivalence relation on A, then we can define the equivalence classes

$$[x]_\text{R} = \{y \in A \mid x\text{R}y\}.$$

It is clear that the family of equivalence classes form a *partition of A*, that is, a family of disjoint sets whose union is A.

We say that the relation R is *antisymmetric* if $x\mathrm{R}y$ and $y\mathrm{R}x$ imply $x = y$. A relation R that is reflexive over X, transitive, and antisymmetric is called a *partial ordering over* X. If a partial ordering R over X has the additional property that for every x, $y \in X$ we have that $x\mathrm{R}y$ or $y\mathrm{R}x$ (i.e., that R is *connected on* X), then R is a *total ordering*, or, simply, an ordering.

As is usual, we use the notations $\sum$ and $\prod$ for finite or infinite sums and products of numbers.

Finally, we sometimes use $\implies$ as abbreviation for 'implies' and $\iff$ as an abbreviation for 'if and only if'.

2. Probability

We shall give an informal account of the main ideas of the mathematical theory of probability, which will be used in Part 1. Part 2 contains a more extensive, but still brief, formal development of the mathematical theory.

There are two primitive elements in probability theory: the algebra of events and the probability measure defined on this algebra. Events can always be represented as subsets of a set Ω, usually called the *sample* or *probability space*. Thus, the algebra of events $\mathcal{A}$ can be considered as a family of subsets of Ω with the following properties:

(1) $\Omega \in \mathcal{A}$; i.e., Ω belongs to $\mathcal{A}$.
(2) If A belongs to $\mathcal{A}$, the complement of A in Ω, A^c or $\Omega - A$, belongs to $\mathcal{A}$.
(3) If A and B belong to $\mathcal{A}$, their union, $A \cup B$, belongs to $\mathcal{A}$.

It is usual to require that the union of a countable family of elements of $\mathcal{A}$, belongs to $\mathcal{A}$. As we shall see later, this is not accepted by everybody. In any case, we shall deal with countable unions using nonstandard analysis, so that we do not need this strengthening of the axiom. By the use of nonstandard analysis, we can also restrict the axioms to the case where Ω is finite.

A few examples follow:

EXAMPLE 1. If the experiment consists on the flipping of a coin, then we may take
$$\Omega = \{H, T\}$$
where H means heads and T, tails. Any subset of Ω is an event. Thus, $\mathcal{A}$ is the set of all subsets of Ω, i.e., the power set of Ω.

EXAMPLE 2. If the experiment consists of casting a die, then the sample space may be
$$\Omega = \{1, 2, 3, 4, 5, 6\}$$
where the outcome i means that i appeared on the die. The algebra of events, $\mathcal{A}$, is the power set of Ω.

EXAMPLE 3. If the experiment consists of the casting two dice, then we may take the sample space to consists of the following 36 points

$$\Omega = \left\{ \begin{array}{l} \langle 1,1\rangle, \langle 1,2\rangle, \langle 1,3\rangle, \langle 1,4\rangle, \langle 1,5\rangle, \langle 1,6\rangle, \\ \langle 2,1\rangle, \langle 2,2\rangle, \langle 2,3\rangle, \langle 2,4\rangle, \langle 2,5\rangle, \langle 2,6\rangle, \\ \langle 3,1\rangle, \langle 3,2\rangle, \langle 3,3\rangle, \langle 3,4\rangle, \langle 3,5\rangle, \langle 3,6\rangle, \\ \langle 4,1\rangle, \langle 4,2\rangle, \langle 4,3\rangle, \langle 4,4\rangle, \langle 4,5\rangle, \langle 4,6\rangle, \\ \langle 5,1\rangle, \langle 5,2\rangle, \langle 5,3\rangle, \langle 5,4\rangle, \langle 5,5\rangle, \langle 5,6\rangle, \\ \langle 6,1\rangle, \langle 6,2\rangle, \langle 6,3\rangle, \langle 6,4\rangle, \langle 6,5\rangle, \langle 6,6\rangle \end{array} \right\},$$

where the outcome $\langle i,j \rangle$ occurs if i appears on the first die and j on the second die.

Suppose that we are not interested in what appears on the dice, but only on the sum of the upper faces. Then, we can also model this experiment with the sample space

$$\Omega_1 = \{2, 3, 4, 5, 6, 7, 8, 9, 10, 11, 12\}.$$

Thus, we can see that the sample space is not determined by the experiment.

It is easy to prove from these axioms that if A and B belong to $\mathcal{A}$, their intersection also belongs to $\mathcal{A}$. It also can be proved that the union and intersection of a finite family of elements of $\mathcal{A}$ is in $\mathcal{A}$, and that $\emptyset \in \mathcal{A}$.

The interpretation of the set-theoretical operations in terms of events are the following. The event $A \cup B$ occurs if A or B (or both) occurs; A^c occurs if A does not occur; and AB occurs if both A and B occur. We say that A and B are *mutually exclusive* if both cannot occur, that is, if their intersection is empty, in symbols, $AB = \emptyset$.

The properties of the probability function Pr, defined over all members of $\mathcal{A}$, are given by the three axioms:

(1) $\Pr \Omega = 1$.
(2) If $A \in \mathcal{A}$, then $\Pr A \geq 0$.
(3) (Additivity) If A and B are mutually exclusive members of $\mathcal{A}$, then

$$\Pr(A \cup B) = \Pr A + \Pr B.$$

It is usual to strengthen additivity to countable additivity, but, as it was mentioned above, we shall deal with the infinite case via nonstandard analysis. The axioms presented here, plus countable additivity, are Kolmogorov's axioms, [56].

If the coin of Example 1 is a fair coin, then we would assign $\Pr\{H\} = 1/2 = \Pr\{T\}$. If it is not a fair coin, however, the assignment might be different. When the event consists of one point in Ω, as above, we usually write $\Pr H$ for $\Pr\{H\}$.

If the dice of Example 3 are fair, then we would assign

$$\Pr \langle i,j \rangle = \frac{1}{36}$$

for every pair, if we take Ω as our sample space. If we take Ω_1, however, the assignments are different. For instance

$$\Pr 2 = \frac{1}{36},$$
$$\Pr 3 = \frac{2}{36},$$
$$\vdots$$
$$\Pr 7 = \frac{6}{36}$$
$$\vdots$$
$$\Pr 12 = \frac{1}{36}$$

It is easy to prove the following:
(1) $\Pr A^c = 1 - \Pr A$.
(2) $\Pr \emptyset = 0$.
(3) If $A_1, A_2, \ldots, A_n$ are pairwise mutually exclusive members of $\mathcal{A}$, then

$$\Pr(A_1 \cup A_2 \cup \cdots \cup A_n) = \Pr A_1 + \Pr A_2 + \cdots + \Pr A_n.$$

(4) $\Pr(A \cup B) = \Pr A + \Pr B - \Pr AB$.

Conditional probability. Suppose that we are in the case of Example 3 with sample space Ω and fair dice. Suppose that we observe that the first die is three. Then, given this information, what is the probability that the sum of the two dice is seven? To calculate this probability, we reduce our sample space to those outcomes where the first die gives three, that is, to $\{\langle 3, 1\rangle, \langle 3, 2\rangle, \langle 3, 3\rangle, \langle 3, 4\rangle, \langle 3, 5\rangle, \langle 3, 6\rangle\}$. Since each of the outcomes in this sample space had originally the same probability of occurring, they should still have equal probabilities. That is, given that the first die is a three, then the (conditional) probability of each of these outcomes is $1/6$, while the (conditional) probability of the outcomes not in this set is 0. Hence, the desired probability is $1/6$.

If we let E and F denote the events of the sum being seven and the first die being three, then the conditional probability just obtained is denoted

$$\Pr(E|F).$$

Thus, we define in general

$$\Pr(E|F) = \frac{\Pr EF}{\Pr F}.$$

Notice that this equation is only well defined when $\Pr F > 0$. It is not difficult to prove that $\Pr(\ |F)$ has all the properties described above for a probability.

The equation can also be written in the useful form

$$\Pr EF = \Pr(E|F) \Pr F.$$

It is not difficult to prove:

Total probability formula: *If $A_1, A_2, \ldots, A_n$ are mutually exclusive events such that their union, $\bigcup_{i=1}^{n} A_i$, is the sure event, Ω, then for any event B*

$$\Pr B = \sum_{i=1}^{n} \Pr(B|A_i) \Pr A_i.$$

Bayes formula: *If $A_1, A_2, \ldots, A_n$ are mutually exclusive events such that their union, $\bigcup_{i=1}^{n} A_i$, is the sure event, Ω, then for any event B*

$$\Pr(A_i|B) = \frac{\Pr(B|A_i) \Pr A_i}{\sum_{j=1}^{n} \Pr(B|A_j) \Pr A_j}.$$

For the case of two events, A_1 and A_2, Bayes' formula reads

$$\Pr(A_i|B) = \frac{\Pr(B|A_i) \Pr A_i}{\Pr(B|A_1) \Pr A_1 + \Pr(B|A_2) \Pr A_2}.$$

EXAMPLE 4. A laboratory blood test is positive in 95% of the patients who, in fact, have a certain disease. The test also yields a false positive result, however, for 1% of the healthy persons tested. Suppose that 0.05% of the population has the disease. What is the probability that a person actually has the disease, given that he tested positive?

Let D be the event that the tested person has the disease, and E the event that the test is positive. The desired probability $\Pr(D|E)$ is obtained by Bayes formula

$$\Pr(D|E) = \frac{\Pr(E|D) \Pr D}{\Pr(E|D) \Pr D + \Pr(E|D^c) \Pr D^c}$$
$$= \frac{0.95 \cdot 0.005}{0.95 \cdot 0.005 + 0.01 \cdot 0.995}$$
$$= \frac{95}{294} \approx 0.323$$

Independent events. The notion of independence introduced here may be called more precisely *probabilistic independence*.

Two events E and F are independent if the occurrence of one of them, say F, does not affect the probability of the other, i.e., E. That is, if

$$\Pr(E|F) = \Pr E.$$

This is equivalent to $\Pr EF = \Pr E \Pr F$. Since this last formula does not have the restriction that $\Pr F > 0$, we adopt it as a definition. The definition of independence can be extended to more than two events. Thus, we have:

The events $E_1, E_2, \ldots, E_n$ are said to be *(probabilistically) independent* if

$$\Pr E_{i_1} E_{i_2} \cdots E_{i_m} = \prod_{j=1}^{m} \Pr E_{i_j},$$

for every $i_1, i_2, \ldots, i_m$ between 1 and n, such that $i_p \neq i_q$ for $p \neq q$. We give an example of pairwise independent events that are not independent:

EXAMPLE 5. Let a ball be drawn from an urn containing four balls, numbered 1, 2, 3, 4. Let $E = \{1, 2\}$, $F = \{1, 3\}$ and $G = \{1, 4\}$. If the drawing of the balls are equally likely, then it is easy to check that

$$\Pr EF = \Pr E \Pr F = \tfrac{1}{4},$$

$$\Pr EG = \Pr E \Pr G = \tfrac{1}{4},$$

$$\Pr FG = \Pr F \Pr G = \tfrac{1}{4}.$$

On the other hand, we have

$$\tfrac{1}{4} = \Pr EFG \neq \tfrac{1}{8} = \Pr E \Pr F \Pr G.$$

Random variables. Most frequently, when performing an experiment, we are interested in a property of the outcome and not in the outcome itself, although we may need the particular sample space, in order to determine the probabilities. For instance, in casting two dice as in Example 3, we would usually adopt the sample space Ω and not Ω_1, because the function Pr is easily determined for it. We may only be interested, however, in knowing the sum of the numbers that appear on the dice, and not on the particular outcome in Ω. In order to deal with this situation we introduce random variables:

A *random variable* or *random quantity* is a function from Ω to $\mathbb{R}$. For example, if the set Ω is the set of tosses of a coin, the function X, defined on Ω, whose value is 0, if the toss results in heads, and 1, if the toss results in tails, is a random variable.

For any set of real numbers S, we write

$$[X \in S] = \{\omega \in \Omega \mid X(\omega) \in S\}.$$

For a real number r we have

$$[X = r] = \{\omega \in \Omega \mid X(\omega) = r\},$$

and for real numbers a and b

$$[a \leq X \leq b] = \{\omega \in \Omega \mid a \leq X(\omega) \leq b\}.$$

For instance, in the case of Example 3, we could introduce the random variables

$$X(\langle i,j \rangle) = i, \quad Y(\langle i,j \rangle) = j, \quad Z(\langle i,j \rangle) = i+j,$$

for every $\langle i,j \rangle \in \Omega$. We can determine the probabilities of the values of Z, for instance:

$$
\begin{aligned}
\Pr[Z=2] &= \Pr\{\langle 1,1\rangle\} &&= \tfrac{1}{36},\\
\Pr[Z=3] &= \Pr\{\langle 1,2\rangle,\langle 2,1\rangle\} &&= \tfrac{2}{36},\\
\Pr[Z=4] &= \Pr\{\langle 1,3\rangle,\langle 3,1\rangle,\langle 2,2\rangle\} &&= \tfrac{3}{36},\\
\Pr[Z=5] &= \Pr\{\langle 1,4\rangle,\langle 4,1\rangle,\langle 2,3\rangle,\langle 3,2\rangle\} &&= \tfrac{4}{36},\\
\Pr[Z=6] &= \Pr\{\langle 1,5\rangle,\langle 5,1\rangle,\langle 2,4\rangle,\langle 4,2\rangle,\langle 3,3\rangle\} &&= \tfrac{5}{36},\\
\Pr[Z=7] &= \Pr\{\langle 1,6\rangle,\langle 6,1\rangle,\langle 2,5\rangle,\langle 5,2\rangle,\langle 3,4\rangle,\langle 4,3\rangle\} &&= \tfrac{6}{36},\\
\Pr[Z=8] &= \Pr\{\langle 2,6\rangle,\langle 6,2\rangle,\langle 3,5\rangle,\langle 5,3\rangle,\langle 4,4\rangle\} &&= \tfrac{5}{36},\\
\Pr[Z=9] &= \Pr\{\langle 3,6\rangle,\langle 6,3\rangle,\langle 4,5\rangle,\langle 5,4\rangle\} &&= \tfrac{4}{36},\\
\Pr[Z=10] &= \Pr\{\langle 4,6\rangle,\langle 6,4\rangle,\langle 5,5\rangle\} &&= \tfrac{3}{36},\\
\Pr[Z=11] &= \Pr\{\langle 5,6\rangle,\langle 6,5\rangle\} &&= \tfrac{2}{36},\\
\Pr[Z=12] &= \Pr\{\langle 6,6\rangle\} &&= \tfrac{1}{36}.
\end{aligned}
$$

Let X be, now, an arbitrary random variable. Once we have the probabilities $\Pr[X=i]$, we can usually forget about the sample space. In fact, the sample space is only used for determining what is called the *mass function* or *probability distribution* of X, namely

$$\mathrm{pr}_X(i) = \Pr[X=i].$$

Another important real funtion associated with a random variable X is the *cumulative distribution function* (or simply the distribution function) of X:

$$F_X(x) = \Pr[X \le x].$$

If X_1 and X_2 are random variables and a and b are real numbers, then we define

$$(aX_1 + bX_2)(\omega) = aX_1(\omega) + bX_2(\omega),$$

$$(X_1 X_2)(\omega) = X_1(\omega) X_2(\omega).$$

For instance, in the case of the two fair dice, with the definitions introduced above, we have $Z = X + Y$.

Next, we introduce the notion of independence for random variables.

The random variables $X_1, X_2, \ldots, X_n$, defined over the same sample space, are said to be *independent*, if for any real numbers $i_{k_1}, i_{k_2}, \ldots, i_{k_m}$, and any $k_1, \ldots, k_m$ between 0 and n

$$\Pr[X_{k_1} = i_{k_1}, X_{k_2} = i_{k_2}, \ldots, X_{k_m} = i_{k_m}] = \\ \Pr[X_{k_1} = i_{k_1}] \Pr[X_{k_2} = i_{k_2}] \cdots \Pr[X_{k_m} = i_{k_m}].$$

For instance, in the case of two fair dice, X and Y are independent, but Z and X are not.

We now define some functions on random variables.

(1) The *mean* or *expectation* of the random variable X is
$$\mathbf{E}\,X = \sum_{\omega \in \Omega} X(\omega)\Pr(\omega) = \sum_{x \in \mathbb{R}} x \Pr\nolimits_X(x).$$

(2) The *variance* of X is
$$\operatorname{Var} X = \mathbf{E}(X - \mathbf{E}\,X)^2 = \sum_{x \in \mathbb{R}}(x - \mathbf{E}\,X)^2 \Pr\nolimits_X(x).$$

The square root of the variance of X is called the *standard deviation* of X.

(3) The *covariance* of the random variables X and Y is
$$\operatorname{Cov}(X, Y) = \mathbf{E}(X - \mathbf{E}\,X)\,\mathbf{E}(Y - \mathbf{E}\,Y).$$

The *correlation coefficient of X and Y* is
$$\frac{\operatorname{Cov}(X,Y)}{\sqrt{\operatorname{Var} X}\sqrt{\operatorname{Var} Y}}.$$

The mean is the weighted average of the values of the random variable, weighted according to their probabilities. The variance is the weighted average of the squared differences to the mean.

The expectation of the variable Z in the case of two fair dice is

$\mathbf{E}\,Z = 2\cdot\frac{1}{36}+3\cdot\frac{2}{36}+4\cdot\frac{3}{36}+5\cdot\frac{4}{36}+6\cdot\frac{5}{36}+7\cdot\frac{6}{36}+8\cdot\frac{5}{36}+9\cdot\frac{4}{36}+10\cdot\frac{3}{36}+11\cdot\frac{2}{36}+12\cdot\frac{1}{36}.$

We can easily get the following formulas (see Chapter VI, Section 4, for a proof):

$$\mathbf{E}(aX_1 + bX_2) = a\,\mathbf{E}\,X_1 + b\,\mathbf{E}\,X_2,$$
$$\operatorname{Var} aX = a^2 \operatorname{Var} X,$$
$$\operatorname{Var} X = \mathbf{E}\,X^2 - (\mathbf{E}\,X)^2,$$
$$\operatorname{Cov}(X, Y) = \mathbf{E}\,XY - \mathbf{E}\,X\,\mathbf{E}\,Y,$$
$$\operatorname{Var}(X + Y) = \operatorname{Var} X + \operatorname{Var} Y + 2\operatorname{Cov}(X, Y).[0]$$

If X and Y are independent, then we have:
$$\mathbf{E}\,XY = \mathbf{E}\,X \cdot \mathbf{E}\,Y.[0]$$

Thus, still assuming independence, we get:
$$\operatorname{Cov}(XY) = 0,[0]$$

and
$$\operatorname{Var}(X + Y) = \operatorname{Var} X + \operatorname{Var} Y.$$

Where it makes sense, the formulas can be extended by induction to n variables.

2. PROBABILITY

We mention here a few important types of random variables.

Constant random variables. These are variables X such that $X(\omega) = c$, for every $\omega \in \Omega$, where c is a real number. We clearly have $\Pr[X = c] = 1$ and $\Pr[X = r] = 0$ for any $r \neq c$. Also $\mathbf{E} X = c$ and $\text{Var } X = 0$. We identify this random variable X with the real number c.

Indicator or Bernoulli random variables. If we have any event E, the *indicator of E* is the function defined by

$$\chi_E(\omega) = \begin{cases} 1, & \text{if } \omega \in E, \\ 0, & \text{if } \omega \notin E. \end{cases}$$

For these random variables we have

$$\Pr[\chi_E = r] = \begin{cases} \Pr E, & \text{if } r = 1, \\ \Pr E^c, & \text{if } r = 0, \\ 0, & \text{otherwise.} \end{cases}$$

In statistical practice, an outcome of an experiment or *trial* in E is called a *success* and one in E^c, a *failure*. So that we define *a Bernouilli random variable* as the indicator of a set, that is, a variable that is 1, in case of success, and 0, in case of failure.

The following equalities are calculated in Chapter VI, Section 4:

$$\mathbf{E} \chi_E = \Pr E,$$
$$\text{Var } \chi_E = \Pr E \Pr E^c.$$

Binomial random variables. Suppose that one performs n independent Bernoulli trials, each of which results in a "success" with probability p and in a "failure" with probability $1 - p$. Let the random variable X measure the number of successes in n trials. Then X is called a *binomial random variable having parameters $\langle n, p \rangle$*. The distribution of X is given by

$$\Pr_X(i) = \binom{n}{i} p^i (1-p)^{n-i}, \qquad i = 0, 1, \ldots, n,$$

where $\binom{n}{i} = n!/(n-i)!i!$ equals the number of sets (called, in this case, combinations) of i objects that can be chosen from n objects. This equation is proved in Chapter VI, Section 4.

EXAMPLE 6. Suppose that a coin is flipped five times. If the flippings are supposed to be independent, what is the probability that three tails and two heads are obtained?

Let X be the number of tails ("successes") that occur in the five flippings. Then X is a binomial random variable with parameters $\langle 5, \frac{1}{2}\rangle$. Hence

$$\Pr[X = 3] = \binom{5}{3}(\frac{1}{2})^3(\frac{1}{2})^2 = \frac{5}{16}.$$

The expectation and the variance of a binomial random variable X are calculated in Chapter VI, Section 4:

$$\mathbf{E}\,X = np,$$
$$\operatorname{Var} X = np(1-p).$$

Inequalities. We shall use the following inequalities, which we shall prove in Chapter VI, Section 5.

Jensen's inequality. Let X be a random variable and $p \geq 1$. Then

$$|\mathbf{E}\,X|^p \leq \mathbf{E}\,|X|^p.$$

In particular we have

$$|\mathbf{E}\,X| \leq \mathbf{E}\,|X|.$$

Markov's inequality. Let f be a positive real function (i.e., $f(x) \geq 0$ for every x) and let $\varepsilon > 0$. Then

$$\Pr[f(X) \geq \varepsilon] \leq \frac{\mathbf{E}\,f(X)}{\varepsilon}.$$

Chebyshev's inequality. Let $p > 0$, $\varepsilon > 0$, and X be a random variable. Then

$$\Pr[|X| \geq \varepsilon] \leq \frac{\mathbf{E}\,|X|^p}{\varepsilon^p}.$$

In particular, we have

$$\Pr[|X - \mathbf{E}\,X| \geq \varepsilon] \leq \frac{\operatorname{Var} X}{\varepsilon^2}.$$

Law of large numbers. Let X be a random variable of finite mean μ and finite variance σ^2. Consider ν independent observations $X_1, X_2, \ldots, X_\nu$ of X. The law of large numbers can be fomulated intuitively by saying that

> if ν is a large number, then almost surely for all large $n \leq \nu$, the average $(X_1 + \cdots + X_n)/n$ is approximately equal to μ.

In order to make precise this statement in usual mathematics, one replaces the finite sequence $X_1, \ldots, X_\nu$ by an actually infinite sequence and constructs the infinite Cartesian product of the probability space Ω with itself, and, also, replaces 'approximately equal' by the appropriate ε, δ statement. The approach that we shall take is different. We retain the finite sequence $X_1, \ldots, X_\nu$ but we let ν be an hyperfinite (infinite nonstandard) natural number, that is, a natural number that is larger than any finite natural number. We also replace 'x is approximately equal to y' by '$x - y$ is infinitesimal', that is, '$|x - y|$ is a positive number which is less than any real positive number'. We symbolize x is approximately equal to y, in this sense, by $x \approx y$.

In order to obtain the usual infinite probability spaces from finite spaces, and to work with infinite and infinitesimal numbers consistently, we shall use nonstandard analysis, which will be discussed formally in Chapter VIII.

In working with infinitesimal analysis, our probability space, Ω, is usually hyperfinite. In this case, not all subsets of Ω are events, that is, there are subsets of Ω where Pr is not defined. Thus, it is convenient to define null events in the following fashion:

(1) A subset A of Ω is called a *null set* if for every real $\varepsilon > 0$ there is a set $B \supseteq A$ such that $\Pr B < \varepsilon$.
(2) Let $\varphi(\omega)$ be any statement about $\omega \in \Omega$. We say that $\varphi(\omega)$ *holds almost everywhere (a.e.)*, or *almost surely (a.s.)* on $\langle \Omega, \Pr \rangle$ if the set $[\varphi]^c$ is a null set.

We shall need the strong law of large numbers. We, first define convergence. Let ν be an infinite natural number. We say that the sequence of hyperreals $x_1, x_2, \ldots, x_\nu$ is *nearly convergent* if there is a hyperreal number x such that $x_\mu \approx x$ for all infinite $\mu \leq \nu$. We also say in this case that the sequence $x_1, \ldots, x_\nu$ nearly converges to x. In Chapter VIII, Section 3 convergence is discussed, and it will be proved that the sequence nearly converges to x if and only if for every noninfinitesimal $\varepsilon > 0$, there is a finite n, such that for every $m \geq n$, we have that $|x_m - x| \leq \varepsilon$.

Similarly, a sequence of random variables $X_1, \ldots, X_\nu$ nearly converges to the random variable X, if $X_n(\omega)$ nearly converges to $X(\omega)$, for every $\omega \in \Omega$. The sequence nearly converges a.s. to X, if the subset of Ω where it does not converge is a null set.

We now give a version of the strong law of large numbers, which is proved as Theorem XI.7.

Strong law of large numbers: *Let $\langle X_n \mid 1 \leq n \leq \nu \rangle$ be a sequence of independent random variables of finite mean μ and finite variance σ^2, where ν is an infinite natural number. Let*

$$Y_n = \frac{X_1 + \cdots + X_n}{n},$$

for $1 \leq n \leq \nu$. Then the sequence $\langle Y_n \mid n \leq \nu \rangle$ nearly converges a.s. to μ.

Product measures and average measures. We shall introduce two constructions of probability measures that we shall use in Part 1 and later. A more thorough discussion appears in Chapter VI, Section 9.

Let $\Pr_1, \Pr_2, \ldots, \Pr_n$ be probabilities defined on the power sets of $\Omega_1, \Omega_2, \ldots, \Omega_n$. We then define the *product probability* Pr on $\Omega_1 \times \Omega_2 \times \cdots \times \Omega_n = \prod_{i=1}^n \Omega_i$, by

$$\Pr \langle \omega_1, \ldots, \omega_n \rangle = \prod_{i=1}^n \Pr \omega_i,$$

for $\langle\omega_1,\ldots,\omega_n\rangle \in \prod_{i=1}^n \Omega_i$, and, for $A \subseteq \prod_{i=1}^n \Omega_i$

$$\Pr A = \sum_{\omega \in A} \Pr \omega.$$

It is clear that if $A_i \subseteq \Omega_i$ for $i = 1, \ldots, n$, then

$$\Pr \prod_{i=1}^n A_i = \prod_{i=1}^n \Pr_i A_i.$$

Let, now, $\bar{A}_i = \Omega_1 \times \cdots \times A_i \times \cdots \Omega_n$, for $i = 1, \ldots n$. Then, by the above

$$\Pr \bar{A}_i = \Pr_i A_i.$$

Thus, the events $\bar{A}_1, \bar{A}_2, \ldots, \bar{A}_n$ are independent.

We turn, now, to the definition of the average probability measure. Let $\Pr_0$ be a probability defined on the power set of Ω_0, and, for each $\omega_0 \in \Omega_0$, let $\Pr_{\omega_0}$ be a probability defined on the power set of Ω_{ω_0}. We define the *average probability measure*, Pr, *of the* $\Pr_{\omega_0}$ *with respect to* $\Pr_0$ on the set

$$\Omega = \{\langle\omega_0,\omega_1\rangle \mid \omega_0 \in \Omega_0, \omega_1 \in \Omega_{\omega_0}\}$$

by

$$\Pr \langle\omega_0,\omega_1\rangle = \Pr_0 \omega_0 \cdot \Pr_{\omega_0} \omega_1,$$

for $\langle\omega_0,\omega_1\rangle \in \Omega$, and, for $A \subseteq \Omega$, we define

$$\Pr A = \sum_{\omega \in \Omega} \Pr \omega.$$

Thus, if we define for $A \subseteq \Omega$ and $\omega_0 \in \Omega_0$

$$A_0 = \{\omega_0 \in \Omega_0 \mid \langle\omega_0,\omega_1\rangle \in A, \text{ for some } \omega_1 \in \Omega_{\omega_0}\}$$

and

$$A_{\omega_0} = \{\omega_1 \mid \langle\omega_0,\omega_1\rangle \in A\},$$

we have

$$\Pr A = \sum_{\omega_0 \in A_0} \Pr_0 \omega_0 \Pr A_{\omega_0}.$$

Hence, Pr is the average probability of the A_{ω_0} weighted according to $\Pr_0 \omega_0$.

Part 1

The concept of probability

CHAPTER I

The problem of foundations

1. The uses of probability

This section is mainly concerned with the analysis of the different uses of the calculus of probability. I shall begin, however, with a brief discussion of the meaning of words related to probability in ordinary language. I do not think that all ordinary language uses of words such as 'probably' or 'probable', can be assimilated to the calculus, but in any case their analysis is helpful to delimit the concept of probability. This analysis is useful in spite of the fact that it may be the case that the present meaning of these words was only fixed after the emergence of the calculus in the 17th century.[1]

The main uses of these words[2] is in statements such as 'it is probable that p', where p is a proposition, e.g., 'it will rain tomorrow'. The phrase 'it is probable that...' serves a similar purpose as statements of the form 'it is true that...', 'it is certain that...', or 'it is possible that...'. When I say 'it is true that p', I am just asserting p. 'It is true that...' only qualifies how my listener is to understand the support that I am giving to p; in this case I confer my complete support. Thus, if p turns out to be false, I am accountable. On the other hand, in saying 'it is probable that p', I am not so accountable, because I indicated that p was not fully supported. If I am challenged, however, I should be able to provide the justification for the support that I did give.

Let us analyze three characteristics of the expression 'it is probable' in ordinary language that must be accounted for in any interpretation of probability. The first characteristic is that the expression 'it is probable' qualifies propositions with respect to the degrees of support that we are willing to adscribe to them. Since a proposition may refer to the single occurrence of a phenomenon, such as 'it will rain tomorrow' or 'heads occurred in the last toss of this coin', probability is naturally applied to single occurrences (what are called *single cases*), and not just to sequences of occurrences, as it is sometimes asserted.

---

[1] See [40].
[2] A similar analysis of common language usages is given in [63].

Second, an assertion 'it is probable that p' may be justified at a certain time, but unjustified later. E.g., at noon today, we may be justified in asserting 'it will probably rain tomorrow', while at five o'clock, this might not be the case anymore. This change in status is usually attributed to shifts in the conditions that provide support to the statement.

Third, we most often believe that there are objective grounds that justify assertions of this type, and, in most cases, there is agreement about these statements. When there is disagreement between two persons, each one usually claims the other to be wrong. For instance, if I say 'it will probably rain tomorrow', and you say the opposite, we may discuss the issue and examine the grounds for our assertions. If there is neither disagreement about facts, nor about the general laws that govern the meteorological phenomena, then usually an agreement is reached.

Even though ordinary language usage of probability has implicit a qualitative scale of degree of support, i.e., true, probable, possible, and false, the quantification of this scale seems not to have arisen from this usage. The numerical assignment of probabilities only began in the 17th century with the study of games of chance. The expressions '... was a chance phenomenon', 'there are great chances that A will occur', 'the chance of A's obtaining is small' also occur in ordinary language.

Chance appears as a property of nature, and not as a qualifier of propositions. It is clear, however, that if the chance of a six obtaining in the next roll of a die is high, then we would be justified in asserting that it is probable that a six will occur in the next roll. Thus, there is a connection between chance and probability as a qualifier of propositions. This connection seems to have been present from the beginnings of the work in the calculus of probability. The word probability was used both for chance and as qualifier of propositions.

Games of chance have existed from great antiquity. Until the 17th century, however, there was no mathematical theory for them. The calculus of probability began to be studied by, among others, Pascal and Fermat in an attempt to solve some problems arising from these games.[3] The theory that was developed during the 17th and 18th centuries is very simple. It is supposed that there are n possible results, and that all events are combinations of these results. The elementary results are equiprobable. Hence, the probability of an event is the number of results favorable to it, divided by the total number of results. This simple model seems intuitively correct for games of chance. As support for this intuitive correctness, one can mention that most elementary textbooks on probability begin, even at present, with examples of games of chance treated with this simple model. Since I think that we should respect common sense as much as possible, I also believe that any interpretation of probability should make sense of this traditional model for games of chance. As a matter of fact, one of the original motivations for my theory was an attempt to understand this traditional model, which seemed to me to be accepted implicitly by most elementary textbooks on probability, although sometimes it was explicitly rejected.

---

[3] See [40].

There are many cases, however, where what appears to be the natural elementary results should not be considered equiprobable. For instance, this is the case when dealing with biased dice or coins. Very early in the studies of probability, it was realized that the biasedness of a coin, for instance, can be discovered from the long run properties of sequences of tosses of the coin. If the proportion of heads in 100 tosses is, say, 70, we conclude that the coin is biased towards heads. An issue connected with this testing procedure for biasedness is the stability of long run frequencies in many situations. I think that any interpretation of probability has to be able to explain this stability, and justify the soundness of the testing procedure just described.

I will, also, call chance *factual probability*, and the use of the calculus of probability for chance, a *factual use* of probability. There are also *epistemic* or *cognitive* uses of probability,[4] that is, uses that relate to our knowledge. With the development of probability, important epistemic and factual uses arose. The principal technical epistemic uses are two applications related to statistics. The first one arises from considering probability as the "very guide of life".[5] This is the epistemic use of probability as the basis for decision theory. The need for a theory of decisions comes about from our uncertainty as to how we ought to behave under circumstances where we are ignorant concerning the state of the world. As Kyburg says in [59, p. 26]

> we attempt to develop rules of behavior which we may follow, whatever the state of the world, in the expectation that we are following rules whose characteristics are generally (1) desirable, and (2) attainable.

The most desirable rule is one that tells us how to discover the true state of the world; the most attainable and simple is to forget the arithmetic and act as we feel like acting. A compromise between these two extremes, is to assign degrees of belief to the occurrence of the different states of the world, based on what we may know about the probabilities of these occurrences, i.e., about the measure of the degree of support of each possible state of the world.

The second epistemic type of use is in the evaluation of scientific hypotheses. Since in these cases we have to decide whether to accept or reject a scientific hypothesis, these uses are assimilated by some people to processes of the kind discussed above, namely decision-theoretic. However, I believe with Fisher, [34, p. 5], that

> ... such processes (i.e., decision-theoretic processes) have a logical basis very different from those of a scientist engaged in gaining from his observations an improved understanding of reality.

I think that Fisher would agree with my opinion that statistical inference, as used in the evaluation and estimation of hypotheses, has the same general purpose as logical inference. It conveys rules of rationality for starting from some propositions accepted as true, arriving at propositions that are to be accepted as

---

[4] See Hacking, [39], for this distinction, which was also introduced in [18].
[5] Bishop Butler in The Analogy of Religion.

true. A typical case of the application of statistical inference is in the acceptance of the hypothesis that smoking is a causative factor for lung cancer. At present, the official statement in the U.S. is: "The Surgeon General has determined that cigarette smoking is harmful to your health", and not just that it is probable that smoking cigarettes is harmful. Thus, we have arrived at the acceptance of a proposition as true based on statistical data and by the use of statistical inference.

Decision theory,[6] on the other hand, studies rules of rationality for guiding our actions in the face of uncertainty. There are two fundamental differences between the evaluation of hypotheses and decision theory. First, in evaluating hypotheses one should not consider the desirabilities (or "goodness") of the consequences of accepting or rejecting the hypotheses, while on deciding on an action, desirabilities of consequences of the action are of primary importance. That is, on deciding on an action, we must take into consideration the advantages and disadvantages of the consequences of our action, what are called, the *utilities* of the consequences. On the other hand, we would be mistaken in considering utilities when deciding whether to accept or not a hypothesis.[7]

Second, decisions on actions can be based on the probabilities of the possible states of the world (i.e., the 'hypotheses' on which the action is based), while accepting or rejecting hypotheses cannot be based on the probabilities of the hypotheses themselves: There is an important difference between accepting a proposition and assigning it a high (even very high) probability. If one accepts a set of propositions, one is committed to accepting all of its logical consequences. For instance, if I accept $p_1, p_2, \ldots, p_n$, then I should also accept their conjunction. On the other hand, each of $p_1, \ldots, p_n$ can be assigned high probability and their conjunction may have a low probability. For example, if one judges $p_1, \ldots, p_n$ to be independent, the probability of the conjunction is the product of the probabilities, and, hence, with n sufficiently large, the probability of the conjunction could be quite low.

In summary, when we accept a proposition as true, we act as if it were true, and, hence, use it as premise in applications of all the normal rules of logic to obtain consequences, which must also be accepted as true. Although this property cannot be considered as a complete characterization of "accepting propositions", it allows one to distinguish between acceptance and the assignment of high probability.[8] When making decisions, on the other hand, we do not need to use the hypotheses in normal logical inferences. If we have accepted a proposition, we can obviously use it for our decisions, but it is not necessary to accept a hypothesis in order to use it for deciding on an action. I shall argue in Chapter IV, Section 1, that it is enough for making decisions to assign probabilities.

Another minor difference is the fact that when deciding on an action we must act, while in accepting hypotheses we can delay our acceptance. So, in this case,

---

[6] Hacking has an excellent discussion of the difference between accepting or rejecting hypotheses and decision theory in [39, pp. 27–32].

[7] A brief discussion of utilities appears in Chapter XX.

[8] For a further discussion of this matter see Chapter IV, Section 2, where Kyburg's lottery paradox, which is highly relevant, is discussed.

2. THE INTERPRETATIONS OF PROBABILITY

if there is no sufficient evidence, we just may say we don't know, and suspend judgement. In decision theory, on the other hand, we must act, although our evidence may not be enough to ascertain truth or falsity.

In short, we may say that in decision theory we pass from probability to probability in order to decide on actions, while in statistical inference proper, we use probability to proceed from truth to truth. Although truth (and not just probability) is a guide for action, truth also has other functions, such as helping us in understanding, explaining, and formulating purposes.

One of the main objectives of this book is to give a coherent account of acceptance and rejection of hypotheses based on probabilistic evidence. This account will be extensively developed in Chapters IV, XVIII, and XIX. I shall also discuss decision theory in Chapters IV and XX, but my discussion will not be as extensive.

Fig. I.1 summarizes our previous discussion. The model of the world is a probabilistic model. That is, it assigns probabilities to propositions. This assignment is thought as logical, in the sense that the model determines the probabilities of propositions without further reference to empirical knowledge. The probabilities thus assigned are used for two purposes: the validation of the model, i.e., its acceptance or rejection, and the use of the model for decision-theoretic purposes.

Besides the discussion just completed about the main epistemic uses of probability, a few words are in order for the newly acquired importance of the study of chance as a part of scientific theories themselves, and not just as the basis for statistical applications for the evaluation of the theories. Although in some of these theories the word 'probability' itself might not occur, probability concepts are present as general statements expressing factual stochastic relations among random quantities. For instance, there may be functions expressing distributions or densities of certain quantities under certain circumstances. In any case, probabilities of events are obtained from them and used in applications.

Examples of this factual application of probability that have been of the greatest importance in the last two centuries, are the uses of probability in physical theories such as statistical mechanics and quantum mechanics. Although I think that any interpretation of probability has to make understandable these uses, I shall not discuss these theories in this book. This analysis will be left for a second publication. I will introduce in this book, however, but not develop extensively, the basis for the theory of stochastic processes, which is another occurrence of probability in scientific theories, such as the theories of Brownian motion and radioactive decay.

2. The interpretations of probability

We shall briefly discuss in this section some of the different schools of interpretation of probability. The description of the different interpretations of probability will be far from complete, both because I shall not include all interpretations that have been offered and because I shall only discuss aspects of

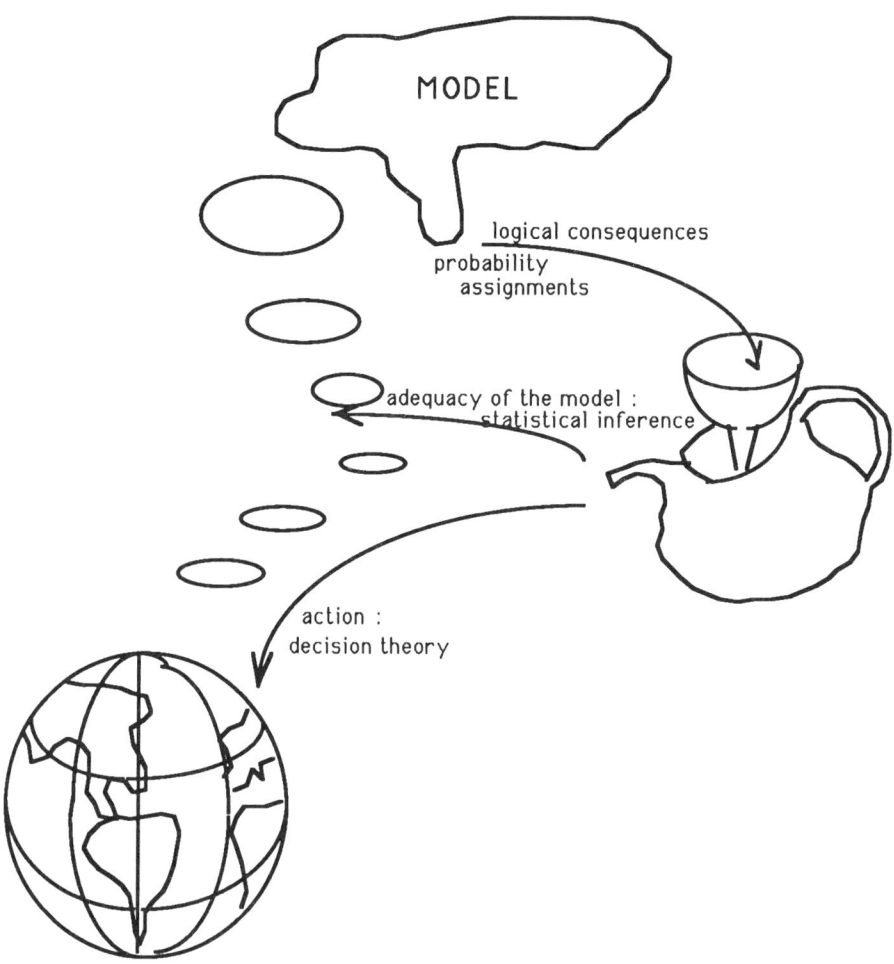

FIGURE I.1. Model of inference and decision

other theories that are relevant to my theory.[9]

The classical definition of probability. Although in all the earliest approaches to probability by Pascal, Fermat, and others, it was implicit a framework of equally likely outcomes, an attemp at an explicit definition of probability seems to have been only offered by de Moivre in 1718. The definition was given more explicitly by Laplace at the beginning of the 19th century (see [31]). Laplace's definition of probability of an event *as the ratio of outcomes favorable to the event to the total number of possible outcomes, each assumed to be equally likely*, was the accepted view until early this century. Thus, the model, in this 'classical view', is constituted by a set of equally likely outcomes, the family of events is an algebra of subsets of this set of outcomes (in most cases, the algebra is the power set), and the probability measure is defined as above.

This 'classical view' is rejected by most modern authors, in part because of the 'compelling reasons', which were given in full in [98]. We shall give a summary of these objections, following, in part, [4, pp. 73–75], and show, later, how they are met in my definition. This is important, since my views are, in a way, similar to the classical position.

(1) *What is meant by 'equally likely'?* Since 'equally likely' seems to be synonymous with 'equally probable' or 'equiprobable' the definition of probability based on this relation seems, prima facie, a vicious circle. We can consider, however, 'equiprobable' to be defined in terms of other notions or as being a primitive notion. Borel, for instance, in [7], claims that everyone has a primitive notion of what is meant by equiprobable. In the definition proposed in this book, 'equiprobable', is defined in terms of notions independent of probability.

(2) *How do we recognize 'equally likely' outcomes?* There are two related principles that have been generally accepted for this purpose. The first is the so-called 'principle of insufficient reason' advanced by Bayes. This principle says, for instance in the case of the casting of a die, that if we have no reason to believe that one or another of the faces is more likely to arise, then we should assume that all are 'equally likely'. The concept, thus defined, is subjective, as a quantitative description of the state of being ignorant. The principle as it stands is contradictory, as it will be shown and discussed in some detail in Chapter III, Section 2.

The second way is an appeal to symmetry or homogeneity of the empirical situation. For instance, if a coin is physically symmetric, why should a tail or a head be favored? This is essentially the basis of the definition proposed in this book. This matter will be taken up in detail in Chapter III, Section 2.

(3) *How restrictive are equally likely outcomes?* Suppose that we assume, on one of the grounds invoked in (2), that a particular die is equally likely to show any of its six faces. As von Mises remarks, [98, p. 69]

---

[9]In particular, I shall not deal with Fisher's fiducial probability, [34], which has been extensively discussed in the literature and that I believe to be a wrong interpretation. For an extensive discussion, see [83].

It is obvious that a slight filing away of one corner of an unbiased die will destroy the equal distribution of chances. Are we to say that now there is no longer a probability of throwing a 3 with such a die, or that the probability of throwing an even number is no longer the sum of the probabilities of throwing a 2, 4, or 6?... (since there is no probability without equally likely cases.)

I believe that the objection presented in this item is by far the most important against the classical conception. On the classical approach, probability seems to find application only in a small range of artificial problems related to unbiased coins or dice, or fair deals of cards, and the like. The wide gamut of enquiries where probability is now employed seem to be outside the realm or probability theory. Some attempts to extend the range of application of the classical definition (for instance, that of Laplace) are not convincing. Fine, in [33], examines appeals to invariance principles, or use of entropy arguments, as bases of support for the classical definition, but finds them unsatisfactory.

One of the purposes of this book is to show that the whole gamut of applications of probability can be encompassed by a modification of the classical definition. In particular, in Chapter XV, Section 4, we prove a general representation theorem that shows that any stochastic process (and, hence, any probability space) can be approximated (in a sense to be made precise later) by a probability measure defined in terms of equiprobability.

(4) It has also been objected that a complete development of the theory has never been attempted. That is, a development complete so as to include all elements of Fig. I.1. For instance, von Mises in [98, p. 61] says:

> ... we shall see that a complete logical development of the theory on the basis of the classical definition has never been attempted. Authors start with the 'equally likely cases', only to abandon this point of view at a suitable moment and turn to a notion of probability based on the frequency definition.

In this book, a complete development is attempted

A few words are in order about what Popper, [76, p. 374], calls the *neo-classical theory*. This is essentially Kolmogorov's axiomatic theory of probability, where probability is anything that satisfies the axioms of the calculus. In general, however, probabilities are taken as generalized measures of possibilities. For instance, in [56, p. 3, 4], Kolmogorov says:

(1) There is assumed a complex set of conditions, $\mathfrak{S}$, which allows any number of repetitions.

(2) ... We include in the set E (which is Kolmogorov's sure event) all the variants which we regard as *a priori* possible.

Example: Let the complex $\mathfrak{S}$ of conditions be the tossing of a coin two times. The set of events mentioned in Paragraph 2 consists of the fact that at each toss either a head or tail may come up. From this it follows that only four different variants (elementary events) are possible, namely: HH, HT, TH, TT.

More precisely, the points of E are outcomes of the experimental setup, $\mathfrak{S}$, as completely specified as we require, which Kolmogorov calls *elementary events*. Sets of elementary events, i.e., subsets of E, give possible outcomes of $\mathfrak{S}$ in a more generalized sense.

Thus, as Popper says, [76, p. 386]:

Thus the interpretation of probabilities as measures of possibilities is rooted in the very structure of the neo-classical theory.

This measure of possibilities is not based, in general, on equal possibilities. The reason for this is that not all probability spaces have equiprobable elementary events. It will be shown in Chapter XV, Section 4, however, that all probability spaces can be approximated (up to an infinitesimal) by spaces based on equiprobability. Moreover, I shall show with many examples, how to obtain naturally the usual spaces that are not based on equal probability, from spaces with equiprobable elementary events.

Frequency theory. Some form of the frequency view of probability is adopted as the right view by most present day statisticians. Although some of the classical authors (for instance, Laplace in [31]) implicitly adopted frequentists or subjectivistic views when trying to analyze applications of probability, one of the earliest attempts at an explicit frequency definition of probability is found in the work of Venn in 1866 (see [97]). Venn defined probability as a ratio in an infinitely large class. The probability of heads is half, because in the infinitely long run of tosses, exactly half will yield heads. The motivation for demanding an infinitely long run is clear: in any finite run, the actual proportion of heads will not be exactly half, except accidentally. But the longer the sequence, the closer will be the proportion of heads among tosses of a fair coin to the number $\frac{1}{2}$.

One of the problems, here, is that we cannot easily make sense of ratios in infinite classes. What we have is limits in ordered infinite classes. Venn's infinite class can be considered an infinite sequence ordered in a natural way: the first toss, the second toss, the third toss, ..., the nth toss, ..., the νth toss, where ν is an infinite natural number. After any number of tosses the relative frequency of heads, i.e., the ratio of heads to the total number of tosses, is well defined. Thus, we also have an infinite sequence of relative frequencies, say $f_1, f_2, \ldots, f_n, \ldots, f_\nu$. It may be the case that $f_\nu \approx f_\mu$, for any infinite μ. If it is, the real number approximately equal to f_μ, for any infinite μ, would seem to be what Venn had in mind for the probability of heads. If there are two infinite numbers, μ and η such that $f_\mu \not\approx f_\eta$, then the probability would not be defined.

A formal definition of probability along these lines was offered by Richard von Mises (see [98]). Von Mises formulated his definition in terms of limits. We can translate it using infinite and infinitesimal numbers, by noting that a sequence a_1, $a_2, \ldots$ tends to the real number L if and only if there is an infinite number ν such that $a_\nu \approx a_\mu \approx L$, for every infinite number $\mu \leq \nu$. Thus, we can rephrase Venn's definition by saying that the probability of heads is the limit of the sequence f_1, $f_2, \ldots$, if the limit exists.

The model, for von Mises, is constituted by a *collective* that is an ordered infinite class of events with additional properties, which we shall discuss a little later. These events are possible results of the repetitions, or trials, of an experiment (tosses of a coin, trials of an experiment, measurements of a quantity, etc.)

The order of the elements in a collective is important, because the same class of events, ordered in a different way, may constitute a class in which the limiting frequency has a different value or is not defined at all. For instance, consider a sequence in which every even-numbered toss yields heads. The limit of the relative frequency of heads is $\frac{1}{2}$. We can reorder the sequence as follows: toss 1, tosses 2 and 4, toss 3, tosses 6 and 8, toss 5, tosses 10 and 12, etc. The limit of the relative frequencies is now $\frac{2}{3}$ instead of $\frac{1}{2}$, although the new sequence contains the same tosses as the old one. In fact, it is possible to rearrange the sequence so that the limit is any value we wish.

We now discuss the extra property that a collective must have. If we had a sequence of tosses of a coin in which all the even-numbered tosses yielded heads and the odd numbered, tails, then we would not, according to von Mises, have a collective, and, hence, we would not have a probability. This is so, because knowing where we are in the sequence, we could predict with perfect certainty whether a head would occur in the next toss or not. This is why, von Mises took as an essential characteristic of a collective, that the outcomes in it occur in an unpredictable, what he called *random*, way. This has been a difficult idea to spell out clearly, and only recently there has been a definition free of contradictions.

The intuitive idea is the following. Consider an ordered sequence of trials of a certain experiment which may result either in success or failure. We assume that the limit of the relative frequencies exists and is a certain number L. Now consider the subsequence consisting of the third trial after a success. For a collective to be random, the limit of relative frequencies in this subsequence should also exist and be L. Any subsequence picked in a similar way, should also have the same limit L. For instance, the subsequences of the trials following a success and a failure, or the subsequence of even numbered trials. All this type of selections were called by von Mises *place selections*. Von Mises required that all subsequences obtained by a place selection have a limiting relative frequency and the limit be L. Although intuitively the idea of a random sequence seems to be clear, it has been very difficult to give a precise mathematical definition of it. It has been done, however.[10]

---

[10] See, for intance, the review article [65], or [96].

2. THE INTERPRETATIONS OF PROBABILITY

Some of the standard objections to von Mises' frequency theory are the following:

(1) Given an infinite sequence of events, it is always an open question whether it is a collective or not. We can never know whether the limit of the relative frequency exists or not, or whether the limit is insensitive to place selections or not.
(2) Most of the sequences that occur in real life are known not to be collectives, since they are not and cannot be infinite. The tosses of a given coin, for instance, can never be considered infinite.

Although von Mises was perfectly aware of these objections, he answered them by saying that in speaking of collectives and limits of relative frequencies in them, he was making an idealization, just as the astronomer, in considering heavenly bodies to be mass points, is also idealizing the situation. I believe, however, that the parallel is not quite right. The astronomer builds a model of reality and in the model he considers the heavenly bodies to be mass points. Thus, the model represents some elements of reality, although not exactly. Infinite sequences, however, cannot be said to represent reality in any reasonable sense. There is no counterpart to an infinite sequence in reality.

A simple alternative to von Mises' theory has been offered by Neyman (see, for instance, [72]). In Neyman's view, probability is defined as the relative frequency in a finite class, called the Fundamental Probability Set or FPS, for short. The class can be any size we wish, but it does not need to be infinite. This proposal avoids some of the criticism mentioned above. It is not clear, however, how it is possible to assign a definite value to the probability of an event. Suppose that we are tossing a coin. To say that the probability of heads is $\frac{1}{2}$ would mean that in a large set, the FPS, the proportion of heads is $\frac{1}{2}$. But this statement is almost always, strictly speaking, false.

I believe, however, that there are more fundamental objections to the frequency theory. The main objections, according to my view, stem from the analysis of probability given in the previous section. In the first place, it is not clear, according to the frequency view, how probability can be applied to single cases. The frequentist has to say that, when talking about single occurrences of a phenomenon, we are really referring elliptically to a sequence of occurrences. An example given by Popper, [75, pp. 31–34], highlights this problem:

> Let us assume that we have a loaded die, and that we have satisfied ourselves, after long sequences of experiments, that the probability of getting a six with this loaded die very nearly equals $\frac{1}{4}$. Now consider a sequence b, say, consisting of throws with this loaded die, but including a few throws (two or perhaps three) with a homogeneous and symmetrical die. Clearly, we shall have to say, with respect to each of these few throws with this correct die, that the probability of a six is $\frac{1}{6}$ rather than $\frac{1}{4}$, in spite of the fact that these throws are, according to our assumptions, *members of a sequence* of throws with the statistical frequency $\frac{1}{4}$.

In this example, we see that the single occurrences of the throw of the correct die, cannot be considered as part of the sequence of throws of the loaded die. I believe, with Popper, that the simple objection presented by this example is decisive. Popper, in the cited article, discusses the possible rejoinders. I shall not go into details here, but what seems to be the best possible answer, from the frequentist point of view, is the following. He might say that the throws of the two dice belong to different sequences: a virtual sequence of throws of the correct die, and the sequence of throws of the loaded die. But, then, he would have to say that the sequence b is not a proper collective, and add, that in order to have a collective the sequence of events must always be a sequence of repetitions of the *same* experiment. Or, as Popper says

> he will say that admissible sequences must be either virtual or actual sequences which are *characterized by a set of generating conditions*—by a set of conditions whose repeated realisation produces the elements of the sequence.

Finally Popper says

> The frequency interpretation always takes probability as relative to a sequence which is assumed as given; and it works on the assumption that a probability is *a property of some given sequence*. But with our modification, the sequence in its turn is defined by its set of *generating conditions*; and in such a way that probability may now be said to be *a property of generating conditions*.

This leads us to the propensity interpretation, which has been proposed by Popper, and that we shall discuss next.

The other objections arise from the way the frequentist has to relate the model to the world. Here, I shall only give a few ideas about the subject. In the course of the book, there will be many comments on this matter. The relationship between the model (a collective, in the case of von Mises, an FPS, in the case of Neyman) and the assignment of probabilities is, as it is expressed in the figure, logical. That is, given that we accept that the model is a collective for a fair coin, then probability of heads is $\frac{1}{2}$, because of the definition of probability itself. One of the major objections, according to my view, is that probability, in the usual sense, simply is not what the frequentists say it is. Thus, the identification of probability with relative frequency is not warranted.

One of the problems we must discuss is how the frequentist validates and uses his model. Many frequentist, such as Neyman, do not talk about accepting or rejecting models, but only as probability and statistical inference as a guide for 'inductive behaviour'. Thus, they only have to explain how to use the model, and not how to validate it. The reason given for using the techniques of statistical inference is that, in the long run, by using the right techniques, the decision will be correct most of the time. The problem with this idea is that there are cases where we are not interested 'in the long run'. For instance, a scientist trying to decide whether or not to accept a hypothesis is not interested in being right most of the time, but only in the particular case he is considering at the moment.

Propensity theory. The most recent empirical interpretation of probability, given in somewhat different forms by Karl Popper, [75] and [76], and Ian Hacking, [39], is the propensity interpretation. In these interpretations, probability is determined by the *propensity* or *chance* of an event to occur. For instance, a biased die might have the propensity to fall $\frac{1}{4}$ of the time with an ace up. This propensity is regarded by Popper and Hacking, as a property of the die together with the mechanism for tossing it, what Hacking calls a *chance setup*. Propensity manifests itself in certain situations. For instance, if a die had the propensity indicated above, and it were rolled a large number of times, then the propensity to come up an ace would manifest itself in the proportion of aces, which would be about $\frac{1}{4}$.

Some authors favorable to propensity theory, such as Popper, [75], [76], seem to identify this propensity with the tendency toward the stability of long run frequencies. Although it is clear that there is a relation between factual probabilities and long run frequencies, a theory such as Popper's, although better in this sense than strict frequentists theories, make it difficult to provide sense directly to single case probabilities, that is, to probabilities applied to single events, without reference to sequences of repeated experiments.

Giere, [36], has proposed a single case propensity theory, but considers propensity as an undefined primitive notion, giving only heuristic explanations, such as saying that propensity is a real kind of power, force, or tendency. I believe that propensity is too mysterious a notion to leave undefined. Hence, a definition of chance in terms of other notions is in order. Suppes, [90], has given axiomatizations of chance for different specific cases where it occurs, for instance for radioactive decay. This is a step in the right direction, but I think that a definition of chance in terms of other concepts, or an axiomatization that works for all cases, is preferable to a different axiomatization for each particular case. My proposal is to define chance (which can be called factual probability) using the notion of real physical possibility, which is understood to be a property of things that manifests itself in certain circumstances. Possibility seems to be more intuitive than propensity. This notion, as well as other notions of possibility, will be discussed in the next chapter.

Many propensity theorists seem to believe that the only validation of probability statements is through statistical inference methods. This is clear in Popper, and, I believe, it is also assumed by other proponents of the theory. In some cases, however, I believe that there are other possible validations of probability statements. Suppose that we have a die and that we measure it carefully establishing that it is symmetric. Then we would be justified in saying that the probability of $\frac{1}{6}$ for an ace is well justified. A propensity theorists might reply that this is so because with previous experiences with symmetric dice indicate that the relative frequency of aces for these dice tends to $\frac{1}{6}$. But I believe that physical laws support the statement that the probability of an ace is $\frac{1}{6}$. This possibility is even clearer in the case of a biased die. Be measuring carefully the distribution of weight, we could arrive at a conclusion about the probability distribution of the faces. We shall discuss the matter further in the sequel.

For propensity theorists, we have a model in which propensity is a primitive concept. This model is validated by the usual statistical procedures. For example, if we assume that the propensity of an ace is $\frac{1}{4}$, then the result of a proportion of $\frac{1}{10}$ of aces in 1,000 castings, would go against the assumption. The problem here is to explain why one should reject the model, if such a result would obtain. A parallel with a deterministic model might help to understand the problem. Suppose that the model asserted that the frequency of an ace was exactly $\frac{1}{4}$. Then the occurrence of a frequency of $\frac{1}{10}$ would invalidate the model, because the statement 'the frequency of an ace is $\frac{1}{10}$' is false according to the model. Let us return, now, to the propensity model. This model only asserts that the *propensity* of the die is to have an ace $\frac{1}{4}$ of the times. The model does not imply that a relative frequency of $\frac{1}{10}$ is false, but only that it is improbable. That is to say, that the propensity of getting 100 aces in 1,000 castings of the die is low. I do not believe that the justification for rejection of such a model, in cases the relative frequency is very different from the probability, which have been given so far by the proponents of propensity theory are enough.

Another problem is the justification of the use of probability statements for decision purposes. We shall discuss further this matter when we introduce a principle of direct inference.

The theory that is proposed in this book is related to propensity theory, in the sense that I believe real possibilities to have a certain tendency to occur, and that I define probability in terms of possibility. Besides being related to propensity, my theory has an additional feature: possibility (and, hence, probability) are related to truth, so that the usual process of validation of models is justified.

Personalistic or subjectivistic theories. Subjectivistic theories take probability statements to assert something about the degree of belief of a particular person at a particular time. A degree of belief is interpreted as an inclination, of the person at the given time, to take certain actions, make certain choices, or accept certain bets. A statement of probability such as "the probability of rain this afternoon is $\frac{1}{4}$" is always understood as having two implicit parameters: a person, say p, and a time, say t. Such a degree of belief is given an interpretation such as: at time t, p will offer three to one odds against the occurrence of rain this afternoon, or, p will be indifferent at t between receiving a prize of one dollar whatever happens, and receiving a prize of four dollars, if it rains this afternoon.

It is not asserted by subjectivists that everybody assigns numbers to their degrees of belief in such a way that they obey the laws of the calculus of probability. It is asserted, however, that a rational person should assign her degrees of beliefs as probabilities. As it is stated in [4, p. 67], a person is considered rational, if she

(1) appreciates that all alternatives can be ordered in terms of their relative positions on the inevitability—impossibility scale,
(2) is *consistent* in her judgements; for example, if A is felt to be more likely than B, and B than C, then C cannot be felt more likely than A,
(3) is *coherent* in not being prepared to accept a series of bets under which she must lose. (Some would say 'may lose'.)

Subjective probability represents the behavior of a conceptual 'rational human being'. These requirements lead logically to a system of axioms for probability theory which are essentially those of Kolmogorov.[11] The theorem of de Finetti, [29], and Ramsey, [77], shows that a subject who is rational, in the sense indicated above, assigns her degrees of belief according to the axioms of probability. Savage, [79], has extended the work of de Finetti and Ramsey.

By the consideration of conditional bets or actions, it is also proved in the theorem mentioned, that conditional probability has to obey

$$\Pr(A|B) = \frac{\Pr AB}{\Pr B}.$$

According to subjectivists, probabilities, or, what for this theory is the same, degrees of belief, can only change, for a rational subject, by what is called *conditionalization*, which I proceed to explain. It is assumed that the subject assigns conditional probabilities, $\Pr_{t_0}(A|B)$, at a certain time t_0, to every pair of event A and B, or at least, to every pair in which she is interested. If at a time t_1, subsequent to t_0, she has observed B to be true (and she has no other relevant knowledge), then the probability assigned at t_1 to A should be

$$\Pr_{t_1} A = \Pr_{t_0}(A|B).$$

Since the subject is supposed to be rational, the only way to change probabilities seems to be by this process of conditionalization. Thus, degrees of belief are not considered as purely subjective, but they must obey certain minimum criteria of rationality.

I think that it is clear that in many cases numerical rational degrees of belief can be assigned to propositions. The theorem of Ramsey-de Finetti,[29], [77], shows, at least in situations where betting makes sense, that if we assign numbers to degrees of belief, they must obey the calculus of probability, if we are at all rational. This restriction to betting situations is important. The only propositions to which this theorem and its extensions apply are those that are decidable: that is, to propositions whose truth or falsity can be decided in a finite amount of time. We have to be able to decide who won a bet.

Although Ramsey and others have eliminated bets in favor of utilities, this restriction to decidable propositions remains in force. Thus, it is not certain, and for me it is very doubtful, that rational degrees of belief for general propositions, such as scientific laws, should obey the calculus. It is not even clear that we may be able to assign numbers to this kind of belief. Furthermore, as we shall see in Chapter IV, Section 3, it is not even certain that it always makes sense to assign numerical degrees of belief, even for decidable propositions. Subjectivists, such as de Finetti or Savage, assume that we can, and do, always assign numerical degrees to our beliefs in decidable propositions, and that the only way in which these degrees change is by conditionalization. The only limitations that these subjectivists put on the assignment of probabilities is that it obey the calculus.

---

[11] De Finetti, however, rejects countable additivity as a plausible axiom.

Strict subjectivists, such as de Finetti, reject objective probability altogether. As de Finetti says, [30, p. X]

> My thesis ... is simply this:
>
> Probability does not exit.
>
> The abandonment of superstitious beliefs about ... Fairies and Witches was an essential step along the road to scientific thinking. Probability, too, if regarded as something endowed with some kind of objective existence, is no less a misleading misconception, an illusory attempt to exteriorize or materialize our true probabilistic beliefs.
>
> In investigating the reasonableness of our modes of thought and behavior under uncertainty, all we require, and all we are reasonably entitled to, is consistency among these beliefs, and their reasonable relation to any kind of relevant objective data ('relevant' in as much as subjectively deemed to be so). This is Probability Theory.

There are subjectivist, such as D. Lewis, [62], Suppes, [90], da Costa, [27], that accept both objective and subjective probabilities.

The main problem, in my judgement, about an extreme subjectivist view, such as de Finetti's, is that it seems clear that objective probability does exist. The attempts of subjectivists to explain psychologically our belief in objective probability (see, for instance, [29]) seem unconvincing. On the other hand, I do not think that subjectivists that accept objective probability have dealt successfully with the relationship between the two. We shall say more about this subject later.

da Costa's pragmatic probability. da Costa, in his paper [27], has proposed a novel theory of subjective probability, which he calls *pragmatic probability*. His main tenet is that epistemic probability can be identified, not with the rational degree of belief in the truth of a proposition, but with the rational degree of belief in the *pragmatic* truth or quasi-truth of a proposition. Informally, he says that a proposition is quasi-true in a certain domain of knowledge D, if things occur in D as if it were true. Thus, scientific theories are, while they are accepted, pragmatically true.

da Costa has been able to formalize the notion of pragmatic truth (see [27] and [68]) and pragmatic probability, showing that the latter obeys the usual axioms of the theory of probability. He gives sense to betting situations for propositions that are not, strictly speaking, decidable, such as scientific laws, by considering bets in the pragmatic truth of the propositions for a limited amount of time, say t. Thus, an individual betting on the quasi truth of a proposition, wins, if the proposition remains quasi-true for time t. In this way, the bet is decidable in a finite amount of time.

Also, general propositions, although almost surely will be shown to be false, and, thus, must receive probability zero, may not be shown to be false during t, and, hence, may be assigned (rationally) pragmatic probability larger than zero.

The main problem with this view, I believe, is the fact that the time t in which a proposition must be quasi-true is entirely arbitrary, and the probabilities depend crucially on it. Thus, the assignment of probabilities has a hidden parameter, t, and different periods of time may yield different assignments. I do not believe that this is in agreement with usual practice.

Logical theories. Logical interpretations take the probability of a statement or event to be a function of a body of evidence. Thus, the statement 'The probability of rain this afternoon is $\frac{1}{4}$' is understood as having one suppressed parameter: the body of evidence e. According to logical theories, the truth conditions for a probability statement are logical. Given the statement 'It will rain this afternoon' and given a body of evidence e, there is exactly one real number r, determined only by logical conditions, such that the probability of 'It will rain this afternoon', relative to the evidence e, is r. A probability statement is either logically true (if it mentions the right number) or logically false (if it mentions the wrong number). For instance, Jeffreys, [49], says:

> The idea of a reasonable degree of belief intermediate between proof and disproof is fundamental. It is an extension of ordinary logic, which deals only with the extreme case... The problem of probability theory is to find out whether a formal theory is possible. Such a theory would be impersonal in the same sense as in ordinary logic: that different people starting from the same data would get the same answers if they followed the rules. It is often said that no such theory is possible. I believe that it is possible, and that inattention to the progress that has actually been made is due to an active preference to muddle.

In the logical view, probability is regarded as a concept which extends the applicability of formal logic. The basic logical relation between two propositions φ and ψ, of 'φ logically implies (or refutes) ψ' is extended to a notion of *degree of implication*, called by Carnap, [9], *degree of confirmation*: when applied to a body of knowledge e, in general expressed by a proposition or a set of propositions, and a potential outcome, expressed also by a proposition, h, probability indicates the extent to which e implies h. Thus, probability is always conditional. It can also be paraphrased by the *rational* degree of belief in h afforded by e. Carnap also calls this rational degree of belief *credibility*, [10].

The main exponents of this type of logical theory are Keynes, [55], Jeffreys, [48], and Carnap, [9] and [10].[12] Carnap has developed an extensive logical theory, without being able, however, to select a unique measure, as a logical concept demands. Carnap's theory will be discussed in several places in Part 1 of the book, especially in Chapter III, Section 2, so that I shall give, here, only a brief outline of the main objections.

---

[12] Kyburg, [59], has presented a different type of logical theory, which does not lead to probabilities in Kolmogorov's sense, but to the assignment of intervals instead of real numbers. This interpretation, although very interesting, is not related to mine, so I shall not discuss it in the book.

The main objection to the usual logical theories is that nobody has succeded in selecting a unique logical measure of probability in this sense. I believe that the principal reason for this to be the fact that probability is placed between logical truth and logical contradiction, and not, between truth and falsehood. The solution for this problem given in this book is the following. Instead of considering logical probability as a relation between two propositions or sets of propositions, it is taken to be the relation between a proposition and a possible model of the proposition. Thus, probability is a semantical concept, instead of being syntactical. This relation between a proposition, φ, and a possible model, $\mathcal{K}$, is what I call *the degree of support of φ given $\mathcal{K}$*.

It is to be noted that in all empirical interpretations, such as the frequency or propensity theories, there is a similar logical connection between a model and a proposition or event. According to the propensity interpretation, for instance, if one assumes a certain model for a chance setup, then the probabilities are determined by the propensities of the model. The empirical connection is given by the relationship between model and reality.

3. The Principle of Direct Inference

I believe that there are three different, but related, aspects of probability, as it occurs in the calculus of probability, and that each one of the three has been taken as the basic (and in some cases the unique) notion of probability by the different schools on the foundations of probability. The objective of this section is to explain how these different aspects of probability are related to each other.

The first is the factual notion of chance. Chance is a property of the objects and the experimental conditions. As we have seen, the chance of an event is interpreted by some authors as its propensity to occur. The propensity of an event A is generally estimated by the long run frequency of the occurrence of A in a sequence of trials of the experiment in question.

Besides the factual interpretation just described, two versions of probability have been put forward: a logical notion of degree of support or confirmation, and a subjective notion of degree of belief. I think that for each of these two notions there are situations in which it makes sense to use the calculus of probability. Degree of support, as a logical concept, assigns a number to the support given to a proposition (call it the hypothesis) by the assumption that a certain state of affairs (call it the support) obtains in the real world. As a logical concept, it is independent of empirical facts and of our knowledge. It depends only on the supposition of a certain state, and not on the actual realization of this state. Because of the logical character of the notion, there should be a measure uniquely determined by hypothesis and support. Thus, such a logical concept accounts very well for the apparent objectivity and dependence on the support of probability that was mentioned in Section 1, i.e., for the fact that if two persons agree on the evidence, they generally agree on the probabilities assigned.

As we briefly discussed in the previous section, there are two possible notions of logical probability. One notion is to consider logical probability as a relation between a proposition and another proposition or set of propositions. This is the version of Carnap, [9] and [10], Kyburg, [59], and others, Carnap, however, has failed to define such a unique measure. In the case of Kyburg, he does not obtain a theory which obeys Kolmogorov's axioms. The main reason for a failure of this kind of theory, as I mentioned before, is that Carnap's degree of confirmation is set between logical truth and contradiction. I believe that it is really between truth and falsity.

The logical theory that I shall present in this book avoids the problem by putting degree of support between truth and falsity. In this way, it is possible to obtain unique measures. Logical probability, in my case, is taken to be a relation between a proposition and a structure, which is a possible model of the proposition. These models, instead of assigning truth and falsehood to propositions, they assign probabilities. Thus, probability turns out to be a semantical concept. I shall discuss informally in the next chapter my conception.

The second notion is that of probability as the measure of the degree of belief in the truth of a proposition that a person has at a certain time. The important conception of degree of belief in our context is that of *rational degree of belief*, that is, the degree of belief that an individual *should* (or, more properly *may*) have, given her knowledge of the situation involved. It may be the case, that knowledge does not determine a unique rational degree of belief, but that there are several that are rational for the situation involved. Since the rational degree of belief of a person depends on her knowledge, this form of probability may be properly called *epistemic probability*. As we saw in the previous section, this is the basic notion (and for some, the only notion) of the subjectivist school of foundations of probability.

Extreme objectivists, such as von Mises, Popper, and, probably, Neyman, reject probability as degree of belief. For instance, Popper seems to think that if there is such a thing as a numerical degree of belief, it should not obey the laws of probability. Hacking, [39], on the other hand, accepts both factual and epistemic probabilities.

On the extreme objectivist view, where no epistemic notion is possible, it is difficult to make sense of the epistemic uses of probability, for instance, to justify statistical inference. One way out, that I believe has been adopted at least by Neyman, is to say that we should act according to the results of statistical tests and estimations, because 'in the long run' we would be successful most of the time. As it was mentioned before, this justification of statistical inference is not enough, for instance, for the cases when we are dealing with the validation of scientific hypotheses: the problem with a scientific hypothesis is whether to accept it or not, and the fact that we are successful 'in the long run' does not help a scientist to decide whether or not a particular hypothesis is correct.

We have seen that extreme subjectivists, such as de Finetti and Savage, reject objective probability altogether. This view of extreme subjectivists, who reject objective probability, contradicts the apparent objectivity of probability and its relation to chance, and makes it very difficult to understand the factual use of probability in scientific theories.

Although I disagree with interpretations based exclusively on only one of these notions, I believe that the three, namely, chance, degree of support, and rational degree of belief, make sense, at least in some cases. I also believe that any interpretation of probability should show that these three notions and their connection are reasonable.

There are many theorists that agree that at least two of the three notions make sense. Fon instance Carnap, [9], accepts objective probability, in a frequency interpretation, and degree of confirmation, which is a logical notion, as legitimate concepts. The connection between the two is given in the following dictum, [9, p. 173]:

> ... probability$_1$ may be regarded as an estimate of probability$_2$.

Probability$_1$ is for Carnap degree of confirmation, and probability$_2$ is for him chance, for which he accepts a frequentist interpretation. An "estimate" can be considered a degree of belief. So he may be paraphrased as saying that a degree of confirmation can lead to a degree of belief.

D. Lewis in [62] and N. C. A. da Costa in [27] also accept both epistemic and factual notions. Lewis says, [62, p. 263]:

> Carnap did well to distinguish two concepts of probability, insisting that both were legitimate and useful and that neither was at fault because it was not the other. I do not think Carnap chose quite the right two concepts, however. In place of his "degree of confirmation", I would put *credence* or *degree of belief*; in place of his "relative frequency in the long run", I would put *chance* or *propensity*, understood as making sense in the single case. The division of labor between the two concepts will be little changed by these replacements. Credence is well suited to play the role of Carnap's probability$_1$, and chance to play the role of probability$_2$.

For those who believe in two or three notions, it is necessary to formulate a principle to connect them. These principles are usually called *principles of direct inference*. Since I believe that the three notions, i.e., chance, degree of support, and rational degree of belief, are useful, I shall formulate my principle divided into two: one part connecting chance with degree of support, and the second, connecting chance with degree of belief.

We state the following provisional version of the Principle of Direct Inference. I shall state strengthened versions at the end of Section 1 in Chapter IV.

Let A be an event. Then:

(1) *The degree of support of A's holding, given that the chance of A is r (where r is a real number between 0 and 1), is also r.*

(2) *Hence, if a person believes at time t that the chance of A is r, then, if he is rational, his degree of belief at time t of A's holding, should also be r.*

This is a principle that allows us to obtain epistemic probabilities from chance; i.e., from the belief that the chance of an event is r, we arrive at the rational degree of belief (i.e., the epistemic probability) in the event.

The principle as stated has two parts. The first part connects chance with degree of support. This is a logical principle, since it does not give the degree of support of A's holding absolutely, but only indicates a relation between two propositions: the proposition that asserts that A holds, and, its support, i.e., the proposition that asserts that the chance of A is r. This first part is not properly a principle of inference, because nothing is inferred: the principle only asserts which is the proper degree of support of an event, given the chance of the event. Degree of support is similar to Carnap's degree of confirmation that connects a hypothesis with its evidence. Degree of support also connects a hypothesis (i.e., that A holds) with its evidence (i.e., that the chance of A is r). The main difference with Carnap's notion is that we only allow as evidence statements about chances.

The second part connects chance and degree of belief, and is appropriately a principle of inference: from the belief that the chance of A holding is r, we infer A's holding with degree of belief r. I think that this second part is a consequence of the first, since the connection between degree of support and degree of belief seems to me obvious. The deduction of the second part from the first goes as follows. Assume that the person p (at time t) believes that the chance of A is r. By the first part, we have that the degree of support of A's holding, given that the chance of A is r, is also r. Since this statement is true and p is rational, p should believe it to be true. From these two beliefs, p should deduce that the absolute degree of support of A's holding is r, and, thus, p should believe that the absolute degree of support of A's holding is r. It seems clear to me that if one is rational, and believes that the degree of support of a proposition is r, then one's degree of belief in the given proposition should also be r. Thus, we get that, since p is rational, p believes with degree r in A's holding.

The first part also has, what I believe, an intuitive appeal. This intuitive appeal, however, should be justified or at least explained in any definition of probability, since the principle does not seem self evident. My definition of probability will provide such a justification.

It is to be noted that the occurrences of r in both parts of the principle are in intensional contexts. That is, although the 'chance of A' is equal to r, we cannot substitute r by the 'the chance of A'.[13] I shall give later a reformulation of the first part that is wholly extensional. For the second part, I do not think there is hope of an extensional reformulation.

The principle, to which I have given what I believe to be its traditional name, is accepted by almost all authors who accept the different concepts involved in it. For instance, we have seen that Carnap accepts a similar principle. Lewis, [62], on the other hand, agrees with the second part. The same is true of da Costa, [27], who is also a moderate subjectivist. Kyburg, [61] and [60], and Hacking, [39], have also defended similar principles.

Even the classical statisticians who believe in a frequency interpretation, accept a sort of Direct Inference Principle: they say that we should act in accordance with the estimates of objective probability obtained by their techniques.

---

[13] This peculiarity has been pointed out by Miller in [69].

Since our actions should be based on our degrees of belief, I think that I am justified in considering this frequentist principle as a form of direct inference. As I mentioned before, the frequentists even justify their principle by saying that following it we would be successful in the long run most of the time.

In summary, I believe that any interpretation of probability should justify both the factual and cognitive uses of probability, plus the Principle of Direct Inference. I attempt to provide such an account.

4. Foundations, an a priori inquiry

The inquiry about the meaning of probability begins as an a priori inquiry.[14] I shall argue, in this section, in favor of this view.

There are many instances of philosophical ideas influencing the development of a science. But statistics is the only case I know of where philosophical ideas are not just relevant for the development of the science, but also have immediate practical implications. It is well-known that the different interpretations of the concept of probability, briefly discussed in the previous sections, which are based on competing philosophical ideas, lead to different statistical techniques. For instance, those who favor a subjectivist or logical interpretation are Bayesian; the believers in a frequency interpretation choose the methods of Neyman-Pearson or Fisher.[15] Although sometimes the application of the different techniques may steer us to similar results, it is also not infrequent that they may yield mutually contradictory consequences.

An attitude that seems reasonable is to accept one technique or another according to the specific state of affairs to which the method is applied. Thus in a given situation there would be no conflict. Coupled with this idea, some believe that the statistician just with his experience can (and must) decide which is the best technique for each situation. This experience is such that it cannot be expressed in words. The statistician, just as the physician before a patient, knows which method to apply without being able to say why.

This comparison, however, is not adequate. If two physicians with very good reputations are studying the same patient, it is almost sure that they will reach the same conclusion. On the other hand, among statisticians this is certainly not the case. There are many instances where statisticians of the best credentials disagree completely, not only on the mathematical techniques to apply, but also on ways of collecting the data.

These examples do not refute the assertion that different techniques sometimes work for different problems, but it does suggest that a theory is necessary to characterize in some way the different types of problems, and to justify the application of the different techniques to different problems. This book presents such a theory.

It is also possible to think that, just as in the natural sciences, experience will enable us to decide between the different statistical methods. It has been usual in

---

[14] As can be easily seen, the influence of [59] is considerable in this section. Most of the content of this section appeared, in a somewhat different form, in [19].

[15] Part 4 of the book is concerned with the analysis of the different statistical techniques.

4. FOUNDATIONS, AN A PRIORI INQUIRY

other sciences that competing theories exist side by side, until by experiments or observations one theory is accepted and the others rejected. Similarly, experience might enable us to decide which statistical techniques work in general, or work in particular situations. Thus, the methods that work well in general or in each situation will be retained, and the others abandoned or restricted in their use. The difficulty here is to understand what we mean when we say that a certain method works. The following analysis, taken from Kyburg, [59, pp. 148–149], is illuminating:

> Let us suppose (what is no doubt contrary to fact) that we have a clear idea of what it is for a statistical technique to work in a given instance. Quite clearly it is not in the nature of a statistical technique to work all the time, or to work in the same degree all the time. It is conceivable, however, that we could collect a number of instances i of the application of the statistical technique, and associate with each instance a number $D(i)$ representing the degree to which that technique worked in that instance. This provides a picture of how well that technique worked in the past. But when we talk about how well a statistical technique works, we are concerned not merely with how well it has worked in the past, but how well it works in general: that is, what concerns us is not the sample of values of the random quantity D that we have happened to observe in the past, but with the general distribution of D. How do we arrive at knowledge of this distribution? Why, by statistical inference of course.

Let us look more carefully at Kyburg's analysis. Suppose that we have a collection of possible techniques T, and that for each t in T we have a random quantity D_t, such that $D_t(i)$ represents the degree to which t worked in instance i. That is, the higher is $D_t(i)$ the better worked the technique t in instance i. Thus, a technique t_0 would be preferable, if the estimation of D_{t_0} is higher than D_t, for all other $t \in T$. The problem is that the estimation of D_t has to be done using one of the techniques in T. Let us call the estimate of D_t using technique s, $\mathbf{E}_s(D_t)$.

Although it is extremely unlikely, it may be that applying any technique t in T, the estimation of D_{t_0}, $\mathbf{F}_t(D_{t_0})$, for a certain t_0, is higher than $\mathbf{E}_t(D_s)$ for all $s \neq t_0$. Then clearly we should conclude that t_0 is the best technique. Thus, it is possible, although very unlikely, that a decision based on experience could be reached.

There is another problem, however, that I believe is insurmountable for a strictly empirical procedure: how are we going to choose our set T of methods. It is clearly impossible to take T to be the set of all logically possible techniques, since we can always design a silly method that only approves of itself. For instance, a silly technique t_1 such that, by definition, $\mathbf{E}_{t_1}(D_{t_1}) = 1$ and $\mathbf{E}_{t_1}(D_s) = 0$, for all $s \neq t_1$. Hence, we must select a reasonable T. But, on what empirical basis can this be done? We would again have to consider a set F of reasonable

T's, and maybe again apply statistical techniques to it. We can conclude with Kyburg, [59, pp. 149–150]:

> This supports the intuition that what is at issue in conflicts among approaches to statistics is a profound and pervasive conception of the nature of statistical inference and the nature of probability... We are offered what appear to be ultimate principles — principles so basic that it is difficult to see how to defend them or argue about them.

Thus, we arrive at the basic difficulty. It seems that in dealing with the foundations of probability and statistical inference, we are confronted with fundamental epistemological questions. Since it is obvious that statistical inference has practical importance, we are in the presence of a philosophical problem with an immediate practical application: how to decide between alternative statistical techniques.

CHAPTER II

Truth, possibility, and probability

This chapter is devoted to an analysis of truth and possibility, which are the two main notions that serve as a basis for my semantic definition of probability. I shall also give an informal account of this definition.

1. Truth as related to probability

This section studies the relationship between truth and epistemic probability. This discussion is not an exhaustive analysis of the notion of truth, but only of its aspects which are relevant to the understanding of epistemic probability. In this section, when talking about probability, I mean mainly epistemic probability.

There has been a tendency among some philosophers, especially among those who work in the foundations of probability, to consider epistemic probability as prior to truth. That is, to think that truth is unattainable, at least in empirical matters, and that, in practice, truth can be identified with high or very high probability. I believe, on the other hand, that truth is prior to probability; that is, that in order to be able to assign probabilities, we must accept some contingent propositions as being true, and not just contingent particular propositions, but also propositions of a general nature.[1]

As it was argued in Chapter I, Section 1, and will be further argued in Chapter IV, Section 2, there is an important difference between accepting a proposition and assigning it a very high probability. This difference comes from the fact that in accepting a set of propositions we are also committed to all of its logical consequences, while on assigning high probability to each proposition of a set we cannot automatically assign high probability to the consequences of the set of propositions. For instance, we may assign high probabilities to two propositions,

---

[1] I also believe that we can know some contingent general propositions as being true, but only acceptance of propositions is needed for my arguments here. In fact, my arguments in this section will only support the view that in order to define probabilities we have to accept some general propositions as being true, not that we have to know them to be true. Actually, the whole construction of my definition of probability will further support the position that the acceptance of some propositions as true is necessary for any assignment of probabilities.

while their conjunction, which is a consequence of the set of the two propositions, may be assigned low probability.

If propositions are never accepted, but only assigned probability, then it is natural to eliminate truth from consideration and place probability between logical truth and contradiction, instead of between truth and falsehood. This position is especially clear in the usual logical theories, but I also think it is true of most of the subjectivist theories.[2] More precisely, as we saw earlier, Carnap, [9], for instance, believes the following: if e logically implies h (i.e., e implies h, is logically true), then the degree of confirmation of h given e is one; if e logically implies not h, this degree is zero; and in all other cases, it is in between. As I have mentioned earlier, I think that probability is really between truth and falsity. Hence, the scale should be: logical truth, truth, probability, possibility, falsity, and logical contradiction. Thus, if e implies h is (factually) true, then the degree of support of h given e, should be one. By taking into account the truth of certain general propositions, we limit the possible results that we have to consider for defining probability to those that are really possible given the truths that we are accepting, and excluding many that, although logically possible, are incompatible with what we accept as true.

The main kind of propositions that must be accepted as true in order to assign probabilities are propositions that assert which are the possible outcomes of a certain setup. As it will be discussed in the next section, these possibilities are factual possibilities (not just logical possibilities) and are determined by the physical situation involved. Thus, the empirical situation determines the factual possibilities, which, in their turn, determine the probabilities.

Tarski, [92],[3] in his semantic definition of truth for formalized languages, replaced the talk of propositions being true or false, by sentences being true or false under an interpretation. These interpretations are represented by set-theoretical systems that, loosely speaking, stand for possible worlds. In my theory, I shall do something similar. Probability will be assigned to sentences of a formalized language under an interpretation which, instead of fixing truth or falsity, determines probability. Probability, in its turn, will be based on a measure of the set of possibilities where sentences are true.

In this and the next few paragraphs, I will argue that it is necessary to accept general contingent propositions, in order to determine and to change, at least some, probabilities. It is clear that everybody accepts that we can change our probabilities, if we accept certain propositions as true. For instance, in the process of conditionalization, used by subjectivists and accepted by most schools in foundations, probabilities of some propositions are changed when we accept further propositions as true. We shall repeat, here, with a few additions, the explanation, given in Chapter I, Section 2 when discussing the subjectivist school, of how conditionalization works. It is assumed that the subject assigns conditional probabilities, $\Pr_{t_0}(A|B)$, at a certain time t_0, to every pair of events A and

---

[2] An exception again, seems to be da Costa, [27].
[3] For the definition of truth, [32] or [11] may be consulted. An excellent informal explanation appears in [94].

B, or at least, to every pair in which he is interested. Suppose that the agent has assigned $\Pr_{t_0}(A|BC)$. If at a time t_1, after t_0, the agent has observed B to be true (and he has no other relevant knowledge), then the probability assigned at t_1 to A, given C, should be

$$\Pr_{t_1}(A|C) = \Pr_{t_0}(A|BC).$$

My idea about the necessity of truth for probability goes further: there has to be an assignment of probabilities based on the acceptance as true of general features of the situation, in particular, which are the factual possibilities for the setup under consideration. Suppose that we were to base our changes of probability only on the process of conditionalization, which is only based on the acceptance as true of particular propositions. As it was seen above, the probabilities at a certain time t_1, depend on the probabilities at an earlier time, t_0. Thus, there must be, in order to avoid an infinite regress, a first general assignment of conditional probabilities. If the first assignment of probabilities is not based on general contingent truths, then there are only two possibilities for this assignment. Either this first assignment is not justified at all, but only irrationally accepted (as by the subjectivists)[4] or a justification is attempted on purely logical grounds.[5]

The first alternative seems too irrational for me to agree with, especially because it involves an arbitrary assignment of conditional probabilities to all pairs of propositions (each and every one of them), whatever "all propositions" may mean. On the other hand, all attempts to justify the first assignment of probabilities on purely logical grounds depend crucially on the language chosen, and have not succeeded in providing criteria for determining the probabilities uniquely.

Since these two alternatives are unacceptable, we are forced to base our first assignment of probabilities on the acceptance of some contingent propositions as true. Actually, I don't believe that there is one first general assignment of probabilities, but one assignment for each particular situation in which we use probability. I do not see how these first assignments could be based upon particular propositions. Thus, they should be based on the general features of the phenomenon, i.e., on general contingent propositions, such as which are the factually possible outcomes of the setup.[6]

If probabilities are not first assigned for particular situations, but to all possible states of affairs, reference to "all propositions" or "all events" is unavoidable. The problem of having to talk about "all propositions" or "all events", seems to me to be one of the main stumbling blocks of the usual logical and subjectivists positions. It is difficult to imagine what "all propositions" are. Even if we knew what all propositions are, the capability of the agent to assign probabilities to all propositions seems very doubtful.

---

[4] Such as de Finetti, [29], Savage, [79] and, also, Lewis, [62].

[5] As by Carnap, [9] and [10].

[6] For another situation where probabilities seem not to be assigned by conditionalization, see [47, Example 11.1]. I believe that also in this case, general features of the situation determine the assignment of probabilities.

For the views that must assign first probabilities to all propositions, an added complication arises, because, in all attempts that have so far been made, the probabilities assigned to particular propositions always depend on the totality of propositions that we are considering (in Carnap's version, it depends on the language chosen). This is not true of deductive logic, where the fact that a proposition is deducible from others depends just on the propositions involved. But in order to define probabilities, we have first to decide on the class of propositions that we consider relevant for the situation. Hence, in order to avoid an infinite regress, and having to refer to all propositions, there must be a first selection of relevant propositions based, not on probability, but on what we accept as true. This first selection must again be based on general features of the situation in question.

In the next two sections, I shall try to show how the probability of a proposition can be based on the truth of the proposition in the different factual possibilities of the setup considered. As it was mentioned above, truth of propositions in possible outcomes, will be replaced by truth (in the sense of Tarski) of sentence in models for these possibilities. Hence, as we shall see, in the definition of probability, truth is involved in two places: we must decide which possibilities we accept as the true factual possibilities, and we also must be able to determine in which of these possibilities each sentence is true.

2. Possibility

There are several notions of possibility. In order to clarify the matter, we shall distinguish with Bunge, [8, Chapter 4], *real* and *conceptual possibility*. Real possibility is a property of nature, while conceptual possibility is always relative to a body of propositions. I begin by considering three varieties of conceptual possibility: technological, physical, and logical possibility. The distinction between the three is clearly stated in Mates' book [66]:

> ... we attempt (with perhaps questionable success) to distinguish technological, physical, and logical possibility. Something is technologically possible if, in the present state of technology, there is a way of doing it. It is physically possible if the hypothesis that it occurs is compatible with the (ideal) laws of nature. As technology advances, more and more things become technologically possible, but this advance always takes place within the domain of physical possibility. Many feats that are physically possible are now technologically impossible and perhaps always will be. Logical possibility is wider even than physical possibility. An event is logically possible if the hypothesis that it occurs is compatible with the laws of logic. If something is technologically impossible but physically possible, we may hope that eventually some clever man may find a way of getting it done. If something is physically impossible, however, no businessman, engineer, or scientist, no matter how ingenious, will

ever bring it off. If something is logically impossible, even the Deity joins this list.

We can see from this description that possibility, in the sense discussed by Mates, applies to propositions and is relative to a certain body of propositions. Logical possibility is relative to the laws of logic. Physical possibility, also, to the laws of physics. Thus, conceptual possibility is the relation of compatibility between a proposition and a body of propositions. The notions of logical and physical possibility discussed by Mates are what may be called, forms of objective conceptual possibility, since they are independent of our knowledge.

The notion of conceptual possibility that is important for the theory presented in this book, especially for statistical inference (cf. Chapter IV), is that of epistemic possibility. *Epistemic possibility* is compatibility not with the actual laws of logic or of nature, but with what we know, or, more appropriately, with what we accept as true. Thus, it is a notion that is relative to the state of our knowledge. Something may be epistemically possible at one time and not at another. Epistemic possibility is also relative to a person or group of persons. Epistemic possibility is similar to physical possibility, but it is possibility not according to the actual laws of nature, but only to what we suppose to be these laws. In fact, we shall be more concerned in this book with epistemic possibility than with an ideal notion of physical possibility, although it is clear that both notions of possibility are closely related.

The notion of possibility, however, that is basic for the theory presented here, is not conceptual possibility but *real* or *factual (physical) possibility*. This notion is the objective basis of probability: real possibilities determine chance, and chance determines degree of support (using the Principle of Direct Inference), the objective notion of conceptual probability.

Physical possibility, real and conceptual, is determined by what is true for the situation we are dealing with. For each given situation, the laws of nature plus the relevant particular facts determine what is physically possible. For something to be possible, according to the notion of real possibility, is for it to be possible to happen. For instance, when rolling a die, its disappearance is not physically possible (not even conceptually), although it is logically possible, but that six obtain in the upper face is not only conceptually, but, also, really possible, in the sense that it is possible for it to happen.

As I mentioned before, following Bunge, I shall call physical possibility in the sense of possibility of happening or occurring, real or factual possibility. For something to be factually possible, it not only has to be compatible with the laws of nature, but also it has to be possible for it to occur. At the time something is really possible, it is not determined whether it is true or false; it is only determined to be false when it ceases to be really possible, i.e., when an alternative real possibility occurs.

The other notion of possibility that is important for us, epistemic possibility, is the possibility of a hypothesis being true, given what we accept as true. That is, the simple compatibility of a proposition, considered as a hypothesis, with the body of propositions that we accept as true. A hypothesis that is epistemically possible, but not really possible, may be determinedly true or false at the time it is epistemically possible, but we may not know which is the case.

An example of conceptual possibility is the following. Suppose that we are interested in the distribution of errors in a certain physical measurement. Suppose that according to the laws for this form of measurement, the errors are distributed according to the normal distribution, but that the mean and standard deviation are not determined by these laws. Then different normal distributions are conceptually (physically) possible, given the laws of nature as they are. Now the distributions that are compatible not with the laws of nature as they are, but with what we suppose them to be, are epistemically possible. It is not that the several distributions are factually possible outcomes, but that each of them is compatible with the laws for the phenomenon (physical possibility), or with what we accept to be these laws (epistemic possibility). It is clear that if something is factually possible, then the proposition expressing it is conceptually possible (i.e., it must be compatible with the actual laws of nature), but the converse is not, in general, true, as we can see from the examples.

One might think that, just as the space of real possibilities determines chance, the space of epistemic possibilities may determine a form of epistemic probability. Real possibilities, however, are endowed with a certain force or tendency to occur, and this force is what determines chance. As can be seen from the next example, a certain type of epistemic possibilities, namely those which come from real possibilities, do determine epistemic probabilities.

An example that deals with different times may help to clarify the difference between factual and epistemic possibility. Suppose that we are rolling a die at 12 noon. At 11 A.M., it is factually possible for a six to come up on the roll of the die at 12 noon. Suppose that at 1 P.M. the die has been rolled but that we don't know the result. Then the result is already determined and, hence, the only really possible result is that which has actually come up. All propositions expressing that one of the particular faces has come up, however, are still epistemically possible for us, since, according to our knowledge, any one of them could have happened. Thus, at 1 P.M., the proposition that expresses that a particular face has obtained is an epistemic possibility that comes from a real possibility. I shall call these possibilities antecedently real epistemic possibilities.[7] That is, *antecedently real epistemic possibilities* are epistemic possibilities that were real at a certain point in the past, and that are still epistemic possibilities, because we don't know what has happened. Antecedently real epistemic possibilities do determine epistemic probabilities, as it will be explained in the next section.

We can see that factual possibility is a property of the objective situation and is independent of our knowledge, while epistemic possibility depends on our knowledge. Real possibility is a dispositional property of things. For instance, when rolling a die, it is a property of the die that when it is rolled any one of the faces can come up. Thus, it is a property of the die that when the die is rolled in a certain way, some possibilities arise. This is what dispositional means: the property needs the obtaining of certain circumstances for it to become evident.

---

[7] The introduction of this notion explicitly was suggested by W. N. Reinhardt.

In the case we are considering, however, the property of having certain real possibilities is not immediately apparent, but only is noticeable through other properties, such as frequency in the long run.

When dealing with epistemic probability per se, assuming some propositions to be true, we obtain probabilities for other propositions. We may not know what is actually true, but we must assume some propositions to be true. Hence, we shall often be working not with what is factually possible according to what is actually the case in reality, but in what would be really possible if what we accept as true were the case. This does not mean that I assume that real possibilities depend on our knowledge, but it is clear that what we believe to be really possible does depend on our knowledge.

Although we cannot hope to give a formal definition of the different notions of possibility, it might be useful to characterize heuristically the notions of possibility in terms of possible worlds, in the style of Leibniz. According to the doctrine of Leibniz, the actual world in which we find ourselves is only one of infinitely many possible worlds that could have existed. To say that a proposition is true is to say that it is true in the actual world. A proposition that is true not only in the actual world but also in all logically possible worlds, is logically true. Propositions that are true in the actual world but not in all logically possible worlds are contingent truths. Scientific laws are of the latter kind; for, although they describe the actual world in considerable generality, there are numerous logically possible worlds in which they do not hold. There may be other worlds, however, besides the actual world, in which they might hold. These worlds in which scientific laws hold, are physically possible worlds. The truths of logic, on the other hand, are true in all logically possible worlds, including of course the actual world (and all physically possible worlds). What we accept as true determines which of the logically possible worlds are epistemically possible.

It is clear that we cannot mathematically represent a possible world completely, with all the objects of the universe in it, and all the properties and relations that hold of them. Rather, for each situation we construct a mathematical approximation to the world (what may be called a partial world), including in its universe just the objects that are relevant for the situation, and distinguishing the relevant properties and relations among these objects.

In general, the laws of nature plus the particular conditions that we accept as being true (this particular conditions determine the relevant universe, properties, and relations), define the range of what is epistemically possible, the epistemically possible worlds. A world in which there are real possibilities is not represented by a possible world, in this sense. Possible worlds, in the sense we have been discussing them, are completely determined. They are, what may be called, completed possible worlds. That is, we look at the possible world as completed in time, i.e., as God is supposed to look at it. But the actual world develops in time, and is not determined. Thus real possibilities, strictly speaking, belong to the actual world while it is still developing, although not to the completed actual world.

A situation where there are several real possibilities is necessary for having a chance phenomenon. The propositions expressing each of the real possibilities are conceptually possible, relative to the description of the chance phenomenon. Thus, we can consider them as represented by possible completed worlds. Abusing the language, we may say that the real possibilities themselves are represented by these completed worlds. Thus, in my theory I shall use the same mathematical device that is used for describing completed possible worlds, for representing real possibilities. Strictly speaking, what we represent in this way, are not the real possibilities, but the propositions expressing them.

The set-theoretic possible world that describes a conceptual possibility, should include all relevant information about how this possibility came about. For real possibilities, the mathematical worlds should include, then, all conditions that might make the possibility occur. In the case of the roll of a die, for instance, the description of each real possibility includes the occurrence of exactly one of the possible results, plus the description of the roll setup. That is, it includes all that is relevant for the understanding of the chance phenomenon, in the case of the example, the roll of a die. After the die is rolled, if we don't know what was the result, then these worlds that were really possible are now only epistemically possible (in fact, antecedently really epistemically possible). If we have some information, for example, if we know that the result was even, then only some of the results that were factually possible remain epistemically possible; the rest are now epistemically impossible.

I believe that this view is similar to that of Bunge, [8], for whom scientific theories determine a range of possibilities, which he also calls real possibilities.

What is really possible is part of the description of the actual world, because it is determined by the objective situation. Hence, possible worlds that are possible models of the actual world in a moment of time, are not completed worlds, and hence, they should also include factual possibilities. Thus, a world that is epistemically possible at a certain time, may be represented by a set of possible completed worlds, and not just one of these worlds, if the chance phenomenon in question admits a set of real possibilities. For instance, before rolling a die, the representation of the actual world still admits all the possible results of the roll, without determining which one is realized. Thus, possible results are mathematically represented by completed possible worlds. Factual probability lives on this set of completed possible worlds, i.e., the set of really possible worlds.

For Carnap, [9], on the other hand, probability lives on the set of logically possible worlds. I believe that the main reason for his inability to find a unique probability measure is that the sample space of all logically possible worlds is too large.

As I shall explain in the next section, in my version, the semantic conception of probability, the sample space is limited to the really possible worlds, and later we shall see that unique measures may exist under this condition. In summary, for me truth determines the really possible worlds, and the set of these worlds determines probability.

3. Probability based on possibility

I shall try, in this section, to state briefly and informally the ideas behind my formalization of factual probability or chance, logical probability, and epistemic probability or rational degree of belief. The next chapter will provide a still informal, but more detailed analysis.

The epistemic probability of a proposition φ depends basically on the degree of support of φ given that we accept a model K as the right model for the phenomenon involved. Thus, the analysis of rational degree of belief that will be given here is derivative from the conception of degree of support via the Principle of Direct Inference.

Factual probability is an objective property of the objects and the experimental conditions, determined by the real possibilities of the objects in these conditions. The degree of support that we attribute to a proposition, on the other hand, depends on what we accept as true, namely it depends on which model K we accept for the situation at hand. The theory presented in this book also attempts to make sense of the Principle of Direct Inference stated in Chapter I, Section 3, which connects chance, degree of support, and rational degree of belief.

Chance is a property of what I shall call, following Hacking, [39, pp. 13–15], chance setups. Trials of a chance setups may be experiments, observations, or, simply, natural phenomena. The only essential characteristic for setup to be subject to chance, is that each of its trials must have a unique result, which is a member of a fixed set of really possible results. In essence, a chance setup consists of a set of objects plus a set of conditions under which these objects have certain real possibilities.

An outcome of a trial of a chance setup (we shall shorten the expression as simply an outcome of a chance setup) will include, in my theory, a description of the chance setup plus a particular result, one of the factually possible results of the setup. Hence, outcome has a technical meaning, which I shall discuss more fully in Chapter III, especially in Sections 1 and 4, and define formally in Part 3. An outcome represents a really possible completed world, including the mechanism of the setup and the specific result that occurs in the outcome, thus building into the outcome the chance mechanism. A chance setup can then be identified with its set of possible outcomes, one possible outcome for each possible result. Since 'possible' is here 'really possible', the set of possible outcomes is determined by the objective conditions of the setup.

Each possible outcome is thought of as indivisible, i.e., as a particular. As such, it has many properties, only some of which are relevant for the setup involved. For instance, in the roll of a die, the position of the die on the ground is not relevant, while the number in the upper face is. These relevant properties determine the family of the events in which we are interested. Events are sets of outcomes. An event occurs, if one of its outcome elements obtains. Since each outcome is indivisible, it is natural to think that an event has a greater chance of occurrence if it contains 'more' outcomes. In order to give a precise meaning to this 'more', it is necessary to introduce a measure on the family of sets of possible outcomes that are events. The chance of an event is then the measure of the set of possible outcomes that constitute the event.

Each possible outcome in an event represents an indivisible real possibility for the event to happen, with each outcome having the same possibility as any other. Each real possibility has, so to speak, a certain force to occur, and we assume that the ultimate possibilities represented by the outcomes have the same force. In other words, each outcome has the same tendency to occur.[8] Thus, I believe that my analysis coincides with Leibniz's ideas expressed in the following dictum:[9]

> Probability is degree of possibility.

A feature of my approach is that chance applies to single events, not necessarily as a part of a sequence of similar experiments. It is possible to look at this definition of factual probability as an explanation for single-case propensities. As was mentioned in Chapter I, Section 2, propensity is taken to be a real kind of power, force, or tendency. Each possibility for an event to occur represents added power for its occurrence. Thus, if there are more possibilities, there is more power; and equal possibilities determine equal power. As J. Bernoulli says, [5, p. 219], referring to equiprobable events:

> All cases are equally possible, that is to say each can come about as easily as any other.

Thus, at least heuristically, propensity is reduced to real possibility, which is, I believe, a clearer notion. Probability is, thus, the measure of possibility. An event is more probable than another (i.e., its chance is greater), if it has more possibilities of happening. Real possibility is the primitive notion from which probability is derived. In order for real possibility to determine probability, the set of possible outcomes for a particular setup (which is determined by real possibility) should determine the family of events and the probability measure that are adequate for the setup. Because of the incorporation of the mechanism of the setup in each outcome, the set of possible outcomes is all we need to know of a chance setup for deciding which probability space is appropriate for it.

This determination of the probability based on the factual possibilities is achieved, in my theory, by performing purely logical operations on the set of possible outcomes. The first step for the determination of probabilities is a detailed analysis of what an outcome is, and the delimitation of the properties and relations that are relevant for the setup. Outcomes can have many different properties. Returning to the example of the roll of a die, we are usually interested in the number that obtains in the upper face, but not in the distance to a certain wall or its color. These last two properties are not considered relevant for the chance setup. We must be able, in formalizing the set of possible outcomes, to include all the relevant properties, and exclude, if possible, all those that are irrelevant.

---

[8] It will be clear after the introduction of the technical notions of an outcome, that in case of a biased die, the fact that a certain face comes up is not an outcome in my technical sense. So the different faces do not have to be equiprobable, according to my definition.

[9] Quoted in [39].

Some properties may turn out to be relevant in my account, although in the usual probabilistic formulations they are not so considered. For instance, in the roll of a biased die, the distribution of the weights in the die is relevant.

As it was said before, and will be developed later, it is possible to include the relevant properties (from now on, when speaking of properties I shall include relations) by thinking of the outcome as a completed possible world and, hence, representing it by a set-theoretical system or model, in the sense of the logical theory of models. The family of events is then determined by the properties that occur in the systems which represent the possible outcomes, i.e., those properties that are relevant for the particular chance setup. One has also to distinguish between those properties that are essential for the experimental setup itself, and those that are just properties of interest for the description of the objects involved in the setup. For instance, if the chance setup consists in the choosing of a sample at random from an urn containing balls of different colors, the property of being in the sample for a particular outcome is the property of the balls that is essential for the specific experiment, but the colors of the balls are properties that are independent of the choosing of the sample itself, although we may be interested in considering them. Properties of the latter kind are intrinsic to the objects of the setup, and do not vary with the different outcomes. Thus, a ball always has a particular color, although it belongs to the sample only in some outcomes. We shall represent properties that are not essential for the description of the mechanism of the chance setup by functions similar to random variables.

Another element that should be included in the description of the setup, and hence in the structure that represents the outcomes, is the relations of stochastic dependence and independence among the different parts of the results, when these results are compounded of simpler ones. Suppose, for instance, that the setup consists of the choosing of an urn at random, and then choosing a ball from this urn. Then the choosing of the urn determines the range of possibilities for the choosing of the balls. So we see that dependence is dependence on the range of possibilities. On the other hand, when we toss a coin twice, the range of possibilities for the second toss does not depend on the result of the first toss; in this case we have independence.

The description of a chance setup is then given by a set of systems, each representing a possible outcome. Each system indicates the relations of dependence and independence, the objects, and the properties of the objects that are essential for the chance mechanism. The probability measure is defined to be the measure (if it exists and is unique) that satisfies the following two conditions. In the first place, it should be invariant under the transformations that preserve the setup. These transformations are the permutations of the objects involved in the setup that transform a possible outcome into another possible outcome. That is, they do not change an outcome that is possible (i.e., one that is one of the really possible outcomes of the setup) into one that is impossible (i.e., one that is not part of the set of really possible outcomes). Invariance under these transformations is a precise rendering of an objective principle of symmetry. This principle is objective, because the group of transformations that determines the symmetries is obtained from the set of possible outcomes, which, in its turn, depends just on the chance setup, which is objective. The principle of symmetry mentioned here will be called the *Principle of Internal Invariance.*

In the second place, the measure should preserve the relations of dependence and independence included in the outcomes. This preservation of dependence is, in fact, another invariance principle: two chance setups which have the same "form" should assign the same probabilities to corresponding events. The models for two chance setups K_1 and K_2 are called homomorphic, i.e., they have the same "form", if there is a function (the homomorphism) that transforms K_1 onto K_2 preserving the dependence and internal invariance structure of the models of the setups. If an event **A** of K_1 contains all the outcomes whose images by the homomorphism form an event **B** of K_2 (i.e., **A** is the inverse image of **B** by the homomorphism), then the probability of **B** in K_2 should be the same as the probability of **A** in K_1. This second principle is called the *Principle of External Invariance.*

The family of events is also determined uniquely by the set of systems representing the possible outcomes of the setup. The family of events is the family generated by those sets of possible outcomes whose measure is determined by the Principles of Internal and External Invariance. In order to be able to derive a logical notion of probability from chance, the probability measure defined on all events should be uniquely determined by the two principles. A more precise explanation of this family will be given in Chapter III, Sections 3 and 4.

Games of chance can, as it will be seen later, easily be modeled by this method, and the resulting probability space is the classical space. It is clear from the account I have given, that it may not be possible in some chance setups to obtain an adequate family of events with its probability measure, following the procedure just outlined. In this respect, my theory is similar to the frequency interpretation, where probability is defined only under certain restrictive conditions. I believe this to be a good feature of the theory. As I plan to show in future chapters, however, all setups where a probability measure has been obtained and used in the past, can be modeled by my methods. In particular, I shall present several examples (especially in Chapter XVI) of reasonable models of natural phenomena, constructed on the basis of these principles. I shall also prove a general representation theorem for stochastic processes (see Chapter XV, Section 4), which asserts that any stochastic process can be approximated, in a sense to be made precise later, by one of my models.

Another side of probability, one of special importance for epistemic concerns, is the degree of support that a proposition has relative to a certain evidence.[10] The basic sort of evidence or, better, support, from my point of view, is that we are in the presence of a certain particular chance setup, which determines a set, say **K**, of possible outcomes. So the basic notion is the degree of support of a proposition φ, given that **K** is the set of possible outcomes. I write this degree as $\Pr_K(\varphi)$, and not $\Pr(\varphi|K)$, i.e., the conditional probability, since we do not, in general, assign a probability to **K**. The degree of support, $\Pr_K(\varphi)$, is

---

[10] In my first major paper on probability, [13], I stressed almost exclusively this side.

3. PROBABILITY BASED ON POSSIBILITY

understood as the degree of possibility of φ being true in $\mathbf{K}$ (which I shall also express, sometimes, abusing the language, as the degree of partial truth of φ in $\mathbf{K}$).

In order to define this notion, we must relate probability to truth. The function $\mathrm{Pr}_{\mathbf{K}}(\varphi)$ is defined to be the (invariant under the two principles) probability measure (if it exists and is unique) of the set of outcomes in $\mathbf{K}$ that make φ true; i.e., the measure of the degree of support of φ given $\mathbf{K}$, is the same as the measure of the chance of the event consisting of the outcomes in $\mathbf{K}$ where φ is true. If φ is true whenever an outcome in $\mathbf{K}$ occurs, we say that φ is true in $\mathbf{K}$, the set of outcomes that make φ true is $\mathbf{K}$ itself, and, hence, $\mathrm{Pr}_{\mathbf{K}}(\varphi) = 1$. The other extreme case obtains when φ is false whenever an outcome in $\mathbf{K}$ occurs: then we say that φ is false in $\mathbf{K}$, the set of outcomes that make φ true is empty, and, hence, $\mathrm{Pr}_{\mathbf{K}}(\varphi) = 0$. For the cases in between, we have that $\mathrm{Pr}_{\mathbf{K}}(\varphi)$ is a number between 0 and 1. In this way, the degree of support of φ given $\mathbf{K}$ is rendered as the degree of "partial truth" of φ in $\mathbf{K}$. So a proposition φ is "more true" (or, better, has a "greater possibility of truth") in $\mathbf{K}$ than another proposition ψ, if the set of possibilities where φ is true is larger (in the sense of the measure) than the set of possibilities where ψ is true.

Since, if $\mathbf{K}$ is the model of the setup, the chance of φ is the degree of possibility of truth of φ in $\mathbf{K}$, this justifies, as will be shown in more detail in the next section, the Principle of Direct Inference, which indicates that the degree of support of φ in $\mathbf{K}$ is the chance of the event determined by φ in $\mathbf{K}$.

Since that which is true or false are propositions, the degree of possibility of truth should also be applied to them. A proposition is true or false according to what obtains in the actual world. Logic studies what is true independently of our knowledge of the real world. Thus, we relativize truth to a possible state of the world, formally a set-theoretical system. In studying pure logic, we don't need to assume any world as the real world. On the other hand, in applying logic, we assume that one of the logically possible worlds $\mathfrak{A}$ is the actual world. Now, if in the real world a certain proposition obtains that is false in $\mathfrak{A}$, then $\mathfrak{A}$ is rejected as an accurate description of the world, and another is sought after. If many propositions obtain that are true in $\mathfrak{A}$, and none that is false in $\mathfrak{A}$, the structure is, at least provisionally, accepted.

Similarly, when studying pure probability (what may be called probabilistic logic), we relativize probability to different sets $\mathbf{K}$ of possible outcomes. Each $\mathbf{K}$ is here a possible world (but not a completed world), which, instead of determining the truth-value of a proposition, spans its possibilities, and, in so doing, defines its probability. Again, in order to apply the theory, we must assume that $\mathbf{K}$ is an accurate description of the factual possibilities in the real world. That is, we must accept $\mathbf{K}$ as true to the world.

Truth is, therefore, prior to probability: in order to assign probabilities we must accept as true that the possible outcomes are those in $\mathbf{K}$. Just as for the models that determine the truth or falsehood of all relevant propositions, there are rules for rejection and acceptance of $\mathbf{K}$ as a representation of the real world. But, since most propositions are only assigned probabilities by $\mathbf{K}$, the rules have a more complicated character than those that apply to the usual models $\mathfrak{A}$ of logic. These rules, which are an important and relatively independent element in my analysis, will be studied in Chapter IV and Part 4.

As was mentioned in Section 1, in usual logic propositions are represented by the sentences of a formal language. We have available, as was also mentioned in Section 1, Tarski's semantic definition of the truth of a sentence φ under an interpretation $\mathfrak{A}$, where $\mathfrak{A}$ is a relational system for the language of φ standing for a possible world. I consider my definition of degree of support also to be semantic. $\Pr_{\mathbf{K}}(\varphi)$, where φ is a sentence, is the probability of φ under the interpretation $\mathbf{K}$. The set $\mathbf{K}$ is not here a relational system, but a set of relational systems for the language of φ, each one representing a possible outcome.

4. The Principle of Direct Inference

The two parts of the Principle of Direct Inference, which is stated in page 36, can now easily be justified with my account. As was mentioned before, the degree of support of φ given that the chance setup is modeled by $\mathbf{K}$ is identified with the degree of possibility of truth of φ in $\mathbf{K}$, i.e, with $\Pr_{\mathbf{K}}(\varphi)$. Thus, the first part of the principle becomes clear. The degree of the possibility of truth of φ in $\mathbf{K}$ is the same as the chance of the event $\mathbf{A}$ that consists of the outcomes in $\mathbf{K}$ where φ is true. Since φ is, in this case, the proposition that $\mathbf{A}$ holds, we have that the degree of support of $\mathbf{A}$'s holding (i.e., of φ) given that the chance of $\mathbf{A}$ is r (this chance is obtained from $\mathbf{K}$), is also r.

We are now ready for an extensional reformulation of the first part of the Principle. Let the chance of an event $\mathbf{A}$, given that $\mathbf{K}$ is the right model of the chance setup, be $\widetilde{\Pr}_{\mathbf{K}}(\mathbf{A})$, and let the degree of support of the sentence φ, given $\mathbf{K}$, be, as earlier, $\Pr_{\mathbf{K}}(\varphi)$. Define $\mathrm{Mod}_{\mathbf{K}}(\varphi)$ as the set of outcomes in $\mathbf{K}$ where φ is true. Then, the extensional formulation of the Principle is

$$\Pr_{\mathbf{K}}(\varphi) = \widetilde{\Pr}_{\mathbf{K}}(\mathrm{Mod}_{\mathbf{K}}(\varphi)).$$

One can also connect rational degree of belief with degree of support, and thus justify the second part of the Principle of Direct Inference. If one is rational, one should believe that φ in the same measure that one has support for φ being true. Hence, if one believes that $\mathbf{K}$ is the set of possible outcomes of the given setup, and that $\Pr_{\mathbf{K}}(\varphi) = r$ (i.e., that the degree of possibility of truth of φ in $\mathbf{K}$ is r), then one should believe φ to the degree r.

This is the basis of the second part of the Principle of Direct Inference, which can be considered as an extension of the following obvious principle: if one believes that $\mathbf{K}$ is the set of possible outcomes (perhaps containing exactly one outcome), and that φ is true in $\mathbf{K}$, then one should believe that φ (or that φ is true).

An example may help in understanding the relations between the three concepts of probability. Suppose that we toss a fair coin at 12 noon of a certain day. At 11 A.M., the chance of heads obtaining at 12 noon is 1/2. Suppose that for this case when the coin is fair, the set of possible outcomes of the chance setup is $\mathbf{K}$. We have, by the Principle of Direct Inference, that the degree of support

4. THE PRINCIPLE OF DIRECT INFERENCE

of heads obtaining, given that **K** is the right model of the chance setup, is also $1/2$, i.e., $\Pr_\mathbf{K}(heads) = 1/2$, and hence, by the second part of the Principle of Direct Inference, that our degree of belief at 11 A.M. in heads obtaining should be $1/2$. Now, at 1 P.M., the coin has already been tossed, and hence there is no chance setup. Heads or tails have obtained. If we don't know which of the two has occurred, heads and tails are still epistemic possibilities, but not real possibilities; in fact, they are antecedently real epistemic possibilities. So the chance at 1 P.M. of heads obtaining at 12 noon is either one or zero. The degree of support is a logical notion, however, and so it is independent of the circumstances. Hence, $\Pr_\mathbf{K}(heads)$ is still $1/2$, since the degree of support depends only on the proposition, i.e., heads obtains, and the support, i.e., **K**. Suppose that at 1 P.M. we don't know the result of the toss. Then all that we know at 1 P.M. about the situation is that the chance setup of the coin toss was modeled by **K**, since the knowledge that it is 1 P.M. does not add information about the chance setup. Therefore, our degree of belief at 1 P.M. in the proposition that heads obtained at 12 should be the same as the degree of support of heads given **K**, i.e., it should still be $1/2$.

Thus, we see that the Principle of Direct Inference can be extended to cases of antecedently real possibilities: in the case of the coin tossed at 12 noon and considered at 1 P.M., we have that the chance of heads at 1 P.M. is not $1/2$, but zero or one. We don't know, however, which of these two numbers is right. Rational degrees of belief should be assigned based on our knowledge. All relevant knowledge is that at 12 noon the chance of heads was $1/2$. Therefore, the rational degree of belief in heads must also be $1/2$.

A general formulation of the Principle for antecedent chances is complicated, because of the difficulty of defining "relevant knowledge". Without attempting full rigor, I shall give a more precise formulation at the end of Chapter IV, Section 1. A good discussion of this subject is found in [62].

CHAPTER III

Probability models

In this chapter, the different elements that constitute the probability models presented in this book and their use for the formalization of the logical notion of probability, i.e., degree of support, will be analyzed. The first four sections deal with the construction of models for chance setups, and the last one explains their use for the formalization of degree of support. The first section deals with the formal representation of simple outcomes, namely, those of the chance setups where there is no need to consider the notions of stochastic dependence or independence. The second section contains an analysis of the symmetry principles that are basic for the definition of the probability measure. The third section analyzes the family of events and the construction of the basic probability space for these models with simple outcomes. The fourth section discusses informally the notions of stochastic dependence and independence that are included in the description of the outcomes for the more complicated chance setups, and the construction of the corresponding probability spaces.

1. Outcomes

An outcome of a chance setup contains, in the formalization presented in this book, a description of the chance mechanism of the setup plus one of the particular results of the setup, i.e., each outcome describes one of the real possibilities for the setup in question.

In order to complete the description of the setup, we should also indicate which additional properties (besides those necessary for the description of the chance mechanism) are relevant for the experiment or observation with which we are concerned. For different experiments that involve the same chance mechanism, there may be different additional properties. These additional properties are not necessary for the description of the chance setup itself, but only for determining which events are considered relevant. They will become important when discussing the notion of degree of support.

Each outcome can be seen as a completed possible world, i.e., a relational system. So its first element is a universe of objects, the objects to which the mechanism of the setup is applied. We can see this more clearly with an example. Suppose the setup is the choosing at random of a sample of size m from a population of size n. In order to be more specific, we may assume that it is a population of balls in an urn. The universe of the outcomes in this case is the set of balls, i.e., the population of balls, say P. Each possible outcome has a different sample, a subset S of P containing m elements. Hence, the outcomes can be represented by all the pairs $\langle P, S \rangle$, with S a subset of m elements of P. Thus, the set $\mathbf{K}$ of all possible outcomes is the set of all $\langle P, S \rangle$, for S a subset of size m of P. But we may be also interested in some properties of the samples. For instance, the number of white and black balls in the sample.

In order to deal with these properties, we introduce *measurement functions*, which are akin to random variables. For the properties mentioned, we could introduce the functions $W : \mathbf{K} \to \mathbb{R}$ and $B : \mathbf{K} \to \mathbb{R}$, such that $W(\langle P, S \rangle)$ and $B(\langle P, S \rangle)$ are, respectively, the number of white and black balls in the sample S. In summary, the chance setup in question is completely described by the triple $\langle \mathbf{K}, W, B \rangle$.

In another experiment, we may be interested in other properties of the sample, for instance, the number of rough or smooth balls. Then we would add these two measurement functions to the set of basic outcomes $\mathbf{K}$. We could also be interested, for example, in whether or not a particular ball, say b, is in the sample S. For this purpose, we can introduce the function $X_b : \mathbf{K} \to \mathbb{R}$ such that

$$X_b(\langle P, S \rangle) = \begin{cases} 1 & \text{if } b \in S, \\ 0 & \text{if } b \notin S. \end{cases}$$

The next example will give a first glimpse of how I deal with situations in which the apparently natural elementary events are not equiprobable. Suppose we have a circular roulette. The only variable element in the mechanism is the random choosing of the initial position. A constant force is applied for a fixed time and the roulette stops at a certain point. The roulette may be biased. The chance mechanism of this setup is the selection of an initial position.

The chance structure, i.e., the set of possible outcomes of this setup, call it **C**, consists of all relational systems of the form $\omega_c = \langle C, r, c \rangle$, where C is the set of points in the circle considered as complex numbers in the unit circle, i.e., $C = \{e^{i\theta} : 0 \leq \theta < 2\pi\}$, r is multiplication between complex numbers, and c is an element of C. It is well known that $r(u_1, u_2)$ represents a rotation of u_1 in the arc of length θ, where $u_2 = e^{i\theta}$. The system $\langle C, r, c \rangle$ stands for the outcome with initial position c.

We need the structure $\mathcal{C} = \langle C, r \rangle$ and not just the set C, since we must have a circle, and not only a set. If we only had $\langle C, c \rangle$, we would be selecting the element c from the set C and not from the circle $\langle C, r \rangle$. The chance structure **C** is then the set of all systems $\langle C, r, c \rangle$.

What we usually observe is not the initial position of the roulette, but its final position. So in order to complete the model for the setup, we add a measurement function X, such that $X(\omega_c)$ is the final position when the initial position is c. In order to have the values of X in the real numbers, we define

$X(\omega_c) = a$, if the final position is e^{ia}, when the initial position is c.

For the moment, X is not a random variable in the usual sense of the word, because we do not yet have a measure. After the introduction of the natural measure, however, X, in this case, will become a random variable in the normal sense, i.e., a measurable function.

By the use of these measurement functions, we can deal with properties of the outcomes. Since these functions are not necessarily measurable, I shall call them measurement functions instead of random variables. This method of using real-valued functions for describing properties of outcomes is customary in probability theory, and it will actually be convenient for us.

We may have different measurement functions, depending on the bias of the roulette. The only restriction is that the function $f(c) = e^{iX(\omega_c)}$ be continuous. The probability structure for the roulette is not just $\mathbf{C}$, but the pair $\langle \mathbf{C}, X \rangle$. If we do not know the bias of the mechanism, we have to assume that the model for the setup is one among the set of structures $\langle \mathbf{C}, X \rangle$, where X can be any of the allowable measurement functions.

The complete description of the model for the roulette is, then, the pair $\langle \mathbf{C}, X \rangle$. If the roulette is, in fact, biased, then a set $\mathbf{A}$ of systems $\langle \mathcal{C}, c \rangle$ with c in an arc A, corresponds to the set $X''\mathbf{A} = \mathbf{B}$ containing the systems $\langle \mathcal{C}, c \rangle$ with $c \in B$, which may be an arc of different length than A. Thus, with the natural definition of the measure, which will be discussed in Section 2, we shall obtain that there are arcs A and B of the same length, such that the measure of $[X \in \mathbf{A}]$ is different from the measure of $[X \in \mathbf{B}]$, and, thus, considering the bias represented by X, we don't get a uniform distribution. That is, Pr_X is not uniform.

Events, which will be treated in Section 3, are subsets of the set of possible outcomes $\mathbf{C}$. As it has been explained earlier, the elements of $\mathbf{C}$ are the relational systems representing the outcomes, without including the additional measurement functions. Thus, in the first example, $\mathbf{K}$ is the set of all $\langle P, S \rangle$ for S a subset of m elements of P. In the second example, $\mathbf{C}$ is the set of $\langle \mathcal{C}, c \rangle$ where c is an element of the circle.

2. The symmetry or first invariance principle

A principle of symmetry was accepted from the beginnings of the calculus of probability. Symmetries were used to calculate the probabilities for games of chance. As was discussed in Chapter I, Section 1, the basic element in the computations of probabilities for games of chance is the assumption of a finite number of equiprobable (called very frequently in the early times equipossible) elementary results, such that all events are combinations of these results. One of the problems with this definition, mentioned in Chapter I, Sections 1 and 2, is that we should have a criterion for equiprobability that does not depend on probability. It seems that for this purpose, a factual principle of symmetry was accepted form the start by Pascal, Fermat, and the other founders of the calculus. For instance, the statement of J. Bernoulli already quoted supports this idea, [5, p. 219]:

> All cases are equally possible that is to say each can come about
> as easily as any other.

In this statement, it seems that Bernoulli had in mind a factual principle, i.e., equally possible does not depend on our knowledge but on the physical facts of the situation involved.

Although in the case of Bernoulli it is not absolutely clear whether he meant a factual or an epistemic principle of symmetry, later, in the 18th and 19th centuries, an epistemic principle came clearly to the fore, especially in the works of Laplace, [31], and his successors. This principle has been called the "Principle of Insufficient Reason" or "Principle of Indifference". Keynes, [55, p. 42], states it in the following form:

> The Principle of Indifference asserts that if there is no known reason for predicating of our subject one rather than another of several alternatives then relatively to such knowledge the assertions of each of these alternatives have an equal probability.

It is easy to change such an epistemic principle into a factual principle of symmetry: we can say that equal probabilities are established when there are in fact no reasons, instead of when one knows of no reasons. I shall call the factual principle the Principle of Symmetry, and reserve the name Principle of Indifference for the epistemic principle. The problem that remains in order to make the Principle of Symmetry precise, is to determine under which conditions there are no reasons. The principle that I shall use to obtain the probability measure is an attempt to state these conditions clearly. It is also of interest to discuss principles of symmetry and indifference more fully, because they are known to lead to contradictions, if one is not careful.

It is clear that if the set of alternatives is not uniquely determined, contradictions might occur. For instance, if the possible results are for this book to be red or not red, then the principle should assign 1/2 to the probability of it being red; but if the possibilities are for the book to be red, or blue, or neither red nor blue, then the same event would be assigned probability 1/3. In the probability models we are discussing, each chance setup determines which are its possible outcomes, and hence, for each setup the Principle of Symmetry should be applied to these indivisible possible outcomes. Thus, no contradiction can arise on this count.

The problem may be, for instance in the case of the book, that it may not be determined which are the elementary possible outcomes. In this case, I would say that we cannot apply the Principle of Symmetry. In any case, if we have determined that there are a finite number of possible results, then the Principle would be enough to yield the probabilities, although one should analyze further why two possible results are equiprobable. In case there are infinitely many possible results, however, the Principle as stated is not by itself enough to establish probabilities. This can be seen from some paradoxes that might arise. I shall present two paradoxes due to Bertrand (1889) to illuminate the point.

I will state the paradoxes in their epistemic form, as Bertrand stated them. I will then show how they can be removed when using a factual version of the

Principle.

Bertrand's mixture paradox. A bottle contains a mixture of water and wine. All that is known about the mixture is that the ratio of water to wine is not less than one, and not more than two. Extending the Principle of Indifference to infinitely many possible results in a continuous way, we could assume that the probability of the mixture lying anywhere between a ratio of one and two of water to wine, arises from a uniform distribution on this interval, that is, to a distribution which assigns equal probabilities to intervals of the same length. The probability of the ratio lying in an interval is, then, the length of the interval divided by the total length.[1] Thus, the probability of the mixture lying between 1 and 3/2 is 1/2.

Now examine the other way of looking at the problem: consider the ratio of wine to water, i.e., the inverse of the previous ratio. From the information given at the beginning of the problem, we know that the ratio of wine to water lies between 1/2 and 1. By the Principle of Indifference, we should assume, once again, a uniform distribution on this interval. Thus, the probability that the ratio of wine to water lies in the interval from 2/3 to 1 is 2/3. Now the wine-to-water ratio 2/3 corresponds to the water-to-wine ratio 3/2. Thus we have arrived at a contradiction:

$$\frac{2}{3} = \Pr(\frac{2}{3} \leq \frac{wine}{water} \leq 1) = \Pr(1 \leq \frac{water}{wine} \leq \frac{3}{2}) = \frac{1}{2},$$

because it is a truth of arithmetic that

$$\frac{2}{3} \leq \frac{wine}{water} \leq 1 \text{ if and only if } 1 \leq \frac{water}{wine} \leq \frac{3}{2}.$$

The source of the difficulty in this paradox is that there is not enough information about the setup to decide which is the chance mechanism involved. Suppose, for instance, that the experiment is conducted as follows. We begin by having in the bottle one volume of water, and then we fill it with a constant flow of wine with the time for stopping chosen at random between one and two volumes of the wine. Then the ratio of wine to water is distributed uniformly. On the other hand, suppose we begin by having in the bottle one volume of wine, and then fill it with a constant flow of water with the time for stopping chose at random between one and two volumes. Then the ratio of water to wine is uniformly distributed. Finally, suppose we start by filling the bottle with a constant flow of water, and the time for stopping is chosen randomly. The bottle is then filled with wine. In this case, neither ratio has a uniform distribution, but what is uniformly distributed is the absolute amount of water in the bottle. Thus different mechanisms explain the different distributions. In the next example, I shall analyze a similar phenomenon more carefully.

---

[1] I shall show in Chapter X, Section 3, using nonstandard analysis, that the uniform distribution is the natural continuous counterpart of a distribution in which all elementary events are equiprobable. In Chapter X, Section 3, I shall analyze further Bertrand's mixture paradox.

Bertrand's random chord paradox. We choose, in this case, a chord at random from a circle of unit radius. The problem is to find the probability that the "random chord" has a length greater than $\sqrt{3}$ (where, of course, $\sqrt{3}$ is the length of the side of an inscribed equilateral triangle). There are several natural assignment of probability for this situation:

(1) Because any chord intersects the circle in two points, we may suppose these two points to be independently distributed on the circumference of the circle. Without loss of generality, we may assume that one of the two points is at the vertex of an inscribed equilateral triangle. The three vertices of the triangle divide the circle into three equal arcs, so there is just 1/3 of the circle in which the other point can lie in order that the length of the resulting chord is greater than $\sqrt{3}$. Thus, the probability of choosing such a chord is 1/3.

(2) The length of the chord is determined by its distance from the center of the circle, and is independent of its direction. Thus, we may suppose that the chord has a fixed direction perpendicular to a given diameter of the circle, with the distance along the diameter measuring its distance from the center. We may suppose that its point of intersection with its diameter has a uniform distribution. For the chord to have a length greater that $\sqrt{3}$, the distance of the point of intersection from the center of the circle must be less than 1/2. So the probability of choosing such a chord is 1/2.

(3) Any chord is uniquely determined by the point of intersection of the chord and a radius perpendicular to the chord (i.e., by the middle point of the chord). So one natural assumption is that this point of intersection is distributed uniformly over the interior of the circle. Then the probability of its lying in any region of area A is A/π, since the total area of the circle is π. For the chord to have length greater than $\sqrt{3}$, this point of intersection must be in a smaller circle with the same center and radius 1/2. Hence, the probability of finding a chord of the required length is 1/4.

Other natural properties of chords can be used to generate distributions other that these three.

The difficulty in the chord paradox is that each solution represents a different setup. That is, each of the probability distributions (1), (2), and (3) models different possible experiments. For instance, the following might be experimental setups modeled by (1), (2), and (3), respectively.

(1) Along a circle of radius one a needle circles, spinning also around the needle's center, which lies on the circumference of the circle. Where the needles stop, it determines a chord on the circle.

(2) Draw parallel lines on the floor at a distance of two (the diameter of the circle). Throw a circle of radius one on top of them. One of the lines has to intersect the circle, and, thus, determine a chord.

(3) Trace parallel horizontal lines on a wall at a distance of two, and insert spikes in the wall on the lines, also at a distance of two. Throw a disc

2. THE SYMMETRY OR FIRST INVARIANCE PRINCIPLE

of radius one so that it is implanted on one of the spikes. When the disc reaches its equilibrium position by the action of gravity, the line will intersect the circle on a chord, with the spike on the middle point of the chord.

Different groups of symmetries determine different probability measures for these cases. In Case 1, the group of transformations under which the chance setup is invariant (i.e., a possible outcome is transformed into another possible outcome) is the group of rotations of the circle. This is so, because the only possible outcomes are those where points from the circle are chosen. That is, any adequate transformation has to transform the circle into itself. Similarly, for Case 2, the diameter where the point is chosen has to be transformed into itself, and hence, the appropriate group consists of translations of this diameter. A similar argument shows that for Case 3, the group is that of translations of the plane. In each case, the invariant measure is as expected. A more detailed analysis showing that these are the right measures, will be presented as examples of the formal models in Chapter XVI.

Principles of Symmetry. A general principle of symmetry based on groups of transformations will be used in my development as a fundamental element in the determination of the probability measure. The measure is a measure invariant under one of these groups. I believe that in all cases where numerical probabilities have been assigned to chance phenomena, it is possible to set up a model of possible outcomes, so that the probability measure is the measure invariant under a group of transformations determined by the set of possible outcomes.

Similar principles of symmetry, but of an epistemic character, have been accepted by all proponents of logical theories of probability. As an example, I shall discuss Carnap's principle, [9, 10]. As we have seen, the basic concept for Carnap is that of the degree of confirmation of a sentence h given another sentence e, considered as evidence for h. Recall that the degree of confirmation of h given e is considered, heuristically, as the degree to which e partially logically implies h.

This degree of confirmation is applied to a language with a fixed set of constants that represent the objects of a fixed (countable) universe. Carnap's principle can be paraphrased as follows:

CARNAP'S PRINCIPLE 1. *If f is a permutation of the objects of the universe, h, e, h', and e' are sentences, and h' and e' are obtained from h and e by substituting a for $f(a)$, for all objects a that occur in h and e, then the degree of confirmation of h given e is the same as the degree of confirmation of h' given e'.*

This is a logical principle, because the degree of confirmation is a logical notion. I consider my principle as factual, because it applies to chance.

I shall now state my Principle of Symmetry more precisely and compare it with Carnap's. We first need some definitions. Suppose that $\omega = \langle A, S \rangle$ is an outcome; for simplicity we assume that it has just one property, namely S. The set A is the universe of the outcome. Let f be a permutation of A. Let $f''(\omega)$ be

$\langle A, f(S)\rangle$, where $f(S)$ is the image of S under f. Let now **K** be a set of possible outcomes with universe A, and let **B** be a subset of **K**. Then we define $\mathbf{B}^f$ to be the set of $f''(\omega)$ for $\omega \in \mathbf{B}$. The set $\mathbf{B}^f$ is the result of the action of f on **B**. My Principle of Symmetry can now be stated.

> Let **K** be the set of possible outcomes of a chance setup, and **B** an event (and, hence, a subset of **K**). If f is a permutation of the universe of **K** such that $\mathbf{K}^f = \mathbf{K}$, then the chance of **B** in the chance setup with **K** as its set of possible outcomes (briefly, the chance of **B** in **K**) is the same as the chance of $\mathbf{B}^f$ in **K**.

This principle of symmetry may be called the *Principle of Internal Invariance*, since the invariance of the probability measure depends only on the internal structure of one setup. In the next section, I shall describe a Principle of External Invariance that relates probability in two setups with the "same chance structure". This second principle is not required, however, for setups with simple outcomes.

In order to illustrate the principle, I shall discuss a couple of examples. Suppose, first, that we have the structure **K**, introduced in Section 1, for the selection of a sample S of elements of a population P. We call the group of permutations of P, f, that satisfy $\mathbf{K}^f = \mathbf{K}$, $G_\mathbf{K}$. It is easy to check, that, in this case, $G_\mathbf{K}$ is the set of all permutations of P.

We now discuss the group of invariance of the roulette setup. The group of invariance of the circle, **C**, call it $G_\mathbf{C}$, is the set of all isomorphisms g, such that $g(\omega_c) \in \mathbf{C}$, where $g(\langle C, r, c\rangle) = \langle g(C), g(r), g(c)\rangle$. That is, for $g(\omega_c) \in \mathbf{C}$, we must have that $g(\omega_c) = \omega_{c'}$ for a certain c' in C, and that g must be an automorphism of the circle, $\langle C, r\rangle$, i.e., $g(\langle C, r\rangle) = \langle C, r\rangle$. Hence, g has to be a rotation of the circle, and not all permutations of C are in $G_\mathbf{C}$.

We can obtain a form of Carnap's principle similar to mine, by replacing sentences by sets of models of sentences. This approach is taken by Carnap in [10], but it was already implicit in [9]. In order to obtain this reformulation, we begin by explaining Carnap's basic ideas. He defines the degree of confirmantion of h, given e, in symbols $\mathfrak{c}(h, e)$, on the basis of an initial logical measure $\mathfrak{m}$, which may be understood as: $\mathfrak{m}(h)$ is the degree of confirmation of h given no evidence (or given a tautology, as evidence). Then, he defines $\mathfrak{c}$, by

$$\mathfrak{c}(h, e) = \frac{\mathfrak{m}(h \wedge e)}{\mathfrak{m}(e)} \qquad (1)$$

where $h \wedge e$ is the conjunction of h and e.

Carnap considers $\mathfrak{m}(h)$ as the measure of the size of the set of possible models of h, which Carnap calls the *range of* h, and writes $\mathfrak{R}(h)$, and we shall call $\mathrm{Mod}(h)$. That is, $\mathrm{Mod}(h)$ is the set of systems where h is true. In [9, p. 297], Carnap has the picture in Fig. III.2. In the picture, the outermost rectangle represents the set of all logically possible models, and the other rectangles the ranges of h and e. The areas represent the measures of the ranges.

All the systems Carnap considers as possible models for his sentences have the same universe, say A. Each element of A is denoted by a constant of the

2. THE SYMMETRY OR FIRST INVARIANCE PRINCIPLE

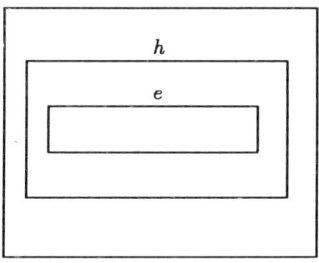
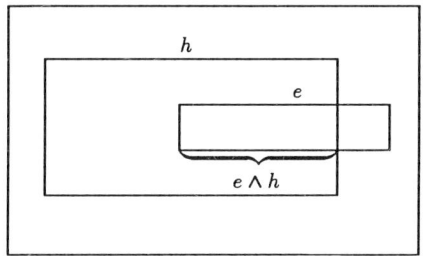

Deductive Logic

'e L-implies h' means that the range of e is entirely contained in that of h.

Inductive Logic

'$c(h,e) = 3/4$' means that three-fourths of the range of e is contained in that of h.

FIGURE III.2. Carnap's model of logic

language, with two different constants denoting different elements. Thus, we can identify the elements of A with constants in the language.

Assume, for simplicity, as before, that we have just one property S. Thus, our systems are all systems of the form $\langle A, S \rangle$, where S is any subset of A (of any number of elements). Since $\mathfrak{m}(h)$ measures the size of $\mathrm{Mod}(h)$, we can write

$$\mathfrak{m}(h) = \bar{\mathfrak{m}}(\mathrm{Mod}(h)) \qquad (2)$$

for a suitable $\bar{\mathfrak{m}}$ which measures this size. That is, $\bar{\mathfrak{m}}$ is defined over sets of possible models. Carnap's principle of symmetry can, now, be stated:

CARNAP'S PRINCIPLE 2. *If f is a permutation of A, and $\mathbf{A}$ and $\mathbf{B}$ are sets of models such that $\mathbf{A}^f = \mathbf{B}$, then $\bar{\mathfrak{m}}(\mathbf{A}) = \bar{\mathfrak{m}}(\mathbf{B})$.*

Let us analyze further the differences between Carnap's approach and mine. In Carnap's approach, $\mathrm{Mod}(h)$ is the set of all logically possible models of h. For instance, in the case we are considering of systems $\langle A, S \rangle$, with S a subset of A, the logically possible S are all subsets of A of any number of elements, including the empty set. Thus, the measure $\mathfrak{m}$ has to be defined over the probability space consisting of systems $\langle A, S \rangle$, for S an arbitrary subset of A. I believe that in most cases there are too many measures invariant under permutations.

In my approach, we only consider the factually possible systems. Which systems are factually possible depends on the particualr setup we are modeling. That is, in the second picture of Fig. III.2, the outermost rectangle should represent the set of factually possible systems, and not all those that are logically possible. For instance, if the setup consists on the choosing at random of a sample S of m elements out of a population A, then the systems $\langle A, S \rangle$ are such that S must have m elements. Thus, not all subsets of A a really possible, but only those with m elements. In this case, as in many others that I will introduce later, there is a unique invariant measure. Of course, there are also cases where there are several invariant measures. In these latter cases, I would say that the factual probability is not defined.

Carnap's Principle 1 can be reformulated, given (2), by: if f is a permutation of A and h' is obtained from h by substituting a by $f(a)$ for all constants a occurring in h, then $\mathfrak{m}(h) = \mathfrak{m}(h')$. Thus, we obtain the equivalence of Principle 1 with Principle 2 restricted to sets of models of sentences, by noticing that if h and h' are as above, then

$$\operatorname{Mod}(h') = \operatorname{Mod}(h)^f.$$

If we take

$$\bar{\mathfrak{c}}(\operatorname{Mod}(h)|\operatorname{Mod}(e)) = \mathfrak{c}(h|e),$$

that is, by (2)

$$\bar{\mathfrak{c}}(\mathbf{B}|\mathbf{K}) = \frac{\bar{\mathfrak{m}}(\mathbf{BK})}{\bar{\mathfrak{m}}(\mathbf{K})},$$

then we could consider $\bar{\mathfrak{c}}(\mathbf{B}|\mathbf{K})$ as the chance of $\mathbf{B}$ in $\mathbf{K}$. By Carnap's Principle 2,

$$\bar{\mathfrak{c}}(\mathbf{B}|\mathbf{K}) = \bar{\mathfrak{c}}(\mathbf{B}^f|\mathbf{K}^f).$$

This form of the principle can be paraphrased as "the chance of $\mathbf{B}$ in $\mathbf{K}$ is the same as the chance of $\mathbf{B}^f$ in $\mathbf{K}^f$". This last form is quite similar to my principle, but omits the requirement that $\mathbf{K}^f = \mathbf{K}$. This difference is very important because it is the reason for my principle being factual, while Carnap's is logical: the facts of the setup are represented by $\mathbf{K}$, and, hence, in my Principle we just consider the factual group of symmetries or group of invariance, consisting of those permutations f of the universe of $\mathbf{K}$ with $\mathbf{K}^f = \mathbf{K}$. That is, this group of invariance is the group that contains all permutations of the universe of $\mathbf{K}$ that transform really possible outcomes into really possible outcomes. Since the laws of the setup are represented by $\mathbf{K}$, these functions can be considered as those under which the laws of the phenomenon are invariant. I believe this to be the main justification for my Principle of Internal Invariance: *if two events are transformed into each other by transformations which preserve the laws of the phenomenon, then they should be assigned the same probability.* In a way, it can be said that $\mathbf{B}$ and $\mathbf{B}^f$ are indistinguishable in the chance setup $\mathbf{K}$.

A few words about randomness are now in order. It seems to me that there is an objective notion of randomness. When we say that we are choosing a ball at random from an urn, we are stating a property of the chance setup consisting of the selection of a ball. It means that it is just as likely for one ball to be chosen as for any other. In general, an outcome is random, if it is an outcome of a setup with equiprobable outcomes, i.e., outcomes which are symmetric with respect to the group of symmetries of the setup. Since the group of invariance is determined by objective properties of the setup, randomness is also objective. In a similar way, a sequence is random, if it is obtained in a chance setup with sequences as outcomes and where all outcomes are equiprobable, i.e., symmetric under the group of symmetries. The definitions of random sequences that have been proposed in the literature,[2] although interesting mathematically and even useful in some cases, do not take into account the essential point that randomness is relative to a chance setup.

As a conclusion of this discussion, I will indicate how the Principle of Indifference should be amended in order to get a reasonable factual Principle of Symmetry. In the first place, it should be interpreted factually. In the second place, there should be a fixed set of possible outcomes, which is determined by the particular chance setup under consideration. In the third place, for two events to be equiprobable there must be a reason, namely, that the events be mutually transformable by transformations from a group determined by the set of possible outcomes of the setup. This group consists of all permutations of the universe such that a possible outcome is not transformed into an impossible one by them. Thus, it is not that there are no reasons, but that there is a good reason, for the events to be equiprobable.

3. The family of events

I now turn to an informal discussion of the construction of the family of events from the set of possible outcomes modeling a certain chance setup.

Recall that a simple chance setup is modeled by a set of possible outcomes **K**. Events are certain subsets of **K**. The probability measure obtained from **K** should be a measure invariant under the group of symmetries described above.

The family of events is generated from certain subsets of **K** whose measure is completely determined by the group of invariance. That is, we define a subset **A** of **K** to be *measure-determined* if there is a real number r such that every measure invariant under the basic group of invariance of **K**, and that is defined on **A** and all its images under functions of the group, assigns r to **A**. For instance, if **A** has n disjoint images by functions of the group, say $\mathbf{A}_1, \ldots, \mathbf{A}_n$, whose union is **K**, then **A** should have measure $1/n$, and hence is measure-determined.

Since the essential property of the measure in these simple cases is its invariance, the measure assigned to measure-determined sets measures their actual degree of possibility.

In order to simplify the mathematics, we close the measure-determined events under certain set operations. Which operations to choose depends on whether we are satisfied with a finitely additive measure or require countable additivity. Since we shall be working with finite (or hyperfinite) structures, every finitely additive measure is automatically countably additive. The process of passing from a nonstandard to a standard structure gives countably additive measures, so we shall always obtain standard countably additive measures.

I happen to believe, but will not argue about it here, that finitely additive measures are not, in general, easier to get than countably additive measures, and I do not see any special advantage in using the finitely additive kind. So I will assume that all our probability measures, if defined on infinite stets, are

---

[2] See, for instance, [65] or [96].

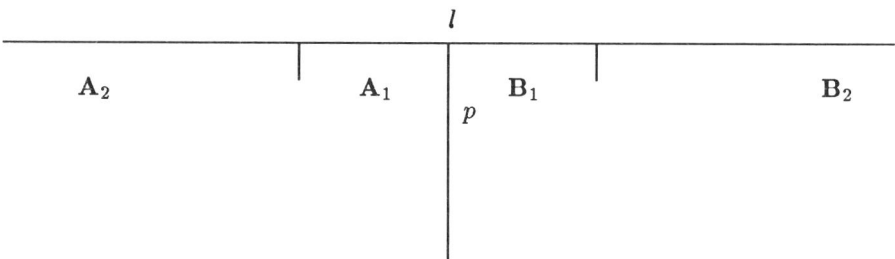

FIGURE III.3. A probability structure with a disjunctive algebra of events that is not an algebra

countably additive. Since we shall be dealing essentially with finite sets, the problem is not important.

The family of events should at least be closed under complements and unions of pairwise disjoint sets (if infinite, also closed under unions of countable pairwise disjoint sets). So the family of events is what may be called a disjunctive algebra (or, if infinite, a disjunctive countably complete or σ-algebra) of sets. We could also require that the family of events be closed under unions in general, and thus get an algebra (or σ-algebra) of sets, but there are examples where this would not be adequate, because of nonuniqueness of the measure.

Disjunctive σ-algebras of sets have been proposed by Suppes, [87], as the natural structures that occur in the applications of probability to quantum mechanics, and called by him quantum mechanical algebras of sets. I shall give, here, an example, unrelated to quantum mechanics, where the natural structure of events is a disjunctive algebra of sets and not an algebra of sets.

Suppose that we throw a very small marble towards a wall trying to run along a line p on the floor, perpendicular to the wall (Fig. III.3). The marble makes a mark on the intersection of the wall with the floor, say the line l. Hence the probabilities of the mark lying in any two subsets of l that are symmetrical with respect to p should be equal, but we can't say anything about the probabilities of subsets that are not symmetrical.

In Fig. III.3, the event $\mathbf{A}_1 \cup \mathbf{A}_2$ should have the same probability as $\mathbf{B}_1 \cup \mathbf{B}_2$, i.e., 1/2, and also $\mathbf{A}_1 \cup \mathbf{B}_2$ should have the same probability of 1/2 as $\mathbf{B}_1 \cup \mathbf{A}_2$, but we don't have information about the probabilities of $\mathbf{A}_1$, $\mathbf{A}_2$, $\mathbf{B}_1$, or $\mathbf{B}_2$.

Hence, the probability of two sets may be determined by the group of symmetries, but not that of their intersection. That is, the natural family of events is not an algebra, but only a disjunctive algebra. It is clear that the measure on this disjunctive algebra is unique. We are able, in this case, to extend the measure to the generated σ-algebra, but the invariant extension is not unique. Since we want to obtain a logical probability relation, i.e., the degree of support, from chance, it is important that the measure be unique. There may be other cases, especially in relation to quantum mechanics, where it may be impossible to extend the measure from the disjunctive algebra to the algebra.

As an example where the measure-determined sets generate an algebra and not only a disjunctive algebra, we return to the case of the roulette setup. Recall that the group of invariance is $G_\mathbf{C}$, the group of rotations. The measure-determined sets include the arcs. It can be proved that the disjuntive σ-algebra generated by the arcs is the family of Borel sets. It is clear that the measure that is invariant under this group, $G_\mathbf{C}$, and defined on the Borel sets, is Lebesgue measure. That is, the sample space for this setup is $\langle \mathbf{C}, \mathcal{F}_\mathbf{C}, \Pr_\mathbf{C} \rangle$, where $\mathcal{F}_\mathbf{C}$ is the family of Borel subsets of the circle, and $\Pr_\mathbf{C}$ is Lebesgue measure. We shall discuss this case with hyperfinite approximations in later chapters.

4. The External Invariance Principle

Dependence and independence. In usual probability theory, the notions of dependence and independence are defined using the probability measure. Thus, we say that the event $\mathbf{A}$ is independent of $\mathbf{B}$, if $\Pr(\mathbf{A}|\mathbf{B}) = \Pr(\mathbf{A})$. Suppes, in [89], has even proposed a probabilistic theory of causality, defining the notion of cause in terms of probability. On the other hand, causal dependence and independence are considered as a primitive notions in the models of this book.

I believe that there are also areas of probability theory where the same is done. For instance, in the theory of stochastic processes, it is assumed that we have a totally ordered set T, and at each t in T we put a probability distribution (or a random variable), such that the distribution at each t in T depends on the distributions at some of the previous s in T. A simple example of this type of process is the urn model. Here, T may be taken to consist of two elements, say s and t with s before t. At s we choose an urn, and from this urn, we choose, at t, a ball.

Another case where these notions, in this case independence, are considered primitive, is in the independent combination of two or more probability spaces. The product space of the given spaces is the natural independent combination. We can put these cases in the framework of the processes discussed above. For instance, assume that we toss a coin twice. The set T consists again of two elements s and t, but now s and t are incomparable, i.e., neither is before the other. At both points in T we have the same probability space. Then the space for the two tosses is the product space.

The possible outcomes for a chance setup may include, in my models, a primitive notion of causal dependence (or independence), represented by an ordered set T. In order to consider both dependence and independence, T, in general, is partially ordered, and not necessarily totally ordered. Since probability is reduced in these models to real possibility, here what has happened up to a certain point s determines the real possibilities at t's that are after s in the ordering. For instance, the choosing of a particular urn determines which are the balls that are possible to choose, i.e., the possibilities at t are determined by what has happened up to t, but not including t. In the case that s and t are incomparable, as in the two tosses of a coin, the possibilities at t are independent of what happens at s. Notice that the model for two tosses of a coin is the same as that of the simultaneous tossing of two coins. Hence, s and t need not be one after the other in time.

An example of the combination of dependence and independence is shown in the following setup. Suppose that we have two sets of two urns, each with balls in the urns. Then we choose independently one urn at random from each set, mix the balls of the chosen urns, and choose a ball at random from the mixture. The set T, for this example, consists of three elements, say s, t, and u. The points s and t are incomparable, but both are before u. At s we choose from one set of urns and at t from the other. The possibilities at s are independent of those at t, but what occurs at s and t jointly determine the possibilities at u, i.e., the possibilities for the choosing of the ball.

Thus, I agree with Suppes, [89], that the notion of causality is not deterministic in nature. Instead of defining it in terms of probability, however, as Suppes does, I consider it as based on possibilities. In my account, what has happened at a certain causal moment t, "causes" the range of possibilities at moments s that come after t. Hence, what is caused is not a particular effect, but a range of possible effects.

It turns out that not every partial ordering T can be taken as a set of causal moments. We shall require further properties that will be spelled out in Chapter XIII.

Compound outcomes. The compound outcomes for composite setups, are two-sorted, i.e., they have two universes, the set T of what may be called causal moments, and the universe of the objects properly involved in the setup, say urns and balls. The outcomes also include as one of their relations, the partial ordering of T. The group of symmetries, or group of invariance, for these chance setups consists of transformations that independently permute both universes. Since they must transform possible outcomes into possible outcomes, and the ordering of T is fixed in all the outcomes for a setup, these transformations restricted to T should be automorphisms of the ordering.

From these two sorted structures, another representation of outcomes can be obtained. Under this second representation, each outcome is represented by a function ξ with domain T, such that at each t in T, $\xi(t)$ represents what is selected at t. For instance, in the first urn example, an outcome is represented by a function ξ, such that $\xi(s)$ is the selected urn and $\xi(t)$ is the selected ball from the urn $\xi(s)$. One can get uniquely from the functions to the two-sorted structures and vice-versa. So we shall use either representation as seems fit.

Probability space. I turn, now, to the characterization of the probability measure and the algebra of events. We begin with an example that will clarify why the Principle of Internal Invariance is not enough to determine the measure in many cases of setups with compound outcomes. Suppose that we have a probability measure Pr defined on a field of sets $\mathcal{F}$ that is invariant under a group of functions G. That is, if A, $B \in \mathcal{F}$, $f \in G$ and $A = f''(B)$ then $\Pr A = \Pr B$. In order for Pr to reflect more accurately the structure of the group G, we should also have that $\Pr A = \Pr B$ should imply that there is an

4. THE EXTERNAL INVARIANCE PRINCIPLE

$f \in G$ such that $A = f''(B)$.[3] It can be easily proved, in this case, that if $\Pr A = \Pr B$ and $A = A_1 \cup A_2$ with A_1 and A_2 disjoint, then there are disjoint B_1 and B_2 such that $B = B_1 \cup B_2$, $\Pr A_1 = \Pr B_1$, and $\Pr A_2 = \Pr B_2$. At least basic events should obey this stronger invariance requirement.

Consider, now, the urn model with two urns, one with two white balls, and the other with one black ball. The event "white ball" should be equiprobable with the event "black ball". The first event, however, can be divided into two events, while the second cannot. These two events must be basic, because they cannot be obtained from other events by Boolean operations. Here, the possible outcomes are three: to obtain white ball 1, to obtain white ball 2, and to obtain the black ball. It is clear that these outcomes are not equiprobable, and hence its probability cannot be obtained through the First Invariance Principle. Thus, we must find another principle to help us obtain the measure for these cases.

A more elaborate example, which also indicates why the Internal Invariance Principle is not enough, is the example communicated by Janina Hosiasson to Jeffreys as a criticism to the classical definition of probability. This example also shows that simple-minded equiprobability does not suffice for a definition of probability. The example is reported in [48, p. 342], but I shall give the version of Popper in [76, pp. 376–378]. We consider two games of chance. In the first game (Fig. III.4), we have one bag and two urns, I and II. In the bag, there are three counters, two of them marked 'I' and one marked 'II'. In urns I and II, there are three balls each: in urn I, two balls are white and one is black; in urn II, one ball is white and two are black.

We draw a counter at random from the bag. If it is marked 'I', we next draw a ball at random from urn I; while if it is marked II, we next draw a ball at random from urn II. The game ends when we have drawn a ball from either urn I or urn II. The question is to determine the probability of drawing a white ball. The answer is, of course, $(\frac{2}{3} \cdot \frac{2}{3}) + (\frac{1}{3} \cdot \frac{1}{3}) = \frac{5}{9}$.

The second game (Fig. III.5) is exactly like the first except that urn II contains two balls only, one white and one black. In this second game, the answer to the question of the probability of drawing a white ball is, of course, $(\frac{2}{3} \cdot \frac{2}{3}) + (\frac{1}{3} \cdot \frac{1}{2}) = \frac{11}{18}$. Our calculations of the result of the second game may be represented by Fig. III.6, or, rather, by the assertion that the game represented by this figure is 'equivalent' to the game in Fig. III.5.

The point made by Janina Hosiasson is put by Popper as follows, [76, p. 378]:

> In the first game we have, essentially, nine equal possibilities. (As the diagram clearly indicates, we might as well have no bags and, instead of the one urn numbered I, two different urns, both numbered I, and thus three urns with nine balls altogether, five of them white.) And we can represent our result by saying that we first *count all* the equal ways or possibilities—nine altogether— of bringing the game to an end, and then *count* the number of

---

[3] In case $\mathcal{F}$ is infinite, we should only require that $A = f''(B)$ except for a set of measure 0, but since we are dealing with finite algebras, we can have the stronger requirement.

III. PROBABILITY MODELS

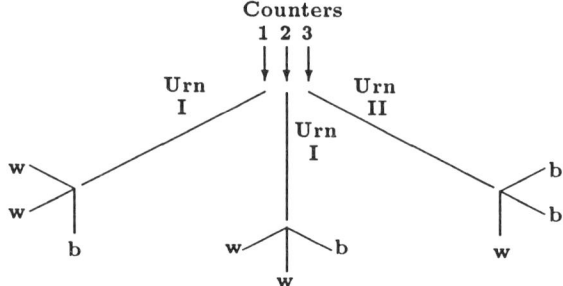

FIGURE III.4. Hosiasson's first game

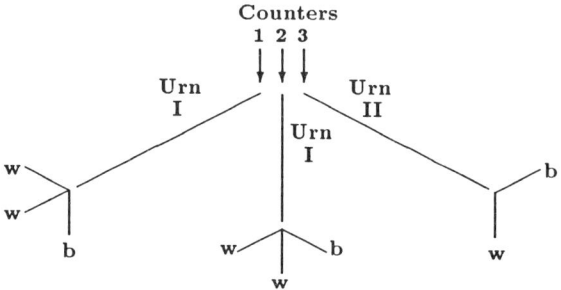

FIGURE III.5. Hosiasson's second game

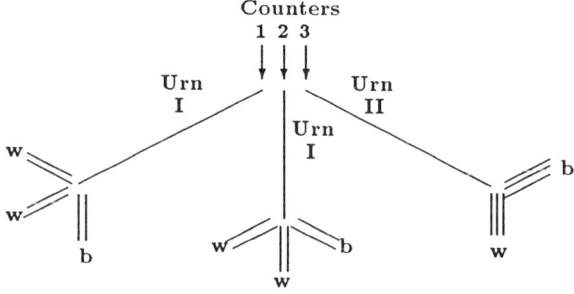

FIGURE III.6. Hosiasson's third game

4. THE EXTERNAL INVARIANCE PRINCIPLE

those which end with the drawing of a *white ball* (i.e., five). The ratio, 5/9, is the solution.

But this method is inapplicable to the second case. Its result is 11/18. Yet it is simply untrue that we have eighteen possibilities to consider of which eleven are favorable: no *counting* of possibilities will yield this result. Although the second game is very simple (there is not the slightest difficulty about its calculation) and although it operates at every stage with equal probabilities or equal possibilities, its result cannot be represented within a theory that *counts* (equal) possibilities. It consists, rather, in a kind of fictitious *construction* of eighteen equal probabilites which we can *calculate* no doubt, but which do not really occur in the second game at all.

Thus, I think that it is clear that the probabilities in Hosiasson's second game (Fig. III.5) cannot be obtained by equally probable possibilities, and, hence, they cannot be obtained by the First Invariance Principle. The game pictured in the third diagram (Fig. III.6), although different from the second game, is equivalent or *homomorphic* (i.e, it has the same form) to this latter game, in a sense to be made precise later. Therefore, as we shall see, it is natural to calculate the probabilities in the second game using Fig. III.6. The precise statement of this fact is what will be called the Second or External Invariance Principle.

Second Invariance Principle. We now begin the discussion of this second principle. The adequate probability measure for a setup, besides being invariant under the group of symmetries, should respect the causal structure of the setup. That is, the measures for two setups with the same causal structure should be obtained in the same way. This second property turns out to be another invariance property, which will be called the *Principle of External Invariance*, since it is invariance between setups and not inside just one setup.

A complete description of this principle is impossible without a formal development. I shall attempt here only an informal and, certainly, very imprecise discussion, and illustrate it with an example. In order to state the principle, we begin by defining when two probability structures of the form we have been discussing have the "same chance structure". For this purpose, a notion of homomorphism between setups will be defined and the measures for the different setups should preserve these homomorphisms. This notion of homomorphism turns out to imply that if there is a homomorphism from **K** onto **H**, then **K** and **H** have essentially the same structure. That is, **K** and **H** must have the same causal structure, and the groups of invariance of the two models must also have the same "structure". The Principle of External Invariance requires that if f is a homomorphism from **K** onto **H**, and **A** and **B** are events in **H** with the same probability measure, then their inverse images under f, $f^{-1}(\mathbf{A})$ and $f^{-1}(\mathbf{B})$ also have the same probability measure in **K**.

In the case of Hosiasson's games, those represented by Fig. III.5 and III.6 are homomorphic in the following sense. We consider the game of Fig. III.6 as a setup having six balls in each urn, each ball represented by a line. The homomorphism f associates each possible outcome of Fig. III.6, with the corresponding outcome in Fig. III.5 labeled by the letters w or b, as indicated in the figure. In a sense to be explained precisely later, this homomorphism preserves the relation of dependence.

The Principle of External Invariance is justified, because the relations of dependence and independence, plus the group of symmetries of the setups are, in my construction, the only elements of the setups which determine the probability measure. Hence, if two setups have the same causal structure (which is determined by the relations of dependence and independence) and groups of symmetries of the "same form", then they should assign the same probabilities to corresponding events.

The notion of measure-determined subset of a model $\mathbf{H}$ has to be widened. Suppose we have an invariant measure Pr (according to the Principle of Internal Invariance) on the subsets of a probability structure $\mathbf{K}$. If we have a homomorphism f from $\mathbf{K}$ onto $\mathbf{H}$,[4] we can define the measure $\mathrm{Pr}_{\mathbf{K}}$ on subsets $\mathbf{A}$ of $\mathbf{H}$, by $\mathrm{Pr}_{\mathbf{K}}(\mathbf{A}) = \mathrm{Pr}(f^{-1}(\mathbf{A}))$. We now call $\mathbf{A}$ measure-determined, if for all such $\mathbf{K}$ and $\mathbf{K}'$ (with the corresponding homomorphisms f and f'), $\mathrm{Pr}_{\mathbf{K}}(\mathbf{A}) = \mathrm{Pr}_{\mathbf{K}'}(\mathbf{A})$. It can be shown that our old measure-determined sets are also measure-determined in the new sense.

The family of events $\mathcal{F}_{\mathbf{H}}$ for $\mathbf{H}$ is again the disjunctive algebra (or disjunctive σ-algebra) of subsets of $\mathbf{H}$ generated by these measure-determined sets. The probability measure $\mathrm{Pr}_{\mathbf{H}}$ on measure-determined sets $\mathbf{A}$, is defined as $\mathrm{Pr}_{\mathbf{H}}(\mathbf{A}) = \mathrm{Pr}_{\mathbf{K}}(\mathbf{A})$, for some $\mathbf{K}$ homomorphic to $\mathbf{H}$, where $\mathrm{Pr}_{\mathbf{K}}$ is obtained as above.[5] This measure $\mathrm{Pr}_{\mathbf{H}}$ is uniquely extended to the generated disjunctive algebra (or disjunctive σ-algebra). The extension to the generated algebra, however, is not unique.

The precise description of the models must wait for the formal development in Part 3. I will just discuss carefully some examples. We begin with urn models similar to Hosiasson's games, but simpler. Let us suppose that we have two urns, u_1, with two white balls, and u_2, with three black balls. We choose one of the urns at random, and then a ball from the urn chosen. As was discussed above, we need here a causal structure of two elements, say t and s, with $t \prec s$. Each possible outcome can be represented by a function ξ with domain $\{t, s\}$ such that $\xi(t) = \langle U, u \rangle$, where U is the set of urns and u the urn selected, and $\xi(s) = \langle B_u, b \rangle$, where B_u is the set of balls in urn u, and b is the ball selected. The chance structure is, then, $\mathbf{U}$, the set of all such ξ. We also denote by $\mathbf{Ur}$ the simple chance structure for the selection of the urn, and $\mathbf{B}_u$ the simple chance structure for the selection of the ball in urn u.

An isomorphism between two such outcomes ξ and η should be obtained from functions g_t and g_s of the following form. The first one, g_t should be a permu-

---

[4] It can be proved that for every finite $\mathbf{H}$ there is $\mathbf{K}$ and a homomorphism f of $\mathbf{K}$ onto $\mathbf{H}$, such that all subsets of $\mathbf{K}$ are measure-determined and that there is an internally invariant measure Pr defined on all subsets of $\mathbf{K}$. This allows for the measure $\mathrm{Pr}_{\mathbf{K}}$ to be always defined. For the mathematical details see Chapter XIV.

[5] We can always choose a $\mathbf{K}$ as explained in footnote 4.

4. THE EXTERNAL INVARIANCE PRINCIPLE

tation of the urns and $g_t(\xi(t)) = \eta(t)$. That is, if u is the urn selected in $\xi(t)$ then $g_t(u)$ is the urn selected at $\eta(t)$. Now, g_s should transform the balls in u into the balls in $g_t(u)$, and if b is the ball selected in ξ then $g_s(b)$ should be the ball selected in η. In our case, since the number of balls in u_1 is different from the number of balls in u_2, there is no function g_s which transforms the balls in one urn into the balls in the other. Thus, g_t must be the identity mapping. Thus, there are very few isomorphisms in the group of invariance, and, hence the invariant measure is underdetermined.

It is intuitively clear, however, that the selection of a white ball should have the same probability as the selection of a black ball. There is no isomorphism, however, that sends a white ball into a black ball. Thus, we must use another principle, besides invariance under the group, to determine the right measure. This is what has been called before, the External Invariance Principle. I shall explain now how it works for this simple example.

Let us construct systems for a different urn model, call it $\mathbf{U}'$. Suppose, now, that we also have two urns, but that each of the urns, call them v_1 and v_2, contains six balls, v_1 contains six white balls and v_2, six black balls. In this case, there are isomorphisms that transform white into black balls, because the two urns have the same number of elements. In fact, for any two outcomes, there will be an isomorphism that transforms one into the other. Hence, the invariant measure is the counting measure. That is, the probability of any event is the number of outcomes in the event divided by the total number of outcomes.

The two chance structures $\mathbf{U}'$ and $\mathbf{U}$ are related by the function f defined in the following way. Let $U' = \{v_1, v_2\}$ and let f_0 be a one-to-one function from U' to U, say $f_0(v_1) = u_1$ and $f_0(v_2) = u_2$. Let f_{v_1} transform the balls in v_1 onto the balls of u_1 in such a way that three balls of v_1 are sent onto each ball of u_1. Similarly, f_{v_2} sends a pair of balls of v_2 onto each of the balls of u_2. Suppose that ξ is an outcome in $\mathbf{U}'$ with $\xi(t) = v$. Then ξ is sent into the outcome $f(\xi)$ of $\mathbf{U}$ if $f_0(v) = f(\xi)(t)$ and $f_v(\xi(s)) = f(\xi)(s)$.

This function f preserves the equiprobability structure of $\mathbf{U}$ in the following sense. We write $\mathbf{A} \sim \mathbf{B}$ if $\mathbf{A}$ and $\mathbf{B}$ are events that are equivalent by a function of the group of invariance of the corresponding chance structure. We then have that if $\mathbf{A}$ and $\mathbf{B}$ are subsets of $\mathbf{U}$ and $\mathbf{A} \sim \mathbf{B}$, then $f^{-1}(\mathbf{A}) \sim f^{-1}(\mathbf{B})$ in $\mathbf{U}'$; if $\mathbf{A}$ and $\mathbf{B}$ are subsets of $\mathbf{Ur}$ and $\mathbf{A} \sim \mathbf{B}$, then $f_0^{-1}(\mathbf{A}) \sim f_0^{-1}(\mathbf{B})$ in $\mathbf{Ur}'$; and if $\mathbf{A}$ and $\mathbf{B}$ are subsets of $\mathbf{B}_u$ with $f_0(v) = u$ and $\mathbf{A} \sim \mathbf{B}$, then $f_v^{-1}(\mathbf{A}) \sim f_v^{-1}(\mathbf{B})$ in $\mathbf{B}'_v$. We call such functions, *homomorphisms*.

The structure $\mathbf{U}'$ and the homomorphism f induce a probability measure Pr on all subsets of $\mathbf{U}$ as follows

$$\Pr \mathbf{A} = \Pr_{\mathbf{U}'}(f^{-1}(\mathbf{A})),$$

where, as we saw above, $\Pr_{\mathbf{U}'}$ is the counting measure, i.e., the invariant measure on $\mathbf{U}'$.

It will be proved in Chapter XIV, that any chance structure of the form of $\mathbf{U}'$ and any homomorphism onto $\mathbf{U}$ induce the same measure on subsets of $\mathbf{U}$. For instance, if instead of six balls we put twelve in the urns, the measure induced is the same. Thus, we define the probability measure Pr_U as the measure thus induced.

It is almost immediate from the definition that this measure is invariant under the internal group of $\mathbf{U}$, and also, under the internal groups of $\mathbf{U}_r$ and $\mathbf{B}_u$. It can also be proved that in general the measure has the right properties.

Models for inference and decision. I would like to discuss here two examples that can be applied to more general cases than those already mentioned. I shall use these examples in the applications to statistical inference in Chapter IV and Part 4. For the first of these two examples, the compound setup consists of n independent trials of an experiment, where n is a natural number, finite or infinite. Assume that the set of possible outcomes of each trial of the experiment is $\mathbf{K}$. The causal ordering T can be taken as any set of n members, all of its elements being mutually incomparable. The outcomes that are obtained in the n trials can be represented by n-termed sequences of elements of $\mathbf{K}$. Let us denote the set of these sequences by $^n\mathbf{K}$. Compound events are subsets of $^n\mathbf{K}$, and it will be shown in Chapter XIV that the measure invariant according to the two principles of invariance is the product measure of the invariant measure for $\mathbf{K}$ itself.

The other example is a generalization of the urn model. Here, the set T consists of two elements t and s, with t preceding s. At t we have a set of possible simple outcomes $\mathbf{K}_0$. For each ω_0 in $\mathbf{K}_0$, we have a set of possible outcomes $\mathbf{K}_{\omega_0}$. The compound outcomes can be represented as functions ξ with domain T and such that $\xi(t)$ is in $\mathbf{K}_0$, and $\xi(s)$ is in $\mathbf{K}_{\xi(t)}$. Let us call $\mathbf{K}$ the set of these compound outcomes. Assume that we have an invariant measure Pr_0 for $\mathbf{K}_0$, and one, Pr_{ω_0}, for each $\mathbf{K}_{\omega_0}$. The invariant dependence preserving measure Pr is that which is expected:

Let $\mathbf{B}$ be a subset of $\mathbf{K}$, i.e., an event. Let $\mathbf{B}_0$ be the set of $\xi(t)$ for ξ in $\mathbf{B}$; and let $\mathbf{B}_{\omega_0}$ be the set of $\xi(s)$ for ξ in $\mathbf{B}$ and $\xi(t) = \omega_0$. It is clear that $\mathbf{B}_0$ is a subset of $\mathbf{K}_0$, and $\mathbf{B}_{\omega_0}$ is a subset of $\mathbf{K}_{\omega_0}$. It will be proved in Chapter XIV that the probability measure Pr invariant according to the two principles has the following property:

$$\mathrm{Pr}\,\mathbf{B} = \int_{\mathbf{B}_0} \mathrm{Pr}_{\omega_0}(\mathbf{B}_{\omega_0})\,d\mathrm{Pr}_0, [0]$$

which in the finite case is

$$\mathrm{Pr}\,\mathbf{B} = \sum_{\omega_0 \in \mathbf{B}_0} \mathrm{Pr}(\omega_0)\,\mathrm{Pr}_{\omega_0}(\mathbf{B}_{\omega_0}).$$

That is, $\mathrm{Pr}\,\mathbf{B}$ is the weighted average of the measures $\mathrm{Pr}_{\omega_0}(\mathbf{B}_{\omega_0})$ according to Pr_0.

As it was mentioned above, it will be proved in Chapter XIV (and discussed in more details in Chapter XVI) that the two invariance requirements, imply that the measures for these two cases must be what has been stated.

I have, thus, completed the informal description of the probability spaces, composed of a basic set $\mathbf{K}$, a disjunctive algebra of events, say $\mathcal{F}_\mathbf{K}$, and a probability measure on this algebra, $\Pr_\mathbf{K}$. These probability spaces bear a resemblance with the usual sample spaces $\langle \Omega, \mathcal{F}, \Pr \rangle$ of Kolmogorov style axiomatic probability theory. There is, however, an important difference. In the probability spaces described here, the set of outcomes $\mathbf{K}$ determines the algebra of events and the probability measure. In the usual sample spaces, the set Ω is just a support for the algebra of events and the measure, and does not play an important role. As a matter of fact, probability theory can be developed without sample spaces, by using directly Boolean algebras of events.[6]

5. Probability models

Models are used both in logic and the natural sciences.[7] In the natural sciences, a model is supposed to be a mathematical representation of the phenomenon that is investigated. Since mathematics can be formalized in set theory, this mathematical representation can be taken to be a set-theoretical system. On the other hand, in mathematical logic, a model of a sentence of a formal language is also a set-theoretical system, in which the language of the sentence is interpreted, and the sentence is true or false according with the interpretation. In these systems, which represent completed possible worlds, the truth or falsity of all sentences of the language is determined. I believe that models in both applications can be represented by the same sort of set-theoretical system. In this section, I shall explain this dual use of models[8] and describe informally the dual use of the probability structures described earlier.

I shall begin with models in the natural sciences. The first step in order to set up a mathematical model of a physical situation, is to replace the physical objects, relations between this objects, properties of the objects, etc. by abstract objects that, so to say, mirror them. That is, to form an abstract world that reflects the physical world. Hence, we must replace the set of material objects that form the physical world by an abstract set A, which should be nonempty; the binary relations between these objects, by abstract set-theoretical relations R, i.e., $R \subseteq {}^2A$; properties, by subsets of A, etc. Thus, in the simplest situations, an abstract world may be thought of as a relational system of the form $\langle A, R_i, a_j \rangle_{i \in I, j \in J}$, where the R_i's are unary (identified with subsets of A), binary, ternary, or, in general, n ary relations (i.e., subsets of ${}^nA$, the set of n-tuples of elements of A), and the a_j's are distinguished elements of A, which are objects we are especially interested in and that we can identify. Each relational system is a logically possible model of reality, that is, it represents a possible world, as was discussed in Chapter II, Section 2 and Section 1 in this chapter.

Not all situations can be modeled by this simple type of relational system. It may be necessary to have two or more distinct universes, functions to the real numbers or to other mathematical structures, or other complications. The

---

[6] See [85], for instance.
[7] My views in this section are influenced by [88]. See also [28].
[8] See also [17].

systems mentioned in Section 4 for compound events are examples of systems with two universes.

Although my wording seems to restrict the models to physical situations, the same type of structures can be used to model economic, social or other phenomena. The objects involved in these may be other than physical, but the abstract systems are similar to those explained above.

A relational system $\omega = \langle A, R_i, a_j \rangle_{i \in I, j \in J}$ can then be considered as an abstract model of nature. It is clearly impossible, however, to describe completely a physical situation, and hence ω can be considered only as an approximate description. Many features of nature are idealized or simply disregarded, since it would be impossible to deal with them all; only what is thought relevant is taken into account.

An example may help to clarify the matter. Suppose that the situation involved is the roll of a die. The universe A of our relational system contains the objects we want to talk about. In this case, A may consist just of the faces of the die, disregarding all other features of the die. The only property of the faces we need is the property of coming up in the roll, call it C, which is a one element subset of A. Thus, our relational systems for the roll of a die may be taken to be $\langle A, C \rangle$.

Suppose now that the die is biased. Then, in order to have a clear picture of the situation, it may be necessary to consider the distribution of weights in the die, and our systems would be much more complicated. They may have to include the real numbers and functions (weights, for instance) into the reals. Thus, there may be different relational systems appropriate for the same situation, depending on what we acknowledge as relevant for the explanation we are seeking.

There is another way of looking at our relation with nature. We use a language in order to describe reality. In this language, we mention objects and speak about relations among them. When our statements match reality, we say they are true. Words in a natural language, such as English, may refer to objects in the world. We also understand what it means for a proposition expressed in English to be true or false: loosely speaking, a proposition is true when it 'corresponds' to reality. Or, as Aristotle puts it:[9]

> To say of what is that it is not, or what is not that it is, is false, while to say of what is that it is, and of what is not that it is not, is true.

This conception of truth as correspondence with reality, has been represented mathematically by Tarski, [92]. The first part of his representation is to mathematize reality via relational systems, as has been explained above. The next step is to mathematize the language. A natural language is approximated by certain mathematical structures called formal (or artificial) languages. These formal languages are approximations to natural languages just as relational systems approximate the real world: some features of natural languages are idealized or

---

[9]Aristotle, Metaphysics, Book Γ 1011b26. For a full discussion of the definition of truth, see [94].

disregarded. Since formal languages are mathematical structures, sentences and other linguistic objects are defined mathematically and represented by mathematical objects.

Just as the interpretation of natural languages is in reality, that of the formal languages is in the mathematical representation of reality, i.e., in relational systems. For each system, we need an appropriate formal language, i.e., one that has symbols for the properties, relations and distinguished elements of the system. For instance, for $\langle A, C \rangle$ we should have a symbol for C. A mathematical definition of truth is then possible. Tarski defined a relation between a sentence φ and a relational system ω: the relation of φ being true in ω (or according to the interpretation ω). This relation is usually expressed symbolically by $\omega \models \varphi$, and is also read, ω is a model of φ.[10]

When studying pure logic, which relational system matches reality is not relevant: for logic, all that is relevant is the abstract relation of truth between a sentence of a formal language and a mathematically represented possible world.

Random variables as part of a language. Probabilists and statisticians have traditionally used, to describe mathematically properties of a situation, what are called random variables or random quantities. As used in probability theory, random variables are functions from a sample space to the real numbers. Since our sample space consists of relational systems, we shall use the following analog of random variables.

Suppose we still consider the roll of a die. The model for the situation is $\langle A, C \rangle$, where C contains exactly one face, the face that obtains in the roll. In order to express that the face that obtains is six, we can use a function with value six. For the roll of a die we would have the function X, such that $X(\langle A, C \rangle) = c$ if and only if $C = \{c\}$. If ω is a system of the form indicated above, then $X(\omega)$ is the face that comes up when the result of the toss is described by ω. In general, X is a function from relational systems (which, as was explained in Section 1, represent outcomes) to real numbers (or to another appropriate set) that measures some feature of the system. Since these functions are introduced here independently of any probability measure, and also due to the fact that the functions will not always turn out to be measurable when the measure is defined, the name random variable, for these functions, is misleading. So I shall call X a *measurement function* or simply a measurement.

Statisticians, on the other hand, use random variables, as logicians use an artificial language. When performing statistical tests, for instance, a random variable X can have different probability distributions. In fact, this means that it can have different interpretations. Suppose that we are measuring a quantity X and that we know that it is normally distributed, but that we do not know the mean or variance. Thus, for each mean μ and standard deviation σ, X can be normally distributed with these parameters. These different distributions are, in fact, possible interpretations of X which, thus, acquires the characteristics of the symbol of a language.

---

[10]For an elementary explanation of this definition see [66]. A less elementary account is in [32]. An excellent informal exposition appears in [94].

We proceed now to explain the use of measurement functions as symbols of a language. We can express properties of the outcomes by the use of measurements. For instance, if we want to express that the result of the roll of a die is 6, we can do it by the sentence $[X = 6]$. Other types of possible sentences are $[X \leq r]$ and $[X \geq r]$, where r is a real number. It is clear what it means for a sentence of one of these forms to be true in a system ω with the measurement function X, namely

$[X \leq r]$ is true in ω if and only if $X(\omega) \leq r$.

This is the relationship between the two uses of measurement functions. We symbolize φ is true in ω with X as measurement function by

$$\langle \omega, X \rangle \models \varphi.$$

We may have more than one measurement function in our language. For instance, we may have X_1, X_2, ..., X_n. It is easy to see how to build a language with these symbols: We include, also, continuous functions of these variables. More precisely, if g is a continuous function of n real variables, then we have the sentence $[g(X_1, \ldots, X_n) \leq r]$ in our language. For example, $g(X_1, \ldots, X_n)$ may be

$$g(X_1, \ldots, X_n) = \frac{1}{n} \sum_{k=1}^{n} X_k$$

Now, we also have

$\langle \omega, X_1, \ldots, X_n \rangle \models [g(X_1, \ldots, X_n) \leq r]$ if and only if $g(X_1(\omega), \ldots, X_n(\omega)) \leq r$.

We can now visualize the dual role of relational systems. On one hand, they are possible models of the world. On the other hand, they are possible models of sentences of a language. Notice that although the systems remain the same, they are models in different senses. In the first case, a relational system may be a model of reality because the properties and relations of the system represent faithfully the real relations which are intended as their interpretations. In the second case, the sentences are interpreted in the relational system, and the system is a possible model of the sentences. That is, for a sentence φ, the system ω may be a model of φ (i.e., φ is true in ω) or not (i.e., φ is false in ω).

The relationship between sentences, models, and reality can be summarized as follows: a sentence is interpreted in a possible model, and it may be true or false in the possible model. The model, in its turn is interpreted in reality, and it may be true or false to reality. But sentences are also interpreted in reality, and they may be true or false in it. The connection between the model and reality can be analyzed through this dual interpretation of sentences: if a sentence true in the model is false in reality, then the model is rejected a model of reality.

I shall introduce formally in Part 3, and have described informally in this chapter, probability structures that play the same dual role. On one hand, they are possible models of reality, when reality is a chance setup. On the other hand, as will be seen in the rest of this section, probability structures serve as interpretations of a language, which will be, in our case, a language of measurements. These interpretations do not determine the truth or falsity of sentences, but only

5. PROBABILITY MODELS 81

their probabilities. These use of measurement functions is, as was mentioned above, similar to the use of random variables in statistical inference. In statistics, random variables are interpreted in sample spaces. Usually there are several sample spaces, i.e., distributions of the random variable, under consideration.

Probability structures were defined in previous sections as sets of relational systems, and the definition of probability for sentences will be based on the definition of truth in a relational system.

Probability models and languages. We have informally seen in the previous sections of this chapter, how a probability structure $\langle \mathbf{K}, \mathcal{F}_{\mathbf{K}}, \mathrm{Pr}_{\mathbf{K}} \rangle$ is obtained, where $\mathbf{K}$ is a set of outcomes represented by relational systems, $\mathcal{F}_{\mathbf{K}}$ is an algebra (or at least a disjunctive algebra) of subsets of $\mathbf{K}$, and $\mathrm{Pr}_{\mathbf{K}}$ is a probability measure on $\mathcal{F}_{\mathbf{K}}$.

As we saw earlier, these probability structures are thought of as possible models, in the natural sciences sense, of chance setups. We can perform many different measurements on $\mathbf{K}$. If $\mathbf{K}$ is a simple structure, that is, with no causal ordering T, a measurement X for $\mathbf{K}$ is just a function from $\mathbf{K}$ to the real numbers. When $\mathbf{K}$ is a possible model of a compound setup, that is, it includes a causal ordering T, then a measurement Y for $\mathbf{K}$ is an function from $^n T \times \mathbf{K}$ into the reals, for some natural number n. That is, for each $t_1, \ldots, t_n \in T$ and $\omega \in \mathbf{K}$, $Y(t_1, \ldots, t_n, \omega)$ is a real number. For simplicity, we take in this section $n = 1$. Thus, $Y(t, \omega)$ gives the measurement of the compound outcome ω at the causal moment t. Instead of writing $Y(t, \omega)$, we shall sometimes write, as it is customary, $Y_t(\omega)$.

For instance, if our compound structure represents n independent repetitions of the same experiment with model $\mathbf{K}$, which is denoted by $^n \mathbf{K}$, and Y is a measurement function on $\mathbf{K}$, then we can define the function $\bar{Y}$ on $^n \mathbf{K}$ by $\bar{Y}(m, \omega) = Y(\omega(m))$, for $1 \leq m \leq n$ and $\omega \in {^n \mathbf{K}}$. We also write $\bar{Y}(m, \omega) = Y_m(\omega)$. Here, $Y_m(\omega)$ is the value of ω at the mth repetition.

As another example, the compound structure for the urn model might have, if we number the urns, a measurement function Z, such that $Z(t, \omega) =$ the number of the urn chosen at $\omega(t)$, and $Z(s, \omega) =$ the number of, say, red balls chosen at $\omega(s)$.

Hence, our possible probability models are pairs $\langle \mathbf{K}, Y \rangle$, where $\mathbf{K}$ is a probability structure and Y is a measurement function appropriate for $\mathbf{K}$ that is measurable in $\mathbf{K}$. The same probability structure may support different measurement functions, and we get thus different possible probability models. We allow, also, more than one measurement function in a possible probability model. For simplicity, I shall only describe here models with one measurement function.

The language to express properties of a model of the form $\langle \mathbf{K}, Y \rangle$, where $\mathbf{K}$ is a compound structure with causal ordering T, consists of expressions of the form $[Y_t \leq r]$ or $[Y_t = r]$, where $t \in T$ and r is a real number, and is closed under negations, conjunctions and disjunctions.

We also include continuous functions of the variables Y_t. Thus, if g is a continuous function of n real variables, then $[g(Y_{t_1}, \ldots, Y_{t_n}) \leq r]$ and $[g(Y_{t_1}, \ldots, Y_{t_n}) = r]$ are sentences. The negation of a sentence φ is denoted $\neg \varphi$, the conjunction of the set of sentences Φ, by $\bigwedge \Phi$, and the disjunction of the set Φ, by $\bigvee \Phi$. The conjunction of two sentences φ and ψ, by $\varphi \wedge \psi$, and its disjunction by $\varphi \vee \psi$. A conjunction is true if all of its conjuncts are true, a disjunction, if at least one disjunct is true, and a negation, if the sentence that is being negated is not true.

By combining these expressions, we can say that Y is in certain set of real numbers B, for some sets of reals B. Thus, we can have expressions that can be abbreviated, with the obvious meaning, by

$$[Y_1 \in B_1] \wedge [Y_2 \in B_2] \wedge \cdots \wedge [Y_n \in B_n],$$

and also

$$[g(Y_{t_1}, \ldots, Y_{t_n}) \in B],$$

where B, B_1, ..., B_n are sets of reals.

It is clear what it means for a sentence $[g(Y_{t_1}, \ldots, Y_{t_n}) \in B]$ to be true in a system ω with the measurement function Y; namely

$$\langle \omega, Y \rangle \models [g(Y_{t_1}, \ldots, Y_{t_n}) \in B] \text{ if and only if } g(Y_{t_1}(\omega), \ldots, Y_{t_n}(\omega)) \in B.$$

For instance, suppose that the model **K** is a possible model for a toss of a coin and $X(\omega) = 1$, if ω is an outcome where the result is heads, and $X(\omega) = 0$, otherwise. Then ${}^n\mathbf{K}$ represents n independent tosses of a coin. Let $\bar{X}$ be defined as above by $\bar{X}(m, \omega) = X(\omega(m))$ for $1 \leq m \leq n$ and $\omega \in {}^n\mathbf{K}$. Let also $X_m(\omega) = \bar{X}(m, \omega)$ and $Y_m(\omega) = \sum_{k=1}^{m} X_k(\omega)$. Then $Y_m(\omega)$ is the number of heads in the m first tosses, for $1 \leq m \leq n$ and $\omega \in {}^n\mathbf{K}$. Then $\langle \omega, Y \rangle \models [Y_m \leq 6]$ if and only if the number of heads in the sequence ω restricted to m is less that or equal to 6.

Also, the variable $\mathrm{Fr}_m = (1/m)Y_m$ represents the relative frequency of heads in the first m tosses, that is, for instance

$$\langle \omega, Y \rangle \models [|\mathrm{Fr}_m - \frac{1}{2}| \geq \delta] \text{ if and only if } |\mathrm{Fr}_m(\omega) - \frac{1}{2}| \geq \delta.$$

This definition of truth fixes an interpretation of the sentences in the relational systems (or outcomes) ω. But we need an interpretation in the possible probability model $\langle \mathbf{K}, Y \rangle$. Each sentence φ is interpreted in $\langle \mathbf{K}, Y \rangle$ by the set of systems in **K** in which φ is true. That is, we define

$$\mathrm{Mod}_{\langle \mathbf{K}, Y \rangle}(\varphi) = \text{ the set of } \omega \in \mathbf{K} \text{ such that } \langle \omega, Y \rangle \models \varphi.$$

Now, $\langle \mathbf{K}, Y \rangle$ does not, in general, associate truth or falsity to sentences, but only probability. We require that $\mathrm{Mod}_{\langle \mathbf{K}, Y \rangle}(\varphi) \in \mathcal{F}_{\mathbf{K}}$ and define

$$\mathrm{Pr}_{\langle \mathbf{K}, Y \rangle}(\varphi) = \mathrm{Pr}_{\mathbf{K}}(\mathrm{Mod}_{\langle \mathbf{K}, Y \rangle}(\varphi)).$$

We use the same symbol Pr for the probability on sentences and the probability of events, since there is no danger of confusion.

Thus, a sentence is assigned as probability the measure of the set of outcomes where it is true, i.e., the probability of a proposition is the meassure of the set of possibilities that make it true.

As an example, we shall discuss the roulette setup. In order to deal with the different models of roulettes with different biases, we introduce a language with one measurement function, say X. Let u and v be two real numbers between

5. PROBABILITY MODELS

0 and 2π. The sentence $[u \leq X \leq v]$ expresses the fact that the final position of the roulette is in the arc between e^{iu} and e^{iv}. The probability assigned to this sentence depends of the structure. Suppose that $\langle \mathbf{C}, X_1 \rangle$ is any of these structures. We first recall the definition of truth. We say that

$$\langle \omega_c, X_1 \rangle \models [u \leq X \leq v], \text{ if } u \leq X_1(\omega_c) \leq v.$$

We have, according to our definition

$$\Pr_{\langle \mathbf{C}, X_1 \rangle}([u \leq X \leq v]) = \Pr_{\mathbf{C}}(\text{Mod}_{\langle \mathbf{C}, X_1 \rangle}([u \leq X \leq v])),$$

where $\text{Mod}_{\langle \mathbf{C}, X_1 \rangle}([u \leq X \leq v])$ is the set of ω_c that satisfy $[u \leq X \leq v]$ (i.e., such that $\langle \omega_c, X_1 \rangle \models [u \leq X \leq v]$).

That is

$$\Pr_{\langle \mathbf{C}, X_1 \rangle}([u \leq X \leq v]) = \Pr_{\mathbf{C}}(\{c : u \leq X_1(\omega_c) \leq v)\},$$

where, as we have already mentioned, $\Pr_{\mathbf{C}}$ is Lebesgue measure.

Thus, if X_1 is the measurement for a symmetrical roulette, the probability of this sentence is $(v-u)/2\pi$, because the set of c such that $u \leq X_1(\omega_c) \leq v$ has length $v - u$.

On the other hand, if $\langle \mathbf{C}, X_2 \rangle$ represents an asymmetrical roulette, then the length of the interval consisting of the c with $u \leq X_2(\omega_c) \leq v$ may be different from the length of the interval $[u, v]$, and hence,

$$\Pr_{\langle \mathbf{C}, X_2 \rangle}([u \leq X \leq v])$$

might not be $(v - u)/2\pi$.

Thus, the same sentence is assigned very different probabilities by different models. I believe that this procedure is similar to that used in statistics, where a random variable is assigned different probabilities by the different possible distributions.

As another example, suppose that $\mathbf{K}$ is a model for infinitely many tosses of a fair coin, and let $\text{Fr}_m(\omega)$ be, as above, the relative frequency of heads in the first m tosses. Then, the weak law of large numbers (which is proved as Theorem XI.5) shows that

$$\lim_{n \to \infty} \Pr_{\langle \mathbf{K}, Y \rangle}([|\text{Fr}_n - \frac{1}{2}| \geq a]) = 0,$$

for any $a > 0$.

We see that two different possible probability models which can be interpretations of the same sentence φ, can assign very different probabilities to φ. For instance, if $\mathbf{J}$ is a model for infinitely many tosses of a biased coin the

$$\lim_{n \to \infty} \Pr_{\langle \mathbf{J}, Y \rangle}([|\text{Fr}_n - \frac{1}{2}| \geq a]) \neq 0.$$

This is the usual situation in classical statistical inference. We have several alternative models that assign different probabilities to the same sentences. As it will be discussed in Chapter IV and Part 4, these different probabilities are used to decide between the different possible models. For instance, suppose that we are tossing a coin and that we don't know whether it is biased or not. Then the possible models represent coins with different biases. These models assign different probabilities to the sentences $[a \leq |\text{Fr}_n - p|]$, for instance, where p is the probability of heads of a certain coin. The probabilities of these sentences help to decide which is the bias of the coin.

As a last example, let's get back to the urn model, **U**, and describe the language that we may interpret in it. We could have a simple measurement function X, such that, for instance, $X(\xi) = 1$ if the ball selected in the outcome ξ is white, and $X(\xi) = 0$, if the ball is black. In this language, we have, then, sentences of the form $[X = 1]$ and $[X = 0]$. Different probability structures $\langle \mathbf{U}_1, X_1 \rangle$ and $\langle \mathbf{U}_2, X_2 \rangle$ might assign different probabilities to these sentences.

But suppose that we would like to talk about the urn selected and the ball selected. We then need a measurement process, say Y, which is a function of two variables. If $\xi \in \mathbf{U}$, then we could have $Y(t, \xi) = 1$ if urn u_1 is selected in ξ, and $Y(t, \xi) = 0$, if urn u_2 is selected. For the balls, we can have $Y(s, \xi) = 1$ for a white ball and $Y(s, \xi) = 0$, otherwise. The language, now, besides Y includes the constants t and s. The basic sentences are of the forms $[Y_t = i]$ and $[Y_s = i]$ for $i = 1$ or 0. The probabilities are assigned as before. For instance

$$\Pr_{\langle \mathbf{U}, Y \rangle}([Y_t = 1]) = \Pr_{\mathbf{U}}(\{\xi \in \mathbf{U} : Y(t, \xi) = 1\}).$$

As in the other examples, there may be several structures which assign different probabilities to each sentence.

CHAPTER IV

The basic principles of inference and decision

Decision theory and statistical inference have a different basis. As discussed in Chapter I, Section 1, in decision theory we use probabilities for deciding on an action, while in statistical inference proper, we use probability in order to accept or reject hypotheses. In this chapter, I shall analyze two principles that I think are fundamental for these epistemic uses of probability. I shall also give some examples of their applications. In the last two sections of this chapter, I shall study their application to the study of the relations between probabilities and the determination of degrees of belief, and between probabilities and frequencies. The next chapter contains an example of the application of the two principles discussed in this one to medical diagnosis.

Part 4 discusses different statistical techniques in the light of these two principles. The second principle, to be discussed in the next section, will play in my system the role of a principle of inverse inference, i.e., a principle that is concerned with arguing from knowledge (or, more generally, the acceptance of the occurrence) of an event or a set of events, to the probability or knowledge (or simply acceptance) of a hypothesis.

As it will be seen in Chapters XVIII and XIX, I accept most of the Neyman-Pearson theory of statistical inference, but my justification for this theory is different from that of Neyman and Pearson themselves. Some techniques that have been advocated by "classical" statisticians are not sound according to the principles laid out in Section 2. They are discussed in Chapter XIX. I believe that my general approach is implicit in current statistical practice and is ultimately derived from Fisher's significance tests (but not his fiducial probability), and, in fact, similar to Birnbaum's, [6], although the principles advanced in the present chapter are different from theirs. In particular, the notion of an "experiment" introduced in Section 2 was inspired on Birnbaum's work. On the ohter hand, the notion of rejection set introduced in the same section seems to be related to Fisher's justification of significance tests via "tail probabilities".

For decision theory, however, I accept a Bayesian approach. The principle for decision theory is discussed in the first section of this chapter and the techiniques themselves, with a few examples, in Chapter XX. I believe that this division of attitude towards these two branches of statistics is, in practice, reflected in the bent of statisticians: those who work in the testing of scientific hypotheses use mainly the classical techniques; those in areas where decision theory is more important, such as economics, are mainly Bayesian.

1. The basis for decisions

In this section, I will discuss what I think is the main basis for decision theory, which is a variant of the already mentioned (Chapter I, Section 3 and Chapter II, Section 4) Direct Inference Principle.

When we have to decide on a certain action, and the states of the world relevant for this action are uncertain, it seems more rational to assign probabilities to the different states and base our decision on these probabilities, than to decide by a simple guess. These probabilities represent our degrees of belief in the occurrence of the different states. As the authors of the Port-Royal Logic say, [3, IV, 16]:

> to decide what one should do to obtain a good or avoid an evil, it is necessary to consider not only the good and the evil in themselves, but also the probability that they happen or not happen; and to view geometrically the proportion that all these things have together.

This is essentially the Bayesian account in which an agent performs an act of maximum expected utility.[1] The expected utility of an act can be taken to be the weighted sum of the utilities of the consequences that it would have in the various possible contingencies, and the weights are the probabilities that the agent assigns to those contingencies conditionally upon the act's being performed.

In order to use a Bayesian technique, the agent needs to assign numerical degrees of belief to the possible contingencies, and desirabilities to the possible consequences of his acts. In this chapter, I shall not be concerned about how numerical desirabilities, what are called utilities, are assigned; I shall concentrate on probabilities representing degrees of belief. As it was discussed in Chapter I, Sections 1 and 2, numerical degrees of belief on decidable propositions must obey the axioms of the calculus of probability. In this section, unless explicitly stated, probability will stand for degree of belief, and I shall restrict its application to decidable propositions.

One way of assigning degrees of belief, which I think is the fundamental method for decisions, is to accept that the possible contingencies constitute a chance setup with a set **K** of possible outcomes, calculate the degrees of support of the different propositions in **K**, use the Principle of Direct Inference to obtain the corresponding degrees of belief, and act accordingly. The structure **K** is, thus, considered a model of reality. The fitness of **K** to this reality is not in question. That is, the determination of the truth or falsity of the proposition that asserts that **K** is an adequate model of the chance setup is not a relevant problem in decision theory. Since we must act, we have to accept a certain **K**, although our reasons for accepting it may not be strong.

---

[1] In the first couple of paragraphs of this section, I am more or less following the introduction to [46].

1. THE BASIS FOR DECISIONS

Since we have to weigh the desirability of a consequence with its possibility of happening, we need to assign probabilities to these consequences and not only to accept them or reject them. In order to assign probabilities, we must accept a certain model for the setup in question. The reasons for accepting the model may be various, but they don't constitute a problem for decision theory proper.

Our first example will be a lottery. Suppose that I have to decide whether or not to buy a ticket for a certain drawing. I set up a model for the drawing, and calculate the probabilities of winning according to the model. For instance, the model may be that all tickets are equally likely. I may have different sorts of evidence for this model, but once I decide to accept it, I compute the probabilities without questioning it anymore. Here, in order to make a decision, I don't consider the proposition that the ticket A wins as a hypothesis to be rejected or accepted, but just assign a probability (i.e., in this case, a degree of belief) to it and act accordingly. Thus, although the probability of the particular ticket A winning may be very low, I don't jump to the conclusion that A will not be drawn. Having the probabilities, we may combine them with utilities (desirabilities) in order to decide on an action.

Suppose, now, that I have two possible models for the chance setup representing the lottery, say $\mathbf{K}_1$ and $\mathbf{K}_2$, and that I can't decide between the two. For instance, according to $\mathbf{K}_1$, all numbers may be equally probable, while, according to $\mathbf{K}_2$, the probability of a number is proportional to its size. I think that then it would be proper to assign probabilities to $\mathbf{K}_1$ and $\mathbf{K}_2$ themselves, and join them in one setup. That is, to consider that there is one compound setup, in which I first choose between $\mathbf{K}_1$ and $\mathbf{K}_2$, and then choose the ticket according to one of them. Here, I am considering that $\mathbf{K}_1$ and $\mathbf{K}_2$ are (antecedently) real possibilities, not just epistemic possibilities. That is, I imagine that somebody actually chose one of these alternatives to happen. Since the occurrences of $\mathbf{K}_1$ or $\mathbf{K}_2$ are decidable propositions, we may assign numerical degrees of belief to them.

The compound model thus obtained is similar, in the framework discussed in Chapter III, Section 4, to the urn model, i.e., we have a set of causal moments of two elements t and s, t preceding s; at moment t, we choose $\mathbf{K}_1$ or $\mathbf{K}_2$, and at moment s, we choose the lottery number according to the mechanism we have chosen at t. The model for reality is now this compound setup. The probabilities that we assign to the two lottery mechanisms, may be based on very flimsy evidence. For instance, we may think that about half of the lottery mechanisms in existence are of type $\mathbf{K}_1$ and half of type $\mathbf{K}_2$, and that the mechanisms are chosen at random. In this case, we can assign with some justification a probability of $1/2$ to the proposition that the lottery is the chance setup described by $\mathbf{K}_i$, for each i. In this case, we believe that this probability is the same as that of coming up heads in the toss of a fair coin. On the other hand, we may have very good evidence for accepting the hypothesis. For instance, we may have based the acceptance on statistical inference techniques.[2]

---

[2] The basis for these techniques will be discussed in Section 2 and further analyzed in Chapter XVIII.

Since we must decide, it is better to do it according to probabilities, however uncertain they may be, than completely irrationally. Here, we can compute the probability of A winning by the total probability formula:

$$\Pr A = \Pr(\mathbf{K}_1) \cdot \Pr(A|\mathbf{K}_1) + \Pr(\mathbf{K}_2) \cdot \Pr(A|\mathbf{K}_2).$$

From the evidence of successive drawings, we may improve the probabilities (i.e., in this case, degrees of belief) we assign to $\mathbf{K}_1$ and $\mathbf{K}_2$. Suppose we call the result of n successive drawings E. Then we can compute the probability of E given $\mathbf{K}_i$, for $i = 1$ and $i = 2$. Now, what we want to use in the total probability formula for A is the probability of $\mathbf{K}_i$ given E. This can be obtained by the use of Bayes formula:

$$\Pr(\mathbf{K}_i|E) = \frac{\Pr(\mathbf{K}_i) \cdot \Pr(E|\mathbf{K}_i)}{\Pr(\mathbf{K}_1) \cdot \Pr(E|\mathbf{K}_1) + \Pr(\mathbf{K}_2) \cdot \Pr(E|\mathbf{K}_2)}.$$

Now, using the total probability formula, replacing $\Pr \mathbf{K}_i$ by $\Pr(\mathbf{K}_i|E)$, we get the probability of A conditional on E, $\Pr(A|E)$. Since we are making a decision that should be guided by our degree of belief in A obtaining, we are not particularly interested whether $\mathbf{K}_i$ is true or not, for $i = 1$ or 2: it is enough to know its probability. The truth of the hypotheses "$\mathbf{K}_i$ is the mechanism of the lottery" is not itself in question. We just need its probability.

This use of Bayes formula is a form of inverse inference: from data, E, to infer the probability of $\mathbf{K}_1$ and $\mathbf{K}_2$. This inverse inference, however, needs a previous use of the Principle of Direct Inference, since we must have $\Pr \mathbf{K}_i$ and $\Pr(E|\mathbf{K}_i)$, which are obtained through this last principle.

Principle of Direct Inference. Thus, the basic principle actually used is a widened form of the Principle of Direct Inference. According to this principle, we accept a particular model for a chance setup, and calculate the probabilities with respect to it. These probabilities are taken to be the rational degrees of belief that we should have, given that we accept the particular model as true to reality. The different propositions we entertain, are not rejected as false or accepted as true, but only assigned probabilities, because they are considered as really possible (or, at least, antecedently really epistemically possible), and not just epistemically possible. We may consider them as hypotheses, but not in order to accept them or reject them, but for guiding our actions, by assigning to them degrees of belief. The model of the chance setup we are working in is not in question, i.e., it is not treated as a hypothesis. We only use it in order to obtain the probabilities and act accordingly. The application of this principle to the assignment of degrees of belief, will be further analyzed in Section 3.

The Principle of Direct Inference should be widened in two steps. As it was stated in Page 36, in its second part, it says:

> Let A be an event. Then if a person believes at time t that the chance of A is r, then, if he is rational, his degree of belief at time t of A's holding, should also be r.

We first widen it to included antecedently real possibilities. This strengthening of the principle, I shall call the

Principle of Direct Inference. Let A be an event. Then if a person believes at time t that the chance of A is r, and at a later time s he has no more relevant knowledge, then, if he is rational, his degree of belief at time s of A's holding, should also be r.

One of the problems with this form of the principle, which is similar to Lewis' formulation in [62], is what is understood by "relevant" knowledge. I shall not discuss here this problem, which is extensively treated in [62].

We further widen the principle, by accepting other ways of assigning degree of belief. For instance, if we decide that our degree of belief in the occurrence of a certain event A is the same as the chance of heads in the toss of a fair coin, then we should assign a degree of belief of $1/2$ to A.[3] Thus, another, widened form of the Principle of Direct Inference, could be stated as saying:

Let A be an event. Then if a person believes that her degree of belief in A's holding is the same as her belief in an event B's holding, and she believes that the chance of B is r, then her degree of belief in A's holding should also be r.

2. The basis for statistical inference

Many writers on the foundations of statistics, probability, or inductive logic, deny that we should ever accept hypotheses, even provisionally. According to Carnap, for instance, a scientific hypothesis h is rendered probable to such and such a degree by evidence e. We ought to have a degree of belief in h corresponding to its probability, but we should never accept h, not even provisionally.

A subjectivist can talk about the acceptance of a hypothesis, but only in a very uninteresting way. To accept a hypothesis h would be to assign to it a subjective probability of one. But since for a subjectivist, the only way probabilities can change is by conditionalization,[4] the only way that we can assign a conditional probability of unity to h on evidence e is to assign an a priori probability of unity to h (the probability of e must be the same as the probability of the conjunction of h and e).

Many non-Bayesians (calling Carnap and the subjectivists Bayesian) have an even greater behavioristic bias. They formulate the problem of statistical inference as a problem of choosing between alternative courses of action. Neyman, for instance, talks only of inductive behavior, not of inductive inference. I have already discussed these views briefly in Chapter I, Section 1, but I would like to add a few remarks here.[5]

The main objection to this behavioristic view of science, is that it seems to be in accordance neither with the practice nor the theoretical views of scientists (and other people): scientists (and people in common situations), simply do *accept* (and *reject*) hypotheses, when the evidence in their favor (or against them) is overwhelming. It seems difficult, given the enormous importance that scientific

---

[3] For another example of the assignment of degrees of belief, see [47, Example 11.1].
[4] See page 31, for a definition of conditionalization.
[5] See, also, [58].

theories have acquired in modern times, to admit that scientific hypotheses serve just as summaries of a large amount of data, and that their only purpose is for simplifying our calculations or guiding our actions. Thus, I believe that statistical inference is inference, that is, that it contains rules to proceed from propositions accepted as true to other propositions that are to be accepted as true.

Kyburg's lottery paradox. I think that Kyburg's lottery paradox, [58], shows that it is impossible to give a rule for the acceptance (or rejection) of hypotheses, based only on their high (or low) probabilities. A discussion of this paradox may be helpful to illuminate the point.

In order to discussed the paradox, it is useful to have in mind the following conditions, given by Hempel, [41],[6] that the body K of accepted propositions should satisfy.

(1) Any logical consequence of a set of accepted statements is likewise an accepted statement; or K contains all logical consequences of any of its subclasses.

(2) The set K of accepted statements is logically consistent.

It seems to me that these conditions are obvious enough as conditions of rationality, when the evidence is deterministic. There are some difficulties, however, with these conditions for the acceptance of hypotheses based on statistical evidence. For instance, it may be reasonable to accept the hypothesis that smoking is a contributing cause of cancer, without accepting that smoking is a contributing cause of cancer in a certain class of men, such as men over 90 years old. I believe that these difficulties can be overcome by a careful stating of the accepted hypothesis. For the purpose of this book, however, these difficulties disappear. Our hypotheses will be statements of the form "**K** is a model of the given chance setup", where **K** is a probabilistic model which assigns probability to propositions. For this type of hypothesis the difficulties mentioned do not arise.

In the discussion of this chapter, I shall write **K** for the possible probability models, although to be more precise it would require us to write $\langle \mathbf{K}, Y \rangle$, where Y is a measurement function. Unless there is danger of confusion, I shall omit the Y, which will be clear from the context.

For example, in the case of smoking and cancer, the hypothesis would be that the structure **K** which assigns a probability distribution to smokers and people with cancer is the right model for the situation. This type of statistical hypothesis can be combined with other accepted hypotheses and one can obtain consequences from them. For instance, suppose that **K** implies that people who smoke for more than 40 years are certain to get emphysema. Then, if we accept the hypothesis that a certain person has smoked for over 40 years, then we must also accept that he will get emphysema. If we had just assigned probabilities to both hypotheses, then we would not be forced to accept the conclusion. If, on the other hand, **K** implied that the probability of getting emphysema is very high for people smoking for over 40 years, then the conclusion of the two hypotheses

---

[6]I am only quoting two of Hempel's three conditions. The third one is not interesting in my context.

would be that the probability of this person having emphysema is high, and we would have to accept this conclusion, namely that the probability of this person having emphysema is high. On the other hand, a high probability for **K** would not necessarily entail a high probability for this person, who has smoked for 40 years, having emphysema.

Kyburg's lottery paradox shows that any rule of acceptance of propositions based on their high probabilities violates Hempel's requirements for the body of accepted propositions. Suppose that we decide to accept h on evidence e when the probability of h given e is greater than 0.999, and, hence, to reject h when this probability is less than 0.001. Take, now, an urn with more than 1,000 successively numbered counters. Suppose that one counter is chosen at random from the urn. Then we must accept, according with our rule, the propositions,

(a) "counter n will not be chosen", for every n.

Also, the setup of the lottery forces us to accept,

(b) "a counter will be chosen".

(a) and (b) are contradictory, and hence we have been forced to accept two contradictory statements, which is impossible by Hempel's clause (2). It is clear that a similar example can be given with a hypothesis of the form "K_n is not the model for the setup", instead of "counter n will not be chosen".

Kyburg's lottery paradox can be modified to take care of the following type of rule of acceptance, which is similar to one that will be adopted later in this section. Accept h, if we have evidence e such that the probability of h given e is greater than r, and it is possible to find a sequence of possible evidences e_n such that the probability of h given e_n tends to one as n tends to infinity. This rule would have the advantage that we need only to accept h provisionally, when its probability is high. Then if we are challenged in our acceptance, we can ask the challenger for a level of probability with which he would be satisfied, and find the corresponding evidence.

In order to take care of this rule of acceptance, Kyburg's lottery may be modified in the following way. Suppose that the results of the lottery are infinite sequences of results of tossing a coin. Then, according to our rule involving infinite sequences of propositions, the following propositions should be accepted:

(a) "sequence s will not obtain", for every sequence s.

Also, the setup of the lottery forces us to accept:

(b) "a sequence will obtain".

Again, (a) and (b) are contradictory.

The following is another, even simpler, example which shows that even a rule that would tell us to accept propositions with probability one fails, if we have an infinite number of possible results: suppose that we are considering the setup of a circular roulette in which any point on a circle may occur. In this case, the probability of any particular point is zero, and thus, our rule allows us to reject the hypothesis "p will occur" for every point p on the circle, while we must accept, "a point on the circle will obtain".

It is clear, even without using Kyburg's paradox, that acceptance of hypotheses by their probabilities violates Hempel's requirements: a set of hypotheses $h_1, \ldots,$

h_n, can all have high probabilities, with their conjunction having low probability. Thus, each member of the set has to be accepted without accepting one of its consequences, namely its conjunction.

I believe that the reason behind the fact that we cannot accept hypotheses based on their probabilities is that once one assigns probabilities to a hypothesis h, one considers it as a proposition expressing a real possibility (or at least an antecedently real epistemic possibility); hence, one cannot reject it as possibly true. If an alternative hypothesis is assigned probability, then this alternative expresses a factual possibility (or at least an antecedently factual epistemic possibility), and, hence, our original hypothesis h cannot be accepted as true, no matter how much higher the probability of h is than that of the alternative. Thus, I believe that it is in principle impossible (and that Kyburg's paradox is a reflection of this fact) to accept or reject hypotheses solely on the basis of their probabilities.

One matter that should be clarified is the following. The hypotheses we shall be considering will be mainly of the form "**K** is the right model for a given chance setup". We have to analyze what it means, for instance, to accept a conjunction of this type of hypotheses or a conjunction of some of this type with regular deterministic ones. An example was given above in the relation of smoking with a certain disease. As another example, suppose that we accept that **K** is a model for a chance setup $\mathfrak{S}_1$ and that **J** is a model for another setup $\mathfrak{S}_2$. The acceptance of both hypotheses as true may not be enough, however, for assigning probabilities to propositions about both setups. If we assume that both setups are at work, we would have to calculate the probabilities of events in a certain way, if we consider them independent, and, in another way, if they are considered dependent. For instance, if we accept that **K** is a model for setup $\mathfrak{S}_1$, **J**, for $\mathfrak{S}_2$, *and* that $\mathfrak{S}_1$ and $\mathfrak{S}_2$ are independent, then the model which covers both setups is **K** × **J**. The mere acceptance **K** for $\mathfrak{S}_1$ and **J** for $\mathfrak{S}_2$, however, does not determine the joint model.

Acceptance and rejection of hypotheses. In statistical inference, the main question is the acceptance of hypotheses as true or their rejection as false without taking into account the desirability of the consequences of acceptance or rejection. Classical statistical inference, in fact, gives rules for the rejection or acceptance of hypotheses, but these rules are not based on the probabilities of the hypotheses. I consider that the principle that will be stated in this section is the basis for these rules. Similarly to classical statistical inference, this principle bases acceptance and rejection of a hypothesis, not on its probability, but on the probability that is assigned according to the hypothesis to other propositions.

I shall begin by analyzing rules for rejection. The rule of rejection of a deterministic hypothesis h (i.e., one that determines a unique outcome) is very simple. Suppose that an experiment is performed. If a result obtains that would be false if the hypothesis were true (i.e., that is false under the hypothesis), then we should reject h.[7] In Popper's view, [74], which is certainly oversimplified here,

---

[7] As is well-known, I am here oversimplifying the rejection of hypotheses. It may be possible to save a hypothesis, in spite of the falsity of one of its consequences e, by changing some

2. THE BASIS FOR STATISTICAL INFERENCE

the basic aim of science is the falsification of hypotheses in this way.

Instead of deterministic hypotheses, we now consider probabilistic hypotheses. The falsification of probabilistic hypotheses has been a problem for views such as Popper's mentioned above. I believe that I am presenting, here, a solution to this problem, which is consistent with Popper's ideas about falsification. The problem for strict falsification of probabilistic hypotheses is that from such a hypothesis we cannot deduce interesting propositions that can be verified empirically, but only assign them probabilites. For instance, from the assumption that the probability of heads in the toss of a coin is 1/2, one cannot deduce any statement about frequencies, but only assign them probabilities. Thus, for instance, one cannot deduce that the relative frequency in 100 tosses is 1/2, but only that this frequency has a certain high probability.[8] Thus, if the relative frequency really obtained is not 1/2, we cannot conclude that the hypothesis is not valid. In my account, a statement having low probability means that its possiblity of truth is low. Thus, a statement with low probability according to the hypothesis occuring in reality is evidence that the hypothesis is false.[9] Popper's account of probability as propensity does not include a connection with truth, and, hence, he does not have available this justification of falsification.

A probabilistic hypothesis is, in my account, the proposition that a set of possible outcomes, say **K**, is an adequate model of reality, i.e., of the chance setup that is in question. For briefness of expression, I shall call this hypothesis just **K**. This hypothesis **K** determines a probability distribution over the possible results of an experiment (or, simply, over the events) we are interested in.[10] This distribution represents the degree of support of the different propositions, given **K**.

It is clear that if a proposition false in **K** were true in reality, we should reject **K**. That is, if the result of the experiment were impossible according to **K**, we would reject **K**. But very seldom the experiments we consider do admit results that are impossible according to a probabilistic hypothesis **K**. So a natural procedure would be to introduce the notion of approximately false and then to reject **K**, if the next best thing to a proposition false in **K** obtained in reality, namely, an approximately false proposition. For convenience of expression, I shall call an event *unlikely*, if the proposition expressing its occurrence is approximately false in this sense. Hence, we may paraphrase the principle "reject **K**, if an approximately false proposition obtains" by "unlikely events don't occur".[11] But, what is an unlikely event? Since I consider probability as a measure of the

---

other aspect of the relation of the hypothesis with experience, and thus e may cease to be a consequence. Here, and in the rest of the book, I shall continue to make this simplification. The account could be modified in order to take into account the complication mentioned in this note.

[8] For a more thorough discussion about the reationship between probabilities and frequencies see Section 4.

[9] The situation is more complicated than this, as will be seen later.

[10] The occurrence of the different possible results is expressed by propositions. So we also talk about a proposition being assigned a certain probability by a model.

[11] See [7].

possibility of truth, it is tempting to say that a proposition φ is approximately false in **K**, if φ has low probability in **K**, i.e., if its degree of support given **K** is low. Hence, we would assume that if a result for which **K** determined low probability occurred, then we should reject **K**. This rule, however, does not work.

Suppose that the chance setup for which **K** is a model is a lottery similar to those discussed before. The drawing of any particular ticket has very low probability according to this model **K**, but such a drawing does not constitute a reason at all for rejecting **K**.

The next few paragraphs constitute an analysis of the notion of unlikely results. The first element that we shall discuss is the consideration that, in order to reject **K**, we must have some grounds to question it, namely, we must have some alternative models in mind for the chance setup. Thus, even the talk about approximately false propositions under **K** requires the consideration of alternatives to **K**. We are forced to deal with classes of alternative hypotheses $\mathbf{K}_i$, for each i in a certain index set I. These are all the hypotheses we consider epistemically possible in the situation involved. All these hypotheses are compatible with what we accept as the laws of the phenomenon in question. It is clear that it is not that $\mathbf{K}_i$ may happen, but only that it is compatible with the accepted laws of nature that $\mathbf{K}_i$ is the right model for the situation we are considering. How to decide which is the right class of alternative hypotheses is not determined, in general, by statistical or probabilistic considerations. There may be general natural laws or particular facts relevant for this decision. In any case, we must accept as true that the class of the $\mathbf{K}_i$ for the i in I, contains all the possible models.

Now we can improve the definition of an approximately false proposition. A proposition φ is approximately false in $\mathbf{K}_j$, relative to the class $\mathbf{K}_i$, for i in I, if φ has much lower probability in $\mathbf{K}_j$, than in a certain $\mathbf{K}_i$, with i in I and $i \neq j$. This definition, however, as we shall see later, does not quite work, so we only accept it provisionally, as a basis for discussion.

It is not, in general, possible to take the family of all logically possible hypotheses as our set of alternatives. Let us suppose that we toss a coin ten times.[12] A sequence $\vec{x} = 0100101110$ (where 0 is for heads, and 1, for tails), counts in favor of the hypothesis, say $\mathbf{K}_{1/2}$, of the coin being fair, if the alternatives are the coins being biased with different degrees of biasedness. It is approximately false in $\mathbf{K}_{1/2}$, however, if one of the possibilities is "the sequence is generated by a machine that produces exactly $\vec{x}$".

Besides being necessary for a natural rule of rejection, the idea that the consideration of a set of alternatives affects the acceptability of a hypothesis, has intuitive appeal. For instance, before the formulation of relativity theory by Einstein, it was very difficult to reject Newtonian mechanics as the right physical theory. After the formulation of its alternative, however, classical mechanics was almost immediately rejected.

In order to declare $\mathbf{K}_j$ false, however, just the occurrence of an unlikely event is not enough. We would not be satisfied, in general, with only one unlikely

---

[12] This example was suggested to me by W. N. Reinhardt.

event. We may, in case an unlikely event occurred, provisionally reject $\mathbf{K}_j$, but we should be able to revise this judgement. Thus, we must have at hand a sequence of unlikely events, whose probabilities approach zero. If we have such a sequence, then, in principle, we could get, with enough effort or time, the occurrence of a result such that its probability is as low as we want, and, thus, we would also be able to revise our judgement. But, as before, we also need to consider the alternatives. Hence, we need to be able to perform[13] a sequence E_n of experiments, or simply observations, such that for every $j \in I$, if $\mathbf{K}_j$ is true, a sequence a_n of results of these experiments almost surely eventually obtains, satisfying that for every $i \in I$ with $i \neq j$, the probability in $\mathbf{K}_i$ of the result a_n tends to zero. We *must* require that if $\mathbf{K}_j$ is true, then such a sequence a_n would almost surely eventually occur. This sequence of experiments, which will be defined more precisely later, is called a *discriminating experiment for* the system $\mathbf{K}_i$ with $i \in I$. We shall abbreviate the proposition 'the result of E_n is a_n' by $[E_n = a_n]$.

The point in dealing with sequences of results of experiments whose probabilities approximate zero, is not that we can get in this way propositions with low probability, but with arbitrarily low probabilities. That is, if we wait enough time or spend enough effort, we can get almost surely a proposition with as low a probability as we want. In this way, we can set in motion a dialectical process: one sets the low probability at which one would be satisfied, and the opponent (which may be the same person) tests the sequence of experiments until she gets to the right level. In order to have a real dialectical process for rejecting $\mathbf{K}_j$, if $\mathbf{K}_m$ is true for any $m \neq j$, a sequence which determines sentences whose probability under $\mathbf{K}_j$ tends to zero must almost surely occur. That is, the set of sequences must have, if $\mathbf{K}_m$ is true, probability infinitely close to one. In actual practice one is able only to get down to a certain finite level, i.e., one can test $[E_n = a_n]$ only up to a certain finite n. This accounts for the fact that, although we may reject a hypothesis and consider it false, this rejection is provisional. If new propositions obtain with high probability, we may be forced to accept it again.[14]

The rules of rejection that we have been considering are asymmetrical in the following sense. For rejecting deterministic hypotheses, we do not need to entertain alternative hypotheses, while for probabilistic ones, we do. For the rules of acceptance of hypotheses, the symmetry is restored. We can't accept a deterministic hypothesis h, just because a proposition φ implied by h has obtained in reality. We are forced to consider alternative hypotheses h_i, for i in an index set I. If a proposition φ obtains that is true according to h_j, but false in all h_m, for $m \neq j$, then h_j should be accepted rather than h_m, with $m \neq j$. Thus, we accept h_j, if we can reject all h_m, with $m \neq j$, but not h_j.

One of the rule for the acceptance of a nondeterministic hypothesis $\mathbf{K}$ is similar. In general, it is not possible, according to this rule, to accept a particular model $\mathbf{K}$, but a class of such models, say the $\mathbf{K}_j$ for $j \in J$, where J is a subset

---

[13] Notice here that we have to be able to perform, and not actually to perform.
[14] Cf. [63, Chapter V].

IV. PRINCIPLES OF INFERENCE AND DECISION

of I. The rule says that we accept provisionally such a class, if we can reject provisionally all models not in the class, i.e., all $\mathbf{K}_i$ for $i \in I - J$.

Just as in the case for rejection, it is not enough for acceptance to have an approximately true proposition. What we need is to have a discriminating sequence of experiments E_n, such that the truth or falsity of φ depends on the result of an element of the sequence. For the dialectical process to be possible, the discriminating experiments should have the two conditions mentioned above. Ideally, the truth in reality of an infinite sequence of propositions $[E_n = a_n]$ would determine truth. In practice, we must be content with a finite sequence approximating truth, and hence, our acceptance is just provisional.

Besides having a discriminating experiment, in order to reject $\mathbf{K}_j$, the experiment E_n, for some n, must result in an outcome that is unlikely according to $\mathbf{K}_j$. The notion of unlikely result for $\mathbf{K}_j$ as that which has low probability according to $\mathbf{K}_j$ and high, according to an alternative, however, does not work: suppose that we are tossing a coin 10 times, and that the sequence 0010000000 occurs. Intuitively, if the alternatives are the models $\mathbf{K}_p$, where p is the probability of heads, this result is unlikely according to the hypothesis $\mathbf{K}_{1/2}$ of the coin being fair. We have, however, that the sequence has low probability under any of the hypotheses.

We proceed, now, to introduce the notion of unlikely result that will be used in this book. Our notion of unlikely result, can be introduced with the following example. Suppose that we are tossing a coin 10 times and recording the frequency of heads. Given that the coin is unbiased, what is an "unlikely result of the experiment"? The result of exactly 5 heads has low probability, but we would not call it unlikely. Suppose that 6 heads are obtained. We call this result unlikely (with respect to an unbiased coin), if the event of obtaining at least 6 heads or at most 4, has low probability. That is, the event of 6 heads or worse (i.e., less likely according to the hypothesis that the coin is unbiased) has low probability. We shall use this idea about an unlikely event for our rule of rejection.

We proceed now to delimit more precisely the notion of an unlikely result. In order to do this, we need to introduce two notions: the notion of 'evidential equivalence', and the notion of 'worse result'. We assume that the alternative hypotheses are $\mathbf{K}_i$, for $i \in I$. We begin with the first of these notions. We notice the following fact, which, as we shall see in Chapter XVIII, is behind the sufficiency principle of statistical inference. Suppose that there are two possible results of E_n, a and b, satisfying the condition: there is a $c > 0$ such that for every $i \in I$

$$\mathrm{Pr}_{\mathbf{K}_i}[E_n = a] = c \cdot \mathrm{Pr}_{\mathbf{K}_i}[E_n = b]. \tag{1}$$

This means that for any $i, j \in I$

$$\frac{\mathrm{Pr}_{\mathbf{K}_i}[E_n = a]}{\mathrm{Pr}_{\mathbf{K}_j}[E_n = a]} = \frac{\mathrm{Pr}_{\mathbf{K}_i}[E_n = b]}{\mathrm{Pr}_{\mathbf{K}_j}[E_n = b]}, \tag{2}$$

provided that $\mathrm{Pr}_{\mathbf{K}_j}[E_n = a] \neq 0$ and $\mathrm{Pr}_{\mathbf{K}_j}[E_n = b] \neq 0$.

An example of this situation is the following. Suppose that we are tossing a coin 60 times and that the hypotheses are coins with different biases. Two

2. THE BASIS FOR STATISTICAL INFERENCE

different sequences with the same number of heads, have the same probability under any of the hypotheses. Thus, (1) is true for this case with $c = 1$.

A natural way of measuring the degree to which a probability is smaller than another is their ratio. Thus, that

$$r = \frac{\Pr_{K_j}(\varphi)}{\Pr_{K_i}(\varphi)}$$

is low for a certain i in I gives evidence against K_j. It is not necessary to require that $\Pr_{K_j}(\varphi)$ be low and $\Pr_{K_i}(\varphi)$ be high, but only that their ratio be low. For instance, if we had that φ was impossible under K_j, but possible under K_i, we would reject K_j, although the probability under K_i may be very low.

Thus, in a situation such as (1), a and b do not discriminate between the different models. That is, the evidence they give with respect to the models is the same. We say, in case (1) is satisfied, that a and b are *evidentially equivalent with respect to E_n and the systems* K_i, *for $i \in I$*. In the example of the coin, then, any two sequences with the same number of heads are evidentially equivalent. It is clear that the relation of evidential equivalence is an equivalence relation between possible results. We write the equivalence class of a, $[a]$. Since all elements of an equivalence class have the same evidential import, we have to consider them together. Thus, we could replace the experiment E_n by another E'_n such that $E'_n = [a]$ instead of $E_n = a$. In the case of the coin, again, the new experiment, instead of having as result the actual sequences, it would have, the number of heads in the sequence. Notice that $[E'_n = [a]]$ can also be written as $[E_n \in [a]]$. In statistical practice, this replacement of E by E' is usually done.

We turn, now, to the second notion mentioned above. We call a possible result b *as least as bad for K_j as a possible result* a, in symbols $b \preceq a$, if $\Pr_{K_j}[E_n \in [b]] \leq \Pr_{K_j}[E_n \in [a]]$ and there is an $i \in I$ where the inequality is reversed. Then, for rejection of K_j, a partial experiment E_n must have a result a such that the probability under K_j of E_n having the value a or any result as least as bad as a is small, say less than a certain α. Thus, we arrive at the notion of "unlikely result": a result, a, of an experiment E_n is *unlikely for K_j*, if the probability of the event of E_n having value a or any value at least as bad for K_j as a, is low under K_j. I think this is a natural characterization of unlikely results.

In case that for any results a and b, if $\Pr_{K_i}[E_n = b] \leq \Pr_{K_i}[E_n = a]$ there is always another K_j where the inequality is reversed, then we can simplify the definition of $b \preceq a$ by eliminating the clause that the inequality be reversed. In fact, as we shall see with examples later, in these cases, it is as if the hypothesis K_i is tested in isolation, i.e., without considering alternatives. These cases are not infrequent.

The Principle of Inverse Inference. The principle that we have been discussing in this section is a principle of inverse inference, in the sense, that we argue from evidence to hypothesis (i.e., to the acceptance of a certain model), and not from the model to the particular facts, as in the Direct Inference Principle. The new principle will be called the *Inverse Inference Principle*, in contradistinction to the Direct Inference Principle, which was discussed in the previous

98   IV. PRINCIPLES OF INFERENCE AND DECISION

section. According to the Principle of Inverse Inference, we consider alternative models for the setup in question.[15] Each model is one of the epistemically possible hypotheses that may explain the setup. The alternative models are not considered as real possibilities, so they are not assigned probabilities. One of the hypotheses, say **K**, is rejected (i.e., taken as false), if there is a discriminating experiment E_n for the set of alternative hypotheses that has a result a_n, such that the probability of a_n or worse is low under **K**.

An essential part of the rules discussed is the fact that the propositions in question should be decidable by a sequence of experiments with the properties indicated above. Although we may, in fact, not use this sequence, there must be the possibility of employing it in case our rejection (or acceptance) is challenged. Then it should be possible to ask the challenger to set a probability level α that is low enough for him, then perform the experiment until we find a result such that it or worse has probability less than or equal to α. I think that this is the intuitive principle behind classical statistical inference. An application of this principle to study the relation of frequencies and probabilities will be given in Section 4, and in Chapter XVIII it will be applied to a study of statistical inference.

This section will be concluded with a more precise version of the Inverse Inference Principle. This will be accomplished by defining more precisely the concept of a discriminating experiment for a system of alternative possible probability models $\langle \mathbf{K}_i \mid i \in I \rangle$. We shall use the measurement functions and language introduce in Chapter III, Section 5. Thus, propositions are replaced by interpreted sentences.

The results of the experiments may be anything. For simplicity, we shall consider that all results of experiments are real numbers or n–tuples of real numbers. Thus, the sentence we shall be considering are of the form $[E_n = r]$, where r is a real number, or $[X_1 = r_1] \wedge [X_2 = r_2] \wedge \cdots \wedge [X_n = r_n]$, which may be written $[E_n = \vec{r}]$. Because not all probability spaces are discrete, we would also need sentences where inequalities replace the equality relation. That is, sentences of the forms $[E_n \leq r]$, $[E_n < r]$, $[r \leq E_n \leq s]$, and so on, where r and s may be real numbers or n–tuples of real numbers. These are precisely the basic sentences or conjunctions of basic sentences in the languages introduced in Chapter III, Section 5.

We now discuss more precisely the possible models of chance setups. We are here in a situation where we need to have the possibility of n repetitions for the setup, for an unbounded set of natural numbers n. We shall model these repetitions by ν repetitions, with ν an infinite natural number. For example, if the model for a setup is **K** and the repetitions are independent, the model for n independent repetitions is $^n\mathbf{K}$, as it was discussed in Chapter III, Section 4. Although most of the applications we shall deal with on this book will be

---

[15] In this section, we have been mainly considering just simple hypotheses, i.e., hypotheses that postulate one set of possible outcomes as model for the chance setup. Later, I shall also consider composite hypotheses, i.e., hypotheses that assert that the model for the setup is one of several related sets of possible outcomes.

2. THE BASIS FOR STATISTICAL INFERENCE

independent repetitions, in general, this may not the case. So, for generality's sake, we shall call the model for n repetitions of **K**, $\mathbf{K}^n$. It may be the case, also, that $\mathbf{K}^n$ is not strictly the model for n independent repetitions, but that there is a process going on and, at time n, the model is $\mathbf{K}^n$.

For the interpretation of the sentences $[E_n = r]$ in a possible probability model **K**, we must have that E_n is a measurement over **K**. In fact, in order to assign probabilities to these sentences, E_n has to be a random variable. When there is no danger of confusion, we shall write $\Pr_{\mathbf{K}_j}(\varphi)$ for $\Pr_{\langle \mathbf{K}_j, X \rangle}(\varphi)$. Usually which is the system of random variables X will be clear from the context, in particular, from the sentence φ. When it is clear from the context, we shall also write $\Pr_{\mathbf{K}_j}$ for $\Pr_{\mathbf{K}_j^n}$.

In order to make the definition of an experiment simpler, however, we shall only consider in this definition sentences of the form $[E_n = r]$ with a new interpretation, or Boolean combinations of them: a sentence $[E_n = r]$ is now interpreted as stating $[r - dr \leq E_n \leq r + dr]$, where dr is a positive infinitesimal.[16] For ratios of probabilities, in the case of continuous distributions, the density $f(r)$ can replace the probablity of $[E_n = r]$. It is well known that if f_j is the density associated with the distribution $\Pr_{\mathbf{K}_j}$, then $f_j(r) \cdot \Delta r \approx \Pr_{\mathbf{K}_j}([r - \Delta r \leq E_n \leq r + \Delta r])$. Thus, with the interpretation given above

$$\frac{\Pr_{\mathbf{K}_j}([E_n = r])}{\Pr_{\mathbf{K}_i}([E_n = s])} \approx \frac{f_j(r)}{f_i(s)}$$

and

$$\Pr_{\mathbf{K}_i}([E_n = r]) \leq \Pr_{\mathbf{K}_i}([E_n = s]) \quad \text{if and only if} \quad f_i(r) \leq f_i(s).$$

For discrete sample spaces, a sentence of the form $[E_n = r]$ is the same with the old and new interpretations.

We shall write, as is usual, $[E_n \in A]$, where A is a set of real numbers, for the disjunction of $[E_n = r]$ for $r \in A$. Also, if A is a set of infinite sequences of real numbers of length an infinite natural number ν, $[E \in A]$ stands for the disjunction of the sentences $\bigwedge_{n=1}^{\nu}[E_n = a_n]$, for all sequences $\langle a_n \mid 1 \leq n \leq \nu \rangle$ in A. That is, $[E \in A]$ stands for the proposition that says the sequence of results of the experiment E is in the set A. As is usual, we say, in this case, that E is *almost surely* in A under a model $\mathbf{K}_j$[17] if the complement of the event $[E \in A]$ is a null set, under $\Pr_{\mathbf{K}_j}$.

The precise definition of a discriminating experiment varies with the set of possible hypotheses that the experiment has to discriminate, and with its purpose, i.e., acceptance or rejection. The only common characteristic is that in case the experiment E_μ were performed with an infinite number μ of repetitions and

---

[16]This can be done formally by using nonstandard analysis, as it was mentioned in the Preliminaries and will be done formally in Part 2. We shall be rather informal in this chapter and the next one in the use of infinitesimals and infinite numbers. Several of the definitions would have to be slightly modified, if we wanted to be precise. We assume, here, that all our functions and sequences are internal, in the sense of nonstandard analysis.

[17]More properly, one should say "under $\langle \mathbf{K}_j, E \rangle$," but, as usual, we shall omit the system of random variables.

a result r_μ, which is an element of a sequence of results, $\langle r_n \mid 1 \leq n \leq \nu \rangle$, with ν infinite, were obtained, then almost surely the purpose of the experiment would be accomplished, i.e., a hypothesis would be accepted or rejected. Of course, it is impossible to perform an experiment an infinite number of times, and, hence, the acceptance or rejection is only provisional.

I shall end this section by giving the definition of a type of discriminating experiment, which will be used in the examples of Sections 4 and Chapter V. Different types of discriminating experiments will be presented in Chapter XVIII, when discussing statistical inference.

We repeat now more formally the definition of evidential equivalence between results. In all the rest of this section, we assume that $\langle E_n \mid 1 \leq n \leq \nu \rangle$, with ν an infinite natural number, is a sequence of random variables for the system of models $\langle \mathbf{K}_i \mid i \in I \rangle$.

DEFINITION IV.1 (EVIDENTIAL EQUIVALENCE). Let a and b be possible results of E_n. Then a and b are *evidentially equivalent*, in symbols

$$a \sim_n b,$$

if there is a $c > 0$ such that for every $i \in I$

$$\Pr_{\mathbf{K}_i}[E_n = a] = c \cdot \Pr_{\mathbf{K}_i}[E_n = b].$$

The corresponding equivalence class is

$$[a] = \{b \mid b \sim_n a\}.$$

We now introduce more formally the definition of 'at least as bad as' that was already mentioned before.

DEFINITION IV.2 (WORSE RESULTS). Let a and b be possible results of a partial experiment E_n. Then b is *at least as bad as a for $\mathbf{K}_j$* (relative to $\mathbf{K}_i$ for $i \in I$ at E_n), in symbols

$$b \preceq_{nj} a,$$

if

$$\Pr_{\mathbf{K}_j}[E_n \in [b]] \leq \Pr_{\mathbf{K}_j}[E_n \in [a]]$$

and there is an $i \in I$, $i \neq j$, such that

$$\Pr_{\mathbf{K}_i}[E_n \in [b]] \geq \Pr_{\mathbf{K}_i}[E_n \in [a]].$$

We use the symbol R_{nja} for the set of b that are at least as bad as a for $\mathbf{K}_j$ at n, called *the rejection set determined by a for $\mathbf{K}_j$ at n*, i.e.

$$R_{nja} = \{b \mid b \preceq_{nj} a\}.$$

In order to shorten some of the definitions, we introduce the following expression. Let E_n for $1 \leq n \leq \nu$, with ν infinite, be a sequence of random variables, and let A be a set of sequences of length ν of possible results of E_n. We say that that E_n is *almost surely (a.s.) eventually in A*, if almost surely there is a sequence $\langle r_n \mid 1 \leq n \leq \nu \rangle \in A$, such that $E_\mu \approx r_\mu$, for every infinite $\mu \leq \nu$. That is, the set of possible outcomes where there is no such sequence $\langle r_n \mid 1 \leq n \leq \nu \rangle$

in A is a null set. We also say that E_n is *almost surely evetually not in A*, if almost surely for no sequence of possible results, $\langle r_n \mid 1 \leq n \leq \nu \rangle$, we have that $E_\mu \approx r_\mu$ for all infinite $\mu \leq \nu$. In standard terms, $E_\mu \approx r_\mu$ for every infinite μ means that the limit of E_n, when n tends to infinity is the same as the limit of r_n, when n tends to infinity. We are now ready to define our experiments.

DEFINITION IV.3 (DISCRIMINATING EXPERIMENTS). Suppose that the epistemically possible models for a setup are $\langle \mathbf{K}_i \mid i \in I \rangle$. A *discriminating sequence of experiments* or, simply, a *discriminating experiment*, $E = \langle E_n \mid 1 \leq n \leq \nu \rangle$, for $\langle \mathbf{K}_i : i \in I \rangle$ is a sequence of measurement functions such that E_n is a random variable on the probability spaces of $\mathbf{K}_i^n$, for $i \in I$ and $1 \leq n \leq \nu$, satisfying the following conditions:

(1) For each $j \in I$, there is a set A of sequences of real numbers such that E_n is almost surely (a.s.) eventually in A,[18] according to $\mathbf{K}_j$, and E_n is a.s. eventually not in A, according to $\mathbf{K}_i$ for $i \neq j$, and such that for any $\langle r_n : 1 \leq n \leq \nu \rangle$ in A, any $i \in I$, $i \neq j$, and any $\alpha > 0$, α noninfinitesimal, there is a finite n, such that for every m with $n \leq m \leq \nu$

$$\Pr\nolimits_{\mathbf{K}_i}([E_m \in R_{mir_m}]) \leq \alpha.$$

(2) It is possible to determine in principle whether the sentences $[E_n = r]$ are true in reality or not, for $n \in \mathbb{N}$.

Condition (2) is not mathematical in character, so we shall not take it into account for mathematical considerations.

The idea behind a discriminating experiment, is that if we were able to perform an infinite sequence of experiments, then almost surely it would be decided which hypothesis is the correct one. Since we are able to perform only a finite sequence, we are able only to accept or reject provisionally.

A similar definition could be given for a discriminating experiment for a system of hypotheses of the form $\langle \mathbf{H}_i \mid i \in I \rangle$, where each $\mathbf{H}_i$ is a set of possible probability models. For simplicity, we shall only consider here the case given in the definition. When discussing statistical inference in Chapter XVIII, we shall introduce explicit rules for some of the more general cases.

We shall discuss many examples of discriminating experiments in later sections. A simple example is the following. Suppose that we are tossing a coin. We can take $E_n = \text{Fr}_n$ to be, for each natural number $n \leq \nu$, with ν infinite, the relative frequency of heads in n tosses. This experiment is discriminating for the set of models for coins with different biases. In order to show this fact, let $\mathbf{K}_p$ be the model that assigns probability p to heads. By the strong law of large numbers,[19] if A is the set of sequences of length ν (where ν is an infinite number) that tend to p, then Fr_n is a.s. eventually in A, under $\mathbf{K}_p$, and it is a.s.

---

[18] When introducing the formal notions of nonstandard analysis, we shall make explicit the requirement that sequence E_n be internal and that A is a (possibly external) set of *internal* sequences. A more precise definition of A appears in Chapter XVIII, Section 2.

[19] See Preliminaries, p. 13 and, for a proof, see Chapter XI, Section 2.

eventually not in A, under $\mathbf{K}_q$, for $q \neq p$. This is our set A of the definition, call it A_p. Now, for any real r and q between 0 and 1, the set R_{nqr} is of the form

$$R_{nqr} = [|\mathrm{Fr}_n - q| \geq |r - q|].$$

This is so, because, if $|b - q| \geq |r - q|$, then

$$\Pr{}_{\mathbf{K}_q}[\mathrm{Fr}_n = b] \leq \Pr{}_{\mathbf{K}_q}[\mathrm{Fr}_n = r]$$

and

$$\Pr{}_{\mathbf{K}_b}[\mathrm{Fr}_n = b] \geq \Pr{}_{\mathbf{K}_b}[\mathrm{Fr}_n = r].$$

It is clear that if $\langle r_n \mid 1 \leq n \leq \nu \rangle \in A_p$, then, again by the strong law of large numbers, for every $\alpha > 0$, α noninfinitesimal, there is a finite n such that for every $m \geq n$, $m \leq \nu$

$$\Pr{}_q([\mathrm{Fr}_m \in R_{mqr_m}]) \leq \alpha.$$

Thus, the conditions for a discriminating experiment are satisfied.

We see that the proof of the existence of a set A with the required characteristics, needs the use of the strong law of large numbers. By weakening Condition (1), we would only use some form of the weak law of large numbers.[20] This other condition would also avoid the use of an infinite product space. It is somewhat more complicated than the original condition, and the advantage of requiring a weak law of large numbers is not very important, so I shall not introduce it explicitly here.

An ideal rule acceptance can easily be formulated with our notion of experiment. This rule is ideal in the sense that it is impossible to carry out in practice. It is stated here only as a model that the more practical rules should approach.

RULE IV.1 (IDEAL RULE OF ACCEPTANCE). *Let $\langle \mathbf{K}_i \mid i \in I \rangle$ be a system of possible probability models. We say that a model $\mathbf{K}_j$ should be accepted with respect to $\mathbf{K}_i$ for $i \in I$, if there is a discriminating experiment $\langle E_n \mid 1 \leq n \leq \nu \rangle$, with ν infinite, for $\langle \mathbf{K}_i \mid i \in I \rangle$, such that*

(1) *The sequence of results $[E_n = r_n]$ obtains in reality.*
(2) *This sequence $\langle r_n \mid 1 \leq n \leq \nu \rangle$ has the property that for every $i \in I$, $i \neq j$, for every noninfinitesimal $\alpha > 0$, there is a finite n such that for every $m \geq n$, $m \leq \nu$*

$$\Pr{}_{\mathbf{K}_i}([E_m \in R_{mir_m}]) \leq \alpha.$$

(3) *There is no other experiment such that a certain $\mathbf{K}_i$, for $i \neq j$ is accepted (according to (1) and (2) above).*[21]

The ideal rule of rejection is, now, very simple.

---

[20] See Chapter XI, Section 2.
[21] The addition of this clause and similar ones in other rules was suggested by N. C. A. da Costa.

RULE IV.2 (IDEAL RULE OF REJECTION). *The model $\mathbf{K}_j$ is rejected if there is an experiment such that an alternative $\mathbf{K}_i$ with $i \neq j$ is accepted (according to Rule IV.1).*

We now make more precise the dialectical rule for *provisionally* rejecting a probabilistic hypothesis. The provisional character of rejection is very important.

RULE IV.3 (DIALECTICAL RULE OF REJECTION). *Let $\langle \mathbf{K}_i \mid i \in I \rangle$ be a system of possible probability models. We say that a model $\mathbf{K}_j$ should be provisionally rejected with respect to $\mathbf{K}_i$ for $i \in I$, if there is a discriminating experiment $\langle E_n \mid 1 \leq n \leq \nu \rangle$, with ν infinite, for $\langle \mathbf{K}_i \mid i \in I \rangle$, such that*

(1) *$[E_n = a]$ obtains, for some a and $n \in \mathbb{N}$.*
(2) *$\Pr_{\mathbf{K}_j}[E_n \in R_{nja}]$ is low (i.e., less than α for a certain small α).*

How low the α has to be, is decided dialectically. There has to be an agreement between our opponent and ourselves about the level of the probability that would convince him. It is also necessary to reach an agreement about the experiment to use. With this agreement, if $\mathbf{K}_j$ is actually false, then by testing larger and larger n's one almost surely one would obtain an n, and a result a for E_n such that the probability of $[E_n \in R_{nja}]$ is less than α.

Examples. Let us take as a first example the tossing of a coin to determine whether the coin is biased or not. Here the $\mathbf{K}_j$ we are testing is the model for an unbiased coin. That is, with the terminology introduced above, $\mathbf{K}_j$ is $\mathbf{K}_{1/2}$. Suppose the alternative models are those for coins with different biases. We already pointed out that tossing the coin n times and observing the frequency of heads is a discriminating experiment for this case. Suppose that we toss the coin 100 times, and that we observe 60 heads. We then determine the probability, under the hypothesis that the coin is unbiased, of the result of 60 heads or worse. In this case, this event is the result of more than 60 heads or less than 40 heads. If its probability is low (i.e., less than α for a certain predetermined α), then we reject the hypothesis.

When we have all system $\mathbf{K}_p$, for $0 \leq p \leq 1$, p a real number, as alternatives, and the experiment, for tossing the coin n times, is Fr_n for each n with $1 \leq n \leq \nu$, we can simplify some of the definitions. In the first place, we can see that no two different relative frequency values are evidentially equivalent (ascccording to Definition IV.1). Second, for every pair of different possible values, a and b, of Fr_n there are p and q such that

$$\Pr_{\mathbf{K}_p}[\mathrm{Fr}_n = a] < \Pr_{\mathbf{K}_p}[\mathrm{Fr}_n = b],$$

and

$$\Pr_{\mathbf{K}_q}[\mathrm{Fr}_n = b] < \Pr_{\mathbf{K}_q}[\mathrm{Fr}_n = a].$$

Thus, the definition of $\preceq$ may be simplified to

$$a \preceq_{np} b \text{ if and only if } \Pr_{\mathbf{K}_p}[\mathrm{Fr}_n = a] \leq \Pr_{\mathbf{K}_p}[\mathrm{Fr}_n = b].$$

Therefore, the definition of R_{npa} (according to Definition IV.2) seems not to depend on the alternatives, and we may consider that $\mathbf{K}_p$ is tested in isolation.

But, notice that we are assuming that all the alternative probabilities are epistemically possible, that we are considering a setup with independent repetitions, etc. If these factors were not present, we might have different alternatives and get differnt results. Thus, not all *logically possible* alternatives are considered.

In order to understand the influence of alternatives, let us consider as possible models, just the $\mathbf{K}_p$ with $1 \geq p \geq 1/2$, and p a real number. The experiment, for this case, can again be Fr_n. We also have, in this case, that no two different results are evidentially equivalent. If we obtain 60 heads in 100 tosses, however, the rejection set for $\mathbf{K}_{1/2}$ consists only of results higher than $3/5$. This is so, since a result, a, less than $2/5$ is not worse than $3/5$, because, although

$$\Pr\nolimits_{\mathbf{K}_{1/2}}[\mathrm{Fr}_{100} = a] < \Pr\nolimits_{\mathbf{K}_{1/2}}[\mathrm{Fr}_{100} = \tfrac{3}{5}],$$

there is no alternative where this inequality is reversed.

We now discuss the coin tossing setup with a different experiment. Assume that the experiment, E_n has as results the actual sequence obtained in n independent tossings of the coin. Assume, first, that the alternative hypotheses are the same as in the first case above, namely, $\mathbf{K}_p$ for $0 \leq p \leq 1$, where p is a real number expressing the probability of heads. It can be shown, similarly as above for frequencies, that this new experiment, E_n, is a discriminating experiment.

As it was shown in page 97, two sequences with the same number of heads are evidentially equivalent. Thus, the probabilities that E_n is in the rejection set determined by a sequence, $\vec{x}$, are the same as the probabilities that Fr_n is in the corresponding rejection set determined by the frequency of heads in $\vec{x}$. Suppose that we toss the coin ten times and we obtain $\vec{x} = 0100101110$ (where 0 is for heads, and 1, for tails). Then the probability that E_{10} is in the rejection set determined by $\vec{x}$, according to $\mathbf{K}_{1/2}$ is one, and, thus, $\mathbf{K}_{1/2}$ would surely not be rejected.

Suppose, now, that we are in the situation explained in page 94 and have as one of the alternative hypothesis (besides those considered before) that the tossings are obtained from a machine, which, as its first ten tosses, produces exactly $\vec{x}$. Let us call this alternative, which is certainly logically possible, $\mathbf{H}$. As can be easily shown, the experiment E_n, $1 \leq n \leq \nu$ is also discriminating for the system of alternative hypotheses, $\mathbf{H}$ and $\mathbf{K}_p$ for $0 \leq p \leq 1$, p real: the set $A_{\mathbf{H}}$, for $\mathbf{H}$, is just the singleton consisting of the sequence, say $\vec{y}$, produced by the machine. Then E_n is surely eventually in A, if $\mathbf{H}$ is true, since, in this case, E_μ, for infinite μ, is the infinite μth result produced by the machine. On the other hand, under any $\mathbf{K}_p$, E_μ is a.s. not this sequence, and thus, E_n is a.s. not in $A_{\mathbf{H}}$. For the other hypotheses, as in the previous case, the sets $A_{\mathbf{K}_p}$ consists of the sequences whose limit is p, except, possibly, for $\vec{y}$.

Let us return to the case E_{10}. Then, $\vec{x}$, i.e., the sequence of ten results produced by the machine, is not evidentially equivalent to any other sequence, because, we have under $\mathbf{H}$

$$\Pr\nolimits_{\mathbf{H}}[E_{10} = \vec{x}] = 1 \quad \text{and} \quad \Pr\nolimits_{\mathbf{H}}[E_{10} = \vec{z}] = 0,$$

2. THE BASIS FOR STATISTICAL INFERENCE

for $\vec{z} \neq \vec{x}$. Thus, there is no $c > 0$ satisfying Definition IV.1 of evidential equivalence. On the other hand, if $\vec{u}$ and $\vec{z}$ are sequences different from $\vec{x}$ with the same number of heads, then as before

$$\Pr \mathbf{K}_p[E_{10} = \vec{u}] = \Pr_{\mathbf{K}_p}[E_n = \vec{z}],$$

and, also

$$\Pr_{\mathbf{H}}[E_{10} = \vec{u}] = \Pr_{\mathbf{H}}[E_{10} = \vec{z}] = 0.$$

Hence, $\vec{u}$ and $\vec{z}$ are evidentially equivalent.

Thus, it is easy to calculate that the rejection set determined by $\vec{x}$ for $\mathbf{K}_{1/2}$ contains only $\vec{x}$, and the two sequences with only ones, say $\vec{1}$ and zeros, say $\vec{0}$: we have, according to Definition IV.2, that

$$\vec{y} \preceq_{10,1/2} \vec{x} \iff \Pr[E_{10} \in [\vec{y}]] \leq \Pr[E_{10} = \vec{x}].$$

So, it is clear that only $\vec{x}, \vec{0}, \vec{1} \preceq_{10,1/2} \vec{x}$. Hence

$$R_{10,1/2,\vec{x}} = \{\vec{x}, \vec{1}, \vec{0}\}.$$

Thus,

$$\Pr[E_{10} \in R_{10,1/2,\vec{x}}] = 3 \cdot (1/2)^{10},$$

which is very low. Therefore, according to our rejection rule, we would be justified in rejecting $\mathbf{K}_{1/2}$, as it is intuitively natural.

As another example, consider the same alternative hypotheses of the coin with different biases, $\mathbf{K}_p$, and the machine that produces $\vec{x}$, $\mathbf{H}$, but now with the experiment Fr. Suppose, in order to be definite, that the limit of the relative frequency of tosses in the sequence produced by the machine is $1/2$. This is not a discriminating experiment for these alternatives: there is no set A of sequences of possible results of Fr such that Fr_n is a.s. eventually in A, according to $\mathbf{H}$ and Fr_n is a.s. eventually not in A, according to the alternatives. If we take A to be the set of sequences whose relative frequency converges to $1/2$, then Fr_n is surely (not just a.s.) eventually in A according to $\mathbf{H}$, but it is also a.s. eventually in A, according to $\mathbf{K}_{1/2}$. Let us consider another set as our candidate for the set A. Let A consist of sequence of possible frequencies in n tosses, $\langle r_n \mid 1 \leq n \leq \nu \rangle$, such that r_n is the exact frequency in the first n tosses of $\vec{x}$, the sequence produced by the machine. Here, again Fr_n is a.s. eventually in A, under $\mathbf{H}$, but it is also a.s. eventually in A under $\mathbf{K}_{1/2}$.

As a final example, consider the same hypotheses, $\mathbf{H}$ and $\mathbf{K}_p$, $0 \leq p \leq 1$, p real, but now assume that the sequence produced by the machine has no limit. In the case, the set A considered last satisfies the first condition of the definition of discriminating experiment, because under $\mathbf{H}$, Fr_n is a.s. eventually in A, and under any other hypotheses, it is a.s. eventually not in A. For the second condition, however, the situation is different. Let $\alpha > 0$, noninfinitesimal, be given. As we saw above, for a result r of Fr_m, the rejection set for the hypothesis $\mathbf{K}_{1/2}$, consists of those results s such that $|s - (1/2)| \geq |r - (1/2)|$. Since the sequence $\langle r_n \mid 1 \leq n \leq \nu \rangle$ has no limit, there is no finite n such that for every m with $n \leq m \leq \nu$

$$\Pr\nolimits_{K_p}[\mathrm{Fr}_m \in R_{mpr_m}] \leq \alpha.$$

Thus, the second part of the condition for a discriminating experiment is not satisfied. This is the end of the examples.

We see, thus, that the consideration of alternative hypotheses is essential. It is possible, as we saw above, that we can consider all "reasonable" alternatives, for instance, all alternatives K_p, for $0 \leq p \leq 1$, p real. But, in this case, we are not entertaining all logically possible alternatives, since the alternative produced by the machine we just mentioned, for instance, would be excluded.

The conditions for acceptance are similar in the requirement of a discriminating experiment. In general, however, there is no precise hypothesis about which we must decide whether to accept it or not. We first perform the experiment, and then decide which hypothesis to accept.

A simple rule of acceptance, that is convenient in many cases, is:

RULE IV.4 (DIALECTICAL RULE OF ACCEPTANCE). *We provisionally accept the structure K_j, relative to K_i, for $i \in I$, if all K_i, for $i \neq j$, have been provisionally rejected with respect to j, according to the just formulated rule of rejection.*

Furthermore, there is no discriminating experiment E' such that K_j is provisionally rejected with respect to E' (according to Rule IV.3).

This rule, which will be applied in the next chapter to medical diagnosis, is too strict for general statistical practice, especially when there is an infinite number of possible hypotheses. A similar rule, however, is also behind confidence region estimation, but in this case, the hypotheses involved are composite, i.e., they are classes of probability structures. We shall discuss confidence regions in Chapter XVIII, Section 7. Other rules of acceptance need different forms of discriminating experiments. We shall see one for formalizing point estimation in Chapter XVIII.

As it has been mentioned before, the fact the we use the real numbers as values for the measurement functions is not important. It is just a matter of convenience. In many examples, we shall have as values other sets instead of the real numbers. In order to transform these measurement functions to functions into the real numbers, some correspondence should be assumed between the set of values and the real numbers. This is, in general, not difficult to do, and I shall leave it to the reader.

There is the possibility that every K_i, for $i \in I$, be rejected. Suppose in the case of the coin, for instance, that hypothesis K_p determines that the probability of heads is p, and that we accept as possible all K_p for $0 < p < 1$, p real. The experiment E_n is the sequence of results in n tosses of the coins. Suppose that a sequence of 0's occurs. Then, if the sequence is long enough, all hypotheses should be provisionally rejected. Indeed, if an infinite sequence of 0's occurred, then all hypotheses should be in fact rejected. In this case, we call the system *untenable*.²² Then, we must add hypotheses to the system to make it tenable. Of course, since in general we can only deal with a finite sequence, we can only determine provisional untenability.

There is another way in which we can arrive at the conclusion that a system of possible probability models is untenable. This is when we have two or more experiments and all the possible models are provisionally rejected in at least one of the experiments.

3. Probability and degree of belief

As has been mentioned before, the subjectivist school on the foundations of probability identifies probability with rational degree of belief. According to my views, rational degrees of belief can, in some cases, be measured by probabilities, but probability is not just degree of belief, since there are other proper uses of the concept.

The connection between the different uses of probability is given by the Principle of Direct Inference in page 36, as ammended at the end of Section 1. Thus, we have at least one case in which we are justified in assigning numerical values to degrees of belief. Namely, if we accept that $\mathbf{K}$ is a model for a chance setup, and that φ has degree of support $\Pr_{\mathbf{K}}(\varphi) = r$, then our degree of belief in φ should be r. Degrees of belief are important, because we should base our decisions on these degrees. In fact, the justification for the use of the Direct Inference Principle as a basis for decision theory goes through rational degrees of belief. That is, if we accept $\mathbf{K}$ as the correct model then, as explained above, we obtain our degrees of belief, and then use these degrees as a basis for our decisions.

There may be other cases when we are justified in numerical degrees of belief. I believe that in cases where we do assign numerical degrees even if we do not have a chance model to base them on, we assign them by comparison with a chance model. For instance, when our degree of belief in φ is $1/2$, it is because we have the same degree of belief in φ as we would have in heads obtaining in the toss of a fair coin.²³ In any case, the Ramsey-de Finetti theorem, [77, 29],²⁴ as was mentioned in Chapter I, Section 2, shows that if we assign numerical degrees at all, satisfying some minimal obvious rational requirements now called *coherence*, this numerical assignment should obey the axioms of the calculus of probability. Coherence is based by de Finetti on conditions of rationality for actions relating to betting behavior, and by other authors, for actions in general.

I will not go into the details of their definitions, which have been extensively discussed in the literature and are very well-known. Coherence has been shown to be a rational requirement only when applied to decidable propositions. I think that this last limitation is important, as was also mentioned in Chapter I, Section

---

²²This term was suggested by W. Reinhardt.

²³This is the content of the widened form of the Principle of Direct Inference at the end of Section 1.

²⁴As was mentioned in Chapter I, Section 2, this theorem asserts in substance that if one doesn't assign degrees of belief obeying the rules of the calculus of probability, and one bets according to the degrees thus assigned, then one is liable to accept bets in such a way that one is sure to lose.

2: the theorem of Ramsey and de Finetti applies only to decidable propositions, namely, those whose truth or falsity can be decided in a finite time, so that the bets (or the other actions considered by Ramsey and other authors) can be decided to be favorable or not.[25]

Example of impossibility of assigning degrees. There may be cases, however, when other rational restrictions make it impossible to assign numbers to degrees even for decidable propositions. Suppose, for instance, that we decide on an ordering of degrees of belief for a certain algebra of events. That is, we are able to decide for any A and B in the algebra, whether A is more probable than B or vice-versa. Even if we assume the algebra to be finite, and that the ordering obeys some natural conditions, there may be no compatible measure.

In order to be more precise, let $\precsim$ be the preordering in question of less than or equal probability.[26] Suppose that $\precsim$ satisfies the following intuitive conditions, proposed by de Finetti, for any events A, B, and C:

(1) $A \precsim B$ or $B \precsim A$.
(2) If $A \precsim B$ and $B \precsim C$, then $A \precsim C$.
(3) $\emptyset \precsim A \precsim X$, and $\emptyset \prec X$, where X is the sure event, $\emptyset$ is the impossible event, and $A \prec B$, means that B not $\precsim A$.
(4) If C is disjoint from A and B, then $A \precsim B$ if and only if $A \cup C \precsim B \cup C$.

Kraft, Pratt, and Seidenberg in [57] have shown that there is an algebra with 32 events, and a relation $\precsim$ satisfying (1)–(4), which has no probability measure Pr, such that $\Pr A \leq \Pr B$ if and only if $A \precsim B$. They also propose an extra condition that insures the existence of a compatible measure, but it is not intuitively clear that an ordering of probabilities should obey it. Thus, it may be possible that the ordering intuitively assigned be as in their counterexample. On the other hand, Savage, [79], and others have given other more intuitive conditions that correspond, I believe, to my requirement of comparability with a chance setup.

Another case where, according to my theory, a measure for degrees of belief may not exist is when we accept a model for a certain chance setup that does not admit a natural probability measure (i.e., an invariant and dependence preserving measure).

Comparison with subjectivists' views. My main two differences with (at least some) subjectivists[27] with respect to degrees of belief, is that I don't believe that it is always possible to assign a measure to degrees of belief, nor do I believe that changes in our degrees of belief can be simply explained through conditionalization. The problem of determining rational criteria for changes in our degrees of belief according to our experience is important. In the next paragraphs, I shall discuss this matter.

I shall explain again how conditionalization works. I shall denote the degree of belief of a certain person X at time t in the proposition φ by $\Pr_t(\varphi)$, and

---

[25] da Costa's inclusion as decidable propositions of those general propositions that are accepted as quasi-true for a definite period of time does not affect the argument here.

[26] Conditions (1) and (2) below assert that $\precsim$ is a preordering.

[27] De Finetti, Savage, and Lewis, for instance. da Costa is again an exception.

the degree of φ conditional on ψ, also at time t, by $\Pr_t(\varphi|\psi)$. Changes by conditionalization then proceed as follows. Suppose that s is a time after t. Then, if ψ is all the knowledge obtained by X from time t to s (more precisely, ψ contains all the new facts that were accepted as true by X between times t and s), then $\Pr_s(\varphi) = \Pr_t(\varphi|\psi)$. Since $\Pr_t$ is the rational degree of belief, and these degrees only apply to decidable propositions, φ and ψ should be decidable propositions.[28]

The fundamental way of changing degrees of belief, according to my view, is not by conditionalization, but by the use of the Inverse Inference Principle combined with the Direct Inference Principle. By the Inverse Inference Principle, we may be able to accept that a certain **K** is the appropriate model for a certain chance setup. Then the Direct Inference Principle forces us to assign to the proposition φ a degree of belief that is equal to the chance in **K** of the event consisting of those outcomes where φ is true. I will call this type of change in our degrees, change by the two principles.

Conditionalization proper only works inside a model **K**. If we have accepted that **K** is the right model, then we might be able to assign conditional degrees of belief, and use conditionalization, as explained above, in order to change our degrees. However, when the Inverse Inference Principle makes us change from **K** to another model **K'** as the correct model of the chance setup in question, then our degrees of belief should be computed according to **K'** and not according to conditionalization in **K**. That is, when conditionalization indicates a different change than the combined use of the two principles, the degrees obtained by these two principles should prevail,[29] because they are based in what we accept as true, and truth is prior to probability, i.e., we must use what we accept as true in order to assign probabilities.[30]

Coherence of the two principles. We must now inquire whether changes by the two principles are coherent or not. Coherence, i.e., the natural conditions of rationality imposed by de Finetti and Ramsey, requires that conditional degrees of belief obey the axioms of the Calculus of Probability. That is, at any time t, we must have

$$\Pr_t(\varphi|\psi) \cdot Pr_t(\psi) = \Pr_t(\varphi \text{ and } \psi).$$

Thus, if all our new knowledge is contained in ψ, i.e., the only new proposition that we accept as true at time s is ψ, then we should change the degree of ψ to one, i.e., $\Pr_s(\psi) = 1$. Also, all the other degrees should be adjusted just on the basis of this change, since ψ expresses all our new knowledge. Since we accept ψ as true, the proposition φ and ψ is equivalent to φ. Therefore, we get that

$$\Pr_s(\varphi) = \Pr_t(\varphi|\psi).$$

Hence, changes by conditionalization seem to be required by coherence.

---

[28] Or, at least, decidable in the sense of da Costa.
[29] Kyburg also believes that the Direct Inference Principle has priority over conditionalization, see [60].
[30] The argument presented in the text can be paraphrased as follows. Truth determines probabilities, so belief or knowledge about truth determines belief or knowledge about probabilities.

The problem that remains is to determine whether changes by the two principles are coherent. Suppose that at time t, we accept a certain $\mathbf{K}$ as model for the situation involved. Then we should have $\Pr_t(\varphi|\psi) = \Pr_{\mathbf{K}}(\varphi|\psi)$. Suppose furthermore, that evidence ψ makes us accept, by the Inverse Inference Principle, that $\mathbf{K}'$ is the adequate model, with $\mathbf{K} \neq \mathbf{K}'$. Then if we accept ψ as true at time s, we should have $\Pr_s(\varphi|\psi) = \Pr_{\mathbf{K}'}(\varphi|\psi)$, and hence, $\Pr_s(\varphi) = \Pr_{\mathbf{K}'}(\varphi|\psi)$, which is different, in general, from $\Pr_t(\varphi|\psi)$, i.e., $\Pr_{\mathbf{K}}(\varphi|\psi)$. Thus, changes by the two principles seem to violate conditionalization, and thus seem to be incoherent.

This violation, however, is only apparent. When we accept ψ as true at time s, we are not only accepting ψ, but also (by the Inverse Inference Principle) the proposition that says that $\mathbf{K}'$ is the true model of the setup. Thus, our new knowledge is represented by ψ and $\mathbf{K}'$ (abbreviating, as usual, the proposition that $\mathbf{K}'$ is the correct model, by $\mathbf{K}'$), and not just by ψ. We cannot write the term $\Pr(\varphi|\psi \text{ and } \mathbf{K}')$, because there is no probability assigned to $\mathbf{K}'$, but we write $\Pr_{\mathbf{K}'}(\varphi|\psi)$, and we do have that $\Pr_s(\varphi) = \Pr_{\mathbf{K}'}(\varphi|\psi)$. Hence, changes by the two principles are not incoherent.

It may be convenient, in order to make a decision, to assign degrees of belief for decidable propositions without having a very precise model in mind. Coherence imposes the laws of the calculus of probability as a requirement for the assignment of degrees of belief. In order to insure coherence and simplify matters, we may require that changes proceed only by conditionalization in the traditional sense. In this case, my use of the two principles, as explained above, adds a requirement for the assignment of degrees of belief: conditional degrees of belief should be assigned in such a way that changes by conditionalization coincide with changes by the two principles. That is, conditionalization on the evidence that makes us change from $\mathbf{K}$ to $\mathbf{K}'$ (using the Inverse Inference Principle) should produce the same effect as changes from probabilities computed in $\mathbf{K}$ to probabilities computed in $\mathbf{K}'$.

We could also go one step further, as D. Lewis does, [62], and not just say that we may require that all changes proceed by conditionalization, but to insist that the only proper way to change degrees of belief is by conditionalization. If one accepts this thesis, then one must accept that one should assign conditional probabilities obeying the two principles. I believe that those subjectivists who accept chance, such as Lewis, [62], and da Costa, [27], and the Principle of Direct Inference, have no recourse but to add this additional requirement to the assignment of conditional probabilities. I think, however, that this procedure does not illuminate the real problems. What matters is that changes in rational degrees of belief proceed through the use of the two principles, and only when there is no change of model, through conditionalization.

4. Probabilities and frequencies

Probabilities and long run frequencies have been connected almost from the beginnings of the calculus of probability. The main school of statisticians even

4. PROBABILITIES AND FREQUENCIES

bases their work on the identification of probability with long run frequencies.[31] The explanation of the relationship between probabilities and frequencies, that is, the fact that from certain probabilities we can deduce that certain frequencies will occur in independent repetitions of an experiment, and that from frequencies we can estimate probabilities, is called by Popper *the fundamental problem of the theory of chance*, [74, Section 49]. Thus, although my interpretation of probability is not based on frequencies, the role that frequencies play in the determination of probabilities should be spelled out, and the fundamental problem of the theory of chance should be solved. This is the purpose of this section.

The following is an obvious example of the connection between frequencies and probabilities. Suppose that we would like to know the probability of heads in the toss of a certain coin. A natural way to investigate this, as has been stated in several places, would be to toss the coin a large number of times (say 1,000) and record the frequency of heads in these tosses. If there were 60% heads, we would naturally say that the probability of heads is approximately 0.6. I shall explain in this section the justification for this procedure, according to the view of probability that has been expounded.

Consider an event A, which is a result of an experiment $\mathfrak{S}$. Suppose that $\mathfrak{S}$ is repeated n times and A occurs $m(A)$ times. Everyone agrees that there is a close relation between $\Pr A$ and $m(A)/n$. There are several views about this relationship. We shall mention two.

According to von Mises, $\Pr A$ and $m(A)/n$ are connected by the equation

$$\Pr A = \lim_{n \to \infty} \frac{m(A)}{n}.$$

This equation is, in fact, von Mises' definition of $\Pr A$. We have already discussed the drawbacks of this definition in Chapter I, Section 2.

Kolmogorov, on the other hand, considers, not an infinite sequence of repetitions, but only a 'large number'. He says in [56, p. 4]:

> Under certain conditions, which we shall not discuss here, we may assume that to an event A which may or may not occur under conditions $\mathfrak{S}$, is assigned a real number $\Pr A$, which has the following characteristics:
>
> (a) One can be practically certain that if the complex of conditions $\mathfrak{S}$ is repeated a large number of times, then if m be the occurrence of event A, the ratio m/n will differ very slightly from $\Pr A$.
>
> (b) If $\Pr A$ is very small, one can be practically certain that when conditions $\mathfrak{S}$ are realized only once, the event A would not occur at all.

The main objection to this view, which is also an objection to Popper's propensity view, for instance, in [76], is that if one translates the phrase 'one can be practically certain' into 'there is a high probability' a circularity seems to be

---
[31] For instance, von Mises, [98], and Neyman, [72].

manifest. Since probability is, in my view, 'possibility of truth' or 'partial truth', we can translate the phrase in terms that are not circular. We shall see in this section how to give definite rules to connect probability and frequencies, which make use of probability as possibility of truth.

Rules for the acceptance of frequencies as estimates of probabilities. We begin with a review of the rules of Section 2 in the present setting. Suppose that $\mathbf{K}$ is the set of possible outcomes of a certain chance setup $\mathfrak{S}$, say the toss of a coin. As we saw in Chapter III, Section 4, and Section 2 in this chapter, we can naturally represent an outcome of the setup consisting of n repetitions of $\mathfrak{S}$, by an n-termed sequence of elements of $\mathbf{K}$. That is, the set of possible outcomes of the chance setup consisting of n independent repetitions of $\mathfrak{S}$ is $^n\mathbf{K}$, the set of n-tuples of elements of $\mathbf{K}$. As an idealization, we also consider infinitely many independent repetitions of $\mathfrak{S}$, which may be represented by ν repetitions, with ν an infinite natural number. Then the possible outcomes are the ν–sequences of elements of $\mathbf{K}$.

Recall that events in $\mathfrak{S}$, in my formulation, are subsets of $\mathbf{K}$. As was explained in Chapter III, Section 4, if the probability measure defined in $\mathbf{K}$ is $\Pr_{\mathbf{K}}$, then the natural measure on $^n\mathbf{K}$ is the product measure (here n is a finite or infinite natural number).[32] From these measures we can define the epistemic probabilities, which I shall denote by the same letters, $\Pr_{\mathbf{K}}$ and $\Pr_{^n\mathbf{K}}$ on the propositions that express the occurrences of the appropriate events. For any proposition φ adequate for $\mathbf{K}$, $\Pr_{\mathbf{K}}(\varphi)$ is the measure $\Pr_{\mathbf{K}}$ of the set of outcomes that make φ true. Similarly, for propositions φ adequate for $^n\mathbf{K}$, $\Pr_{^n\mathbf{K}}(\varphi)$ is the measure (the product measure of $\Pr_{\mathbf{K}}$, n times) of the set of outcomes (in this case sequences) that make φ true. As I have already explained, $\Pr_{\mathbf{K}}(\varphi)$ represents the degree of possibility of truth of φ in $\mathbf{K}$.

Suppose that we have a chance setup $\mathfrak{S}$, and that we would like to test which set of possible outcomes is the right model for $\mathfrak{S}$. If $\mathbf{K}$ is such a model, then $^n\mathbf{K}$ is the model for n independent repetitions of $\mathfrak{S}$. I shall apply the Principle of Inverse Inference of Section 2 in order to determine which of the models is an adequate model for $\mathfrak{S}$. Since the probabilities in $^n\mathbf{K}$ are determined by those in $\mathbf{K}$, when referring to a proposition φ concerning n-termed sequences of outcomes, its probability in $\mathbf{K}$ should be understood as that in $^n\mathbf{K}$.

In order to obtain rules of acceptance for our present situation, we first specialize the notion of discriminating experiment defined in Section 2. We say that $E = \langle E_n \mid 1 \leq n \leq \nu \rangle$ is a discriminating experiment for $\mathbf{K}_i$, $i \in I$, if E is a discriminating experiment for $\langle ^n\mathbf{K}_i \mid 1 \leq n \leq \nu \rangle$ with $i \in I$, so that E_n is a random variable over the probability space of $^n\mathbf{K}_i$, for each $i \in I$ and $1 \leq n \leq \nu$.[33]

Our ideal rule of rejection, Rule IV.2, asserts that the effect of the occurrence of an infinite sequence of results, such that the probabilites of the set of results that are at least as bad for $\mathbf{K}_j$ tend to zero is similar to the occurrence of an event that is impossible according to $\mathbf{K}_j$: both have the effect of rejecting the

---

[32] A justification of this product measure from considerations of invariance will be given in Part 3.

[33] As before, we should also require that the sequence E_n be internal.

hypothesis that $\mathbf{K}_j$ is the true model of reality. However, there is a difference. While the occurrence of an event that is impossible according to $\mathbf{K}_j$, rejects this hypothesis outright, the occurrence of the infinite sequence rejects it only relative to the class of alternatives. This is so because the definition of worse result requires not only that the probability of the result be smaller for $\mathbf{K}_j$, but also that it be higher for an alternative $\mathbf{K}_i$. The dual rule of acceptance (IV.1) can be interpreted in the same way.

These two rules (IV.1 and IV.2) are ideal in the sense that they are impossible to apply in practice, because they require an infinite number of realizations of the experiments. They are only stated as idealizations that more practical rules should approximate. The approximating rules that were stated, Rules IV.3 and IV.4, are dialectical in character. This means that these rules give me procedures to convince somebody (this 'somebody' may be myself) to accept or reject a hypothesis provisionally. These procedures are of the following form: I ask him to give me 'reasonable' criteria by which he would be convinced, and then provide him with what he asked for. 'Reasonable' here means 'approximating the ideal rule'.

I shall now apply these rules for the provisional acceptance for the important special case when the evidence consists of frequencies. This application makes the rule stronger, in a certain sense, than our original Rule IV.4. We need acceptance for this case, and not only rejection, in order to explain how long run frequencies approximate probabilities. For instance, we must explain why, if the relative frequency of an event was 60%, we should accept that its probability is approximately 0.6.

Suppose that we have a proposition φ for which we can determine the relative frequency of its occurrence in a series of independent repetitions of a setup $\mathfrak{S}$. The problem is to determine which probability of φ is a reasonable hypothesis to accept. In order to do this, we choose, for each real q in the unit interval, one set of possible outcomes $\mathbf{K}_q$ (representing a possible probability model) with the property that $\Pr_{\mathbf{K}_q}(\varphi) = q$.

Since we are assuming that we only know the probability of φ, the internal structure of the different $\mathbf{K}_q$'s is irrelevant. We only need to assume that $\mathbf{K}_q$ is a set of possible outcomes with $\Pr_{\mathbf{K}_q}(\varphi) = q$. By only using the frequency of φ, we cannot distinguish between two models that assign the same probability to φ. This is why we must pick exactly one model that assigns q to φ, for each real q between zero and one.

The procedure that I shall describe below, can also be applied when we are given the alternative hypotheses in advance and we have to accept one of these hypotheses, if there is a proposition φ that is assigned different probabilities by the different alternatives.

We now return to the case where φ is given, and we want to estimate its probability. Since we are using the frequency of φ, the sequence of experiments is clear: E_n is just the observation of the relative frequency of the truth of φ in n repetitions of the setup. We call this E_n, Fr_n. The proof that $\langle \mathrm{Fr}_n \mid 1 \leq n \leq \nu \rangle$ is a discriminating experiment in the sense of Definition IV.3 is the same as that given in Section 2, page 101, for the case of the setup of the tossing of a coin.

For the case of frequencies that we are discussing, the experiment Fr that we are considering has another stronger property, namely effectivity. Although I shall not give a precise definition of the notion of an effective sequence, a few words of explantion are in order. We say that a sequence $\langle a_n \mid 1 \leq n \leq \nu \rangle$ is *effective* if, given a finite n, there is an algorithm to calculate a_n. We use the expression "for every finite p we can effectively find n_p" to mean that there is an algoritm to calculate n_p from p, when p is finite. We use other similar expressions whose interpretation will be clear from the context.

Let r be a given real number strictly between 0 and 1. We would like to determine whether $\mathbf{K}_r$, call it simply $\mathbf{K}$, is the acceptable hypothesis, by using a dialectical rule. Let $y \in {}^k\mathbf{K}_q$, for some natural number k and some q. Then $\mathrm{Fr}_n(y)$ is the relative frequency of the truth of φ in the first n members of y. We can obtain Fr_n as follows. Let

$$X_i = \begin{cases} 1, & \text{if } \varphi \text{ is true in the } i\text{th repetition,} \\ 0, & \text{otherwise.} \end{cases}$$

Then X_i is an Bernoulli random variable and

$$\sum_{i=1}^{n} X_i$$

is a binomial random variable (see page 11). Now

$$\mathrm{Fr}_n = \frac{1}{n}\sum_{i=1}^{n} X_i.$$

For any $t \in [0,1]$ and s a real number, $[|\mathrm{Fr}_n - t| \geq s]$ is the proposition that says that the sequence y that obtained has the property that $|\mathrm{Fr}_n(y) - t| \geq s$. Similarly, with $<$, $\leq$, or $>$, instead of $\geq$. As we saw in Section 2 for the coin tossing example, the rejection set R_{nrs} of results that are at least as bad as s for the probability to be r is $[|\mathrm{Fr}_n - r| \geq |s - r|]$.

For the rest of this section, we write, for any q between 0 qnd 1, Pr_q, for $\mathrm{Pr}_{\mathbf{K}_q}$, E_q, for $\mathrm{E}_{\mathbf{K}_q}$, and Var_q, for $\mathrm{Var}_{\mathbf{K}_q}$.

As was explained in Section 2, the definition of the rejection set, R_{nrs}, for this case is very simple. We can just say that $t \in R_{nrs}$ if and only if $\mathrm{Pr}_r[\mathrm{Fr}_n = t] \leq \mathrm{Pr}_r[\mathrm{Fr}_n = s]$. This is so, because no two different results are evidentially equivalent, and for every t and s, $t \neq s$

$$\mathrm{Pr}_t[\mathrm{Fr}_n = t] \geq \mathrm{Pr}_t[\mathrm{Fr}_n = s] \text{ and } \mathrm{Pr}_s[\mathrm{Fr}_n = t] \leq \mathrm{Pr}_s[\mathrm{Fr}_n = s].$$

Thus, we easily get, as was asserted above, that

$$R_{nrs} = [|\mathrm{Fr}_n - r| \geq |s - r|].$$

4. PROBABILITIES AND FREQUENCIES

As is mentioned in page 12 and will be shown in page 139

$$\text{Var}_q(\sum_{i=1}^n X_i) = nq(1-q).$$

By the equation on page 10, which is proved on page 137

$$\text{Var}_q(\text{Fr}_n) = \frac{1}{n^2}\text{Var}_q(\sum_{i=1}^n X_i) = \frac{1}{n^2}nq(1-q) = \frac{1}{n}q(1-q).$$

Thus, by Chebyshev's inequality, which is mentioned in page 12 and proved as Theorem VI.3, we have that for any q between 0 and 1

$$\Pr_q[|\text{Fr}_n - q| \geq k] \leq \frac{q(1-q)}{nk^2}.$$

It is easy to prove that $1/2$ is the maximum of the function $g(q) = q(1-q)$ in the interval $[0,1]$. Thus, for any q between 0 and 1

$$\Pr_q[|\text{Fr}_n - q| \geq k] \leq \frac{1}{4nk^2}.$$

Thus, if we put

$$\frac{1}{m} = \frac{1}{4nk^2}$$

we get

$$k = \frac{1}{2}\sqrt{\frac{m}{n}}.$$

Therefore, if we define the computable function

$$f(m,n) = \frac{1}{2}\sqrt{\frac{m}{n}},$$

then for all natural numbers m and n

$$\Pr_q[|\text{Fr}_n - q| \geq f(m,n)] \leq \frac{1}{m}.$$

Hence, if we set $\alpha = 1/m$, and obtain a sequence of length n, y, such that $|\text{Fr}_n(y) - q| \geq f(m,n)$, then, by Rule IV.3, $\mathbf{K}_q$ is rejected. Thus, for any m and n we can effectively determine the q's such that $\mathbf{K}_q$ is rejected at level $\alpha = 1/m$ with a sequence, y, of length n. This means that given that $\mathbf{K}_q$ is the true model for the setup, the possibility of truth of the set of results that are at least as bad for $\mathbf{K}_q$ as y is less than $1/m$. Thus, if $\mathbf{K}_q$ is the true model, y is an unlikely result, that is, y is nearly false. Hence, we are justified in provisionally rejecting $\mathbf{K}_q$ as the true model for the setup. Thus, we say that, given the n-sequence y, the probability, r, of φ is between $\text{Fr}_n(y) - f(m,n)$ and $\text{Fr}_n(y) + f(m,n)$ *with confidence* $1 - 1/m$. Since we can make m as large as we please, $1 - 1/m$ can be as close to one as we please.

It is clear that for a fixed m, $f(m,n)$ tends to zero as n tends to infinity. Since we can let n tend to infinity, frequencies approximate probabilities. This method for accepting frequencies as probabilities is again dialectical, but in a stronger sense, as it was already explained, than our old rules: here, our opponent can give us the n and the m, and, from the given sequence, we can effectively obtain the probabilities that are accepted as possible.

The method explained here for frequencies can be used, for instance, in order to determine, when in doubt, which is the right model for an experiment of choosing a chord at random in a circle, i.e., for Bertrand's chord paradox. Suppose that we perform an experiment where we choose a chord at random in a circle. We want to know which of the models discussed in Chapter III, Section 2 applies. The alternative hypotheses are given by the three models, say $\mathbf{K}_1$, $\mathbf{K}_2$, and $\mathbf{K}_3$. We perform a large number of experiments, and calculate the relative frequency of the chord having length greater than $\sqrt{3}$. The model that assigns a probability closest to this frequency is then accepted.

CHAPTER V

A medical example

The mutual relations between the two principles, i.e., the direct and inverse inference priciples, and between them and statistical inference and decision theory, will be analyzed further in Part 4. At this point, I shall study an example where both principles are used, but only in their intuitive versions, generally without formal statistical developments. The example in question is the analysis of the principles of reasoning behind medical diagnosis.

There is a vast recent literature on the subject of medical diagnosis (see, for instance, [91, 84]), especially devoted to the construction of computer systems, in particular, expert systems, as an aid in diagnosis. We shall not review this literature here, but merely analyze, what I think, are the basic logical principles that should underlie medical diagnosis. Thus, although the discussion in this chapter might be of help in the design of computer-aided diagnostic procedures, it should not be taken as a proposal for implementing these procedures.

1. The disease model

I shall begin with a brief description of the situation involved. A physician is examining a patient, and needs to discover which disease he has in order to treat him. The ascertaining of the disease is done through a study of the symptoms of the patient. I include in the term 'symptom' all relevant findings about the patient: his history, physical signs, laboratory tests, etc.

The theory behind medical diagnosis is that the disease causes the symptoms. More precisely, each disease has a cause that produces different clusters of symptoms in different people. Each disease is, thus, a chance setup determining the possible clusters of symptoms arising in the different patients that have the disease. In fact, diseases should be considered as stochastic processes which evolve in time. The disease is supposed to have a beginning, that is, the patient at a certain instant of time acquires the disease. The disease has then an evolution. That is, at successive moments of time different alterations of the body may occur. Each of these changes in the physiology manifests itself as different clusters of symptoms.

V. A MEDICAL EXAMPLE

The connection between the different alterations of the body physiology is probabilistic, or more precisely, possibilistic, in nature. Thus, what has happened up to a certain time determines the range of possibilities of alterations at successive times. The connection between the alterations of the body and the symptoms observed may also be probabilistic. That is, certain alterations of the body may determine a range of possible clusters of symptoms and not a specific cluster.

For simplicity, we shall assume that the time moments are discrete, that is, that we have a succession of causal moments t_1, t_2, ..., t_ν, with ν an infinite natural number. The causal universe of our model is then $T = \{t_1, t_2, \ldots, t_\nu\}$, with the ordering $t_i \leq t_j$ for $i \leq j$. We can imagine the successive moments as the successive moments of consultation. What has occurred at times t_1, ..., t_n determines the range of possible alterations at time t_{n+1}. These alterations, in their turn, determine the possible clusters of symptoms.

Thus, we see that there are two parallel interconnected processes going on. The internal process is the real causal process: the cause of the disease produces successive alterations of the physiology and anatomy of the body, which, in their turn, produce further alterations: the different alterations influence the range of possibilities for subsequent alterations. The visible process is the sequence of symptoms observable in the patient at the sequence of times, t_1, t_2, ... Thus, this visible process is not a real causal process. The symptoms, in general, do not cause the range of possibilities for subsequent symptoms. The cause of the symptoms is the alterations of the body. So, calling the moments t_1, t_2, ..., causal moments, for the process that has as range of possibilities at each t_n the different clusters of symptoms, is somewhat of an abuse of language. The probabilistic model is, thus, quite complicated. As we shall see, several simplifications are necessary to make the model workable.

As it is clear from the description given, I accept a sort of probabilistic conception of causality. As has been mentioned before, a similar conception has been put forward by Suppes in [89]. The main difference is that Suppes defines causes in terms of conditional probabilities, while I consider the causal structure as primitive. Thus, in my view, there is a causal ordering, and what has happened before a certain moment t, determines the *possibilities at t*. These possibilities, in their turn, determine the probabilities. Thus, I would call my conception of causality *possibilistic* rather than probabilistic.

Diseases are classified according to the type of cause. Examples of these categories are infectious, congenital, neoplastic, or degenerative diseases. In fact, there is a hierarchical organization with categories, subcategories, and so on. The lowest level is constituted by the particular diseases. Very often a symptom gives support to one of the broader categories, without singling out a particular disease in the category. For instance, there are symptoms that are specific for infectious diseases, without distinguishing between the different illnesses of this type.

Besides the symptoms, medical diagnosis is also based on anatomical and physiological knowledge. I shall not be concerned very much with this aspect, so my discussion is certainly oversimplified.

2. The Bayesian model

I shall begin with a simple model that is based on the Direct Inference Principle. This model, which works in accordance with the principles used by Bayesian statisticians, has actually been employed for computerizing medical diagnosis.[1] There is a finite, but maybe very large, number of possible diseases $D_1, \ldots, D_m$. The model assigns a probability distribution to these diseases. In practice, the a priori probability of disease D_i, $\Pr D_i$, is determined by the frequency of D_i among the patients that consult the given clinic. As we saw in Chapter IV, Section 4, assigning probabilities based on frequencies is justified.

I am abusing language somewhat, because what we really are considering is the probability that this patient (namely, the particular patient seeing the physician) has disease D_i, and not the probability of the disease itself. In all this medical analysis, I will not distinguish between the diseases or symptoms and the propositions that this patient has the disease or the symptoms.

For each disease D_i, we also have a probability distribution for the different sets of symptoms, i.e., $\Pr(S|D_i)$, the probability that the patient has symptoms S, given that he has disease D_i. This probability is calculated as the frequency of these symptoms among the patients who have disease D_i.

The Bayesian model that we are considering is similar to the urn model discussed in Chapter III, Section 4. It assumes one chance setup with two components, the second depending on the first. The disease model is, then, simplified as follows. The first component of the urn model, namely, the choosing of the urn, is the chance setup having as possible outcomes the patient having the different diseases. An assumption about the real possibilities of the different diseases leads to a probability distribution. The belief in a certain factual probability distribution at a time when the possibilities are real (i.e., a time before the patient has acquired the disease) determines an epistemic probability, i.e., a degree of belief. And, finally, at a later time (i.e., the time when the physician is examining the patient), although the possibilities are no longer real (but only antecedently real), the epistemic distribution is retained, since no new relevant knowledge has been acquired. Thus, the prior probability measure, $\Pr D_i$, is obtained by the use of the Principle of Direct Inference. In fact, however, we also need a use of the Principle of Inverse Inference: we don't really know the prior distribution, but the assumption about the real possibilities is based on an estimate according to the frequency of each disease, as it was explained in Chapter IV, Section 4.

The probabilities $\Pr(S|D_i)$ are determined in a similar way. Each disease determines the range of real possibilities of the clusters of symptoms. (This is the second component of the urn model, i.e., the choosing of a ball.) These possibilities lead, again, to a factual probability distribution of the different clusters of symptoms at the time when the possibilities of symptoms are real (that is, before they occur), which determines an epistemic probability. At the time when the physician is examining the patient, although the possibilities are no longer real (but only antecedently real), this epistemic distribution is retained, since there is no new relevant knowledge. As above, we use both the Inverse Inference

---

[1] See, for instance, [64], which is an early reference.

Principle to estimate the chance of the symptoms S, given disease D_i, and the Direct Inference Principle to obtain the probabilities $\Pr(S|D_i)$, from chance.

As it is easily seen, the disease model explained in the previous section is greatly oversimplified, when applying the Bayesian model. For this latter model, we only consider two causal moments. At the first one, a disease is chosen, and at the second one, a cluster of symptoms.

Another model, which, I think, represents better the Bayesian mode of thinking is the following. This new model is not concerned with the causal disease model, but imitates more faithfully the urn model. The action of the physician is seen as if he chose at random an urn from a set of urns, each urn representing a disease. Each urn contains all patients that consult the clinic and have the particular disease represented by the urn. Each urn-disease, contains balls representing the different clusters of symptoms. The physician, observes the cluster of symptoms, and, hence, it is as if she selected a ball from the urn. The physician has, then, to determine the probabilities of the different possible urns (i.e., diseases).

According to the Bayesian model (in either of the two forms) we are now discussing, each disease is really possible (or, more precisely, antecedently really possible), and hence we can never consider the hypothesis that the patient has a particular disease as false. The probabilities calculated in this model, however, give the rational degrees of belief that the physician ought to have. The assignment of degrees of belief is permissible, because, at least in principle, the propositions involved are factually determined and indeed decidable in principle, i.e., it can be decided in the long run whether a patient has a disease or not (for instance, in his autopsy), and whether or not he has a certain symptom.

What we are after is $\Pr(D_i|S)$, the probability of the patient having disease D_i when presenting symptoms S. This probability can be obtained from Bayes formula:

$$\Pr(D_i|S) = \frac{\Pr D_i \cdot \Pr(S|D_i)}{\sum_{j=0}^{n} \Pr D_j \cdot \Pr(S|D_j)}.$$

Notice that for the computations using this formula to be useful, S should contain all the symptoms that have been discovered in the patient. This is what has been called the requirement of total evidence. That is, the probability of the disease has to be computed with respect to the total evidence available. Also, diseases $D_1, D_2, \ldots, D_m$ must include all possible diseases, in order that Bayes formula be applicable.

Instead of using Bayes formula, $\Pr(D_i|S)$ may be computed directly as the frequency of disease D_i among the patients with symptoms S. This does not change essentially the type of principle used, since this second method also assumes one chance setup where all the probabilities are calculated.

When using the Direct Inference Principle as above, each disease is not entertained as a hypothesis to be accepted or rejected according to the symptoms observed, but it is just assigned a conditional probability $\Pr(D_i|S)$, i.e., a rational degree of belief in D_i, given S. In fact, this conditional probability should never be 0 or 1, because then the disease D_i or the rest of the diseases would cease to be viable possibilities, and further changes in probability according to Bayes' formula would become impossible.

The probabilities obtained in the model help, in the usual decision-theoretic fashion, to make the right decision about the treatment. This may be done by assigning numerical utilities, but usually it is done more informally.[2] There are several difficulties, however, with this model, even in this decision-theoretic framework. In the first place, groups of diseases always are in fact completely taken out of consideration by the physician. That is, the hypothesis that the patient has one of these diseases is rejected. This may be a mistake on the physician's part, but, actually, it is practically impossible to deal with the full number of diseases. Even the fastest computers can't do it. The computer programs that have been devised assume that the physician decides first on a certain category of diseases, and that the model is then applied with just the diseases in the accepted class as possible, i.e., $D_1, \ldots, D_m$ is limited to this class. The fact that in order ot apply Bayes' formula we need all the alternative diseases, complicates the matter further. Disregarding diseases with positive probability may introduce large errors.

In the second place, it is awkward to deal in the model with the fact that symptoms might give evidence for a category of diseases, and not for a particular disease. In order to avoid this problem, alternative models, based on a different measure of degrees of belief, have been proposed. The most prominent of these is the Dempster-Shafer model, which is discussed in [38]. Since this model does not obey the laws of the Calculus of Probability, it is incoherent, in the sense of de Finetti.

In the third place, it is difficult to accommodate the stochastic nature of the evolution of diseases, which was explained in the previous section. This evolution, in particular, includes anatomo-physiological knowledge, which may be of interest in deciding on a disease. As we saw above, the model is too oversimplified, and, as such, it is certain to lead to mistakes in some cases.

3. The classical statistical inference model

Instead of only applying the Direct Inference Principle, we could also use the Inverse Inference Principle. In this case, we assume that there are alternative hypotheses for the explanation of the patient's symptoms, each hypothesis calling for the patient having a particular disease or combination of diseases. Here, the patient is not one chosen from a set of patients, but he is considered individually. Thus, the hypotheses of the patient having the different diseases, are just considered epistemic possibilities, not even antecedently real epistemic possibilities.

I shall disregard the possibility of a combination of diseases, for simplicity's sake, but it would be no problem to include it here.

With each hypothesis, we associate a model for the corresponding chance setup, which has the structure explained in Section 1. Thus, with the hypothesis

---

[2]See Chapter XX, Section 4, for a discussion of how utilities may be used in medical diagnosis.

of the patient having disease D, we associate the chance setup with possible outcomes that consist of the patient having different clusters of symptoms, supposing that he has D. For each one of these setups we obtain a probability distribution $\Pr_D$ on the different clusters of symptoms. $\Pr_D(S)$ is the probability of symptoms S determined by the patient having D. This is what is called, following Fisher, the likelihood of D, given S. Notice that we don't write the likelihood as the conditional probability $\Pr(S|D)$, because there is no common distribution for all the diseases.

The Inverse Inference Principle then works as follows. We assume that we observe the patient at a sequence of times, t_n, indexed by n. (It would be possible to formulate the rules having a continuous time variable t, instead of t_n.) The discriminating experiment E is defined as $[E_n = S]$ if up to time t_n the cluster of symptoms S has been observed. Thus, it is possible to emphasize the evolutionary character of diseases. For instance, $\Pr_D[E_n = S]$ might be different from $\Pr_D[E_k = S]$, for $k \neq n$.

It is easy to show that this is a discriminating experiment: with each disease D a set A_D of sequences of possible clusters of symptoms $\langle S_n \mid 1 \leq n \leq \nu \rangle$ is associated. That is, there is a set A_D of sequences of symptoms such that, if D is the right disease, E_n is a.s., in this case, in fact, surely, in A_D. In general, if D_1 and D_2 are different diseases, then A_{D_1} and A_{D_2} are disjoint. That is, a complete sequence of clusters of symptoms (through all times, including, maybe the autopsy) discriminates between two diseases.

It is also true that if we wait long enough, new symptoms S' will appear such that $\Pr_{D_i}[E_k \in R_{kiS'}]$ will be low enough for some $k > n$ (in case D_i is actually false). Therefore, by waiting long enough, we can get a proposition with a probability as low as we want.

The sequence of discriminating experiments is, in this case, just the observation of the symptoms, and the sequence of propositions express the cluster of symptoms observed at different times. We shall discuss, now, the notion of evidential equivalence, in the sense of Definition IV.1, for our situation. Suppose that two different clusters of symptoms, S_1 and S_2, are evidentially equivalent, that is, that there is a $c > 0$ such that for every i, with $1 \leq i \leq m$

$$\Pr_{D_i}[E_n = S_1] = c \cdot \Pr_{D_i}[E_n = S_2].$$

Then, we would naturally consider S_1 and S_2 to be different forms of the same cluster of symptoms. Thus, in general, with very few exceptions, no two different clusters of symptoms are evidentially equivalent, in the sense of Definition IV.1. This is so, because the symptoms are chosen so that they discriminate between diseases. That is, medical knowledge is organized so that, as far as possible, different clusters of symptoms are not evidentially equivalent.

In most cases, it is also true that, for any pairs of clusters of symptoms S and S', there are diseases D_i and D_j such that

$$\Pr_{D_i}[E_n = S] < \Pr_{D_i}[E_n = S'] \text{ and } \Pr_{D_j}[E_n = S] > \Pr_{D_j}[E_n = S'],$$

3. THE CLASSICAL STATISTICAL INFERENCE MODEL

so that we can define that S' is worse than S for D_i (Definition IV.2), by only requiring that

$$\Pr{}_{D_i}[E_n = S] > \Pr{}_{D_i}[E_n = S']. \qquad (*)$$

The conditions for provisional rejection of D_i can be met: D_i is rejected if a cluster of symptoms S occurs at E_n and the probability of $[E_n = S]$ or clusters with less probability under D_i, is low. If we can reject all diseases D_i, except for one, say D_j, then D_j is accepted (using Rule IV.4).

Formally, we define the rejection set R_{nDS} as the set of clusters of symptoms containing, at time t_n, S or any cluster that is as bad as S for D. Then, we reject D, if S is observed and

$$\Pr[E_n \in R_{nDS}] \leq \alpha,$$

for a certain low α.

Notice that all diseases could be rejected if a cluster of symptoms S occurred at E_n such that, for any desease D_j, the probability of S or worse is low (i.e., the probability of the rejection set determined by S is low). Then, we should either add another disease to the epistemic possibilities, or say that this is a very unusual case, and suspend judgement.

In applying the Inverse Inference Principle for rejection, the only place where the alternative models are needed is in the definition of "worse results", which depends on "evidential equivalence". Thus, as was explained above, for the definition of worse result, we only need formula (*), which is independent of the alternatives. Hence, it seems that a disease can be tested in isolation. The reason for this fact is that the possible symptoms have been chosen so that they discriminate between the possible diseases. In order to choose the possible symptoms in the right way, however, consideration of the different possible diseases is necessary. Therefore, we have that in the particular decision of a physician, a disease can be considered in isolation for rejection, but in choosing which symptoms should be taken into account (that is, in the data base of medical knowledge), alternative diseases must be considered.

For acceptance, however, one needs to take into account all possible alternatives. One must be able to reject all alternatives, in order to accept a disease. This situation is similar to the deterministic case. Suppose that the alternative hypotheses are deterministic in character. That is, each hypothesis determines truth or falsehood of consequences, and not only probabilities. If we can obtain in reality consequences that are false according to all hypotheses, except one, h, then we have to accept h. But, if we don't know whether the set of alternatives is exhaustive, we cannot accept h, because there may be a hypothesis that has not been considered and which should not be rejected.

The Inverse Inference Principle can account very well for the hierarchical aspect of diseases. Instead of considering the hypothesis that the patient has D_i, we may consider that of his having a certain category of diseases, say D (for instance, an infectious disease). We can compute

$$\Pr{}_D[E_n \in R_{nDS}]$$

where R_{nDS} is obtained with reference to alternative categories, instead of alternative diseases, applying the same method as above. Then D may be rejected or accepted. Again, we must have a sequence of clusters of symptoms. Here, rejected means rejected as false, and taken out of consideration; and accepted means accepted as true, and all diseases outside of D taken out of consideration. As was explained before, this does not mean that the decision might not be reconsidered later.

Accepting a disease means being able to use the proposition that the patient has the disease together with other accepted propositions in order to obtain logical consequences, which also are accepted. Hence, we can use biological knowledge to obtain conclusions about the treatment or predictions about the future course of the patient. If, as in the Bayesian model, we only assigned probabilities to the patient having the different diseases, and did not accept one, and these consequences were just probable and not completely determined by the disease, the probabilities of these consequences may be too low to be of any relevance. On the other hand, when we accept a disease, the probability of a consequence *is exactly* the probability of the consequence given the disease.

I believe that a combination of both principles is the best approach for medical diagnosis. Bayes formula and rejection sets should both be considered at the same time. Those diseases or categories of diseases for which the probability of the rejection set are very low, are rejected and taken out of consideration. That is, the set of $D_1, \ldots, D_m$ in Bayes formula should contain only those diseases whose rejection sets do not have very low probabilities. This method allows us to eliminate whole categories of diseases. For instance, if S obtains and $\Pr_D[E_n \in R_{nDS}]$ is very low for infectious diseases, no infectious disease would appear among the D_i's.

Notice that the level of the probability at which we are prepared to reject D_i depends on several factors. One of them is the frequency of the diseases. If D_i is much more frequent than D_j, then the probability of the rejection set should be very low. Since we have a set of sequences of clusters of symptoms whose rejection sets tend in probability to zero, which is of probability one, in principle, if we wait long enough, we can obtain this low probability.

The ideal situation is when just one disease, say D_j, is not rejected. Then the physician would clearly accept as true that the patient has D_j, and would not bother about probabilities in making his decision. This elimination of all diseases but one is not a rare occurrence, although, as was said above, it is always subject to reconsideration.

In order to analyze further the two inferential models, we shall consider the following situation. Suppose that there is a symptom, S, that occurs in just one disease, say D_i (this is what is called a pathognomonic symptom), and assume that this symptom is observed. In using the Bayesian model, we have that $\Pr(S|D_j) = 0$ for all $j \neq i$, and hence, by Bayes formula

$$\Pr(D_i|S) = \frac{\Pr(S|D_i) \Pr D_i}{\Pr(S|D_i) \Pr D_i} = 1,$$

so that the Bayesian model behaves as it should. Notice, however, that we cannot apply the Bayesian model again, because the disease has probability one, and no further changes are possible with Bayes formula. That is, D_i is definitively accepted.

Apply, now, the inverse inference principle to the same situation. We have that $\Pr_{D_j}[E_n = S] = 0$, for any $j \neq i$. Thus, since all symptoms worse than S for D_j, $j \neq i$, must also have probability zero, and as we assume that there are only finitely many clusters of symptoms, the probability of the rejection set is zero. Hence, it is less than α for any positive α. Therefore, all disease D_j, $j \neq i$, are rejected at any level, and, thus, D_i is accepted. In this case, however, there is a situation in which acceptance may be changed into rejection. Suppose that other symptoms appear such that the corresponding rejection set is very improbable with respect to D_i. This would mean that D_i should also be rejected, but that D_j, for $j \neq i$, also remain rejected. We have, then, two alternatives opened: either say that the alternatives we were considering are not exhaustive, or disregard the new evidence or the pathognomonic symptom. In any case, I believe that, although the Bayesian solution does not seem too inadequate, the classical inference solution appears to be intuitively better.

In order to clarify further the differences between the two principles, let us consider the case where the physician needs to know the disease of the patient for reasons other than his treatment. For instance, let us suppose that the physician is testing a drug to see whether or not it is effective for the treatment of a particular disease. Then he really needs to know whether the patient has the disease or not, and he is not in a decision-theoretic framework. He has to eliminate all alternative diseases as real possibilities (even antecedently real). I believe that in this case, the frequency of the disease in the population, i.e., the a priori probability of the disease, is irrelevant, and, hence, one should not apply the Bayesian model.

Although likelihoods have been used in computer-assisted medical diagnosis, I have not seen the use of rejection sets, which, I think, emulate more faithfully the usual processes of statistical inference. I believe that a diagnostic inference system based on probabilities of rejection sets may be promising.

Part 2

Elements of Probability Theory

CHAPTER VI

Introduction to Probability Theory

1. Probability

This exposition is inspired by Nelson's approach in [71]. Thus, we only discuss finite probability spaces, the continuous case being dealt with via the use of infinitesimal (nonstandard) analysis.

Probability Theory is usually applied when we perform an experiment whose outcome is not known in advance. In order to use the theory, however, we must know which is the set of possible outcomes. This set of possible outcomes is formalized in the theory by a set Ω, called the *sample* or *probability space*. For examples of probability spaces, see Section 2 in the Preliminaries (Examples 1, 2, 3). We shall refer to these examples in the sequel.

We associate a weight $\text{pr}(\omega)$ to the possibility of each outcome, which is a real number between 0 and 1. The only condition we impose on pr is that the total probability be 1. That is

$$\sum_{\omega \in \Omega} \text{pr}(\omega) = 1.$$

If the coin of Example 1 is a fair coin, then we would assign $\text{pr}(H) = 1/2 = \text{pr}(T)$. If it is not a fair coin, however, the assignment might be different.

If the dice of Example 3 are fair, then we would assign

$$\text{pr}(\langle i, j \rangle) = \frac{1}{36}$$

for every pair, if we take Ω as our sample space. If we take Ω_1, however, the assignments are different. For instance

$$\text{pr}(2) = \frac{1}{36},$$
$$\text{pr}(3) = \frac{2}{36},$$
$$\vdots$$
$$\text{pr}(7) = \frac{6}{36}$$
$$\vdots$$
$$\text{pr}(12) = \frac{1}{36}$$

Once we have the distribution of probabilities, pr, defined on elements of Ω, we define a probability measure Pr over all *subsets* of Ω. As it is easily seen, for our finite spaces, Pr is completely determined by pr. We adopt the convention that the distribution of probabilities over Ω is written with lower-case letters and the corresponding probability measure over subsets of Ω begins with a capital letter. For instance, pr and Pr, pr_1 and Pr_1, or pr' and Pr'.

Thus, we define:

DEFINITION VI.1 (PROBABILITY SPACES).

(1) A *(finite) probability space* or *(finite) sample space* is a finite set Ω and a function pr, called a *distribution of probabilities*, or, simply, a *distribution*, on Ω, i.e., a pair $\langle \Omega, \text{pr} \rangle$, such that
   (a) pr is strictly positive, that is, $\text{pr}(\omega) > 0$ for every $\omega \in \Omega$.
   (b)
   $$\sum_{\omega \in \Omega} \text{pr}(\omega) = 1.$$
   We call $\langle \Omega, \text{pr} \rangle$ a *weak* probability space, if conditions (b) and
   (c) pr is positive, that is, $\text{pr}(\omega) \geq 0$ for every $\omega \in \Omega$, are satisfied.
(2) An event A is any subset of Ω, and the *probability* of A is
$$\text{Pr}\, A = \sum_{\omega \in A} \text{pr}(\omega).$$

Since pr determines Pr, in order to check whether something is a probability space, we need only to check (1), in the definition.

For instance, if E is the event $E = \{\langle 1,6 \rangle, \langle 6,1 \rangle, \langle 2,5 \rangle, \langle 5,2 \rangle, \langle 3,4 \rangle, \langle 4,3 \rangle\}$, in Example 3 with Ω as a sample space, then E is the event that the sum of the two dice equals seven, and its probability is

$$\text{Pr}\, E = \sum_{\omega \in E} \text{pr}(\omega) = \frac{6}{36}.$$

1. PROBABILITY

A weak probability space can always be replaced by a probability space that preserves most of its properties. If we have a space $\langle \Omega', \mathrm{pr}' \rangle$ where pr' is only positive, but satisfies (b), we can always replace it by a probability space $\langle \Omega, \mathrm{pr} \rangle$ where pr is strictly positive, as follows. Let

$$\Omega = \{\omega \in \Omega' \mid \mathrm{pr}'(\omega) > 0\} \text{ and } \mathrm{pr} = \mathrm{pr}' \restriction \Omega.$$

Then $\langle \Omega, \mathrm{pr} \rangle$ is a probability space with pr strictly positive. Any event $A' \subseteq \Omega'$ can be replaced by $A = A' \cap \Omega$. Then $\Pr' A' = \Pr A$. So that any experiment that is modeled by $\langle \Omega', \mathrm{pr}' \rangle$ can also be modeled by $\langle \Omega, \mathrm{pr} \rangle$. Thus, there is no loss of generality in requiring pr to be strictly positive, and it simplifies some of the developments.

We say that an event E *occurs*, if an outcome belonging to E occurs. For any two events E and F their union $E \cup F$ occurs if either E occurs or F occurs or both occur. The intersection of E and F, EF, occurs if and only if both E and F occur.

The empty or *impossible* event, $\emptyset$, is the event that never occurs, and the universal or *sure* event is Ω. If $EF = \emptyset$, we say that E and F are *mutually exclusive* or *incompatible*. Finally, for any two events E and F, their difference, $E - F$, occurs when E occurs and F does not occur. In particular, the complement (relative to Ω), of E, E^c, occurs when E does not occur.

From the definition of probability on events, we immediately obtain for any pair of events E and F

(1) $0 \leq \Pr E \leq 1$.
(2) $\Pr \Omega = 1$.
(3) If E and F are mutually exclusive, then

$$\Pr(E \cup F) = \Pr E + \Pr F.$$

(4) $\Pr E > 0$, for every $E \neq \emptyset$.

Kolmogorov's axioms for the finite case, which were introduced in Preliminaries, Section 2, have two main elements: an algebra of events, that is, a family $\mathcal{B}$ of subsets of Ω such that $\Omega \in \mathcal{B}$ and closed under finite unions and complements (and hence, under differences and intersections), and a function $\Pr$ defined on $\mathcal{B}$ satisfying (1), (2), and (3). Thus, a Kolmogorov probability space can be represented by a triple $\langle \Omega, \mathcal{B}, \Pr \rangle$ satisfying the properties indicated above. As for probability spaces, a space which satisfies (1), (2), (3), can be replaced by one satisfying (1), (2), (3), (4). This replacement is done as follows. Let A be the union of all $B \in \mathcal{B}$ such that $\Pr B = 0$. Then we have that A is an event and $\Pr A = 0$. We take $\Omega' = \Omega - A$, $\mathcal{B}'$ to be the algebra consisting of $\Omega' \cap C$ for $C \in \mathcal{B}$, and if $C \in \mathcal{B}$, we take $\Pr'(C \cap \Omega') = \Pr C$. Then $\langle \Omega', \mathcal{B}', \Pr' \rangle$ satisfies (1), (2), (3), and (4). We also have, that $\mathcal{B}' \subseteq \mathcal{B}$ and if $C \in \mathcal{B}'$, then $\Pr' C = \Pr C$. Thus, we define:

DEFINITION VI.2 (KOLMOGOROV PROBABILITY SPACES). A triple $\langle \Omega, \mathcal{B}, \Pr \rangle$, Ω finite, is a *(finite) Kolmogorov probability space* if it satisfies

(1) $\mathcal{B}$ is an algebra of subsets of Ω.

(2) $\Pr : \mathcal{B} \to [0,1]$.
(3) $\Pr \Omega = 1$.
(4) If E and F are mutually exclusive elements of $\mathcal{B}$, then
$$\Pr(E \cup F) = \Pr E + \Pr F.$$
(5) $\Pr E > 0$, for every $E \neq \emptyset$, $E \in \mathcal{B}$.

We say that $\langle \Omega, \mathcal{B}, \Pr \rangle$ is a *weak* Kolmogorov space, if it only satisfies (1), (2), (3), and (4).

It is clear that if we have a (weak) finite probability space, $\langle \Omega, \text{pr} \rangle$ in our sense, then $\langle \Omega, \mathcal{P}(\Omega), \Pr \rangle$ is a (weak) finite Kolmogorov probability space.

It is not difficult to prove (see Section 7, Corollary VI.9) that for any finite (weak) Kolmogorov probability space, $\langle \Omega, \mathcal{B}, \Pr \rangle$, one can find a finite (weak) probability space $\langle \Omega', \text{pr}' \rangle$ and a one-one onto function $f : \mathcal{B} \to \mathcal{P}(\Omega')$ such that $\Pr A = \Pr' f(A)$, for every $A \in \mathcal{B}$. Thus, our axioms are enough for dealing with finite probability spaces. Infinite probability spaces are another matter. We shall deal with them, however, by reducing them to finite (or, more properly, hyperfinite) spaces using nonstandard analysis.

The following properties are not difficult to prove, either from our original axioms or from Kolmogorov's axioms.

(1) $\Pr \emptyset = 0$.
(2) $\Pr E + \Pr E^c = 1$.
(3) $\Pr(E \cup F) = \Pr E + \Pr F - \Pr EF$.

2. Conditional probability

Suppose that we are in the case of Example 3 with sample space Ω and fair dice. Suppose that we observe that the first die is three. Then, given this information, what is the probability that the sum of the two dice is seven? To calculate this probability, we reduce our sample space to those outcomes where the first die gives three, that is, to $\{\langle 3,1\rangle, \langle 3,2\rangle, \langle 3,3\rangle, \langle 3,4\rangle, \langle 3,5\rangle, \langle 3,6\rangle\}$. Since each of the outcomes in this sample space had originally the same probability of occurring, they should still have equal probabilities. That is, given that the first die is a three, then the (conditional) probability of each of these outcomes is 1/6, while the (conditional) probability of the outcomes not in this set is 0. Hence, the desired probability is 1/6.

If we let E and F denote the event of the sum being seven and the first die being three, respectively, then the conditional probability just obtained is denoted
$$\Pr(E|F).$$
In order to obtain a general formula, we reason as follows. If the event F occurs, then E occurs just in case an outcome both in E and F occurs, that is, an outcome in EF occurs. Since we know that F has occurred, we must take F as our new sample space, and define a new distribution pr_F. We must set the

2. CONDITIONAL PROBABILITY

probability of points outside of F to 0, and, hence, $\sum_{\omega \in F} \mathrm{pr}_F(\omega)$ must be equal to 1. So we set, for $\omega \in F$

$$\mathrm{pr}_F(\omega) = \frac{\mathrm{pr}(\omega)}{\Pr F},$$

and then pr_F is a distribution of probabilities over F, i.e., $\langle F, \mathrm{pr}_F \rangle$ is a sample space. Thus, we get

$$\Pr(E|F) = \sum_{\omega \in EF} \mathrm{pr}_F(\omega) = \frac{\Pr EF}{\Pr F}.$$

Notice that this equation is only well defined when $\Pr F > 0$, that is, when $F \neq \emptyset$. The equation defines $\Pr(\ |F)$ as the probability determined by the distribution pr_F. Thus, $\Pr(\ |F)$ has all the properties described obove for a probability. We then define formally:

DEFINITION VI.3 (CONDITIONAL PROBABILITY). Let F be an event such that

$$\Pr F \neq 0.$$

Then we define, for any event E

$$\Pr(E|F) = \frac{\Pr EF}{\Pr F}.$$

The equation can also be written in the useful form

$$\Pr EF = \Pr(E|F) \Pr F.$$

Bayes formula. Let E and F be events. We have

$$E = EF \cup EF^c$$

with EF and EF^c mutually exclusive. Then

$$\begin{aligned} \Pr E &= \Pr EF + \Pr EF^c \\ &= \Pr(E|F) \Pr F + \Pr(E|F^c) \Pr F^c \\ &= \Pr(E|F) \Pr F + \Pr(E|F^c)(1 - \Pr F) \end{aligned}$$

This formula is what is called the *total probability formula*, which can be generalized as follows. Let $F_1, F_2, \ldots, F_n$ be pairwise mutually exclusive events with positive probability such that their union is the whole sample space (i.e., the sure event). Then for any event E

$$\Pr E = \sum_{j=1}^{n} \Pr(E|F_j) \Pr F_j.$$

With the same assumption about the sequence of F's, we can deduce for any i with $1 \leq i \leq n$

$$\Pr(F_i|E) = \frac{\Pr EF_i}{\Pr E}$$
$$= \frac{\Pr(E|F_i)\Pr F_i}{\sum_{j=1}^{n}\Pr(E|F_j)\Pr F_j}.$$

This is *Bayes formula*. In particular, we have

$$\Pr(F|E) = \frac{\Pr(E|F)\Pr F}{\Pr(E|F)\Pr F + \Pr(E|F^c)\Pr F^c}.$$

Example 4 is an example of the use of Bayes formula.

Independent events. We conclude this section with a discussion of independence as is defined inside probability theory. I shall introduce later another notion of independence that does not depend on the probability measure. The notion of independence introduced here may be called more precisely *probabilistic independence*.

Two events E and F are independent if the occurrence of one of them, say F, does not affect the probability of the other, i.e., E. That is, if

$$\Pr(E|F) = \Pr E.$$

This is equivalent to $\Pr EF = \Pr E \Pr F$. Since this last formula does not have the restriction that $\Pr F > 0$, we adopt it as a definition. The definition of independence can be extended to more than two events. Thus, we have:

DEFINITION VI.4 (PROBABILISTIC INDEPENDENCE). The events E_1, E_2, ..., E_n are said to be *(probabilistically) independent* if

$$\Pr E_{i_1} E_{i_2} \cdots E_{i_m} = \prod_{j=1}^{m} \Pr E_{i_j},$$

for every $i_1, i_2, \ldots, i_m$ between 1 and n, such that $i_p \neq i_q$ for $p \neq q$.

Example 5 shows that pairwise independent events might not be independent.

3. Random variables

Most frequently, when performing an experiment, we are interested in a property of the outcome and not on the outcome itself, although we may need the particular sample space, in order to determine the probabilities. For instance, in casting two dice as in Example 3, we would usually adopt the sample space Ω and not Ω_1, because the function pr is easily determined for it. We may only be interested, however, in knowing the sum of the numbers that appear on the dice, and not on the particular outcome in Ω. In order to deal with this situation we introduce random variables.

3. RANDOM VARIABLES

DEFINITION VI.5 (RANDOM VARIABLES). A *random variable* X is a function
$$X : \Omega \to \mathbb{R}.$$
We write, for any property of real numbers, $\varphi(x)$,
$$[\varphi(X)] = \{\omega \in \Omega \mid \varphi(X(\omega))\}.$$

In particular, for any set of real numbers S
$$[X \in S] = \{\omega \in \Omega \mid X(\omega) \in S\}$$
and for a real number r
$$[X = r] = \{\omega \in \Omega \mid X(\omega) = r\}.$$

We shall use for some examples the random variables X, Y, and Z introduced in Preliminaries, Section 2, for the case of the two dice (Example 3). Namely, X is the face occuring in the first die, Y is the face occurring in the second die, and Z, the sum of the two faces.

Let X be an arbitrary random variable. Once we have the probabilities $\Pr[X = i]$, we can usually forget about the sample space. In fact, it is only used for determining the following probability space:

DEFINITION VI.6 (PROBABILITY SPACE OF A RANDOM VARIABLE).

(1) The *range* of the random variable X, Λ_X, is defined by
$$\Lambda_X = \{X(\omega) \mid \omega \in \Omega\}.$$

(2) For $i \in \Lambda_X$, we define
$$\operatorname{pr}_X(i) = \Pr[X = i].$$
The function pr_X is called the *mass function* or *distribution* of X. We shall occasionally use the same symbol pr_X for the function defined on all of $\mathbb{R}$, by the same definition.

(3) The pair $\langle \Lambda_X, \operatorname{pr}_X \rangle$ is called the *probability space of X*. The probability determined by pr_X, namely, $\Pr_X$, is called the *probability distribution of X*.

Since Ω is finite, Λ_X is a finite set of real numbers. We clearly have that pr_X is strictly positive on Λ_X and
$$\sum_{i \in \Lambda_X} \operatorname{pr}_X(i) = 1.$$

Thus, pr_X is a distribution of probabilities over Λ_X, and, hence, $\langle \Lambda_X, \operatorname{pr}_X \rangle$ is a probability space, the space *induced* by X on $\mathbb{R}$. It is also clear that $\operatorname{pr}_X(x) = 0$, for $x \notin \Lambda_X$.

Any function $\operatorname{pr} : S \to \mathbb{R}$ such that S is a finite set of real numbers and
$$\sum_{x \in S} \operatorname{pr}(x) = 1$$

can be considered as the distribution of a random variable. We just take $\Omega = S$ and $X(x) = x$, for every $x \in S$. Then $\text{pr} = \text{pr}_X$.

DEFINITION VI.7 (EQUIVALENCE). We call two random variables X and Y (possibly defined on different sample spaces) *equivalent*, if X and Y have the same probability spaces, that is, if $\langle \Lambda_X, \text{pr}_X \rangle = \langle \Lambda_Y, \text{pr}_Y \rangle$.

Probability theory is concerned only with those properties of random variables that are shared by all equivalent random variables.

Let X be any random variable. Let X' be the identity restricted to Λ_X, i.e., $X'(x) = x$ for $x \in \Lambda_X$. Then X' is a random variable on $\langle \Lambda_X, \text{pr}_X \rangle$, $\Lambda_{X'} = \Lambda_X$, and $\text{pr}_{X'} = \text{pr}_X$. Thus, X' is equivalent to X. Therefore, in studying random variables, it is no loss of generality to assume that they are defined on $\langle \Lambda_X, \text{pr}_X \rangle$.

If X_1 and X_2 are random variables and a and b are real numbers, then we define

$$(aX_1 + bX_2)(\omega) = aX_1(\omega) + bX_2(\omega),$$
$$(X_1 X_2)(\omega) = X_1(\omega) X_2(\omega).$$

For instance, in the case of the two fair dice, with the definitions introduced above, we have $Z = X + Y$.

Next, we introduce the notion of independence for random variables:

DEFINITION VI.8 (INDEPENDENT RANDOM VARIABLES). We say that the random variables X and Y, defined over the same sample space, are *independent* if for any real numbers i and j

$$\Pr[X = i, Y = j] = \Pr[X = i] \Pr[X = j].$$

For instance, in the case of two fair dice, with X, Y and Z as before, X and Y are independent, but Z and X are not.

4. Expectation, variance and covariance

We now define some functions on random variables.

DEFINITION VI.9 (EXPECTATION AND VARIANCE).

(1) The *mean* or *expectation* of the random variable X is

$$\mathbf{E}\, X = \sum_{\omega \in \Omega} X(\omega) \, \text{pr}(\omega) = \sum_{x \in \mathbb{R}} x \, \text{pr}_X(x).$$

(2) The *variance* of X is

$$\text{Var}\, X = \mathbf{E}(X - \mathbf{E}\, X)^2 = \sum_{x \in \mathbb{R}} (x - \mathbf{E}\, X)^2 \, \text{pr}_X(x).$$

The square root of the variance of X is called the *standard deviation* of X.

4. EXPECTATION, VARIANCE AND COVARIANCE

(3) The *covariance* of the random variables X and Y is

$$\mathrm{Cov}(X,Y) = \mathbf{E}((X - \mathbf{E}\,X)(Y - \mathbf{E}\,Y)).$$

The *correlation coefficient of X and Y* is

$$\frac{\mathrm{Cov}(X,Y)}{\sqrt{\mathrm{Var}\,X}\sqrt{\mathrm{Var}\,Y}}.$$

The mean is the weighted average of the values of the random variable, weighted according to their probabilities. The variance is the weighted average of the squared differences to the mean.

We note that

$$\mathbf{E}\,X = \sum_{i \in \Lambda_X} i\,\mathrm{pr}_X(i).$$

The right hand side of this equation is also the expectation of the identity function on $\langle \Lambda_X, \mathrm{pr}_X \rangle$. Hence, the expectation of a random variable depends only on its probability space. The same is true for the variance and covariance.

We can easily get the following formulas

$$\mathbf{E}(aX_1 + bX_2) = a\,\mathbf{E}\,X_1 + b\,\mathbf{E}\,X_2,$$
$$\mathrm{Var}\,X = \mathbf{E}(X - \mathbf{E}\,X)^2$$
$$= \mathbf{E}\,X^2 - 2(\mathbf{E}\,X)^2 + (\mathbf{E}\,X)^2$$
$$= \mathbf{E}\,X^2 - (\mathbf{E}\,X)^2.$$
$$\mathrm{Var}\,aX = \mathbf{E}\,a^2 X^2 - (\mathbf{E}\,aX)^2$$
$$= a^2(\mathbf{E}\,X^2 - (\mathbf{E}\,X)^2)$$
$$= a^2\,\mathrm{Var}\,X,$$
$$\mathrm{Cov}(X,Y) = \mathbf{E}(X - \mathbf{E}\,X)(Y - \mathbf{E}\,Y)$$
$$= \mathbf{E}(XY - Y\,\mathbf{E}\,X - X\,\mathbf{E}\,Y + \mathbf{E}\,X\,\mathbf{E}\,Y)$$
$$= \mathbf{E}\,XY - \mathbf{E}\,Y\,\mathbf{E}\,X - \mathbf{E}\,X\,\mathbf{E}\,Y + \mathbf{E}\,X\,\mathbf{E}\,Y$$
$$= \mathbf{E}\,XY - \mathbf{E}\,X\,\mathbf{E}\,Y,$$

and

$$\mathrm{Var}(X + Y) = \mathbf{E}(X + Y - \mathbf{E}(X+Y))^2$$
$$= \mathbf{E}((X - \mathbf{E}\,X) + (Y - \mathbf{E}\,Y))^2$$
$$= \mathbf{E}((X - \mathbf{E}\,X)^2 + (Y - \mathbf{E}\,Y)^2 + 2(X - \mathbf{E}\,X)(Y - \mathbf{E}\,Y))$$
$$= \mathrm{Var}\,X + \mathrm{Var}\,Y + 2\,\mathrm{Cov}(X,Y).$$

If X and Y are independent, then we have

$$\mathbf{E}\,XY = \sum_{x,y \in \mathbb{R}} xy \Pr[X = x, Y = y]$$
$$= \sum_{x,y \in \mathbb{R}} xy \Pr[X = x] \Pr[Y = y]$$
$$= \sum_{x \in \mathbb{R}} x \Pr[X = x] \cdot \sum_{y \in \mathbb{R}} y \Pr[Y = y]$$
$$= \mathbf{E}\,X \cdot \mathbf{E}\,Y.$$

Thus, still assuming independence, we get

$$\mathrm{Cov}(XY) = \mathbf{E}\,XY - \mathbf{E}\,X\,\mathbf{E}\,Y$$
$$= \mathbf{E}\,X\,\mathbf{E}\,Y - \mathbf{E}\,X\,\mathbf{E}\,Y$$
$$= 0,$$

and

$$\mathrm{Var}(X + Y) = \mathrm{Var}\,X + \mathrm{Var}\,Y + 2\,\mathrm{Cov}(X, Y)$$
$$= \mathrm{Var}\,X + \mathrm{Var}\,Y.$$

Where it makes sense, the formulas can be extended by induction to n variables.

We mention here a few important types of random variables.

Constant random variables. These are variables X such that $X(\omega) = c$, for every $\omega \in \Omega$, where c is a real number. We clearly have $\Pr[X = c] = 1$ and $\Pr[X = r] = 0$ for any $r \neq c$. Also $\mathbf{E}\,X = c$ and $\mathrm{Var}\,X = 0$. We have that $\Lambda_X = \{c\}$ and $\mathrm{pr}_X(c) = 1$, so that we are justified in identifying the random variable X with the real number c.

Indicator or Bernoulli random variables. If we have any event E, the *indicator of E* is the function defined by

$$\chi_E(\omega) = \begin{cases} 1, & \text{if } \omega \in E, \\ 0, & \text{if } \omega \notin E. \end{cases}$$

For these random variables we have

$$\mathrm{pr}_{\chi_E}(r) = \begin{cases} \Pr E, & \text{if } r = 1, \\ \Pr E^c, & \text{if } r = 0, \\ 0, & \text{otherwise.} \end{cases}$$

In statistical practice, an outcome of an experiment or *trial* in E is called a *success* and one in E^c, a *failure*. So that we define *a Bernoulli random variable* as a variable that is 1, in case of success, and 0, in case of failure.

We also have:

$$\mathbf{E}\,\chi_E = 1 \cdot \Pr E + 0 \cdot \Pr E^c = \Pr E,$$
$$\begin{aligned}\operatorname{Var} \chi_E &= \mathbf{E}\,\chi_E^2 - (\mathbf{E}\,\chi_E)^2 \\ &= \Pr E - (\Pr E)^2 \\ &= \Pr E\,(1 - \Pr E) \\ &= \Pr E \, \Pr E^c\end{aligned}$$

Binomial random variables. Suppose that one performs n independent Bernoulli trials, each of which results in a "success" with probability p and in a "failure" with probability $1-p$. Let the random variable X measure the number of successes in n trials. Then X is called a binomial random variable having parameters $\langle n,p \rangle$. The distribution of X is given by

$$\operatorname{pr}_X(i) = \binom{n}{i} p^i (1-p)^{n-i}, \qquad i = 0, 1, \ldots, n,$$

where $\binom{n}{i} = n!/(n-i)!i!$ equals the number of sets (called, in this case, combinations) of i objects that can be chosen from n objects. This equation can be proved by first noticing that, since the trials are supposed to be independent, the probability of any particular sequence containing i successes and $n-i$ failures is $p^i(1-p)^{n-i}$. The equation follows since there are $\binom{n}{i}$ different sequences of the n outcomes with i successes and $n-i$ failures.

Note that, by the binomial theorem, the probabilities $\operatorname{pr}_X(i)$ sum to one, so that this is an adequate distribution.

The expectation and the variance of a binomial random variable X can be calculated as follows. Let X_i be the Bernoulli random variable that is one for success in the ith trial, and zero for failure. Then

$$X = \sum_{i=0}^{n} X_i.$$

Since the sequence $X_1, X_2, \ldots, X_n$ is assumed to be independent, by the formulas obtained above

$$\mathbf{E}\,X = \sum_{i=1}^{n} \mathbf{E}\,X_i = np,$$
$$\operatorname{Var} X = \sum_{i=1}^{n} \operatorname{Var} X_i = np(1-p).$$

Expectation of a function of a random variable. Suppose $f : \mathbb{R} \to \mathbb{R}$ and X a random variable on Ω. Then we can consider the random variable $f(X)$ defined by

$$f(X)(\omega) = f(X(\omega))$$

for every $\omega \in \Omega$. We shall also have occasion later to consider functions of several variables. That is, if $f : {}^n\mathbb{R} \to \mathbb{R}$ and $X_1, X_2, \ldots, X_n$ are random variables, over the same space Ω, then

$$f(X_1, X_2, \ldots, X_n)(\omega) = f(X_1(\omega), X_2(\omega), \ldots, X_n(\omega)),$$

for every $\omega \in \Omega$.

We now want to calculate $\mathbf{E}\, f(X)$. We have

$$\mathbf{E}\, f(X) = \sum_{x \in \mathbb{R}} x \Pr[f(X) = x].[0]$$

But, we have

$$x \Pr[f(X) = x] = \sum_{f(y)=x} f(y) \Pr[X = y].[0]$$

Hence

$$\mathbf{E}\, f(X) = \sum_{x \in \mathbb{R}} f(x) \Pr[X = x].[0]$$

Therefore, we also get

$$\mathbf{E}\, f(X) = \sum_{\omega \in \Omega} f(X(\omega)) \operatorname{pr}(\omega) = \sum_{x \in \Lambda_X} f(x) \operatorname{pr}_X(x).$$

5. Inequalities

We shall use the following inequalities.

THEOREM VI.1 (JENSEN'S INEQUALITY). *Let X be a random variable and $p \geq 1$. Then*

$$|\mathbf{E}\, X|^p \leq \mathbf{E}\, |X|^p.$$

In particular we have

$$|\mathbf{E}\, X| \leq \mathbf{E}\, |X|.$$

PROOF. This is obtained, because for $p \geq 1$ we have

$$\left|\sum_{\omega \in \Omega} X(\omega) \operatorname{pr}(\omega)\right|^p \leq \sum_{\omega \in \Omega} |X(\omega)|^p \operatorname{pr}(\omega).$$

□

THEOREM VI.2 (MARKOV'S INEQUALITY). *Let f be a positive real function (i.e., $f(x) \geq 0$ for every x) and let $\varepsilon > 0$. Then*

$$\Pr[f(X) \geq \varepsilon] \leq \frac{\mathbf{E}\, f(X)}{\varepsilon}.$$

PROOF. We have
$$\mathbf{E} f(X) = \sum_{\omega \in \Omega} f(X(\omega)) \operatorname{pr}(\omega)$$
$$\geq \sum_{\omega \in [f(X) \geq \varepsilon]} f(X(\omega)) \operatorname{pr}(\omega)$$
$$\geq \varepsilon \Pr[f(X) \geq \varepsilon],$$

so that
$$\Pr[f(x) \geq \varepsilon] \leq \frac{\mathbf{E} f(X)}{\varepsilon}.$$
□

THEOREM VI.3 (CHEBYSHEV'S INEQUALITY). *Let $p > 0$, $\varepsilon > 0$, and X be a random variable. Then*
$$\Pr[|X| \geq \varepsilon] \leq \frac{\mathbf{E}|X|^p}{\varepsilon^p}.$$

In particular, we have
$$\Pr[|X - \mathbf{E} X| \geq \varepsilon] \leq \frac{\operatorname{Var} X}{\varepsilon^2}.$$

PROOF. We have for $\varepsilon > 0$, $p > 0$ and $\omega \in \Omega$
$$|X(\omega)| \geq \varepsilon \iff |X(\omega)|^p \geq \varepsilon^p.$$

Hence
$$[|X| \geq \varepsilon] = [|X|^p \geq \varepsilon^p].$$

By Markov's inequality
$$\Pr[|X|^p \geq \varepsilon^p] \leq \frac{\mathbf{E}|X|^p}{\varepsilon^p}.$$

Thus, we obtain the theorem.

In order to obtain the last clause, we apply the theorem with $X - \mathbf{E} X$ instead of X and $p = 2$. □

6. Algebras of random variables

The set of all random variables on Ω is the set ${}^\Omega \mathbb{R}$. We know that this set is closed under linear combinations, i.e., for $a, b \in \mathbb{R}$ and $X, Y \in {}^\Omega \mathbb{R}$, then $aX + bY \in {}^\Omega \mathbb{R}$. That is, ${}^\Omega \mathbb{R}$ is a vector space. But also, $X, Y \in {}^\Omega \mathbb{R}$ imply $XY \in {}^\Omega \mathbb{R}$. Thus, ${}^\Omega \mathbb{R}$ is also an *algebra*, i.e., a set closed under linear combinations and products.

We must consider other algebras of random variables. So we define

DEFINITION VI.10 (ALGEBRAS).

(1) By an *algebra of random variables* $\mathcal{V}$ we mean a subset of ${}^\Omega \mathbb{R}$ satisfying the following properties:
   (a) If $a, b \in \mathbb{R}$, and $X, Y \in \mathcal{V}$, then $aX + bY \in \mathcal{V}$.
   (b) If X and $Y \in \mathcal{V}$, then $XY \in \mathcal{V}$.

(c) If X is constant (i.e., $X(\omega) = c$ for a $c \in \mathbb{R}$ and all $\omega \in \Omega$), then $X \in \mathcal{V}$.

(2) An *atom* of a subset of $^\Omega\mathbb{R}$, $\mathcal{V}$, in particular an algebra, is a maximal event A such that each random variable in $\mathcal{V}$ is constant in A. That is, for every $X \in \mathcal{V}$, there is a real number c such that $X(\omega) = c$ for all $\omega \in A$, and if $\omega \notin A$, then there is an $X \in \mathcal{V}$ such that $X(\omega) \neq X(\omega')$, for any $\omega' \in A$. We denote by at($\mathcal{V}$) the set of all atoms in $\mathcal{V}$.

PROPOSITION VI.4. *Let $\mathcal{V}$ be a set of random variables, subset of $^\Omega\mathbb{R}$. Then Ω is partitioned into atoms, that is, Ω is the union of the atoms and they are pairwise disjoint.*

PROOF. We first prove that for any $\omega \in \Omega$, there is an atom A such that $\omega \in A$. Let $\Omega = \{\omega_1, \omega_2, \ldots, \omega_n\}$. Define the sequence A_i of subsets of Ω by recursion as follows

(1) $A_0 = \{\omega\}$.
(2) Suppose that A_i is defined. Then

$$A_{i+1} = \begin{cases} A_i \cup \{\omega_i\}, & \text{if for every } X \in \mathcal{V}, X(\omega_i) = X(\eta) \text{ for each } \eta \in A_i, \\ A_i, & \text{otherwise.} \end{cases}$$

Then take $A = A_n$. It is clear that A is an atom and that $\omega \in A$.

We prove, now, that if A and B are atoms such that $A \cap B \neq \emptyset$, then $A = B$. Suppose that A and B are atoms with $A \cap B \neq \emptyset$, and let $\omega \in A \cap B$. Let $\omega' \in A$ and $\omega'' \in B$ and let X be any random variable in $\mathcal{V}$. Since $\omega, \omega'' \in B$, we have that $X(\omega) = X(\omega'')$. Since $\omega, \omega' \in A$ we also have $X(\omega) = X(\omega')$. Thus, for any element ω' of A, any element ω'' of B and any $X \in \mathcal{V}$, we have $X(\omega') = X(\omega'')$. Thus, since A and B are atoms, $A = B$. □

The trivial algebra of random variables consists of all the constant random variables. There is only one atom for this algebra, namely Ω.

We define:

DEFINITION VI.11 (GENERATED ALGEBRA). *The algebra $\mathcal{V}$ generated by the random variables $X_1, X_2, \ldots, X_n$ on Ω is the smallest algebra containing these variables and all constant random variables.*

As an example consider the case of two fair dice with the random variable Z defined above, i.e., $Z(\langle i, j \rangle) = i + j$. Consider the algebra $\mathcal{V}$ generated by Z. This algebra contains all constants and linear combinations of these constants with Z. There are 11 atoms, which are:

$$A_2 = \{\langle 1, 1 \rangle\},$$
$$A_3 = \{\langle 1, 2 \rangle, \langle 2, 1 \rangle\},$$
$$A_4 = \{\langle 1, 3 \rangle \langle 3, 1 \rangle, \langle 2, 2 \rangle\},$$
$$A_5 = \{\langle 1, 4 \rangle, \langle 4, 1 \rangle, \langle 2, 3 \rangle, \langle 3, 2 \rangle\},$$
$$A_6 = \{\langle 1, 5 \rangle, \langle 5, 1 \rangle, \langle 2, 4 \rangle, \langle 4, 2 \rangle, \langle 3, 3 \rangle\},$$

6. ALGEBRAS OF RANDOM VARIABLES

$$A_7 = \{\langle 1,6\rangle, \langle 6,1\rangle, \langle 2,5\rangle, \langle 5,2\rangle, \langle 3,4\rangle, \langle 4,3\rangle\},$$
$$A_8 = \{\langle 2,6\rangle, \langle 6,2\rangle, \langle 3,5\rangle, \langle 5,3\rangle, \langle 4,4\rangle\},$$
$$A_9 = \{\langle 3,6\rangle, \langle 6,3\rangle, \langle 4,5\rangle, \langle 5,4\rangle\},$$
$$A_{10} = \{\langle 4,6\rangle, \langle 6,4\rangle, \langle 5,5\rangle\},$$
$$A_{11} = \{\langle 5,6\rangle \langle 6,5\rangle\},$$
$$A_{12} = \{\langle 6,6\rangle\}.$$

The subscript of A denotes the value of Z at the elements of the atom.

If A is an atom of $\mathcal{V}$ and $X \in \mathcal{V}$, we denote by $X(A)$ the value of X at the elements of the atom A. Thus, in the example above, $Z(A_i) = i$. Thus, we see that in the case that the algebra $\mathcal{V}$ is generated by one random variable X, its atoms are the subsets of Ω of the form $[X = i]$, for the numbers $i \in \Lambda_X$.

Algebras of random variables have the following properties:

THEOREM VI.5. *Let $\mathcal{V}$ be an algebra of random variables. Then*

(1) *The indicator of any atom is in $\mathcal{V}$.*
(2) *If $\mathcal{W}$ is any subset of $^\Omega\mathbb{R}$, the algebra generated by $\mathcal{W}$, $\mathcal{V}$, is the set of all linear combinations of indicators of atoms of $\mathcal{W}$, i.e., $X \in \mathcal{V}$ if and only if there are indicators of atoms $Y_1, Y_2, \ldots, Y_m$ and real numbers $a_1, a_2, \ldots, a_m$ such that*

$$X = a_1 Y_1 + a_2 Y_2 + \cdots + a_m Y_m.$$

Moreover, the atoms of $\mathcal{V}$ are the same as the atoms of $\mathcal{W}$.

(3) *$\mathcal{V}$ consists of all random variables which are constant on the atoms of $\mathcal{V}$.*
(4) *Let f be an arbitrary function from $^n\mathbb{R}$ into $\mathbb{R}$ and $X_1, X_2, \ldots, X_n \in \mathcal{V}$. Then $f(X_1, X_2, \ldots, X_n) \in \mathcal{V}$.*

PROOF OF (1). Let A be an atom. If $A = \Omega$ then the indicator of A is the constant function of value 1, and hence, it is in $\mathcal{V}$.

Suppose that there is an $\omega \notin A$. Then, by definition of atom, there is an $X \in \mathcal{V}$ such that $X(A) \neq X(\omega)$. Let

$$X_\omega = \frac{X - X(\omega)}{X(A) - X(\omega)}.$$

Then $X_\omega \in \mathcal{V}$, $X_\omega(A) = 1$, and $X_\omega(\omega) = 0$. We have that

$$Y = \prod_{\omega \notin A} X_\omega$$

is in $\mathcal{V}$. If $\eta \in A$, then $X_\omega(\eta) = 1$ for every $\omega \notin A$. Hence, $Y(\eta) = 1$. On the other hand, if $\eta \notin A$, then $X_\eta(\eta) = 0$. Hence, $Y(\eta) = 0$. Thus, Y is χ_A, the indicator of A. □

PROOF OF (2). First, let $\mathcal{U}$ be the set of all linear combinations of indicators of the atoms of $\mathcal{W}$. We shall prove that $\mathcal{U}$ is an algebra. It is clear that $\mathcal{U}$ is

closed under linear combinations. Suppose, now, that $X, Y \in \mathcal{U}$. Then
$$X = a_1\chi_{A_1} + a_2\chi_{A_2} + \cdots + a_n\chi_{A_n},$$
$$Y = b_1\chi_{B_1} + b_2\chi_{B_2} + \cdots + b_m\chi_{B_m},$$
where $A_1, \ldots, A_n$ and $B_1, \ldots, B_m$ are atoms of $\mathcal{W}$, and, hence
$$X \cdot Y = \sum_{i=1}^{n}\sum_{j=1}^{m} a_i b_j \chi_{A_i}\chi_{B_j}$$
$$= \sum_{i=1}^{n}\sum_{j=1}^{m} a_i b_j \chi_{A_i B_j}.$$

But, since A_i and B_j are atoms, $A_i B_j = \emptyset$ or $A_i B_j = A_i$. Thus, $X \cdot Y$ is a linear combination of indicators of atoms, and, hence, it belongs to $\mathcal{U}$. Therefore, we have proved that $\mathcal{U}$ is an algebra.

Let, now, $Z \in \mathcal{W}$ and let $A_1, A_2, \ldots$, be the atoms of $\mathcal{W}$, and $Z(\omega) = a_i$ for $\omega \in A_i$, $i = 1, \ldots, n$. Then
$$Z = a_1\chi_{A_1} + \cdots + a_n\chi_{A_n},$$
and, hence, $Z \in \mathcal{U}$. Thus, we have proved that $\mathcal{V} \subseteq \mathcal{U}$.

By 1, $\mathcal{V}$ contains all the indicators of atoms in $\mathcal{V}$. Since $\mathcal{W} \subseteq \mathcal{V}$, all the atoms in $\mathcal{W}$ are atoms in $\mathcal{V}$. Thus, $\mathcal{V}$ contains all indicators of atoms in $\mathcal{W}$. Therefore, $\mathcal{U} \subseteq \mathcal{V}$, and, hence, $\mathcal{U} = \mathcal{V}$. □

PROOF OF (3). Now, let X be a function on Ω which is constant on the atoms of $\mathcal{V}$, let $A_1, A_2, \ldots, A_m$ be the atoms of $\mathcal{V}$, and let $X(A_i) = a_i$. Then
$$X = a_1\chi_{A_1} + a_2\chi_{A_2} + \cdots + a_m\chi_{A_m}.$$
Thus, $X \in \mathcal{V}$. □

PROOF OF (4). Let f be a real function of n variables, and $X_1, X_2, \ldots, X_n \in \mathcal{V}$. Then, since $X_1, \ldots, X_n$ are constant on the atoms, $f(X_1, \ldots, X_n)$ is also constant on the atoms. Hence, by (3), we get (4). □

We have the following corollary:

COROLLARY VI.6. *Let $\mathcal{W}$ be a subset of $^{\Omega}\mathbb{R}$. Then if $\mathcal{V}$ is the algebra of random variables over Ω generated by $\mathcal{W}$, then $\mathcal{V}$ can be considered as the algebra of all random variables over* $\mathrm{at}(\mathcal{W})$, $^{\mathrm{at}(\mathcal{W})}\mathbb{R}$, *by identifying the random variable $X \in \mathcal{V}$ with $X_{\mathcal{V}}$ on* $\mathrm{at}(\mathcal{V})$ *defined by*
$$X_{\mathcal{V}}(A) = X(A),$$
where A is an atom of $\mathcal{V}$. That is, the function $F : \mathcal{V} \to {}^{\mathrm{at}(\mathcal{W})}\mathbb{R}$, defined by $F(X) = X_{\mathcal{V}}$, is one-one and onto and satisfies:
  if $X, Y \in \mathcal{V}$ and a, b are real numbers then
$$F(aX + bY) = aF(X) + bF(Y) \text{ and } F(XY) = F(X)F(Y).$$
Moreover, we have

(1) *If $X \in \mathcal{V}$ is an indicator, then $X_\mathcal{V}$ is also an indicator.*
(2) *If $\langle \Omega, \mathrm{pr} \rangle$ is a probability space, then $\langle \mathrm{at}(\mathcal{W}), \mathrm{pr}' \rangle$, where $\mathrm{pr}'(A) = \Pr A$, for any $A \in \mathrm{at}(\mathcal{W})$, is also a probability space (with expectation $\mathbf{E}'_\mathcal{V}$) satisfying*

$$\mathbf{E}\, X = \mathbf{E}'_\mathcal{V}\, X_\mathcal{V}$$

for any $X \in \mathcal{V}$.

PROOF. We must prove that the function $F(X) = X_\mathcal{V}$, defined on $\mathcal{V}$, is one-one and onto $^{\mathrm{at}(\mathcal{W})}\mathbb{R}$, and that it preserves the operations of the algebra. Suppose that $F(X) = F(Y)$. Since, by the previous theorem, the atoms of $\mathcal{V}$ are the same as the atoms of $\mathcal{W}$, $X(A) = Y(A)$ for any atom A of $\mathcal{V}$. Since for any $\omega \in \Omega$, $\omega \in A_\omega$, for a certain atom A_ω, we have that $X = Y$. Thus, F is one-one.

Let now Z be an arbitrary random variable on $^{\mathrm{at}(\mathcal{W})}\mathbb{R}$. Define X on Ω by $X(\omega) = Z(A_\omega)$, where A_ω is the atom to which ω belongs. It is clear that $X \in \mathcal{V}$ and $F(X) = Z$. Thus F is onto.

It is also clear that F preserves the operations of the algebra $\mathcal{V}$.

Proof of (1). It is clear from the definition that an indicator, χ_C, is sent into the indicator χ_D, where $D = \{A \in \mathrm{at}(\mathcal{V}) \mid A \cap C \neq \emptyset\}$. Thus, we have (1).

Proof of (2). It easy to show that $\langle \mathrm{at}(\mathcal{W}), \mathrm{pr}' \rangle) = \langle \mathrm{at}(\mathcal{V}), \mathrm{pr}' \rangle$ is a probability space. We also have for $X \in \mathcal{V}$

$$\begin{aligned}\mathbf{E}\, X &= \sum_{\omega \in \Omega} X(\omega) \mathrm{pr}(\omega) \\ &= \sum_{A \in \mathrm{at}(\mathcal{V})} \sum_{\omega \in A} X(\omega) \mathrm{pr}(\omega) \\ &= \sum_{A \in \mathrm{at}(\mathcal{V})} X_\mathcal{V}(A) \sum_{\omega \in A} \mathrm{pr}(\omega) \\ &= \sum_{A \in \mathrm{at}(\mathcal{V})} X_\mathcal{V}(A) \Pr A \\ &= \sum_{A \in \mathrm{at}(\mathcal{V})} X_\mathcal{V}(A) \mathrm{pr}'(A) \\ &= \mathbf{E}'_\mathcal{V}\, X_\mathcal{V}. \end{aligned}$$

□

7. Algebras of events

The family of all events $\mathcal{P}(\Omega)$ is closed under unions, intersections and complements and it contains Ω. Thus, it is an algebra of events. We define:

DEFINITION VI.12 (ALGEBRAS OF EVENTS). A nonempty subset of $\mathcal{P}(\Omega)$, $\mathcal{B}$, satisfying:
(1) If A and $B \in \mathcal{B}$, then $A \cup B \in \mathcal{A}$,
(2) If $A \in \mathcal{B}$, then $A^c \in \mathcal{B}$,

is called an *algebra of events*.

An element $A \in \mathcal{B}$ such that $A \neq \emptyset$ and there is no $B \in \mathcal{B}$ such that $B \neq \emptyset$ and $B \subset A$, is called an *atom* of $\mathcal{B}$.

Notice that we have two notions of atoms: atom of an algebra of random variables and atom of an algebra of events. The duality between these two notions of algebras, implies, as we shall see later (Theorem VI.8), that, in a sense, the two notions of atoms coincide. As an example of atoms of algebra of events, in the algebra $\mathcal{P}\Omega$, the atoms are the singletons. The trivial algebra of events is the algebra of two elements, $\{\Omega, \emptyset\}$, whose only atom is Ω.

Let $\mathcal{B}$ be an algebra of events. Since

$$AB = (A^c \cup B^c)^c$$

and

$$A - B = AB^c$$

$\mathcal{B}$ is closed under intersections and differences, i.e., $A, B \in \mathcal{B}$ imply $AB, A - B \in \mathcal{B}$. Also $\Omega = A \cup A^c$; thus, $\Omega \in \mathcal{B}$.

We also have the following property:

PROPOSITION VI.7. *Let $\mathcal{B}$ be an algebra of events. Then*
1. *Suppose that $A = \bigcup_{i=1}^m A_i$, where $A_i \in \mathcal{B}$, for $i = 1, \ldots, m$. Then there are pairwise disjoint $A'_1 \subseteq A_1, \ldots, A'_m \subseteq A_m$ in $\mathcal{B}$, such that $A = \bigcup_{i=1}^m A'_i$.*
2. *For every $A_1, \ldots, A_m \in \mathcal{B}$ and $a_1, \ldots, a_m \in \mathbb{R}$, there are pairwise disjoint $B_1, \ldots, B_k$ and $b_1, \ldots, b_k \in \mathbb{R}$ such that*

$$\bigcup_{i=1}^m A_i = \bigcup_{i=1}^k B_i$$

and

$$a_1 \chi_{A_1} + \cdots + a_m \chi_{A_m} = b_1 \chi_{B_1} + \cdots + b_k \chi_{B_k}.$$

3. *If $\chi_A = a_1 \chi_{A_1} + \cdots + a_m \chi_{A_m}$, where $a_i \neq 0$, and $A_i \cap A_j = \emptyset$, for $i, j = 1, \ldots, m$, $i \neq j$, then $A = A_1 \cup \cdots \cup A_m$.*

PROOF OF (1). We just define by recursion the A'_i as follows:
1. $A'_1 = A_1$.
2. Suppose that A'_j is defined for $j = 1, \ldots, i$. Then $A'_{i+1} = A_{i+1} - \bigcup_{j=1}^i A'_j$.

□

PROOF OF (2). The proof is by induction on m. Let $A_1, \ldots, A_{m+1}$, and $a_1, \ldots, a_{m+1}$ be given, and suppose that $B_1, \ldots, B_k$ with $b_1, \ldots, b_k$ work for $A_1, \ldots, A_m$ with $a_1, \ldots, a_m$. Let $A = \bigcup_{i=1}^m A_i$, and let $C_i = (A - A_{m+1}) \cap B_i$, $D_i = A \cap A_{m+1} \cap B_i$, $E = A_{m+1} - A$, $c_i = b_i$, $d_i = b_i + a_{m+1}$ and $e = a_{m+1}$, for $i = 1, \ldots, k$. Then, we have

$$\bigcup_{i=1}^{m+1} A_i = \bigcup_{i=1}^k C_i \cup \bigcup_{i=1}^k D_i \cup E$$

7. ALGEBRAS OF EVENTS

and

$$a_1\chi_{A_1} + \cdots + a_{m+1}\chi_{A_{m+1}} = c_1\chi_{C_1} + \cdots + c_m\chi_{C_m} + d_1\chi_{D_1} + \cdots + d_m\chi_{D_m} + e\chi_E.$$

□

PROOF OF (3). Let

$$\chi_A = a_1\chi_{A_1} + \cdots + a_m\chi_{A_m}.$$

Let $x \in A$. Then $\chi_A(x) = 1$. Since the A_i's are pairwise disjoint, there is at most one A_i such that $x \in A_i$. Thus

$$a_1\chi_{A_1}(x) + \cdots + a_m\chi_{A_m}(x) = a_i\chi_{A_i}(x).$$

Hence, $a_i\chi_{A_i}(x) = 1$, and so $\chi_{A_i}(x) = 1$. Thus, $x \in A_i$ and

$$x \in \bigcup_{i=1}^{m} A_i.$$

Assume, now, that $x \in \bigcup_{i=1}^{m} A_i$. Then there is exactly one $i = 1, \ldots, m$ such that $x \in A_i$. Thus

$$a_1\chi_{A_1}(x) + \cdots + a_m\chi_{A_m}(x) = a_i\chi_{A_i}(x) \neq 0.$$

Hence, $\chi_A(x) \neq 0$, and so $x \in A$. □

We have the following relationship between an algebra of random variables and an algebra of events. We first introduce a definition. Notice, for this definition, that, given any partition of Ω, $A_1, A_2, \ldots, A_n$, the family consisting of all the unions of members of the partition is an algebra of events. Recall, also, that the atoms of an algebra of random variables form a partition of Ω.

DEFINITION VI.13.

(1) Let $\mathcal{V}$ be an algebra of random variables over Ω. We define $\mathcal{V}_e$ to be the algebra of events in Ω consisting of unions of the atoms of $\mathcal{V}$.
(2) Let $\mathcal{B}$ be an algebra of events in Ω. We define $\mathcal{B}_{rv}$ to be the algebra of random variables generated by the indicators of sets in $\mathcal{B}$.

If $\mathcal{B}$ is the trivial algebra of events $\{\Omega, \emptyset\}$, then $\mathcal{B}_{rv}$ is the trivial algebra containing the constant random variables. On the other hand, if $\mathcal{V}$ is the trivial algebra of constant random variables, then $\mathcal{V}_e$ is the trivial algebra of events, $\{\Omega, \emptyset\}$.

THEOREM VI.8 (DUALITY).

(1) Let $\mathcal{V}$ be an algebra of random variables over Ω. Then $\mathcal{V}_e$ is an algebra of events and $(\mathcal{V}_e)_{rv} = \mathcal{V}$.
Therefore, the atoms of $\mathcal{V}$ (as an algebra of random variables) are the same as the atoms of $\mathcal{V}_e$ (as an algebra of events).

(2) Let $\mathcal{B}$ be an algebra of events in Ω. Then $(\mathcal{B}_{rv})_e = \mathcal{B}$.
Therefore, the atoms of $\mathcal{B}$, as an algebra of events, are the same as the atoms of $\mathcal{B}_{rv}$ (as an algebra of random variables) and, hence, there are atoms $A_1, \ldots, A_n$ of $\mathcal{B}$ such that every element of $\mathcal{B}$ is a union of some of these atoms.

PROOF OF (1). The fact that $\mathcal{V}_e$ is an algebra of events is deduced from the easy to prove statement that the unions of elements of a partition of Ω form an algebra of events.

We have that $(\mathcal{V}_e)_{rv}$ contains all indicators of the atoms of $\mathcal{V}$. Then, by Corollary VI.6, $\mathcal{V} \subseteq (\mathcal{V}_e)_{rv}$. On the other hand, if $C \in \mathcal{V}_e$ with $C = A_1 \cup \cdots \cup A_m$, where $A_1, A_2, \ldots, A_m$ are distinct (and hence pairwise disjoint) atoms, then

$$\chi_C = \chi_{A_1} + \chi_{A_2} + \cdots + \chi_{A_m}$$

and, thus, $\chi_C \in \mathcal{V}$. Thus, $(\mathcal{V}_e)_{rv} \subseteq \mathcal{V}$. □

PROOF OF (2). Suppose first that $A \in \mathcal{B}$. Then $\chi_A \in \mathcal{B}_{rv}$. Thus

$$\chi_A = a_1 \chi_{A_1} + \cdots + a_m \chi_{A_m},$$

where $A_1, \ldots, A_m$ are distinct atoms of $\mathcal{B}_{rv}$. By Proposition VI.7, (3), $A = A_1 \cup \cdots \cup A_m$ and, hence, $A \in (\mathcal{B}_{rv})_e$. Thus, $\mathcal{B} \subseteq (\mathcal{B}_{rv})_e$.

Suppose, now, that A is an atom of $\mathcal{B}_{rv}$. Then, by Theorem VI.5, $\chi_A \in \mathcal{B}_{rv}$. Thus,

$$\chi_A = a_1 \chi_{A_1} + \cdots + a_m \chi_{A_m},$$

where $A_1, \ldots, A_m \in \mathcal{B}$. Then, by Proposition VI.7, (2), there are $B_1, \ldots, B_k \in \mathcal{B}$, pairwise disjoint, and $b_1, \ldots, b_k$ (which clearly can be taken $\neq 0$) such that

$$\chi_A = b_1 \chi_{B_1} + \cdots + b_k \chi_{B_k}.$$

By Proposition VI.7, (3), $A = B_1 \cup \cdots \cup B_k$, and, hence, $A \in \mathcal{B}$.
Since every atom of $\mathcal{B}_{rv}$ is in $\mathcal{B}$, $(\mathcal{B}_{rv})_e \subseteq \mathcal{B}$. □

We now prove that every finite Kolmogorov probability space can be replaced by a finite probability space.

COROLLARY VI.9. *For every (weak) Kolmogorov probability space $\langle \Omega_1, \mathcal{B}, \Pr_1 \rangle$, there is a (weak) probability space $\langle \Omega, \text{pr} \rangle$ and a function F such that*

(1) *F is a one-one function of $\mathcal{B}$ onto $\mathcal{P}\Omega$.*
(2) *For every $A \in \mathcal{B}$, we have*

$$\Pr_1 A = \Pr F(A).$$

PROOF. By the previous theorem, there are atoms $A_1, \ldots, A_n \in \mathcal{B}$ such that every element of $\mathcal{B}$ is a union of these atoms. Let $\Omega = \{A_1, \ldots, A_n\}$ and define $\text{pr}(A) = \Pr_1 A$, for $A \in \Omega$. Then $\langle \Omega, \text{pr} \rangle$ is a (weak) probability space.

Let, now, $A \in \mathcal{B}$. Then $A = B_1 \cup \cdots \cup B_m$, where $B_1, \ldots, B_m \in \Omega$. The B's are pairwise disjoint. Let $F(A) = \{B_1, \ldots, B_m\}$. Thus

$$\operatorname{Pr}_1 A = \sum_{i=1}^{m} \operatorname{Pr}_1 B_i$$

$$= \sum_{i=1}^{m} \operatorname{pr}(B_i)$$

$$= \operatorname{Pr} F(B).$$

□

We have the following theorem that extends the duality between algebras of events and algebras of random variables:

THEOREM VI.10. *Let $\mathcal{V}$ and $\mathcal{W}$ be algebras of random variables. Then the following conditions are equivalent:*

(1) $\mathcal{W} \subseteq \mathcal{V}$.
(2) *Every atom in $\mathcal{W}$ is the union of atoms in $\mathcal{V}$.*
(3) $\mathcal{W}_e \subseteq \mathcal{V}_e$.

PROOF. Suppose that $\mathcal{W} \subseteq \mathcal{V}$ and let A be an atom of $\mathcal{V}$. Every random variable in $\mathcal{W}$ is constant on A, and, thus, there is an atom of $\mathcal{W}$, say B, such that $A \subseteq B$. So, if A is an atom in $\mathcal{V}$, and B an atom in $\mathcal{W}$, we have that $A \cap B = \emptyset$ or $A \cap B = A$.

We then get (2), i.e.

$$B = \bigcup_{\substack{A \in \operatorname{at}(\mathcal{V}) \\ A \subseteq B}} A \cap B.$$

Suppose, now, (2). We have that every element if $\mathcal{W}_e$ is the union of atoms in $\mathcal{W}$, and, hence, the union of atoms in $\mathcal{V}$. Thus, we get (3).

Finally, suppose (3). We have that $\mathcal{W} = (\mathcal{W}_e)_{rv}$ and $\mathcal{V} = (\mathcal{V}_e)_{rv}$. Since all the generators of $(\mathcal{W}_e)_{rv}$ are generators of $(\mathcal{V}_e)_{rv}$, we get (1). □

8. Conditional distributions and expectations

Let $\mathcal{V}$ be an algebra of random variables and A an atom of $\mathcal{V}$. As in Section 2, we can take A as a probability space with respect to pr_A defined by

$$\operatorname{pr}_A(\omega) = \frac{1}{\operatorname{Pr} A} \operatorname{pr}(\omega)$$

for all $\omega \in A$. It is clear that $\langle A, \operatorname{pr}_A \rangle$ is a sample space, that is

$$\sum_{\omega \in A} \operatorname{pr}_A(\omega) = 1.$$

The conditional expectation of a random variable X with respect to $\mathcal{V}$ (or to $\mathcal{V}_e$) is a random variable $\mathbf{E}_{\mathcal{V}} X$ that at each $\omega \in \Omega$ gives the expectation of X if the atom A_ω, such that $\omega \in A_\omega$, were to occur. That is, we define

DEFINITION VI.14 (CONDITIONAL EXPECTATION). Let $\mathcal{V}$ be an algebra of random variables over Ω, and X an arbitrary random variable over Ω. For $\omega \in \Omega$, we define

$$\mathbf{E}_{\mathcal{V}} X(\omega) = \mathbf{E}(X|\mathcal{V})(\omega) = \frac{1}{\Pr A_\omega} \sum_{\eta \in A_\omega} X(\eta) \pr(\eta) = \sum_{\eta \in A_\omega} X(\eta) \pr_{A_\omega}(\eta),$$

where A_ω is the atom containing ω.

In case $\mathcal{V}$ is the algebra of random variables generated by the random variables $X_1, X_2, \ldots, X_m$, we also write

$$\mathbf{E}_{\mathcal{V}} X = \mathbf{E}(X|X_1, X_2, \ldots, X_m).$$

When $\mathcal{B}$ is an algebra of events, we write $\mathbf{E}_{\mathcal{B}}$ for $\mathbf{E}_{\mathcal{B}_{rv}}$.

We have that $\mathbf{E}_{\mathcal{V}} X \in \mathcal{V}$, and $\mathbf{E}_{\mathcal{V}} X$ on each atom, A, is the expectation of X with respect to $\pr_A$. Using the convention indicated above that if A is an atom, then, for any $X \in \mathcal{V}$, $X(A) = X(\omega)$, for $\omega \in A$, we can write

$$\mathbf{E}_{\mathcal{V}} X(A) = \sum_{\eta \in A} X(\eta) \pr_A(\eta).$$

As an example, consider the trivial algebra, $\mathcal{T}$, of the constant random variables; then, since its only atom is Ω, it is easy to check that $\mathbf{E}_{\mathcal{T}} = \mathbf{E}$, i.e., $\mathbf{E}_{\mathcal{T}} X$ is the constant random variable with value $\mathbf{E} X$.

From the fact that for each atom A, $\mathbf{E}_{\mathcal{V}}$ is an expectation, we can easily prove the following properties.

(1) If $Y \in \mathcal{V}$, then $\mathbf{E}_{\mathcal{V}} Y = Y$.
(2) The conditional expectation is $\mathcal{V}$-linear, i.e., for $Y_1, Y_2 \in \mathcal{V}$, and X_1, X_2 arbitrary random variables, we have

$$\mathbf{E}_{\mathcal{V}}(Y_1 X_1 + Y_2 X_2) = Y_1 \mathbf{E}_{\mathcal{V}} X_1 + Y_2 \mathbf{E}_{\mathcal{V}} X_2.$$

Thus, the element of $\mathcal{V}$ act as constants for $\mathbf{E}_{\mathcal{V}}$.

(3) If $\mathcal{W}$ is an algebra of random variables, such that $\mathcal{W} \subseteq \mathcal{V}$, then $\mathbf{E}_{\mathcal{W}} \mathbf{E}_{\mathcal{V}} = \mathbf{E}_{\mathcal{W}}$. In particular, since the trivial algebra $\mathcal{T}$ is a subset of any algebra, we have $\mathbf{E} \mathbf{E}_{\mathcal{V}} = \mathbf{E}$, that is, for any random variable X, $\mathbf{E} \mathbf{E}_{\mathcal{V}} X = \mathbf{E} X$.

The proof of (3), depends on Theorem VI.10, (2).

EXAMPLE VI.1. We continue with the example of the two fair dice, and the random variables

$$X(\langle i,j \rangle) = i, \qquad Y(\langle i,j \rangle) = j, \qquad Z(\langle i,j \rangle) = i + j,$$

for every $\langle i,j \rangle \in \Omega$. Let $\mathcal{V}$ be the algebra generated by Z. Let $\omega = \langle 3, 4 \rangle$. Then ω is in the atom

$$A_\omega = \{\langle 1,6 \rangle, \langle 6,1 \rangle, \langle 2,5 \rangle, \langle 5,2 \rangle, \langle 3,4 \rangle, \langle 4,3 \rangle\}.$$

8. CONDITIONAL DISTRIBUTIONS AND EXPECTATIONS

Then
$$\mathbf{E}(X|Z)(\omega) = \frac{1}{\Pr A_\omega} \sum_{\eta \in A_\omega} X(\eta) \Pr(\eta)$$
$$= \frac{1}{6/36} \sum_{i=1}^{6} i \cdot \frac{1}{36}$$
$$= \frac{7}{2}$$

Thus, the expectation of X knowing that seven has appeared is $7/2$. On the other hand, by property (1) mentioned above, $\mathbf{E}(Z|Z) = Z$. By symmetry, we can see that $\mathbf{E}(X|Z) = \mathbf{E}(Y|Z)$. Since, by property (2)
$$\mathbf{E}(X|Z) + \mathbf{E}(Y|Z) = \mathbf{E}(X+Y|Z) = \mathbf{E}(Z|Z) = Z,$$
we have that
$$\mathbf{E}(X|Z) = \mathbf{E}(Y|Z) = \tfrac{1}{2} Z.$$

We can also define the conditional probability of an event B with respect to an algebra of random variables $\mathcal{V}$. This is a random variable, such that, for each $\omega \in \Omega$, its value is the conditional probability of B given that the atom A_ω, with $\omega \in A_\omega$, has occurred. That is:

DEFINITION VI.15 (CONDITIONAL PROBABILITY). Let B be an event, $\mathcal{V}$ an algebra of random variables, and $\omega \in \Omega$. Then
$$\Pr{}_\mathcal{V} B(\omega) = \Pr(B|\mathcal{V})(\omega) = \mathbf{E}_\mathcal{V}(\chi_B)(\omega) = \frac{\Pr(B A_\omega)}{\Pr A_\omega},$$
where A_ω is the atom of $\mathcal{V}$ containing ω.

If $X_1, X_2, \ldots, X_m$ are random variables, we also write
$$\Pr(B|X_1, X_2, \ldots, X_m) = \Pr(B|\mathcal{V}),$$
where $\mathcal{V}$ is the algebra generated by $X_1, X_2, \ldots, X_n$.

We have, then, that $\Pr(B|\mathcal{V})$ is a random variable, which is constant on the atoms of $\mathcal{V}$, and, hence, it belongs to $\mathcal{V}$.

The old definition of conditional probability is a special case of this one. Let E and F be events with $\Pr F > 0$. Consider the indicator of F, χ_F, and let $\mathcal{V}$ be the algebra generated by χ_F. If $\Pr F^c \neq 0$, this algebra has exactly two atoms, namely, F and F^c. Thus
$$\Pr(E|\chi_F)(\omega) = \begin{cases} \dfrac{\Pr(EF)}{\Pr F} = \Pr(E|F), & \text{if } \omega \in F, \\ \dfrac{\Pr(EF^c)}{\Pr F^c} = \Pr(E|F^c), & \text{if } \omega \in F^c. \end{cases}$$

Thus, we have that $\Pr(E|F) = \Pr(E|\chi_F)(F)$.

We can also relativize Jensen's and Markov's inequalities to $\mathcal{V}$. Jensen's inequality is relativized to
$$|\mathbf{E}_\mathcal{V} X|^p \leq \mathbf{E}_\mathcal{V} |X|^p,$$
for $p \geq 1$, and Markov's inequality
$$\Pr_\mathcal{V}[f(X) \geq Y] \leq \frac{\mathbf{E}_\mathcal{V} f(X)}{Y},$$
for f positive and $Y > 0$, $Y \in \mathcal{V}$, that is
$$\Pr_\mathcal{V}[f(X) \geq Y](\omega) \leq \frac{\mathbf{E}_\mathcal{V} f(X)(\omega)}{Y(\omega)},$$
for every $\omega \in \Omega$.

9. Constructions of probability spaces

We first introduce formally the notation used in Corollary VI.6:

DEFINITION VI.16. Let $\langle \Omega, \mathrm{pr} \rangle$ be a probability space and let $\mathcal{V}$ be an algebra of random variables on this space. Then we define the probability space $\langle \mathrm{at}(\mathcal{V}), \mathrm{pr}' \rangle$ where $\mathrm{pr}'(A) = \Pr A$ for any $A \in \mathrm{at}(\mathcal{V})$. We write $\Pr'$ for the probability measure induced by pr'.

We say that the original probability space $\langle \Omega, \mathrm{pr} \rangle$ is *fibered* over $\langle \mathrm{at}(\mathcal{V}), \mathrm{pr}' \rangle$ with *fibers* $\langle A, \mathrm{pr}_A \rangle$, where $A \in \mathrm{at}(\mathcal{V})$.

The two constructions that we shall frequently use in the sequel and that we shall discuss in this section, are special cases of "fibering". We have the following easy proposition:

PROPOSITION VI.11. *Let $\mathcal{V}$ be an algebra of random variables. Then*
1. *For any $\omega \in \Omega$, $\mathrm{pr}(\omega) = \mathrm{pr}'(A_\omega) \, \mathrm{pr}_{A_\omega}(\omega)$, where A_ω is the atom to which ω belongs.*
2. *Let X be a random variable, then*
$$\mathbf{E} X = \mathbf{E}'_\mathcal{V} \mathbf{E}_\mathcal{V} X,$$
where $\mathbf{E}'_\mathcal{V}$ is the conditional expectation relative to $\langle \mathrm{at}(\mathcal{V}), \mathrm{pr}' \rangle$.

PROOF. Part (1) is immediate from the definitions. We now prove (2). We have
$$\begin{aligned}
\mathbf{E}'_\mathcal{V} \mathbf{E}_\mathcal{V} X &= \sum_{A \in \mathrm{at}(\mathcal{V})} (\mathbf{E}_\mathcal{V} X)(A) \, \mathrm{pr}'(A) \\
&= \sum_{A \in \mathrm{at}(\mathcal{V})} \left(\sum_{\omega \in A} X(\omega) \, \mathrm{pr}_A(\omega) \right) \Pr A \\
&= \sum_{\omega \in \Omega} X(\omega) \, \mathrm{pr}(\omega) \\
&= \mathbf{E} X.
\end{aligned}$$

□

9. CONSTRUCTIONS OF PROBABILITY SPACES

Part (2) of this proposition is a generalization of the total probability formula: let $\mathcal{B}$ be the algebra of events generated by the partition of Ω, $F_1, F_2, \ldots, F_n$, and let E be any event. Then the atoms of $\mathcal{B}$ are precisely $F_1, \ldots, F_n$. We have

$$\Pr E = \mathbf{E}\,\chi_E$$
$$= \mathbf{E}'_\mathcal{B}\,\mathbf{E}_\mathcal{B}\,\chi_E$$
$$= \mathbf{E}'_\mathcal{B}(\Pr(E|\mathcal{B}))$$
$$= \sum_{i=1}^{n}(\Pr(E|\mathcal{B}))(F_i)\,\Pr F_i.$$

It is easy to show that $\Pr(E|\mathcal{B})(\omega) = \Pr(E|F_i)$, if $\omega \in F_i$. Thus, $\Pr(E|\mathcal{B})(F_i) = \Pr(E|F_i)$. Therefore, we get the total probability formula

$$\Pr E = \sum_{i=1}^{n} \Pr(E|F_i)\,\Pr F_i.$$

We now proceed to define the average space. This is the general construction of fibered spaces. That is, a space $\langle \Omega_0, \mathrm{pr}_0 \rangle$ is given, and for each $\omega \in \Omega_0$, we have a space $\langle \Omega_\omega, \mathrm{pr}_\omega \rangle$. Then we want to get a space $\langle \Omega, \mathrm{pr} \rangle$ that is fibered over $\langle \Omega_0, \mathrm{pr}_0 \rangle$ (or, more precisely, over a copy of this space) with fibers $\langle \Omega_\omega, \mathrm{pr}_\omega \rangle$.

DEFINITION VI.17 (AVERAGE SPACES). Let $\langle \Omega_0, \mathrm{pr}_0 \rangle$ be a finite sample space, and, for each $\omega \in \Omega_0$, let $\langle \Omega_\omega, \mathrm{pr}_\omega \rangle$ be a finite probability space. The *average probability space of* $\langle \langle \Omega_\omega, \mathrm{pr}_\omega \rangle \mid \omega \in \Omega_0 \rangle$ *with respect to* $\langle \Omega_0, \mathrm{pr}_0 \rangle$, $\langle \Omega, \mathrm{pr} \rangle$, where pr is called the *average measure*, is defined as follows:

$$\Omega = \bigcup \{\{\omega\} \times \Omega_\omega \mid \omega \in \Omega_0\} = \{\langle \omega, \eta \rangle \mid \omega \in \Omega_0,\, \eta \in \Omega_\omega\},$$

and, if $\omega \in \Omega_0$ and $\eta \in \Omega_\omega$, then

$$\mathrm{pr}(\langle \omega, \eta \rangle) = \mathrm{pr}_0(\omega)\,\mathrm{pr}_\omega(\eta).$$

We now obtain the average measure as a fibered measure. Let Ω be as in the definition, and let $\mathcal{V}$ be the algebra of random variables on Ω which depend only on their first argument. That is, $X \in \mathcal{V}$ means that $X(\omega, \eta) = X(\omega, \eta')$, for every $\eta, \eta' \in \Omega_\omega$. The atoms of $\mathcal{V}$ are the sets $\{\omega\} \times \Omega_\omega$ for $\omega \in \Omega_0$. (Instead of this algebra, we could consider just the algebra of events generated by the partition of Ω_0 given by these atoms.) We define $\mathrm{pr}'(\{\omega\} \times \Omega_\omega) = \mathrm{pr}_0(\omega)$ and $\mathrm{pr}_{\{\omega\} \times \Omega_\omega} = \mathrm{pr}_\omega$. Then, it is easy to check that for any random variable X on Ω

$$\mathbf{E}\,X = \mathbf{E}'_\mathcal{V}\,\mathbf{E}_\mathcal{V}\,X,$$

where $\mathbf{E}$ is the expectation determined by the average measure, and $\mathbf{E}'$ by pr'. Thus the average measure can be said to be fibered over $\langle \Omega_0, \mathrm{pr}_0 \rangle$ with fibers $\langle \Omega_\omega, \mathrm{pr}_\omega \rangle$.

It is not difficult to show that the average measure of any subset A of the average space Ω can be obtained as follows. For any $\omega \in \Omega_0$, let $A_\omega = \{\eta \mid$

$\langle \omega, \eta \rangle \in A\}$. Then $A_\omega \subseteq \Omega_\omega$. Let X_A be the random variable defined on Ω_0 by $X_A(\omega) = \Pr_\omega(A_\omega)$. We can calculate

$$\Pr A = \sum_{r \in \mathbb{R}} r \Pr_0[\Pr_\omega(A_\omega) = r] = \mathbf{E}_0 X_A,$$

i.e., $\Pr A$ is the expectation of X_A with respect to pr_0 (or, what is the same, the weighted average with respect to the measure $\Pr_0$) of the probabilities $\Pr_\omega(A_\omega)$ for $\omega \in \Omega$.

We now define the product spaces:

DEFINITION VI.18 (PRODUCT SPACES). Let $\langle \Omega_i, \mathrm{pr}_i \rangle$ be probability spaces for $1 \le i \le n$. Then the *product probability space of* $\langle \langle \Omega_i, \mathrm{pr}_i \rangle \mid 1 \le i \le n \rangle$

$$\langle \Omega, \mathrm{pr} \rangle = \prod_{i=1}^{n} \langle \Omega_i, \mathrm{pr}_i \rangle,$$

is defined as follows:

(1) $\Omega = \prod_{i=1}^{n} \Omega_i$,
(2) $\mathrm{pr}(\langle x_1, \ldots, x_n \rangle) = \prod_{i=1}^{n} \mathrm{pr}_i(x_i)$, for each $\langle x_1, \ldots, x_n \rangle \in \Omega$.

pr is called the *product distribution*. In case we have two spaces $\langle \Omega_1, \mathrm{pr}_1 \rangle$ and $\langle \Omega_2, \mathrm{pr}_2 \rangle$, we also write the product space $\langle \Omega_1 \times \Omega_2, \mathrm{pr}_1 \times \mathrm{pr}_2 \rangle$. When $\langle \Omega_i, \mathrm{pr}_i \rangle = \langle \Omega, \mathrm{pr} \rangle$, for all i with $1 \le i \le n$, we write $\prod_{i=1}^{n} \langle \Omega_i, \mathrm{pr}_i \rangle = \langle {}^n\Omega, {}^n \mathrm{pr} \rangle$.

It is clear that if the $\langle \Omega_i, \mathrm{pr}_i \rangle$ are probability spaces, then their product is also a probability space.

Let us see why the product space is a case of fibering. Consider the case of the product of two spaces, $\langle \Omega_1 \times \Omega_2, \mathrm{pr}_1 \times \mathrm{pr}_2 \rangle$. Let $\mathcal{V}_1$ be the algebra of random variables on $\Omega = \Omega_1 \times \Omega_2$ which depend only on the first variable, that is, the variables X such that $X(\omega_1, \eta) = X(\omega_1, \eta')$ for every $\eta, \eta' \in \Omega_2$. The atoms of this algebra are the sets

$$A_\eta = \{\langle \omega_1, \omega_2 \rangle \mid \omega_1 = \eta, \omega_2 \in \Omega_2\}$$

for a certain $\eta \in \Omega_1$. Then we have the probability distributions $\mathrm{pr}'_{\mathcal{V}_1}$ on $\mathrm{at}(\mathcal{V}_1)$ defined by

$$\mathrm{pr}'_{\mathcal{V}_1}(A_\eta) = \mathrm{pr}_1(\eta),$$

and pr_{A_η} on each atom A_η, by

$$\mathrm{pr}_{A_\eta}(\langle \eta, \omega_2 \rangle) = \mathrm{pr}_2(\omega_2).$$

We then show, for any random variable X on Ω, $\mathbf{E}\,X = \mathbf{E}'_{\nu_1}\mathbf{E}_{\nu_1}X$:

$$\begin{aligned}
\mathbf{E}\,X &= \mathbf{E}'_{\nu_1}(\mathbf{E}_{\nu_1}X) \\
&= \sum_{\omega_1 \in \Omega_1} (\mathbf{E}_{\nu_1}X)(A_{\omega_1})\,\mathrm{pr}'_{\nu_1}(A_{\omega_1}) \\
&= \sum_{\omega_1 \in \Omega_1} \bigl(\sum_{\omega_2 \in A_\eta} X(\langle\omega_1,\omega_2\rangle)\bigr)\,\mathrm{pr}_{A_{\omega_1}}(\omega_2))\,\mathrm{pr}'_{\nu_1}(A_{\omega_1}) \\
&= \sum_{\omega_1 \in \Omega_1}\sum_{\omega_2 \in \Omega_2} X(\langle\omega_1,\omega_2\rangle)\,\mathrm{pr}_1(\omega_1)\,\mathrm{pr}_2(\omega_2)
\end{aligned}$$

So that the product measure is fibered over $\langle\Omega_1,\mathrm{pr}_1\rangle$ with fibers $\langle\Omega_2,\mathrm{pr}_2\rangle$ for each $\omega_1 \in \Omega_1$.

We turn, now, to products with an arbitrary finite number of factors. It is clear that if $A \subseteq \Omega$ with $A = \prod_{i=1}^n A_i$ and $A_i \subseteq \Omega_i$, then

$$\mathrm{Pr}\,A = \prod_{i=1}^n \mathrm{Pr}_i\,A_i.$$

10. Stochastic processes

A stochastic process is a function whose values are random variables:

DEFINITION VI.19 (STOCHASTIC PROCESSES). Let T be a finite set and let $\langle\Omega,\mathrm{pr}\rangle$ be a finite probability space.

(1) A *stochastic process indexed by T and defined over* $\langle\Omega,\mathrm{pr}\rangle$ is a function $\xi : T \to {}^\Omega\mathbb{R}$. We write $\xi(t,\omega)$ for the value of $\xi(t)$ at $\omega \in \Omega$, and we write $\xi(\cdot,\omega)$ for the function from T to $\mathbb{R}$ which associates to each $t \in T$, $\xi(t,\omega) \in \mathbb{R}$. Thus each $\xi(t)$, written also ξ_t, is a random variable, each $\xi(t,\omega)$ is a real number, and each $\xi(\cdot,\omega)$ is a real-valued function on T called a *trajectory* or *sample path of the process*.

(2) Let Λ_ξ be the set of all trajectories of the stochastic process ξ. Then Λ_ξ is a finite subset of ${}^T\mathbb{R}$, the set of all function from T to $\mathbb{R}$. We define the distribution of ξ, pr_ξ, on elements of Λ_ξ by

$$\mathrm{pr}_\xi(\lambda) = \Pr[\xi(t) = \lambda(t) \text{ for all } t \in T].$$

The corresponding probability measure, Pr_ξ, is called the *probability distribution of ξ*.

Decoding the definition, we have that

$$\mathrm{pr}_\xi(\lambda) = \Pr\{\omega \mid \xi(t,\omega) = \lambda(t), \text{ for all } t \in T\}.$$

From the definition, we get that $\langle\Lambda_\xi,\mathrm{pr}_\xi\rangle$ is a probability space. As usual, we have denoted by Pr_ξ the probability measure determined by pr_ξ, that is, for

$A \subseteq \Lambda_\xi$

$$\Pr{}_\xi A = \sum_{\lambda \in A} \mathrm{pr}_\xi(\lambda).$$

DEFINITION VI.20. Two stochastic processes ξ and ξ' indexed by the same finite set T are called *equivalent* in case $\langle \Lambda_\xi, \mathrm{pr}_\xi \rangle = \langle \Lambda_{\xi'}, \mathrm{pr}_{\xi'} \rangle$; that is, in case they have the same trajectories with the same probabilities.

Notice that the evaluation function $\eta : T \to {}^{\Lambda_\xi}\mathbb{R}$, defined by $\eta(t, \lambda) = \lambda(t)$, is a stochastic process defined over $\langle \Lambda_\xi, \mathrm{pr}_\xi \rangle$ which is equivalent to ξ. In order to see this, we must prove that $\Lambda_\xi = \Lambda_\eta$ and $\mathrm{pr}_\xi = \mathrm{pr}_\eta$. We have, $\lambda \in \Lambda_\xi$ if and only if $\lambda = \eta(\cdot, \lambda)$; hence, $\lambda \in \Lambda_\xi$ if and only if $\lambda \in \Lambda_\eta$. On the other hand

$$\begin{aligned}\mathrm{pr}_\eta(\lambda) &= \Pr{}_\xi \{\mu \in \Lambda_\xi \mid \eta(t, \mu) = \lambda(t) \text{ all } t \in T\} \\ &= \Pr{}_\xi \{\mu \in \Lambda_\xi \mid \mu(t) = \lambda(t) \text{ all } t \in T\} \\ &= \mathrm{pr}_\xi(\lambda).\end{aligned}$$

Thus, in studying a stochastic process there is no loss in generality in assuming that it is defined over the space of its trajectories.

If T consists of a single element t, then we can identify a stochastic process ξ indexed by T with the random variable $\xi(t)$. That is, a random variable is a special case of a stochastic process. If $X = \xi(t)$ is such a random variable, then the trajectories of X are functions $\{\langle t, \lambda \rangle\}$ where $\lambda \in \mathbb{R}$. So they can be identified with real numbers, i.e., $\{\langle t, \lambda \rangle\}$ can be identified with the real number λ. Thus, the distribution of X is $\mathrm{pr}_X(\lambda) = \Pr[X = \lambda]$, where λ is in

$$\Lambda_X = \{\lambda \in \mathbb{R} \mid \Pr[X = \lambda] \neq 0\}.$$

This is exactly what we had before in Section 3.

When $T = \{1, 2, \ldots, n\}$, we write a process ξ over T, as $X_1, X_2, \ldots, X_n$, where $\xi(i) = X_i$ for $i = 1, \ldots, n$. We also use the vector notation, writing $\xi = \vec{X}$, and we call $\vec{X}$ a *random vector*. The distribution, pr_ξ, is called the *joint distribution of* $X_1, X_2, \ldots, X_n$ or *of the random vector* $\vec{X}$, and is also written, if $\lambda(1) = x_1, \lambda(2) = x_2, \ldots, \lambda(n) = x_n$

$$\begin{aligned}\mathrm{pr}_{X_1, X_2, \ldots, X_n}(x_1, x_2, \ldots, x_n) &= \mathrm{pr}_{\vec{X}}(\vec{x}) \\ &= \mathrm{pr}_\xi(\lambda) \\ &= \Pr[X_1 = x_1, X_2 = x_2, \ldots, X_n = x_n] \\ &= \Pr[\vec{X} = \vec{x}],\end{aligned}$$

In particular, we write $\mathrm{pr}_{X,Y}$ for the joint distribution of two variables, X, Y. That is, $\mathrm{pr}_{X,Y}(x, y) = \mathrm{pr}_\xi(\lambda) = \Pr[X = x, Y = y]$, where $\xi(1) = X$, $\xi(2) = Y$, $\lambda(1) = x$, and $\lambda(2) = y$.

We now extend the notion of independence to stochastic processes.

10. STOCHASTIC PROCESSES

DEFINITION VI.21. The random variables $\xi(t)$, for $t \in T$, of a stochastic process ξ indexed by T are called independent if

$$\operatorname{pr}_\xi(\lambda) = \prod_{t \in T} \operatorname{pr}_{\xi(t)}(\lambda(t))$$

for all $\lambda \in \Lambda_\xi$.

The construction of the product space gives the standard example of independent variables. Suppose, for instance, that $T = \{1, 2, \ldots, \nu\}$ and that X is a random variable on a space $\langle \Omega, \operatorname{pr} \rangle$, then the random variables X_n defined on the product space $\langle {}^\nu\Omega, {}^\nu\operatorname{pr} \rangle$ by

$$X_n(\omega_1, \ldots, \omega_\nu) = X(\omega_n)$$

are independent. The stochastic process ξ indexed by T and defined by $\xi(n) = X_n$ describes ν independent observations of the random variable X.

If $X_1, \ldots, X_n$ are random variables over the same probability space, then they are called independent, if they are independent as a random process indexed by n. In particular, two random variables X_1 and X_2 over the same probability space $\langle \Omega, \operatorname{pr} \rangle$, considered as a process X defined on $\{1, 2\}$, are independent, if and only if

$$\begin{aligned}\Pr[X_1 = x_1, X_2 = x_2] &= \operatorname{pr}_X(\langle x_1, x_2 \rangle) \\ &= \operatorname{pr}_{X_1}(x_1)\operatorname{pr}_{X_2}(x_2) \\ &= \Pr[X_1 = x_1]\Pr[X_2 = x_2],\end{aligned}$$

for all reals x_1 and x_2. Thus, the new notion coincides with the old one introduced in Section 3.

If $A_1, \ldots, A_n$ are events, then their indicator functions $\chi_{A_1}, \ldots, \chi_{A_n}$ are a stochastic process indexed by n. It is easy to check that the events $A_1, \ldots, A_n$ are independent exactly in case their indicator functions are independent.

CHAPTER VII

Disjunctive probability spaces and invariant measures

1. Disjunctive algebras of sets

As we have seen, probabilities are usually defined on algebras of sets, that is families of sets closed under unions, intersections and complements. For the construction to be given later, in order to insure the uniqueness of the probability measure, we need a more general type of structure. These structures have been also considered by Tarski in [93], and by Suppes in [87].

DEFINITION VII.1 (DISJUNCTIVE ALGEBRAS OF SETS). A nonempty family $\mathcal{A}$ of subsets of a set Ω is called a disjunctive algebra of subsets of Ω, if

(1) If $A \in \mathcal{A}$, then $A^c \in \mathcal{A}$.
(2) If A and $B \in \mathcal{A}$, and A and B are disjoint, then $A \cup B \in \mathcal{A}$.

That is, a disjunctive algebra of subsets of Ω is a family closed under complements and unions of pairwise disjoint sets. It is clear that an algebra of sets is a disjunctive algebra. It is also easy to see that the intersection of a collection of disjunctive algebras of subsets of Ω is also a disjunctive algebra. Hence, there is a smallest disjunctive algebra containing a given family $\mathcal{B}$ of subsets of a set Ω. This disjunctive algebra will be called the *disjunctive algebra generated by* $\mathcal{B}$. Disjunctive algebras enjoy certain closure properties that are of interest. I shall summarize them in the next proposition.

PROPOSITION VII.1. *Let $\mathcal{A}$ be a disjunctive algebra of subsets of a set Ω. Then if A and $B \in \mathcal{A}$, with $A \subseteq B$, we have that $B - A \in \mathcal{A}$. Also, $\Omega \in \mathcal{A}$ and $\emptyset \in \mathcal{A}$.*

PROOF. Suppose that $A, B \in \mathcal{A}$ with $A \subseteq B$. We have that $A - B = (A^c \cup B)^c$. But A^c is disjoint from B, because $A \subseteq B$. Hence, $A^c \cup B \in \mathcal{A}$, and, thus, $A - B \in \mathcal{A}$.

Since $\mathcal{A}$ is nonempty, there is an $A \in \mathcal{A}$. Then $A^c \in \mathcal{A}$, and, hence, $\Omega = A \cup A^c \in \mathcal{A}$. Also, $\emptyset = \Omega^c \in \mathcal{A}$. $\square$

A sort of reciprocal is also true.

PROPOSITION VII.2. *A family $\mathcal{A}$ of subsets of a set Ω is a disjunctive algebra if and only if $\Omega \in \mathcal{A}$, and for every $A, B \in \mathcal{A}$ with $A \subseteq B$ we have $B - A \in \mathcal{A}$.*

PROOF. By the previous proposition, we know that if $\mathcal{A}$ is a disjunctive algebra, then $\mathcal{A}$ contains Ω and it is closed under the difference of B and A, when $A \subseteq B$.

So, assume that $\Omega \in \mathcal{A}$ and that $A, B \in \mathcal{A}$ with $A \subseteq B$ implies that $B - A \in \mathcal{A}$. It is clear that if $A \in \mathcal{A}$, then $A^c = \Omega - A \in \mathcal{A}$. Thus, $\mathcal{A}$ is closed under complements. Suppose, now, that $A, B \in \mathcal{A}$, with $A \cap B = \emptyset$. We have that $A \subseteq B^c$. Hence, $B^c - A \in \mathcal{A}$, and $(B^c - A)^c \in \mathcal{A}$. But, $A \cup B = (B^c - A)^c$. Therefore, $\mathcal{A}$ is closed under disjoint unions. □

The following theorem indicates a relation between disjunctive algebras and algebras.

THEOREM VII.3. *Let $\mathcal{B}$ be a family of subsets of Ω closed under intersections. Then the disjunctive algebra $\mathcal{A}$ generated by $\mathcal{B}$ is an algebra.*

PROOF. Let $\mathcal{A}_A = \{B \in \mathcal{A} \mid A \cap B \in \mathcal{A}\}$. We first prove that for every $A \in \mathcal{B}$, $\mathcal{A}_A = \mathcal{A}$. Since $\mathcal{B}$ is closed under intersections, $\mathcal{B} \subseteq \mathcal{A}_A$. If $B \in \mathcal{A}_A$, then

$$A \cap B^c = A - (A \cap B) \in \mathcal{A}.$$

Hence, $B^c \in \mathcal{A}_A$. Now, let B and C be disjoint elements of $\mathcal{A}_A$. Then,

$$A \cap (B \cup C) = (A \cap B) \cup (A \cap C).$$

Hence, $B \cup C \in \mathcal{A}_A$. Thus, $\mathcal{A}_A$ contains $\mathcal{B}$, $\mathcal{A}_A$ is a subset of $\mathcal{A}$, and is closed under complements and disjoint unions. Hence, $\mathcal{A}_A = \mathcal{A}$.

Let, now, $\mathcal{C} = \{A \in \mathcal{A} \mid \mathcal{A}_A = \mathcal{A}\}$. We show in a similar way that $\mathcal{C} = \mathcal{A}$. We have just proved that $\mathcal{C} \supseteq \mathcal{B}$. Suppose that $A \in \mathcal{C}$, and let $B \in \mathcal{A}$. Then

$$A^c \cap B = B - (A \cap B) \in \mathcal{A}.$$

Hence, $\mathcal{A}_{A^c} = \mathcal{A}$. Similarly, if C and D are disjoint sets in $\mathcal{C}$ and $B \in \mathcal{A}$, then

$$B \cap (C \cup D) = (B \cap C) \cup (B \cap D) \in \mathcal{A}.$$

Hence, $\mathcal{A}_{C \cup D} = \mathcal{A}$. Thus, we have shown that for every $A, B \in \mathcal{A}$, $B \in \mathcal{A}_A$, that is, $A \cap B \in \mathcal{A}$. Therefore, $\mathcal{A}$ is a disjunctive algebra closed under intersections, and, hence, $\mathcal{A}$ is an algebra. □

2. Finite disjunctive probability spaces

Let us now consider finite disjunctive probability spaces.

DEFINITION VII.2 (DISJUNCTIVE PROBABILITY SPACES). A *(finite) disjunctive probability space* is a triple $\langle \Omega, \mathcal{A}, \Pr \rangle$, where Ω is a finite nonempty set, $\mathcal{A}$ is a disjunctive algebra of subsets of Ω, and $\Pr$ satisfies

(1) $\Pr : \mathcal{A} \to [0, 1]$.
(2) $\Pr \Omega = 1$.

2. FINITE DISJUNCTIVE PROBABILITY SPACES

(3) If E and F are mutually exclusive elements of $\mathcal{A}$, then

$$\Pr(E \cup F) = \Pr E + \Pr F.$$

(4) $\Pr E > 0$, for every $E \neq \emptyset$, $E \in \mathcal{A}$.

Since logical probability requires unique measures, one of the reasons for choosing disjunctive algebras instead of algebras is given in the following theorem.

THEOREM VII.4. *Let Ω be a set, $\mathcal{B}$ a family of subsets of Ω and $\Pr$ a function on $\mathcal{B}$ into $[0, 1]$. Then, there is at most one probability measure that is an extension of $\Pr$ to the disjunctive algebra $\mathcal{A}$ generated by $\mathcal{B}$.*

PROOF. Suppose that $\Pr_1$ and $\Pr_2$ are probability measures defined on $\mathcal{A}$, extensions of $\Pr$. Let

$$\mathcal{D} = \{A \in \mathcal{A} \mid \Pr_1 A = \Pr_2 A\}.$$

We prove by induction that $\mathcal{D} = \mathcal{A}$. It is clear that $\mathcal{B} \subseteq \mathcal{D}$.

Suppose that $\Pr_1 A = \Pr_2 A$. Then

$$\Pr_1 A^c = 1 - \Pr_1 A = 1 - \Pr_2 A = \Pr_2 A^c.$$

Hence, $\mathcal{D}$ is closed under complements. Suppose, finally, that B and C are disjoint elements of $\mathcal{D}$. Then

$$\Pr_1(B \cup C) = \Pr_1 B + \Pr_1 C = \Pr_2 B + \Pr_2 C = \Pr_2(B \cup C).$$

Therefore, $\mathcal{D}$ is also closed under pairwise disjoint unions. □

The disjunctive algebra $\mathcal{A}$ is not necessarily the power set of Ω. In fact, if it is a disjunctive algebra that is not an algebra, it cannot be the power set. In this case, not every real function on Ω is a random variable. So we introduce the following definition.

DEFINITION VII.3 (RANDOM VARIABLES). Let $\langle \Omega, \mathcal{A}, \Pr \rangle$ be a finite disjunctive probability space. A real valued function X with domain Ω is a *random variable over* $\langle \Omega, \mathcal{A}, \Pr \rangle$ or, briefly, a $\Pr$-*random variable*, if for every real number λ the set $[X = \lambda] \in \mathcal{A}$.

It is clear that if $A \in \mathcal{A}$, then its indicator function χ_A is a random variable. We now define the product space and the average space of disjunctive spaces.

DEFINITION VII.4.

(1) Let $\mathcal{A}_i$ be finite disjunctive algebras for $1 \leq i \leq n$. Then the *product algebra of* $\langle \mathcal{A}_i \mid 1 \leq i \leq n \rangle$

$$\mathcal{A} = \prod_{i=1}^{n} \mathcal{A}_i,$$

is the disjunctive algebra generated by the family

$$\{\prod_{i=1}^{n} A_i \mid A_i \in \mathcal{A}_i \text{ for } 1 \leq i \leq n\}.$$

(2) Let $\langle \Omega_0, \mathcal{A}_0, \Pr_0 \rangle$ and $\langle \Omega_\omega, \mathcal{A}_\omega, \Pr_\omega \rangle$ be disjunctive probability spaces for each $\omega \in \Omega_0$. The average probability space, $\langle \Omega, \mathcal{A}, \Pr \rangle$, of the system of probability spaces $\langle \langle \Omega_\omega, \mathcal{A}_\omega, \Pr_\omega \rangle \mid \omega \in \Omega_0 \rangle$ with respect to the space $\langle \Omega_0, \mathcal{A}_0, \Pr_0 \rangle$ is defined as follows:

$$\Omega = \bigcup \{\{\omega\} \times \Omega_\omega \mid \omega \in \Omega_0\} = \{\langle \omega, \eta \rangle \mid \omega \in \Omega_0, \eta \in \Omega_\omega\}.$$

For each $A \subseteq \Omega$ and $\omega \in \Omega_0$, we define $A_\omega = \{\eta \mid \langle \omega, \eta \rangle \in A\}$. Then

$$\mathcal{A} = \{A \subseteq \Omega \mid \text{for all real } r, [\Pr_\omega A_\omega = r] \in \mathcal{A}_0\},$$

and, for $A \in \mathcal{A}$, define

$$\Pr A = \sum_{r \in \mathbb{R}} r \Pr_0[\Pr_\omega A_\omega = r].$$

It is clear that if we define $Y_A(\omega) = \Pr_\omega A_\omega$, then

$$\mathcal{A} = \{A \subseteq \Omega \mid Y_A \text{ is } \Pr_0\text{-random variable}\}.$$

and that for $A \in \mathcal{A}$ we have

$$\Pr A = \mathbf{E}_0 Y_A.$$

We have the following theorem:

THEOREM VII.5. *The average probability space of finite disjunctive probability spaces is a disjunctive probability space.*

PROOF. We must prove that $\mathcal{A}$ is a disjunctive algebra and that $\Pr$ is a probability. It is clear that $\Omega \in \mathcal{A}$, since $Y_\Omega = \chi_{\Omega_0}$ and χ_{Ω_0} is a $\Pr_0$-random variable. Now, let $A \in \mathcal{A}$. Then Y_A is $\Pr_0$ random variable. We have, for $\omega \in \Omega$, $A^c_\omega = (\Omega - A)_\omega = \Omega_\omega - A_\omega$. Hence

$$Y_{A^c}(\omega) = \Pr_\omega A^c_\omega = 1 - \Pr_\omega(A_\omega) = 1 - Y_A(\omega).$$

Thus, Y_{A^c} is a $\Pr_0$ random variable and $A^c \in \mathcal{A}$.

Suppose that $A, B \in \mathcal{A}$ with A and B disjoint. Then Y_A and Y_B are $\Pr_0$ random variables, and since $(A \cup B)_\omega = A_\omega \cup B_\omega$, it is easy to show that $Y_{A \cup B} = Y_A + Y_B$. Hence, $Y_{A \cup B}$ is a $\Pr_0$ random variable and $A \cup B \in \mathcal{A}$. From the additive properties of the expectation, which are easy to prove, we get the additivity of $\Pr$. □

The notion of expectation (or integral) can be developed for disjunctive probability spaces, just as for probability spaces. In fact, this notion can be developed for infinite disjunctive spaces. The average measure can also be constructed, even for infinite spaces. I do not know, however, whether the product measure construction can be carried through in general, even for finite disjunctive spaces. Fortunately, our disjunctive probability spaces will always be extendible to the

2. FINITE DISJUNCTIVE PROBABILITY SPACES

usual probability spaces where the field of events is the power set. This is not true in general. It is not difficult to construct an example of an infinite disjunctive probability space which cannot be extended to a probability space.[1] I do not know whether there is a finite disjunctive space which cannot be so extended.

DEFINITION VII.5. A finite disjunctive probability space $\langle \Omega, \mathcal{A}, \Pr \rangle$ is called *extendible*, if the probability measure Pr can be extended to a probability measure $\Pr'$ defined on $\mathcal{P}(\Omega)$.

The following propositions are easy to prove.

PROPOSITION VII.6. *Let $\langle \Omega_i, \mathcal{A}_i, \Pr_i \rangle$ for $1 \leq i \leq n$ be extendible disjunctive probability spaces. Let Ω be the product of the Ω_i for $i < n$. For each subset of Ω, $A = \prod_{i=1}^{n} A_i$ with $A_i \in \mathcal{A}_i$ for $1 \leq i \leq n$, define*

$$\Pr A = \prod_{i=1}^{n} \Pr_i A_i.$$

Then Pr can be extended uniquely to the product algebra $\mathcal{A} = \prod_{i=1}^{n} \mathcal{A}_i$. Furthermore, the disjunctive probability space $\langle \Omega, \mathcal{A}, \Pr \rangle$ is extendible.

PROOF. We extend the disjunctive probability spaces $\langle \Omega_i, \mathcal{A}_i, \Pr_i \rangle$ to the respective power sets and construct the product measure. It is clear that the product measure extends Pr. The uniqueness on the disjunctive product space follows from Theorem VII.4. □

The proof of the next proposition is similar.

PROPOSITION VII.7. *Let $\langle \Omega_0, \mathcal{A}_0, \Pr_0 \rangle$ and $\langle \Omega_\omega, \mathcal{A}_\omega, \Pr_\omega \rangle$ be extendible finite disjunctive probability spaces for each $\omega \in \Omega_0$. Then the average disjunctive probability space $\langle \Omega, \mathcal{A}, \Pr \rangle$ is extendible.*

PROOF. Let $\Pr_0'$ and $\Pr_\omega'$ be the extensions of $\Pr_0$ and $\Pr_\omega$, for $\omega \in \Omega_0$, and let $\Pr'$ be the average measure of $\Pr_0'$ and $\Pr_\omega'$. Then $\Pr'$ is defined on $\mathcal{P}\Omega$. We have for $A \in \mathcal{A}$,

$$\Pr' A = \mathbf{E}_0' Y_A = \mathbf{E}_0 Y_A = \Pr A.$$

So, $\Pr'$ is an extension of Pr. □

---

[1] The construction of this example is as follows. We proceed as in the construction of the average measure. Let both $\langle \Omega_0, \mathcal{A}_0, \Pr_0 \rangle$ and $\langle \Omega_\omega, \mathcal{A}_\omega, \Pr_\omega \rangle$, for $\omega \in \Omega_0$, be the unit interval with Lebesgue measure. Let $\langle \Omega, \mathcal{A}, \Pr \rangle$ be the average probability space, as above. It is easy to see that Pr is a σ-additive measure on $\mathcal{A}$. Let C be a Lebesgue nonmeasurable set. Define subsets of Ω, A and B, such that $\Pr_\omega A_\omega = \Pr_\omega B_\omega < 1/2$, for every $\omega \in \Omega_0$, but $A_\omega = B_\omega$, if $\omega \in C^c$, and $A_\omega \cap B_\omega = \emptyset$, if $\omega \in C$. Then A and B are measurable, but $A \cap B$ cannot be measurable, because then C would be Lebesgue measurable.

3. Probabilities invariant under groups

One of the main notions for the construction of the probability models is that of a measure invariant under a group of transformation. Here we discuss these measures for finite spaces. Many of the properties obtained here can be extended, suitably modified, to infinite spaces, but since we will not need them, we shall concentrate on the finite case.[2]

One of the main concepts to be used in the sequel is the notion of a measure invariant under a group. When we talk of a measure, we always mean a probability measure, that is, one satisfying Kolmogorov's axioms. I proceed to define measures invariant under groups.

DEFINITION VII.6 (INVARIANT MEASURES).

(1) A *group of permutations of a set* Ω is a nonempty set G of permutations of Ω, such that if f and $g \in G$, then $f \circ g$ and $f^{-1} \in G$.
(2) A measure Pr on a family $\mathcal{B}$ of subsets of Ω is *invariant under a group of permutations G of Ω* (or, briefly, is G-*invariant*) if the following two conditions are satisfied.
   (a) $\mathcal{B}$ is G-closed, that is, if $A \in \mathcal{B}$ and $g \in G$, then $g''A \in \mathcal{B}$.
   (b) If $A \in \mathcal{B}$ and $g \in G$, then $\Pr A = \Pr g''A$.

If we have a set Ω and G a group of permutations of Ω, we can define the relation on subsets of Ω, $\sim_G$, by:

DEFINITION VII.7. Let G be a group of permutations of the set Ω, $A, B \subseteq \Omega$. Then $A \sim_G B$ if and only if there is a $g \in G$ such that $B = g''A$.

Since G is a group, it is easy to check that $\sim_G$ is an equivalence relation, which has two important additional properties:

(1) $\sim_G$ is *refining*, that is, if $A \sim_G B$ and $A = A_1 \cup A_2$, with A_1 and A_2 disjoint, then there are B_1 and B_2 disjoint, such that $B = B_1 \cup B_2$, $A_1 \sim_G B_1$, and $A_2 \sim_G B_2$.
(2) $\sim_G$ is *strictly positive*, that is, if $A \sim_G \emptyset$, then $A = \emptyset$.

Condition (2) is obvious, and (1) can easily be proved in the following way. Suppose that $A \sim_G B$ and $A = A_1 \cup A_2$, with A_1 and A_2 disjoint. Then there is a $g \in G$ such that $g''A = B$. Thus if one takes $B_1 = g''A_1$ and $B_2 = g''A_2$, one obtains the desired conclusion.

We extend $\sim_G$ to a relation $\simeq_G$ by defining, with the same hypotheses as in the previous definition.

DEFINITION VII.8. Let $\mathcal{A}$ be a disjunctive algebra of events. Then $A \simeq_{G,\mathcal{A}} B$ if there are pairwise disjoint elements, $A_1, \ldots, A_n \in \mathcal{A}$ and $B_1, \ldots, B_n \in \mathcal{A}$ such that $A_i \sim_G B_i$ for $i = 1, \ldots, n$, $A = \bigcup_{i=1}^{n} A_i$ and $B = \bigcup_{i=1}^{n} B_i$.

We write $A \simeq_G B$ for $A \simeq_{G, \mathcal{P}\Omega} B$.

---

[2] Conditions of an algebraic character for the existence of invariant measures on infinite algebras were obtained in [15] and [14]. Since we shall deal mainly with finite algebras, where invariant measures always exist, we will not use these conditions.

When it is clear from the context, I shall write $\simeq$ and $\sim$ instead of $\simeq_G$ and $\sim_G$. We shall prove later that $\simeq$ is also an equivalence relation which is refining and strictly positive (Corollary VII.10). It also has the additional property, which is easy to check:

(3) The relation $\simeq_G$ is *additive*, that is, if $A_1 \simeq B_1$ and $A_2 \simeq B_2$ with A_1 and A_2, and B_1 and B_2 disjoint, then $A_1 \cup A_2 \simeq B_1 \cup B_2$.

The main equivalence relations that we shall need are $\sim_G$ and $\simeq_G$. Since most of the theorems that we shall prove are also valid for some other equivalence relations, we shall discuss equivalence relations in some generality. Notice that the relation of equiprobability is an equivalence relation, so our discussion is tied up with the classical definition of probability. We now introduce officially the terminology discussed above for equivalence relations.[3]

DEFINITION VII.9. *Let Ω be a finite set, $\mathcal{A}$ a disjunctive algebra of subsets of Ω, and R an equivalence relation defined on subsets of Ω. Then:*

(1) *R is additive (on $\mathcal{A}$) if whenever A_1, A_2, B_1, $B_2 \in \mathcal{A}$, $A_1 \cap A_2 = \emptyset$, $B_1 \cap B_2 = \emptyset$, and $A_i \, \mathrm{R} \, B_i$ for $i = 1, 2$, then $A_1 \cup A_2 \, \mathrm{R} \, B_1 \cup B_2$.*
(2) *R is refining (on $\mathcal{A}$) if whenever A, B, A_1, $A_2 \in \mathcal{A}$, $A \, \mathrm{R} \, B$, $A = A_1 \cup A_2$, and $A_1 \cap A_2 = \emptyset$, we have that there are B_1, $B_2 \in \mathcal{A}$ such that $B = B_1 \cup B_2$, $B_1 \cap B_2 = \emptyset$, and $A_1 \, \mathrm{R} \, B_1$ and $A_2 \, \mathrm{R} \, B_2$.*
(3) *R is strictly positive (on $\mathcal{A}$) if $A \in \mathcal{A}$ and $A \, \mathrm{R} \, \emptyset$ imply that $A = \emptyset$.*
(4) *$\mathcal{A}$ is R–closed if whenever $A \in \mathcal{A}$ and $A \, \mathrm{R} \, B$, we have that $B \in \mathcal{A}$.*
(5) *Let Pr be a probability defined on $\mathcal{A}$. Then Pr is R–invariant if $\mathcal{A}$ is R–closed, and whenever $A \, \mathrm{R} \, B$, with $A, B \in \mathcal{A}$, we have that $\mathrm{Pr} \, A = \mathrm{Pr} \, B$.*
(6) *$\mathrm{R}^\sharp$ is the intersection of all additive equivalence relations on $\mathcal{A}$ containing R.*

We have that $\mathrm{R}^\sharp$ is the smallest additive equivalence relation containing R, because the intersection of additive equivalence relations is an additive equivalence relation. When we do not mention "on $\mathcal{A}$" we mean on the power set of Ω.

It is clear that by induction we extend the additive and refining properties to sequences of n elements. That is if R is additive and $A_1, \ldots, A_n$ and $B_1, \ldots, B_n$ are pairwise disjoint families of subsets of $\mathcal{A}$ such that $A_i \, \mathrm{R} \, B_i$ for $i = 1, \ldots, n$, we have that

$$\bigcup_{i=1}^n A_i \, \mathrm{R} \, \bigcup_{i=1}^n B_i.$$

Also, if R is refining, $A, B \in \mathcal{A}$, $A \, \mathrm{R} \, B$, and $A = \bigcup_{i=1}^n A_i$, where $A_1, \ldots, A_n$ is a pairwise disjoint family of elements of $\mathcal{A}$, then $B = \bigcup_{i=1}^n B_i$, with $B_1, \ldots, B_n$ pairwise disjoint elements of $\mathcal{A}$, and $A_i \, \mathrm{R} \, B_i$ for $i = 1, \ldots, n$. We have the following theorem:

THEOREM VII.8. *Let R be a refining equivalence relation on an algebra $\mathcal{A}$ of subsets of Ω, and $A, B \in \mathcal{A}$. Then $A \, \mathrm{R}^\sharp \, B$ if and only if there are families*

---

[3]Relations of the form defined here are discussed extensively in [93] and [99].

of pairwise disjoint elements of $\mathcal{A}$, A_1, ..., A_n *and* B_1, ..., B_n, *such that* $A = \bigcup_{i=1}^{n} A_i$, $B = \bigcup_{i=1}^{n} B_i$, *and* $A_i \operatorname{R} B_i$ *for* $i = 1$, ..., n.

PROOF. Let S be the relation that holds between $A, B \in \mathcal{A}$ if and only if there are families of pairwise disjoint elements of $\mathcal{A}$, A_1, ..., A_n and B_1, ..., B_n, such that $A = \bigcup_{i=1}^{n} A_i$, $B = \bigcup_{i=1}^{n} B_i$, and $A_i \operatorname{R} B_i$ for $i = 1$, ..., n.

It is clear that S is symmetric, reflexive, and additive, and that $\operatorname{R}^{\sharp} \supseteq \operatorname{S}$. So we only have to prove that S is transitive. Suppose that $A \operatorname{S} B$ and $B \operatorname{S} C$. Then, by definition, there are pairwise disjoint sequences A_1, ..., A_n, B_1, ..., B_n, B'_1, ..., B'_m, C_1, ..., C_m, such that

$$A = \bigcup_{i=1}^{n} A_i,$$

$$B = \bigcup_{i=1}^{n} B_i$$

$$= \bigcup_{j=1}^{m} B'_j,$$

$$C = \bigcup_{j=1}^{m} C_j,$$

$A_i \operatorname{R} B_i$ for $i = 1$, ..., n, and $B'_j \operatorname{R} C_j$ for $j = 1$, ..., m.

Let $B_{ij} = B_i \cap B'_j$. Then the B_{ij} are pairwise disjoint elements of $\mathcal{A}$, such that

$$B = \bigcup_{i=1}^{n} \bigcup_{j=1}^{m} B_{ij},$$

$$B_i = \bigcup_{j=1}^{m} B_{ij},$$

$$B'_j = \bigcup_{i=1}^{n} B_{ij}.$$

So, by refinement, there are pairwise disjoint sequences A_{ij} and C_{ij} in $\mathcal{A}$ such that

$$A_i = \bigcup_{j=1}^{m} A_{ij},$$

$$C_j = \bigcup_{i=1}^{n} C_{ij},$$

and $A_{ij} \operatorname{R} C_{ij}$. But then

$$A = \bigcup_{i=1}^{n} \bigcup_{j=1}^{m} A_{ij}$$

and

$$C = \bigcup_{i=1}^{n} \bigcup_{j=1}^{m} C_{ij}.$$

Thus, $A \mathrel{S} C$, and we have proved that S is transitive. Since S is an additive equivalence relation which obviously contains R, $S \supseteq R^\sharp$. Hence, $S = R^\sharp$ and the theorem is proved. □

THEOREM VII.9. *Let $\langle \Omega, \mathrm{pr} \rangle$ be a finite probability space, and R, a refining equivalence relation on subsets of Ω. Then*

(1) *$R^\sharp$ is also refining.*
(2) *If R is strictly positive, then $R^\sharp$ is also strictly positive.*
(3) *Pr is R–invariant if and only if Pr is $R^\sharp$–invariant.*

PROOF OF (1). Suppose that $A \mathrel{R^\sharp} B$, and $A = C_1 \cup C_2$, with C_1 and C_2 disjoint. Since $A \mathrel{R^\sharp} B$, we have that there are pairwise disjoint sequences A_1, ..., A_n and B_1, ..., B_n such that $A = \bigcup_{i=1}^n A_i$, $B = \bigcup_{i=1}^n B_i$, and $A_i \mathrel{R} B_i$ for $i = 1, \ldots, n$.

Let $C_{i1} = A_i \cap C_1$ and $C_{i2} = A_i \cap C_2$. We have that $A_i = C_{i1} \cup C_{i2}$, with C_{i1}, C_{i2} disjoint. Since R is refining, there are D_{i1} and D_{i2}, such that $B_i = D_{i1} \cup D_{i2}$, D_{i1}, D_{i2} disjoint, and $C_{i1} \mathrel{R} D_{i1}$, $C_{i2} \mathrel{R} D_{i2}$, for $i = 1, \ldots, n$. Then, we have that the sequences $D_{11}, \ldots, D_{n1}$, and $D_{12}, \ldots, D_{n2}$ are pairwise disjoint. Hence, if we define $D_1 = \bigcup_{i=1}^n D_{i1}$, and $D_2 = \bigcup_{i=1}^n D_{i2}$, we get that $D_1 \mathrel{R^\sharp} C_1$ and $D_2 \mathrel{R^\sharp} C_2$. It is easy to see that $B = D_1 \cup D_2$, and D_1 and D_2 are disjoint.

The proof of 2 is left to the reader. □

PROOF OF (3). Since $R^\sharp$ is an extension of R, it is clear that if Pr is $R^\sharp$–invariant, then Pr is R–invariant.

Suppose, now, that Pr is R–invariant, and let $A \mathrel{R^\sharp} B$. Then there are pairwise disjoint sequences $A_1, \ldots, A_n$ and $B_1, \ldots, B_n$ such that $A = \bigcup_{i=1}^n A_i$, $B = \bigcup_{i=1}^n B_i$, and $A_i \mathrel{R} B_i$, for $i = 1, \ldots, n$. Thus, $\mathrm{Pr}\, A_i = \mathrm{Pr}\, B_i$, for $i = 1, \ldots, n$. Since Pr is additive, $\mathrm{Pr}\, A = \mathrm{Pr}\, B$. □

The next corollary collects some simple properties of $\simeq_G$. We could, by suitably modifying the definitions, prove similar theorems for arbitrary disjunctive probability spaces, but since we shall always deal with extendible spaces, we shall not need these stronger theorems.

COROLLARY VII.10. *Let $\langle \Omega, \mathrm{pr} \rangle$ be a finite probability space, and G a group of permutations of Ω. Then*

(1) *$\simeq_G$ is an additive, refining and strictly positive equivalence relation.*
(2) *Pr is G–invariant if and only if Pr is $\simeq_G$–invariant.*

PROOF. We have noted above that $\sim_G$ is refining and strictly positive. Thus, by the previous theorem, we obtain (1).

Condition (2) is an immediate consequence of the previous theorem, noticing that Pr is G–invariant if and only if Pr is $\sim_G$–invariant. □

Many of the theorems true for G–invariant measures, are also true for R–invariant measures, where R is a refining, and strictly positive equivalence relation. The main reason for this fact is given by the following theorem.

THEOREM VII.11. *Let* R *be a refining and strictly positive equivalence relation on subsets of a finite set* Ω. *Then* $A\,\mathrm{R}\,B$ *implies that* A *and* B *have the same number of elements.*

In fact, $A\,\mathrm{R}\,B$ *implies that there are sequences of distinct elements of* Ω, a_1, ..., a_n *and* b_1, ..., b_n, *such that* $A = \{a_1, \ldots, a_n\}$, $B = \{b_1, \ldots, b_n\}$ *and* $\{a_i\}\,\mathrm{R}\,\{b_i\}$ *for* $i = 1, \ldots, n$.[4]

PROOF. Suppose that $A\,\mathrm{R}\,B$ and $A = \{a_1, \ldots, a_n\}$ with $a_1, \ldots, a_n$ a sequence of distinct elements of Ω. We now prove by induction on n that $B = \{b_1, \ldots, b_n\}$ with $b_1, \ldots, b_n$ a sequence of distinct elements of Ω, and $\{a_i\}\,\mathrm{R}\,\{b_i\}$ for $i = 1, \ldots, n$. If $n = 1$, then $A = \{a\}$. If B had more than one element, then $B = B_1 \cup B_2$, with B_1 and B_2 disjoint and nonempty. Then $A = A_1 \cup A_2$, with $A_1\,\mathrm{R}\,B_1$, $A_2\,\mathrm{R}\,B_2$, and A_1, A_2 disjoint. Then, by positivity, A_1 and A_2 are also nonempty. But this is contradiction.

Suppose, now that the statement is true for n and let $A = \{a_1, \ldots, a_{n+1}\}$. Then $A = \{a_1, \ldots, a_n\} \cup \{a_{n+1}\}$. Since R is refining, $B = B_1 \cup B_2$ with $B_1\,\mathrm{R}\,\{a_1, \ldots, a_n\}$ and $B_2\,\mathrm{R}\,\{a_{n+1}\}$. As above, we prove that B_2 has one element. So by the induction hypothesis, we can complete the proof. □

We shall also need the following properties of refining strictly positive equivalence relations.

THEOREM VII.12 (REFINING PROPERTY). *Let* R *be a refining relation on subsets of a finite set* Ω. *If* A, B, C, *and* D *are subsets of* Ω *such that* $A \cap B = \emptyset = C \cap D$ *and* $A \cup B\,\mathrm{R}\,C \cup D$ *then there are pairwise disjoint subsets* E_1, E_2, E_3 *and* E_4 *such that* $A = E_1 \cup E_2$, $B = E_3 \cup E_4$, $C\,\mathrm{R}\,E_1 \cup E_3$ *and* $D\,\mathrm{R}\,E_2 \cup E_4$.[5]

PROOF. By refinement, there are disjoint C_1 and D_1 such that $A \cup B = C_1 \cup D_1$, $C\,\mathrm{R}\,C_1$ and $D\,\mathrm{R}\,D_1$. Let $E_1 = A \cap C_1$, $E_2 = A \cap D_1$, $E_3 = B \cap C_1$, and $E_4 = B \cap D_1$. The result is now clear. □

THEOREM VII.13. *Let* R *be a refining strictly positive equivalence relation on subsets of a finite set* Ω. *If* A, B, C *and* D *are subsets of* Ω, *such that* $A \cap B = \emptyset = C \cap D$, $A\,\mathrm{R}\,C$ *and* $A \cup B\,\mathrm{R}\,C \cup D$, *then* $B\,\mathrm{R}\,D$.

---

[4]This theorem is found in [13, Lemma 4.1]. Other similar theorems are also found in the same paper.

[5]Similar theorems about refining equivalence relations without the restriction to Ω finite are proved in [93].

3. PROBABILITIES INVARIANT UNDER GROUPS

PROOF. The proof is by induction on the number of elements of A. Suppose that the theorem is true for all sets with less elements than A, and that A, B and C are as in the hypothesis of the theorem. By Theorem VII.12, there are pairwise disjoint E_1, E_2, E_3, and E_4 such that $A = E_1 \cup E_2$, $B = E_3 \cup E_4$, and $C \operatorname{R} E_1 \cup E_3$ and $D \operatorname{R} E_2 \cup E_4$. If E_1 has the same number of elements as A, then E_2 is empty. Since we have that $A \operatorname{R} C$, then $A \operatorname{R} E_1 \cup E_3$ and, by strict positivity, E_3 is empty. Hence, $B = E_4$ and $D \operatorname{R} E_4$, i.e., $B \operatorname{R} D$. If E_1 has less elements than A, then we apply the induction hypothesis to $E_1 \cup E_2 \operatorname{R} E_1 \cup E_3$ and get $E_2 \operatorname{R} E_3$. Hence, $E_3 \cup E_4 \operatorname{R} E_2 \cup E_4$, i.e., $B \operatorname{R} D$. □

The following theorem explains the behavior of R–invariance under extensions.

THEOREM VII.14. *Let* R *be a refining and strictly positive equivalence relation on subsets of* Ω. *If a family of subsets of* Ω *is R-closed, then the disjunctive algebra of subsets and the algebra of subsets of* Ω *generated by this family are also R-closed.*

PROOF. Let $\mathcal{A}$ be the disjunctive algebra generated by the R–closed family $\mathcal{B}$. Let
$$\mathcal{D} = \{A \in \mathcal{A} \mid B \in \mathcal{A}, \text{ for every } B \operatorname{R} A\}.$$
We prove by induction that $\mathcal{D} = \mathcal{A}$. It is clear that $\mathcal{B} \subseteq \mathcal{D}$. Suppose that $A \in \mathcal{D}$ and that $A^c \operatorname{R} B$. We have
$$\Omega = A^c \cup A = B^c \cup B.$$
Hence, by Theorem VII.13, $B^c \operatorname{R} A$. Since $A \in \mathcal{D}$, we have that $B^c \in \mathcal{A}$, and, therefore, $B \in \mathcal{A}$. Thus, $A^c \in \mathcal{D}$ and $\mathcal{D}$ is closed under complements.

Suppose now that A_1 and A_2 are disjoint elements of $\mathcal{D}$, and that $A_1 \cup A_2 \operatorname{R} B$. Then $B = B_1 \cup B_2$, with $B_1 \cap B_2 = \emptyset$, $A_1 \operatorname{R} B_1$ and $A_2 \operatorname{R} B_2$. Then $B_1, B_2 \in \mathcal{A}$, and hence $B \in \mathcal{A}$. Thus, $A_1 \cup A_2 \in \mathcal{D}$ and $\mathcal{D}$ is closed under disjoint unions. Therefore $\mathcal{D} = \mathcal{A}$.

A similar proof works for arbitrary unions, and, hence, for the algebra generated by $\mathcal{B}$. □

Finally in this section, we state definitions and properties of extendible measures.

DEFINITION VII.10. Let $\langle \Omega, \mathcal{A}, \Pr \rangle$ be a finite disjunctive probability space and R a refining and strictly positive equivalence relation on subsets of Ω. Then Pr is an *extendible* R*-invariant measure* if

(1) If $A \in \mathcal{A}$ and $A \operatorname{R} B$, then $B \in \mathcal{A}$.
(2) There is an extension $\Pr'$ of $\Pr$ to $\mathcal{P}\Omega$ that is R-invariant.

When R is $\sim_G$, for some group G, we say *extendible* G*-invariant measure*.

THEOREM VII.15. *Let* $\mathcal{A}$ *be a disjunctive algebra of subsets of the finite set* Ω *and let* R *be a strictly positive, refining equivalence relation on subsets of* Ω, *such that* $\mathcal{A}$ *is* R*-closed. Then there is an extendible* R*-invariant measure* $\Pr$ *on* $\mathcal{A}$.

PROOF. Let n be the number of elements Ω. For each $A \subseteq \Omega$, define $\Pr' A = m/n$, where m is the number of elements of A. Then, since whenever $A \mathrel{R} B$, A and B have the same number of elements, and, thus, we have, $\Pr' A = \Pr' B$. Let $\Pr$ be the restriction of $\Pr'$ to $\mathcal{A}$. □

Although it is possible to define stochastic processes on disjunctive spaces, since we shall always deal with extendible disjunctive spaces, we will not need them.

CHAPTER VIII

Elements of infinitesimal analysis

In order to obtain the usual infinite probability spaces from finite spaces, and to work with infinite and infinitesimal numbers consistently, we shall use nonstandard analysis. This chapter is devoted to a brief introduction to its methods.[1] I shall stress the intuitive ideas behind nonstandard (also called infinitesimal) analysis omitting many of the technical details and proofs. We assume no previous knowledge of nonstandard methods. We develop nonstandard analysis axiomatically, so that no deep understanding of mathematical logic is required.[2] This chapter, however, contains a rather intuitive development, without attempting to be completely rigorous. The complete set of axioms, and a sketch of a proof that they are consistent, appears in Appendix A.[3]

Nonstandard analysis is a system for making precise statements that talk about a very large number of objects, and about numbers which are very close to zero. We repeat the example quoted in the Preliminaries. Let X be a random variable of mean μ and finite variance σ^2. Consider ν independent observations X_1, X_2, ..., X_ν of X. The law of large numbers can be fomulated intuitively by saying that

> if ν is a large number, then almost surely for all large $n \leq \nu$, the average $(X_1 + \cdots + X_n)/n$ is approximately equal to μ.

In order to make this statement precise in usual mathematics, one replaces the finite sequence $X_1, \ldots, X_\nu$ by an actually infinite sequence and constructs the infinite Cartesian product of the probability space Ω with itself, and, also, replaces 'approximate equal' by the appropriate ε, δ statement. The approach that uses nonstandard analysis is different. We retain, in this case, the finite

---

[1] For a more complete development see [71], [1], or [26]. My introduction is inspired mainly on [71].

[2] Nonstandard analysis was first developed by A. Robinson in the sixties (see [78]). He used the method of enlargements and provided a construction of the nonstandard numbers. Nelson, [71], uses nonstandard set theory, [70], instead of enlargements. We shall follow Robinson and use the method of enlargements. The main mathematical techniques, however, especially in Chapter XI, are mainly in [71].

[3] This axiomatic development appears, also, in [12].

sequence $X_1, \ldots, X_\nu$ but we let ν be an hyperfinite (infinite nonstandard) natural number, that is, a natural number that is larger than any finite natural number. We also replace 'x is approximately equal to y' by '$x - y$ is infinitesimal', that is, '$|x - y|$ is a positive number which is less than any real positive number'.

Nonstandad analysis shows how to deal consistently with these objects: infinite natural numbers and infinitesimal numbers.

In order to achieve these aims, without constructing it explicitly here, we shall use an extension, $^*\mathbb{R}$ of the real number system satisfying the following conditions:[4]

(1) Every mathematical notion which is meaningful for the system of real numbers is also meaningful for $^*\mathbb{R}$. In particular, $^*\mathbb{R}$ is an ordered field extension of $\mathbb{R}$.

(2) Every mathematical statement formulated with the usual notions (which are these notions will be defined more precisely later) which is meaningful and true for the system of real numbers is meaningful and true also for $^*\mathbb{R}$: provided that we interpret any reference to sets or functions in the system $^*\mathbb{R}$ not in terms of the totality of sets or functions, but in terms of a certain subset, called the family of *internal* sets or functions. (The properties of this families will be spelled out later.) For example, if the statement contains a phrase 'for all sets of numbers', we interpret this as 'for all internal sets of numbers'. Similarly, the phrase 'there exists a set of numbers' as 'there exists an internal set of numbers'. All elements of $^*\mathbb{R}$, however, are internal: the phrase 'for all numbers' is interpreted in $^*\mathbb{R}$ as 'for all elements of $^*\mathbb{R}$'.

(3) The system of internal entities of $^*\mathbb{R}$ has the following property: if S is an internal set, then all elements of S are also internal.

(4) The set $^*\mathbb{R}$ properly contains the set of real numbers, $\mathbb{R}$. In particular, there is a positive element of $^*\mathbb{R}$ which is smaller than all positive real numbers, i.e., an infinitesimal element.

The subset of $^*\mathbb{R}$ satisfying the property of being a natural number is denoted by $^*\mathbb{N}$. It is an internal set which properly contains $\mathbb{N}$. In particular, there is an individual in $^*\mathbb{N}$, which, according to the order relation in $^*\mathbb{N}$, is greater than all numbers of $\mathbb{N}$.

The field $^*\mathbb{R}$ is non-Archimedean, since it contains numbers that are greater than all numbers in $\mathbb{R}$, which is a subfield of $^*\mathbb{R}$. Thus, for some number $a \in {^*\mathbb{R}}$

$$1 < a, \; 2 < a, \; \ldots, n < a, \; \ldots, \quad \text{and so on}$$

where n is any element of $\mathbb{N}$. We have, however, that for any $a \in {^*\mathbb{R}}$, there is a $\nu \in {^*\mathbb{N}}$ such that $a < \nu$. In fact, for any positive element $a \in {^*\mathbb{R}}$, there is a $\nu \in {^*\mathbb{N}}$ such that

$$\nu \leq a < \nu + 1.$$

---

[4]For a construction of $^*\mathbb{R}$, see Appendix A. Similar constructions appear in [78], [52] and [1]. The conditions that follow are paraphrased from [78, pp. 50–56].

1. Infinitesimals

As explained above, we consider an extension of the set of natural numbers, $\mathbb{N}$, and hence of $\mathbb{R}$, which has an infinite number ν. That is, a natural number ν that is larger than all elements of $\mathbb{N}$. We denote the extended natural numbers by *$\mathbb{N}$, and the extended reals by *$\mathbb{R}$, which we call the *hyperreals*. We also extend all functions and relations defined on $\mathbb{R}$ to *$\mathbb{R}$ and assume that *$\mathbb{R}$ is an ordered field under the extended field operations and relations. Thus, $1/\nu$ is smaller than any positive real number, and thus, it is an infinitesimal that is different from 0. We also assume that *$\mathbb{N}$ is discretely ordered, that is, if $n \in$ *$\mathbb{N}$ then there is no element of *$\mathbb{N}$ strictly between n and $n+1$. We must notice that the addition of one infinite natural number, together with the requirement that *$\mathbb{R}$ be a field, means that many other infinite and infinitesimal numbers have to be added.

We shall not construct *$\mathbb{R}$ here,[5] but assume that we have done it and proceed axiomatically giving the main properties of this set that we shall need. A. Robinson, [78], was the first to prove that these additions can be done consistently.

The following definitions introduce the basic concepts of infinitesimal analysis. These definitions make sense for any ordered field extending $\mathbb{R}$. Notice that, since *$\mathbb{R}$ is an ordered field, the absolute value $|x|$ of a number x makes sense and has the usual properties.

DEFINITION VIII.1. Let $x, y \in$ *$\mathbb{R}$. Then
(1) x is *finite* if and only if $|x| < n$ for some $n \in \mathbb{N}$.
(2) x is *infinite* if and only if x is not finite, i.e., if and only if $n < |x|$ for all $n \in \mathbb{N}$. x is infinite and positive is symbolyzed by $x \approx \infty$, and x is infinite and negative by $x \approx -\infty$. We write $x \ll \infty$ to mean $x \not\approx \infty$, and $-\infty \ll x$, to mean $x \not\approx -\infty$.
(3) x is *infinitesimal* if and only if $|x| < 1/n$ for every $n \in \mathbb{N}$.
(4) x is infinitely close to y, symbolized $x \approx y$, if and only if $x - y$ is infinitesimal.
(5) $x \lesssim y$ if and only if $x \leq y + \varepsilon$ for some infinitesimal ε.
(6) $x \ll y$ if and only if $x < y$ and $x \not\approx y$.
(7) The elements of $\mathbb{R}$ are called *standard* and the elements of *$\mathbb{R} - \mathbb{R}$, *nonstandard*.

It is easy to show that $\approx$ is an equivalence relation. It is also clear that the only infinitesimal that is a real number is 0. From these definitions, it follows immediately that $|x| \ll \infty$ if and only if x is finite, and $|x| \approx \infty$ if and only if x is infinite.

By assumption, there is an infinite element ν in *$\mathbb{R}$. It is almost immediate that $1/\nu$ is infinitesimal. Since *$\mathbb{R}$ is a field, $1/\nu \in$ *$\mathbb{R}$ and, hence, *$\mathbb{R}$ contains infinitesimal elements.

The following is a list of properties, which are deduced easily from the definitions.

PROPOSITION VIII.1.

---
[5] For a sketch of the construction, see Appendix A.

(1) $x \approx 0$ if and only if x is infinitesimal.
(2) $x \approx 0$ if and only if for all positive reals r we have $|x| < r$.
(3) Infinitesimals are finite.
(4) Let $x \neq 0$. Then $x \approx 0$ if and only if $1/x$ is infinite.
(5) If x and y are finite, then so are $x + y$ and xy.
(6) If x and y are infinitesimal, then so are $x + y$ and xy.
(7) If $x \approx 0$ and $|y| \ll \infty$, then $xy \approx 0$.
(8) $x \lesssim y$ and $y \lesssim x$ if and only if $x \approx y$.
(9) For all $n \in {}^*\mathbb{N}$, n is standard if and only if n is finite.
(10) The only standard infinitesimal is 0.
(11) If x and y are (standard) real numbers, then

$$x \approx y \iff x = y.$$

From these properties, we deduce:

PROPOSITION VIII.2.
(1) $x \approx y$ and $z \not\approx 0 \implies \dfrac{x}{z} \approx \dfrac{y}{z}$.
(2) $x \approx y$ and $|u| \ll \infty \implies ux \approx uy$.
(3) $x \approx z$, $y \approx z$ and $x \leq u \leq y \implies u \approx z$.
(4) $x \approx 0 \implies x^p \approx 0$, where p is a positive rational standard number.

We also need *relative approximate equality*:

DEFINITION VIII.2. Let $\varepsilon \neq 0$. Then we define

$$x \approx y \ (\varepsilon) \quad \text{if and only if} \quad \dfrac{x}{\varepsilon} \approx \dfrac{y}{\varepsilon}.$$

Relative approximate equality has the same properties as approximate equality. In particular, all properties stated above are true when '$\approx$' is replaced by '$\approx \ (\varepsilon)$'.

We shall occasionally use the standard part of a finite number. In any case, its introduction helps to understand the structure or the hyperreal line. We first prove:

PROPOSITION VIII.3 (STANDARD PART). *Let a be a finite number. Then there is a unique real number r such that $a \approx r$.*

PROOF. Since the only real infinitesimal is 0, if r and $s \in \mathbb{R}$ satisfy $r \approx s$, we have $r = s$. Hence, the unicity follows.

We now show the existence. Let a be finite. Let $r = \sup\{s \in \mathbb{R} \mid s \leq a\}$. This least upper bound exists, because a is finite. Suppose that $r \not\approx a$. Then $|r - a| \not\approx 0$ and, hence, there is a real number s such that $0 < s \leq |r - a|$. If $r \gg a$, then $r - s \geq a$, while if $r \ll a$, $r + s \leq a$ and, thus, in both cases, r is not a least upper bound. Therefore, $r \approx a$. □

In accordance with this theorem we can define the function st on the finite elements of $^*\mathbb{R}$ to the real numbers by

$$\operatorname{st} a = {}^\circ a = \text{ the unique real } r \text{ such that } r \approx a.$$

It is clear that for any finite number a, $a = \operatorname{st} a + \varepsilon$, where $\varepsilon \approx 0$. Thus, it is easy to show that st is a homomorphism from the finite elements of $^*\mathbb{R}$ onto $\mathbb{R}$.

We can also introduce infinitesimal and infinite elements in other sets, besides the real numbers, such as $^n\mathbb{R}$ or the complex numbers, $\mathbb{C}$. As an example, I shall consider $\mathbb{C}$. The set $^*\mathbb{C}$ can be considered as the set of all numbers $x + iy$, where $x, y \in {}^*\mathbb{R}$ and $i = \sqrt{-1}$. A complex number z is infinitesimal, if $|z|$ is infinitesimal as a real number. That is, the complex number $z = x + iy$ is infinitesimal, if and only if both x and y are infinitesimals as real numbers. The complex numbers z and u are infinitely close, in symbols $z \approx u$, if and only if $z - u$ is infinitesimal, i.e., $|z - u| \approx 0$. Thus, if $z_1 = x_1 + iy_1$ and $z_2 = x_2 + iy_2$, then $z_1 \approx z_2$ if and only if $x_1 \approx x_2$ and $y_1 \approx y_2$, i.e., the real and imaginary parts of z_1 and z_2 are approximately equal. We can similarly extend the definition of infinite numbers.

We now introduce the definitions and notations for $^{n*}\mathbb{R}$. The elements of $^{n*}\mathbb{R}$ are referred to as vectors and denoted $\vec{x}$. In general, we adopt the convention that $\vec{x} = \langle x_1, x_2, \ldots, x_n \rangle$, $\vec{y} = \langle y_1, y_2, \ldots, y_n \rangle$, etc. The number n will usually be clear from the context.

We denote by $\|\vec{x}\|$ the length of $\vec{x}$, i.e.

$$\|\vec{x}\| = \sqrt{x_1^2 + x_2^2 + \cdots + x_n^2},$$

where $x_1, \ldots, x_n$ are the components of $\vec{x}$. Thus, the distance between $\vec{x}$ and $\vec{y}$ can be written as $\|\vec{x} - \vec{y}\|$.

We say that $\vec{x}$ is *real* or *standard*, if all its components $x_1, \ldots, x_n$, are real.

We say that $\vec{x}$ is infinitesimal, if $\|\vec{x}\|$ is infinitesimal. We also say that $\vec{x}$ is approximately equal to $\vec{y}$, in symbols, $\vec{x} \approx \vec{y}$, if $\|\vec{x} - \vec{y}\|$ is infinitesimal. We have the following easy proposition:

PROPOSITION VIII.4.

(1) $\vec{x}$ is infinitesimal if and only if its components, $x_1, x_2, \ldots, x_n$ are infinitesimals.
(2) $\vec{x} \approx \vec{y}$ if and only if $x_1 \approx y_1$, $x_2 \approx y_2$, $\ldots$, $x_n \approx y_n$.

2. Standard, internal and external sets

In order to incorporate all of standard mathematics, and prove our main theorems, we must distinguish between standard, internal, and external sets. We shall only introduce here the minimum required for our development; in Appendix A, a complete set of axioms is given, so that all of analysis can be obtained on its basis.

We call the usual mathematical notions, not including those that are obtained from the notions of 'infinitesimal', 'standard', or any of the new notions introduced before, *standard notions*. In particular, the membership relation, $\in$, is standard. In fact, as it is done in Appendix A, we can just have the membership and equality relations as the only primitive standard notions, and the notion of 'being standard', as the only primitive nonstandard notion. All nonstandard notions can be defined with the latter notion. For instance, we can define 'x is infinitesmal' by '$x < 1/n$ for every standard natural number n'.

We need for our development the notions of standard and internal sets and functions. The standard sets we need are subsets of $^{n*}\mathbb{R}$, and the standard funtions which are needed are functions from subsets of $^{n*}\mathbb{R}$ into $^*\mathbb{R}$. We also consider standard and internal sets which are families of these subsets or functions defined on these families into $^*\mathbb{R}$. Instead of defining *standard objects*, we shall give some of their properties. For a more complete study see Appendix A.

In order to codify the sets we need, we introduce the following definition:
For any set X, let

$$V_0(X) = X$$
$$V_{n+1}(X) = V_n(X) \cup \mathcal{P}(V_n(X)).$$

All objects that we need are in one of the sets $V_m(^*\mathbb{R})$, for a certain m. By the usual set-theoretic construction, the functions of interest are also in these sets. As an example we shall determine the level m where a function $f : {}^*\mathbb{R} \to {}^*\mathbb{R}$ is. Such a function is a set of ordered pairs of elements of $^*\mathbb{R}$, so we have to see where ordered pairs are. Let $a, b \in {}^*\mathbb{R}$. We have, using the customary definition of ordered pair

$$\langle a, b \rangle = \{\{a\}\{a, b\}\}.$$

The sets $\{a\}$ and $\{a, b\}$ are subsets of $^*\mathbb{R}$, and, hence, belong to $V_1(^*\mathbb{R})$. Thus, the ordered pair $\langle a, b \rangle$ is a subset of $V_1(^*\mathbb{R})$, and belongs to $V_2(^*\mathbb{R})$. Therefore, f is a subset of $V_2(^*\mathbb{R})$ and belongs to $V_3(^*\mathbb{R})$.

Similarly, n–tuples of elements of $^*\mathbb{R}$, for $n \in \mathbb{N}$, which are functions from $\{1, \ldots, n\}$ to $^*\mathbb{R}$, belong to $V_3(^*\mathbb{R})$, and, thus, $^{n*}\mathbb{R}$, the set of all n–tuples of elements of $^*\mathbb{R}$, is in $V_4(^*\mathbb{R})$.

As another example, suppose that $\Omega \in V_m(^*\mathbb{R})$. A probability measure is a function Pr from subsets of Ω into $^*\mathbb{R}$, and, hence, Pr $\in V_{m+4}(^*\mathbb{R})$. In a similar way, we can see that all sets, families of sets, or functions, can always be assumed to belong to one of these sets $V_m(^*\mathbb{R})$.

We shall need the following properties of standard sets:

(1) Real vectors, i.e., elements of $^n\mathbb{R}$ (but not $^n\mathbb{R}$ itself) are standard.
(2) The sets $^*\mathbb{R}$ and $^*\mathbb{N}$ are standard.
(3) If a set A can be defined as the subset of $^{n*}\mathbb{R}$ or $V_m(^*\mathbb{R})$, for a certain m, that satisfies a condition which only involves standard notions, and other, already defined, standard objects (i.e., real numbers, standard sets or functions), then A or f are also standard. Thus, in order to define standard objects we do not use "infinitesimal", "real", "finite" (in its new sense), "infinite", "$\approx$", etc.
(4) A finite set of standard objects is standard. Unions and intersections of a finite number of standard sets are standard. Sums, differences, products, quotients, compositions, and inverses of standard functions are standard.

2. STANDARD, INTERNAL AND EXTERNAL SETS

(5) If f is standard, then the domain of f is standard, and, for every standard x in its domain, $f(x)$ is also standard. In particular, if $x \in \mathbb{R}$ is in the domain of f, $f(x) \in \mathbb{R}$.

(6) This is the main property of standard sets. Intuitively it says that a standard set is determined by its standard objects. For instance, if A and B are standard subsets of $^*\mathbb{R}$ that contain the same real numbers (i.e., $A \cap \mathbb{R} = B \cap \mathbb{R}$), then $A = B$. A similar property is valid for $^n\mathbb{R}$. In general, if A and B are standard sets that have the same standard objects, then $A = B$.

(7) If f and g are standard functions with the same domain and such that for each real x (or each standard x) in its domain, $f(x) = g(x)$, then $f = g$.

(8) Every subset of $^n\mathbb{R}$ and every function on $^n\mathbb{R}$ has a standard extension. In general, every set S of standard objects has a standard extension, $^*S$, such that the only standard objects in $^*S$ are the elements of S. A similar statement is true for functions.

Intervals with real endpoints are examples of standard sets. For instance, if a, b are standard real numbers, then

$$[a, b] = \{x \in {^*\mathbb{R}} \,|\, a \leq x \leq b\}$$

is standard.

Elementary functions, are also standard. For instance,

$$f(x) = \frac{x^2 - 1}{x^2 + 1}$$

is a standard function defined over all of $^*\mathbb{R}$.

It is clear by (6) that the standard extension of a set of standard objects is unique.

It is easy to show that if a standard set A contains all the reals, then $A = {^*\mathbb{R}}$. Also, if A is a standard set of positive hyperreals which contains all positive reals, then $A = {^*\mathbb{R}^+}$, the set of positive hyperreals.

In general, we have the following proposition:

PROPOSITION VIII.5. *If A and B are standard with $A \subseteq B$ and A contains all the standard objects in B, then $A = B$.*

PROOF. Let A and B be standard, $A \subseteq B$, where A contains all the standard objects in B. Then, if $x \in B$ is standard, then $x \in A$. Thus, all standard objects in B are in A. But $A \subseteq B$, and, hence, all standard objects in A are in B. Therefore, A and B contain the same standard objects. Since A and B are standard, $A = B$. □

PROPOSITION VIII.6. *If g and h are standard functions with the same domain $A \subseteq {^{n*}\mathbb{R}}$ into $^{m*}\mathbb{R}$, for certain n and m, and for every $\vec{x} \in A$, $g(\vec{x}) \approx h(\vec{x})$, then $g = h$.*

PROOF. If $\vec{x}$ is real in A, then $g(\vec{x})$ and $h(\vec{x})$ are real, and, as $g(\vec{x}) \approx h(\vec{x})$, we have $g(\vec{x}) = h(\vec{x})$. Thus, the standard set $\{\vec{x} \in A \,|\, g(\vec{x}) = h(\vec{x})\}$ contains all the real vectors in A. Since A is standard

$$A = \{\vec{x} \in A \,|\, g(\vec{x}) = h(\vec{x})\}.$$

□

The notion of internal set or function is wider than the corresponding standard notion. The family of *internal objects* has the following properties.

(1) Every standard set is internal.
(2) Every element of an internal set is internal.
(3) A set or function is internal if it can be defined using only standard mathematical notions and other, already defined, internal objects (i.e., elements of $^{n*}\mathbb{R}$, internal sets or functions). That is, just as for standard objects, in order to define internal objects we do not use "infinitesimal", "real", "finite", "infinite", "$\approx$", etc. On the other hand, we may use for defining an internal set, elements of $^*\mathbb{R}$ and not just of $\mathbb{R}$.
(4) The union and intersection of an internal family of internal sets is internal and the family of the internal subsets of an internal set, is internal. Sums, differences, products, quotients, composition and inverse of internal functions are internal.

A set or function which is not internal is called *external*.

We can conclude from (1) and (2), that all elements of $^{n*}\mathbb{R}$, and $^{n*}\mathbb{R}$ itself, are internal.

The intervals of $^*\mathbb{R}$ are instances of internal sets. For example, if $a, b \in {}^*\mathbb{R}$, not necessarily real (for instance a and b may be infinitesimals) the interval $[a, b]$ can be defined by

$$[a, b] = \{x \in {}^*\mathbb{R} \,|\, a \leq x \leq b\}.$$

The condition which defines $[a, b]$, $a \leq x \leq b$, only has usual mathematical notions, in this definition a and b occur, which are elements of $^*\mathbb{R}$, and the only set that occurs, in this case $^*\mathbb{R}$, is internal.

On the other hand, as we shall show later, the set of all infinitesimals is not internal. We can see that in its definition we use the notion of "infinitesimal", so that the fact that it is not internal does not violate the rules given above.

All elementary functions are standard, and hence, internal, because they can be defined with standard notions and particular real numbers. For instance, the function

$$f(x) = \frac{1}{x}, \quad \text{if } x \neq 0$$

is standard because it can be defined with standard notions and the number zero.

Another internal function, but this time not standard, is $f(x) = \varepsilon$ for all $x \in {}^*\mathbb{R}$, where $\varepsilon \approx 0$.

On the other hand, the function st defined in the previous section is external. It needs for its definition the notions of "real" (or "standard") and "$\approx$".

We assume the following principle, called the *transfer principle*:

2. STANDARD, INTERNAL AND EXTERNAL SETS

Every standard theorem is true for standard objects.

In Appendix A, we indicate how to prove this assumption from our axioms. As was stated in the introduction to this chapter, every phrase of the form 'there is a set' or 'for every set' should be replaced, in the statement of each standard theorem, by 'there is an internal set' or 'for every internal set'. Similar changes have to be done for functions. Most of the standard theorems we shall use do not contain these types of phrases, so that we don't need to worry about them.

The least upper bound axiom is not true for external subsets of $^*\mathbb{R}$. For instance, $\mathbb{R}$ is a bounded (in $^*\mathbb{R}$) subset of $^*\mathbb{R}$ which does not have a least upper bound. On the other hand, although we shall not use it, the least upper bound principle is valid for internal subsets. Thus, $\mathbb{R}$ is external.

We also need some properties of $^*\mathbb{N}$.

(1) $\mathbb{N} \subseteq {^*\mathbb{N}}$.
(2) The finite elements of $^*\mathbb{N}$ are exactly the elements of $\mathbb{N}$.
(3) $^*\mathbb{N}$ is a proper extension of $\mathbb{N}$. That is, there are elements in $^*\mathbb{N}$ which are not in $\mathbb{N}$. These elements of $^*\mathbb{N} - \mathbb{N}$ are infinite.
(4) For every positive hyperreal number $x \in {^*\mathbb{R}}$, there is a natural number $\nu \in {^*\mathbb{N}}$, such that
$$\nu - 1 < x \leq \nu.$$
(5) $^*\mathbb{N}$ satisfies the Internal Induction Principle, that is:
   Internal Induction Principle: If S is an internal subset of $^*\mathbb{N}$ that satisfies the following conditions:
   (a) $1 \in S$;
   (b) $n \in S \implies n+1 \in S$;
   then we have $S = {^*\mathbb{N}}$.
Note that S must be internal.

With these properties, it is easy to show that internal subsets of $^*\mathbb{N}$ satisfy all standard conditions satisfied by subsets of $\mathbb{N}$ in usual mathematics. For instance, every nonempty internal subset of $^*\mathbb{N}$ has a least element, and every nonempty internal bounded subset of $^*\mathbb{N}$ has a largest element.

The induction principle, without the restriction on internal sets, is, of course, valid for $\mathbb{N}$. We shall call this induction principle for $\mathbb{N}$, the *external induction principle*.

It is also clear that
$$1 \in \mathbb{N}$$
and
$$n \in \mathbb{N} \implies n+1 \in \mathbb{N}.$$

Thus, if $\mathbb{N}$ were internal, by the internal induction principle, $\mathbb{N}$ would be equal to $^*\mathbb{N}$. But we know it is not. Therefore, $\mathbb{N}$ is external. We thus deduce the *overflow principles*, [78].

THEOREM VIII.7 (OVERFLOW PRINCIPLES).

(1) *If an internal set A contains all finite natural numbers larger than a certain finite natural number n, then A also contains all infinite natural numbers less than a certain infinite natural number ν.*

(2) *If an internal set A contains all infinite natural numbers less than a certain $\nu \approx \infty$, then it contains all finite natural number larger than a certain finite number n.*

(3) *If an internal set A contains all positive infinitesimals, then A contains a positive real number d.*

(4) *If an internal set A contains all positive noninfinitesimal numbers less than a certain real number, then A contains all infinitesimals larger than a certain infinitesimal.*

(5) (**Standard overflow**). *If a standard subset A of* $^*\mathbb{R}$ *is nonempty, then it contains a finite (standard) real number.*

The third overflow principle shows that the set of infinitesimals is external.

PROOF OF (1). Suppose that A is internal and let A contains all finite natural numbers larger than the finite natural number n, and $B = A \cup \{m \in {}^*\mathbb{N} \mid m \leq n\}$. Then B is internal. If $^*\mathbb{N} - B$ is empty, then the theorem is obviously true. So, in order to complete the proof, suppose that it is nonempty. The set $^*\mathbb{N} - B$ is internal, and, hence by the internal induction principle, since it is assumed to be nonempty, it has a minimal member. Let $\nu + 1$ be that minimal element, which cannot be finite. Then ν satisfies the conditions of the theorem. □

PROOF OF (2). Suppose that A satisfies the hypothesis of (2). Then the set $B = A \cup \{n \in {}^*\mathbb{N} \mid n \geq \nu\}$ is internal. If $^*\mathbb{N} - B$ is empty, we are done. Let, then, $^*\mathbb{N} - B \neq \emptyset$. Then $^*\mathbb{N} - B$ is an internal set of finite numbers, because all infinite numbers are in B. Thus, $^*\mathbb{N} - B$ is bounded and, hence, it contains a maximal element, n. The number n, which must be finite, is the required number. □

PROOF OF (3). Let B be the internal subset of $^*\mathbb{N}$ defined by

$$B = \{\nu \mid (0, \frac{1}{\nu}) \subseteq A\}.$$

From the hypothesis of (3) we deduce that B contain all infinite natural numbers. Then, by (2), B contains a finite natural number n. Take $d = 1/n$. (4) is obtained from (1), as (3) is from (2). □

PROOF OF (5). Suppose that A is a standard subset of $^*\mathbb{R}$ which does not contain real numbers. The empty set, $\emptyset$, is obviously standard and contains the same standard reals as A, namely none. Hence, $A = \emptyset$. □

As a simple example of the use of overflow, we shall prove the following propositions.

PROPOSITION VIII.8. *A number $x \approx 0$ if and only if $|x| < 1/\nu$ for some nonstandard (infinite) natural number ν.*

2. STANDARD, INTERNAL AND EXTERNAL SETS

PROOF. It is clear that if $|x| < 1/\nu$ for a certain infinite ν, then $|x| < 1/n$, for every finite natural number n, because a finite number is always less than an infinite number.

So, in order to prove the converse, assume that $|x| < 1/n$ for every finite natural number n. Then, the set

$$A = \{n \in {}^*\mathbb{N} \mid |x| < \frac{1}{n}\}$$

is internal and contain all finite numbers. Hence, A contains an infinite number, ν. This is the required number. □

PROPOSITION VIII.9. *Let $f : S \to {}^*\mathbb{R}$ be an internal function such that for every $x \in S$, $f(x)$ is finite. Then there is a finite number m such that $f(x) \leq m$, for every $x \in S$.*

The proposition asserts that an internal function that has only finite values is bounded by a finite number.

PROOF. The set $\{n \in {}^*\mathbb{N} \mid f(x) \leq n, \text{ for every } x \in S\}$ is internal and contains all infinite natural numbers. Hence, by overflow, it contains a finite natural number m. This is a bound of f. □

We also need a lemma proved by A. Robinson, [78].

LEMMA VIII.10 (ROBINSON'S LEMMA). *Let f be an internal function such that for every finite $x \geq a$ in its domain, where a is finite, $f(x) \approx 0$. Then there is an infinite b such that, for all x with $a \leq x \leq b$, we have $f(x) \approx 0$.*
In particular, we have:
Let $x_1, x_2, \ldots, x_n, \ldots, x_\mu$, for $n \leq \mu \approx \infty$, be an internal sequence. If $x_i \approx 0$ for every finite natural number $i \in \mathbb{N}$, then there is an infinite $\nu \leq \mu$ such that $x_i \approx 0$ for every $i \leq \nu$.

Most frequently, we shall use the statement for sequences.

PROOF. We cannot use overflow directly on the set of x such that $f(x) \approx 0$, because this set is not internal. So we consider instead the set S of all x such that $|f(y)| \leq 1/y$ for all $y \leq x$, y in the domain of f. The set S contains all finite natural numbers larger than a, and, hence, by overflow it contains an infinite number ν. Let $y \leq \nu$. If y is finite, then $f(y) \approx 0$, by hypothesis. If y is infinite, then $|f(y)| \leq 1/y$, and hence $f(y) \approx 0$. Thus, $f(y) \approx 0$, for all $y \leq \nu \approx \infty$.

The second assertion, for sequences, follows from the first, or it can be proved directly in a similar way. □

In a similar way, one can prove that for every infinitesimal $\delta > 0$ there is an infinitesimal $\varepsilon > 0$ such that δ/ε is infinitesimal (or ε/δ is infinite), i.e., ε is a much larger infinitesimal than δ. Also, it is easy to prove that if ν is an infinite natural number then there is another infinite natural number $\mu < \nu$ such that $\nu - \mu$ is infinite.

If ν is an infinite natural number, the set $\{1, 2, \ldots, \nu\}$ is internal. This fact motivates the following definition:

DEFINITION VIII.3 (HYPERFINITE SETS). If the number of elements of an internal set S is ν, where ν is a natural number, then S is called *hyperfinite*.

More formally, an internal set S is hyperfinite if there is an internal one-one function f from $\{1, 2, \ldots, \nu\}$ onto S, where $\nu \in {}^*\mathbb{N}$. The number ν is called the *internal cardinality* of S; we shall use $\#S$ to denote the internal cardinality of S.

If ν, the number of elements of a set S, is an infinite natural number, S is not, strictly speaking, finite, but hyperfinite. In particular, $\{1, 2, \ldots, \nu\}$ is hyperfinite.

Hyperfinite sets are useful, because they can approximate standard sets in a sense we shall discuss later, and, although they may be infinite, by internal induction they behave formally as finite sets; thus combinatorial and counting arguments valid for finite sets are also true for hyperfinite sets.

For instance, if $T = \{t_1, t_2, \ldots, t_\nu\}$ is a hyperfinite set of numbers (which contains ν elements with $\nu \approx \infty$), then T has a maximal and a minimal element; the union of a hyperfinite family of hyperfinite sets is hyperfinite; the sum of any hyperfinite subset A of ${}^*\mathbb{R}$, $\sum_{x \in A} x$, always exists. As we shall see later, integration can be obtained from hyperfinite summation.

We shall discuss a few examples of how hyperfinite sets can approximate other sets.

EXAMPLE VIII.1. Let $T = \{0, 1, \ldots, \nu\}$, where ν is an infinite natural number. This is a hyperfinite analogue of $\mathbb{N}$, since the set of standard images of the finite elements of T, $\mathrm{st}'' T$, is $\mathbb{N}$.

EXAMPLE VIII.2. Consider, now, the closed interval $[0, 1]$. We define the discrete time line
$$T = \{0, dt, 2dt, \ldots, 1\},$$
where $dt = 1/\nu$ for a certain infinite $\nu \in {}^*\mathbb{N}$. The set T is clearly internal since it can be defined as the set
$$\{k\, dt \mid 0 \leq k \leq \nu\}.$$
This definition also shows that T is hyperfinite with $\#T = \nu + 1$, because the bijection $f(k) = k\, dt$ is clearly internal. Now, for each real number r in $[0, 1)$, there is exactly one k such that $k\, dt \leq r < (k+1)\, dt$. Just take k to be the largest integer $\leq \nu$ such that $k\, dt \leq r$. This k exists by internal induction, since the set of integers $\leq \nu$ is hyperfinite (see, also, the observation at the end of page 172). For this k, $r - k\, dt \leq dt$, and, hence, $r \approx k\, dt$. Thus, in a sense, T approximates $[0, 1] \cap \mathbb{R}$, since $\mathrm{st}'' T = [0, 1] \cap \mathbb{R}$. If we take $\nu = \mu!$ for a certain infinite $\mu \in {}^*\mathbb{N}$, then T contains all standard rationals between 0 and 1.

EXAMPLE VIII.3. If instead of taking T the set of $k\, dt$ for $k \leq \nu$, we consider
$$P = \{k\, dt \mid 0 \leq k \leq \nu^2\},$$

then P approximates the whole interval $[0, +\infty)$, in the sense that for any $x \in [0, +\infty) \cap \mathbb{R}$ there is a $y \in P$ such that $y \approx x$, i.e., st$''P = [0, +\infty) \cap \mathbb{R}$.

In our development of probability theory, we shall deal almost exclusively with hyperfinite sets and with functions, for instance stochastic processes, from these sets into $^*\mathbb{R}$.

Important hyperfinite sets are the near intervals as in Example VIII.2:

DEFINITION VIII.4 (NEAR INTERVALS). Let $\nu \in {}^*\mathbb{N}$. A hyperfinite set $T = \{t_0, t_2, \ldots, t_\nu\}$ is a near interval for $[a, b]$, if $t_0 = a$, $t_\nu = b$, and the difference $t_{i+1} - t_i = dt_i$ is infinitesimal for every $i = 0, 1, 2, \ldots, \nu - 1$.

As an example, we have that, if T is a near interval for $[a, b]$, then for all $x \in [a, b]$ there is a $t \in T$ such that $x \approx t$. It is enough to take the maximum of the $t \in T$ such that $t \leq x$.

EXAMPLE VIII.4. We now consider a hyperfinite approximation to the unit circle. We take the unit circle $\mathbf{C}$ to be the set of complex numbers u such that $|u| = 1$; that is, the set of numbers $e^{i\theta}$, for $\theta \in [0, 2\pi]$.

Let ν be an infinite natural number, and let $d\theta = 2\pi/\nu$. Then a hyperfinite approximation to $\mathbf{C}$ is the set

$$C = \{e^{ik\,d\theta} \mid 0 \leq k \leq \nu\}.$$

Again we have that every element of $\mathbf{C}$ is approximated by an element of C. It is clear that C can be obtained from a near interval in the following way. Let

$$T = \{0, d\theta, 2\,d\theta, \ldots, 2\pi\}.$$

Then T is a near interval for $[0, 2\pi]$, and $dt = d\theta$, for every $t \in T$. We have

$$C = \{e^{it} \mid t \in T\}.$$

3. Internal analogues of convergence

We shall consider three kinds of sequences of numbers: sequences x_n for $n \in \mathbb{N}$, sequences x_n for $1 \leq n \leq \nu$, where $\nu \approx \infty$, and sequences x_n, for $n \in {}^*\mathbb{N}$. The first type cannot be internal. In order to deal with convergence for this first kind of sequences, we need to add one more principle to our characterization of internal sets:

Denumerable comprehension: *If $x_1, x_2, \ldots, x_n, \ldots$, for $n \in \mathbb{N}$, is a sequence of elements of an internal set A, then there is an internal sequence of elements of A, $x_1, x_2, \ldots, x_\nu, \ldots$, for $\nu \in {}^*\mathbb{N}$, which extends it.*

There are several nonstandard analogues of the notion of a convergent sequence. One of them is near convergence of sequences as in [71, p. 20]. This analogue uses sequences with domain the numbers less than an infinite natural number. In this case, the sequences contain a hyperfinite number of elements.

DEFINITION VIII.5 (NEAR CONVERGENCE). Let ν be an infinite natural number. We say that the sequence of hyperreals $x_1, x_2, \ldots, x_\nu$ is *nearly convergent* if there is a hyperreal number x such that $x_\mu \approx x$ for all infinite $\mu \leq \nu$. We also say in this case that the sequence $x_1, \ldots, x_\nu$ *nearly converges to* x.

We call the usual notion of convergence for sequences with domain $\mathbb{N}$, S-convergence, that is, the sequence $\{x_n\}_{n \in \mathbb{N}}$ *S-converges* to x, if for every $\varepsilon \gg 0$ there is an $n_0 \in \mathbb{N}$ such that $|x_n - x| < \varepsilon$ for every $n \in \mathbb{N}, n \geq n_0$.

Notice that if a sequence $x_1, \ldots, x_\nu$ nearly converges to x, then it also nearly converges to y if and only if $x \approx y$. The following theorem, which is implicit in the discussion of [71, p. 20], may help to understand why the notion of near convergence is an analogue of S-convergence.

THEOREM VIII.11. *Let $x_1, x_2, \ldots, x_n, \ldots$, for $n \in \mathbb{N}$, be a sequence of hyperreal numbers. Then the sequence S-converges to a number x if and only if there is an internal extension $x_1, x_2, \ldots, x_\nu$, for a certain $\nu \approx \infty$, that nearly converges to x.*

In fact, if the sequence S-converges to x, then for every internal extension $x_1, x_2, \ldots, x_\mu$, for $\mu \approx \infty$, there is a $\nu \approx \infty$, $\nu \leq \mu$, such that $x_1, x_2, \ldots, x_\nu$ nearly converges to x.

PROOF. Suppose that the sequence $x_1, \ldots, x_\nu$ nearly converges to x and take the (external) restriction of the sequence to $\mathbb{N}$, i.e., $x_1, x_2, \ldots, x_n, \ldots$ for $n \in \mathbb{N}$. Let $\varepsilon \gg 0$. The internal set $\{m \mid \text{for all } n \geq m, |x_n - x| \leq \varepsilon\}$ contains all infinite numbers $\leq \nu$. Hence, by overflow, Theorem VIII.7, it contains an $n_\varepsilon \ll \infty$. Hence, $x_1, x_2, \ldots,$ S-converges to x.

On the other hand, suppose that $x_1, x_2, \ldots$ is a sequence defined on $\mathbb{N}$, S-convergent to x. Then, by denumerable comprehension, we can extend the sequence to an internal sequence on $^*\mathbb{N}$. Let $A_m = \{n \in {}^*\mathbb{N} \mid |x_n - x| \leq 1/m\}$, for m a finite natural number. Then A_m is internal and contains all finite $n \geq n_m$ for a certain finite n_m. Hence, by overflow, Theorem VIII.7, it contains an infinite μ_m such that $|x_n - x| \leq 1/m$ for all n with $n_m \leq n \leq \mu_m$. Now, make the sequence μ decreasing by taking $\mu'_m = \inf\{\mu_j \mid j \leq m\}$. Then we still have $|x_n - x| \leq 1/m$ for n satisfying $n_m \leq n \leq \mu'_m$. We now need an infinite number less that μ'_m for every finite m. Since we cannot take the minimum of all the μ'_m, for all $m \in \mathbb{N}$, because it is not internal, we consider an internal extension of the sequence μ'_m to $^*\mathbb{N}$. The internal set $S_1 = \{m \in {}^*\mathbb{N} \mid \mu'_j > \mu'_m \text{ for all } j < m\}$ contains all finite natural numbers, and hence by overflow, Theorem VIII.7, it contains an infinite natural number η_1. The reciprocals of the μ'_j are infinitesimal for every finite j. Hence, by Robinson's lemma VIII.10, there is an infinite number η_2 such that μ'_j is infinite for every $j \leq \eta_2$. Let η be the least of η_1 and η_2 and let $\nu = \mu'_\eta$. Then for any finite number m, if n is an infinite number such that $n \leq \nu$, then $n_m \leq n \leq \mu_m$ and, hence, $|x_n - x| \leq 1/m$. Thus, $|x_n - x| \approx 0$, for any infinite $n \leq \nu$. Thus, the sequence $x_1, x_2, \ldots, x_\nu$ nearly converges to x. □

We shall occasionally need another analogue of the ordinary notion of convergence applied to sequences of numbers $\{x_n\}_{n \in {}^*\mathbb{N}}$.

DEFINITION VIII.6. We say that the sequence $\{x_n\}_{n \in {}^*\mathbb{N}}$ *-converges to x, if for every $\varepsilon > 0$ (real or not) there is a $\nu_0 \in {}^*\mathbb{N}$ (finite or not) such that for every $\nu \geq \nu_0$, $|x_\nu - x| < \varepsilon$.

We have the following theorem that relates, for standard sequences, the three notions of convergence. For internal sequences that are not standard the theorem is not true, as it can be seen by the sequence $x_n = \varepsilon \approx 0$, for every $n \in {}^*\mathbb{N}$, which is internal, nearly converges to zero (and, hence, S-converges to zero), but does not *-converge to zero.

THEOREM VIII.12. *Let $x_1, x_2, \ldots, x_n, \ldots$, for $n \in {}^*\mathbb{N}$ be a standard sequence, and let $x \in \mathbb{R}$. Then the following conditions are equivalent:*
  (1) *The restriction of the sequence to $\mathbb{N}$, $\{x_n\}_{n \in \mathbb{N}}$, S-converges to x.*
  (2) *For every $\nu \approx \infty$, the restriction of the sequence to ν, $\{x_n\}_{n \leq \nu}$, nearly converges to x.*
  (3) *The sequence *-converges to x.*

PROOF. We have already proved (Theorem VIII.11) that for internal, and hence standard, sequences (1) and (2) are equivalent. We now prove that (2) implies (3). Suppose (2). Then the standard set A consisting of the $\varepsilon > 0$ satisfying that there is an $n \in {}^*\mathbb{N}$ such that for all $m \geq n$, $|x_m - x| \leq \varepsilon$ contains all positive real numbers. Hence, it contains all positive hyperreals. Thus, we have (3).

Finally, assume (3). Let $\varepsilon > 0$ be a real number. Then the set B consisting of the $n \in {}^*\mathbb{N}$ such that for every $m \geq n$, $|x_m - x| \leq \varepsilon$ is standard and nonempty. Then, by standard overflow, Theorem VIII.7, B contains a finite n. Thus, we have (1). Hence (3) implies (1). □

4. Series

We now introduce the nonstandard analogue to convergent series corresponding to convergent sequences.

DEFINITION VIII.7 (CONVERGENCE OF SERIES).
  (1) Let $x_1, x_2, \ldots, x_\nu$ be a sequence, where ν is an infinite natural number. We say that $\sum_{i=1}^{\nu} x_i$ is *nearly convergent* if the sequence of partial sums, $y_n = \sum_{i=1}^{n} x_i$, for $n = 1, 2, \ldots, \nu$, is nearly convergent.
  (2) Let $\{x_n\}_{n \in \mathbb{N}}$ be a sequence. We say that $\sum_{i \in \mathbb{N}} x_i = \sum_{i=1}^{\infty} x_i$ is *S-convergent* if the sequence of partial sums, $y_n = \sum_{i=1}^{n} x_i$, for $n = 1, 2, \ldots$, is S-convergent.
  (3) Let $\{x_n\}_{n \in {}^*\mathbb{N}}$ be a sequence. We say that $\sum_{i \in {}^*\mathbb{N}} x_i = \sum_{i=1}^{\infty} x_i$ is **-convergent* if the sequence of partial sums, $y_n = \sum_{i=1}^{n} x_i$, for $n \in {}^*\mathbb{N}$, is *-convergent.

The notation $\sum_{i=1}^{\infty} x_i$ is ambiguos. It has a different meaning for a sequence on $\mathbb{N}$ and a sequence on ${}^*\mathbb{N}$.

Since $\sum_{i=1}^{\nu} x_i$ denotes a number and not a sequence, there is an abuse of language here. Notice that the sum, $\sum_{i=1}^{\nu} x_i$, always exists, but the series, written in the same way $\sum_{i=1}^{\nu} x_i$, does not always nearly converge. In order for the series to nearly converge, we must have

$$\sum_{i=1}^{\nu} x_i \approx \sum_{i=1}^{\mu} x_i,$$

for every infinite $\mu \leq \nu$.

It is clear that if the series $\sum_{i=1}^{\nu} x_i$ nearly converges, then it nearly converges to its sum $y = \sum_{i=1}^{\nu} x_i$, and that the series $\sum_{i=1}^{\nu} x_i$ nearly converges if and only if the tails $\sum_{i=\mu}^{\nu} x_i$ are infinitesimal for all infinite $\mu \leq \nu$. Thus, we have the following theorem.

THEOREM VIII.13 (COMPARISON TEST). *Let $x_1, \ldots, x_\nu$, and $y_1, \ldots, y_\nu$ be internal sequences where ν is infinite. Then if $|x_n| \leq |y_n|$ for all infinite $n \leq \nu$ and $\sum_{i=1}^{\nu} |y_i|$ nearly converges, we have that $\sum_{i=1}^{\nu} |x_i|$ also nearly converges.*

We also have:

THEOREM VIII.14. *Let $x_1, x_2, \ldots, x_\nu$ be an internal sequence where ν is infinite, and suppose that $|x_i| \ll \infty$ for all $i \ll \infty$ and that $\sum_{i=1}^{\nu} |x_i|$ nearly converges. Then $\sum_{i=1}^{\nu} |x_i| \ll \infty$.*

PROOF. The set A of all n such that $\sum_{i=n}^{\nu} |x_i| \leq 1$ contains all infinite numbers. Then by overflow (Theorem VIII.7) it must contain a finite number m. But

$$\sum_{i=1}^{m-1} |x_i| \leq (m-1) \max\{|x_i| \mid i < m\}.$$

Hence

$$\sum_{i=1}^{\nu} |x_i| = \sum_{i=1}^{m-1} |x_i| + \sum_{i=m}^{\nu} |x_i|$$

is finite. □

From Theorem VIII.11 for sequences, we immediately obtain:

THEOREM VIII.15. *Let $x_1, x_2, \ldots, x_n, \ldots$, for $n \in \mathbb{N}$, be a sequence of hyperreal numbers. Then the series $\sum_{i=1}^{\infty} x_i$ S-converges to a number x if and only if there is an internal extension $x_1, x_2, \ldots, x_\nu$, for a certain $\nu \approx \infty$, such that the series $\sum_{i=1}^{\nu} x_i$ nearly converges to x.*

In fact, if the sequence S-converges to x, then for every internal extension $x_1, x_2, \ldots, x_\mu$, for $\mu \approx \infty$, there is a $\nu \approx \infty$, $\nu \leq \mu$, such that $\sum_{i=1}^{\nu} x_i$ nearly converges to x.

A theorem similar to VIII.12 is true for standard series.

THEOREM VIII.16. *Let $x_1, x_2, \ldots, x_n, \ldots$, for $n \in {}^*\mathbb{N}$ be a standard sequence. Then the following conditions are equivalent:*

(1) *The restriction of the series, $\sum_{n \in \mathbb{N}} x_n$, to $\mathbb{N}$, S-converges to a real number x.*

(2) *For every $\nu \approx \infty$, the restriction of the series to ν, $\sum_{n=1}^{\nu} x_n$ nearly converges to x.*
(3) *The series $\sum_{n=1}^{\infty} x_n$ *-converges to x.*

5. Convergence of functions

We need a few facts about internal analogues for the notion of limits at infinity for functions.

DEFINITION VIII.8 (CONVERGENCE AT INFINITY).
(1) Let f be a function defined on an interval $[a, b]$ where $b \approx \infty$ (or $a \approx -\infty$). We say that f *nearly converges to c at ∞ ($-\infty$)*, if $f(x) \approx c$ for every $x \approx \infty$, $x \leq b$ ($x \approx -\infty$, $x \geq a$).
(2) Let f be a function defined on the finite numbers of an interval $[a, b]$ where $b \approx \infty$ (or $a \approx -\infty$). We say that f *S-converges to c at ∞ ($-\infty$)*, if for every $\varepsilon \gg 0$ there is a finite $M > 0$ ($M < 0$) such that, for every finite x with $b \geq x > M$ ($a \leq x < M$), $|f(x) - c| < \varepsilon$.
(3) Let f be a function defined on an interval $[a, \infty)$ (or $(-\infty, b]$). We say that f **-converges to c at ∞ ($-\infty$)*, if for every $\varepsilon > 0$ there is an $M > 0$ ($M < 0$) such that for every $x > M$ ($x < M$), $|f(x) - c| < \varepsilon$.

Theorems similar to the theorems on convergence are true for these notions. For instance, we have:

THEOREM VIII.17. *Let f be an internal function defined on an interval $[a, b]$, where $b \approx \infty$ ($a \approx -\infty$). Then f nearly converges to c at infinity if and only if f S-converges to c at infinity.*

THEOREM VIII.18. *Let f be a standard function defined on an interval $[a, \infty)$ ($(-\infty, b]$). Then the following conditions are equivalent:*

(1) *The restriction of f to any interval containing the finite numbers in its domain, S-converges at infinity to a real number c.*
(2) *For every $b \approx \infty$ ($a \approx -\infty$), the restriction of f to $[a, b]$ nearly converges at infinity to c.*
(3) *The function f *-converges at infinity to c.*

The proofs, which are similar to those in the preceding sections, are left to the reader.

Finally, we introduce the notion of convergence of functions:

DEFINITION VIII.9 (CONVERGENCE OF FUNCTIONS). Let $f : A \to {}^*\mathbb{R}$. Then
(1) f *nearly converges to c at t_0*, if for every $t \approx t_0$, $t \in A$, $f(t) \approx c$.
(2) f *S-converges to c at t_0*, if for every $\varepsilon \gg 0$, there is a $\delta \gg 0$ such that $|t - t_0| < \delta$, $t \in A$ implies $|f(t) - f(t_0)| < \varepsilon$.
(3) f **-converges to c at t_0*, if for every $\varepsilon > 0$, there is a $\delta > 0$ such that $|t - t_0| < \delta$, $t \in A$ implies $|f(t) - f(t_0)| < \varepsilon$.

We have:

THEOREM VIII.19. *Let $f : A \to {}^*\mathbb{R}$ be an internal function. Then f nearly converges to c at t_0 if and only if f S-converges to c at t_0.*

For standard functions, the three notions of convergence are equivalent:

THEOREM VIII.20. *Let $f : A \to {}^*\mathbb{R}$ be a standard function and let $t_0 \in \mathbb{R}$. Then the three notions of convergence at t_0 are equivalent.*

The proof are similar to the previous proofs and are left to the reader.

6. Continuity

We now discuss the different notions of continuity, which are analogues of the standard notion of continuity.

DEFINITION VIII.10 (CONTINUITY). Let $f : A \to {}^*\mathbb{R}$, where $A \subseteq {}^{n*}\mathbb{R}$; then

(1) f is *nearly continuous at $\vec{x} \in A$* if whenever $\vec{y} \approx \vec{x}$ with $\vec{y} \in A$, we have $f(\vec{y}) \approx f(\vec{x})$. f is *nearly continuous on A* if it is nearly continuous at each $\vec{x}$ in A.

(2) We say that f is *S-continuous at $\vec{x} \in A$*, if for every $\varepsilon \gg 0$ there is a $\delta \gg 0$ such that if $\|\vec{x} - \vec{y}\| < \delta$ and $\vec{y} \in A$, we have that $|f(\vec{x}) - f(\vec{y})| < \varepsilon$. f is *S-continuous on A* if it is S-continuous at each $\vec{x}$ in A.

(3) Finally, we say that f is **-continuous at $\vec{x} \in A$*, if for every $\varepsilon > 0$ there is a $\delta > 0$ such that if $\|\vec{x} - \vec{y}\| < \delta$ and $\vec{y} \in A$, we have that $|f(\vec{x}) - f(\vec{y})| < \varepsilon$. f is **-continuous on A* if it is *-continuous at each $\vec{x}$ in A.

We shall use the notion of near continuity mainly when A is a near interval T. But the notion makes sense also for other types of sets. In order to see that near continuity at t is a nonstandard analogue of continuity at t, we have the following theorem.

THEOREM VIII.21. *Let $f : A \to {}^*\mathbb{R}$, with $A \subseteq {}^{n*}\mathbb{R}$, be internal. Then f is nearly continuous at $\vec{x} \in A$ if and only if f is S-continuous at $\vec{x}$.*

PROOF. Let f be internal nearly continuous at $\vec{x} \in A$ and let $\varepsilon \gg 0$. Let $\Delta_{\vec{x}}$ be the set of all δ such that for all $\vec{y} \in A$, if $\|\vec{y} - \vec{x}\| \leq \delta$ then $|f(\vec{y}) - f(\vec{x})| \leq \varepsilon$. Since the internal set $\Delta_{\vec{x}}$ contains all positive infinitesimals, by overflow (Theorem VIII.7) it contains a $\delta \gg 0$. Thus, f is S-continuous at $\vec{x}$.

Suppose, now, that f is S-continuous at $\vec{x}$. Then for every $\varepsilon \gg 0$ there is a $\delta \gg 0$ such that if $\vec{y} \in A$ and $\|\vec{y} - \vec{x}\| < \delta$, then $|f(\vec{y}) - f(\vec{x})| < \varepsilon$. Let $\vec{y} \in A$, $\vec{y} \approx \vec{x}$ and let $\varepsilon \gg 0$. Then we have that $\|\vec{y} - \vec{x}\| < \delta$, for the δ that works for ε, because $\delta \gg 0$. So that, since $\varepsilon \gg 0$ was arbitrary, $|f(\vec{y}) - f(\vec{x})| < \varepsilon$, for any $\varepsilon \gg 0$. Thus, $|f(\vec{y}) - f(\vec{x})| \approx 0$. Therefore, f is nearly continuous at $\vec{x}$. □

Now, if T is hyperfinite then near continuity on T is the internal analogue of uniform continuity, since we have:

THEOREM VIII.22. *Let $T \subseteq {}^{n*}\mathbb{R}$ be hyperfinite. Then the internal function $f : T \to {}^*\mathbb{R}$ is nearly continuous on T if and only if for every $\varepsilon \gg 0$ there is a $\delta \gg 0$ such that for any $\vec{t}, \vec{s} \in T$ with $\|\vec{t} - \vec{s}\| \leq \delta$ we have $|f(\vec{t}) - f(\vec{s})| \leq \varepsilon$.*

PROOF. Let $f : T \to {}^*\mathbb{R}$ be nearly continuous on T and let $\varepsilon \gg 0$. For each $\vec{t} \in T$, let $\Delta_{\vec{t}}$ be as in the previous proof, and let

$$A = \{n \in {}^*\mathbb{N} \mid \frac{1}{n} \in \Delta_{\vec{t}}\}.$$

Then A is an internal subset of ${}^*\mathbb{N}$, and, thus, by internal induction, it has a least element n. Let $1/n = \delta_{\vec{t}}$, i.e., $\delta_{\vec{t}}$ is the largest element in $\Delta_{\vec{t}}$ that is the reciprocal of an integer. Then, as we saw above, $0 \ll \delta_{\vec{t}}$. Let

$$\delta = \min\{\delta_{\vec{t}} \mid \vec{t} \in T\}.$$

We have that $0 \ll \delta$, because $\delta = \delta_{\vec{t}}$, for some $\vec{t}$ in the hyperfinite set T. Thus we have proved that if f is nearly continuous on T, then for every $\varepsilon \gg 0$, there is a $\delta \gg 0$, such that if $\vec{s}, \vec{t} \in T$ and $\|\vec{s} - \vec{t}\| \leq \delta$, then $|f(\vec{s}) - f(\vec{t})| \leq \varepsilon$.
The converse implication is obvious. □

Finally, we show that for standard functions the three notions of continuity are equivalent:

THEOREM VIII.23. *Let f be a standard function defined on $A \subseteq {}^{n*}\mathbb{R}$ and $\vec{t}_0 \in A$, $\vec{t}$ real. Then the following conditions are equivalent:*

(1) *f is nearly continuous at $\vec{t}_0$.*
(2) *f is S-continuous at $\vec{t}_0$.*
(3) *f is *-continuous at $\vec{t}_0$.*

PROOF. By Theorem VIII.21, we know that (1) and (2) are equivalent. So, assume (2). The set B of all $\varepsilon > 0$ for which there is a $\delta > 0$ such that for all $\vec{t} \in A$ with $\|\vec{t} - \vec{t}_0\| < \delta$ we have $|f(t) - f(t_0)| < \varepsilon$ is standard and includes all positive reals. Then B includes all positive hyperreals. Thus, we have (3).
Finally, assume (3) and let $\varepsilon > 0$ be real. The set B of all $\delta > 0$ such that for all $\vec{t} \in A$ with $\|\vec{t} - \vec{t}_0\| < \delta$ we have $|f(t) - f(t_0)| < \varepsilon$ is standard and nonempty. By standard overflow, Theorem VIII.7, B contains a positive real δ. Thus, we have (2). □

7. Differentials and derivatives

We need to define when a point is in the interior of a set:

DEFINITION VIII.11. We say that $\vec{a}$ is in the *interior* of a subset A of ${}^{n*}\mathbb{R}$, if $\vec{x} \in A$, for every $\vec{x} \approx \vec{a}$.

We now define differentials as differences:

DEFINITION VIII.12 (DIFFERENTIALS).

(1) We take dx to be a nonzero infinitesimal. Then, for any real function of one variable f such that x and $x + dx$ are in its domain, we define

$$df(x, dx) = f(x + dx) - f(x).$$

(2) The function f is *differentiable* at a point a of its domain if there is a function g, independent of dx, such that

$$df(x, dx) \approx g(x)\, dx \quad (dx),$$

for every infinitesimal dx and $x \approx a$, such that x and $x + dx$ are in the domain of f.

This is the traditional definition of differentials that was adopted in [78], but not in other versions of nonstandard calculus such as [51] and [52].

Notice that the definition of differentiable at a point a is equivalent to saying

$$\frac{f(x + dx) - f(x)}{dx} \approx g(x)$$

for every infinitesimal dx and $x \approx a$.

Recall that the derivative of a standard function f at a real point t_0 is the limit of the differential quotient

$$\frac{f(t_0 + dt) - f(t_0)}{dt}$$

when dt tends to zero. We say that a function f is *derivable* at an interior point t_0 if the derivative exists at t_0. We see, from Theorem VIII.20, that if a standard function f is differentiable, then it is derivable, and

$$f'(t) \approx \frac{df(t, dt)}{dt}$$

for every infinitesimal dt and every $t \approx t_0$. Also, it can be proved in this case that there is exactly one standard function f' which is the derivative of f. We shall call this standard function, f', the *derivative* of f. It is not difficult to show that this definition coincides with the usual standard definition.

We proceed, now, to introduce partial differentials and derivatives for functions of several variables. For simplicity of expression, we define these notions for functions of two variables. The extension to n variables is easy.

DEFINITION VIII.13 (PARTIAL DIFFERENTIALS). Let $z = F(x, y)$ and $\langle x, y \rangle$ be in the domain of F.

(1) If we let y be constant, we define

$$\partial_x z = \partial_x F(x, y) = F(x + dx, y) - F(x, y),$$

for $\langle x + dx, y \rangle$ in the domain of F. Similarly, letting x be constant

$$\partial_y z = \partial_y F(x, y) = F(x, y + dy) - F(x, y),$$

for $\langle x, y + dy \rangle$ in the domain of F. The functions $\partial_x z$ and $\partial_y z$ are called *partial differentials*.

(2) If there is a function F_x, independent of dx, such that

$$\partial_x z \approx F_x(x, y)\, dx \quad (dx),$$

for every infinitesimal dx with $\langle x + dx, y \rangle$ in the domain of F, then $F_x(x, y)$ is called a *partial derivative of F with respect to x at $\langle x, y \rangle$*. Similarly, we define $F_y(x, y)$.

The definition of partial derivative is equivalent to

$$\partial_x z = F_x(x, y)\, dx + \varepsilon\, dx$$

where $\varepsilon \approx 0$. That is

$$\frac{\partial_x z}{dx} \approx F_x(x, y)$$

for every $dx \approx 0$. Suppose, now, that F is standard. Define for x, y real

$$F_x(x, y) = \operatorname{st} \frac{F(x + dx, y) - F(x, y)}{dx}.$$

and take the standard extension, called also F_x. This is *the derivative* of a standard function, which can be shown to be unique.

Alternative notations for the partial derivatives are

$$F_x(x, y) = \frac{\partial z}{\partial x}$$

$$F_y(x, y) = \frac{\partial z}{\partial y}.$$

Similarly as for functions of one variable, these definitions of partial derivatives are equivalent, for standard functions, to the standard definitions.

DEFINITION VIII.14 (TOTAL DIFFERENTIAL). Let $z = F(x, y)$ and $\langle x, y \rangle$ be in the domain of F. Then:

(1) The *total differential* or, simply *differential* of $z = F(x, y)$ is

$$dz = dF(x, y) = F(x + \Delta x, y + \Delta y) - F(x, y),$$

where $\langle x + \Delta x, y + \Delta y \rangle$ is in the domain of F. The function dz is, really, a function of four variables and should be written $dF(x, y, \Delta x, \Delta y)$.

(2) For standard functions F, the *total derivative* or, simply *derivative* of $z = F(x, y)$ is defined by

$$Dz = DF(x, y) = F_x(x, y)\, \Delta x + F_y(x, y)\, \Delta y$$

$$= \frac{\partial z}{\partial x} \Delta x + \frac{\partial z}{\partial y} \Delta y.$$

Again, Dz is a function of four variables.

Although we shall use dF and DF mainly for Δx and Δy infinitesimal, the definition makes sense for any hyperreal number, and, for some proofs, it is necessary to have the functions defined for all hyperreal numbers. We write dx and dy in case we are assuming that they are infinitesimal.

For instance, if $z = xy$

$$dz = (x + dx)(y + dy) - xy$$
$$= y\,dx + x\,dy + dx\,dy$$

and

$$Dz = y\,dx + x\,dy.$$

The length of the vector $\langle dx, dy \rangle$ will be written ds. That is

$$ds = \sqrt{(dx)^2 + (dy)^2}.$$

If we don't assume that dx and dy are infinitesimal, then we write

$$\Delta s = \sqrt{\Delta x^2 + \Delta y^2}.$$

DEFINITION VIII.15. We say that the function F is *differentiable at a point* $\langle x_0, y_0 \rangle$ in its domain, if for every infinitesimal vector $\langle dx, dy \rangle$ such that the domain of F contains $\langle x_0 + dx, y_0 + dy \rangle$

$$dF(x_0, y_0) \approx DF(x_0, y_0) \quad (ds).$$

The last expression can also be written

$$dz \approx Dz \quad (ds).$$

The differentiability of F is, of course, equivalent to

$$dF(x_0, y_0) = DF(x_0, y_0) + \varepsilon\,ds$$

with $\varepsilon \approx 0$.

It is not difficult to show, that for standard functions, the notion of differentiability we have defined is equivalent to the standard notion. That is:

THEOREM VIII.24. *Let F be a standard function and $\langle x_0, y_0 \rangle$ be a real interior point in its domain. Then the following conditions are equivalent:*

(1) *For every dx and dy infinitesimals, there are ε_1, ε_2 infinitesimal, such that*

$$dF(x_0, y_0) = DF(x_0, y_0) + \varepsilon_1\,dx + \varepsilon_2\,dy.$$

(2) *F is differentiable at $\langle x_0, y_0 \rangle$.*

(3) *For every $\varepsilon \gg 0$, there is a $\delta \gg 0$, such that, if $\|(\Delta x, \Delta y)\| < \delta$, then*

$$\left\| \frac{F(x_0 + \Delta x, y_0 + \Delta y) - F(x_0, y_0) - DF(x_0, y_0)}{\sqrt{\Delta x^2 + \Delta y^2}} \right\| < \varepsilon.$$

7. DIFFERENTIALS AND DERIVATIVES

(4) *For every $\varepsilon > 0$, there is a $\delta > 0$ such that, if $\|(\Delta x, \Delta y)\| < \delta$, then*
$$\left\| \frac{F(x_0 + \Delta x, y_0 + \Delta y) - F(x_0, y_0) - DF(x_0, y_0)}{\sqrt{\Delta x^2 + \Delta y^2}} \right\| < \varepsilon.$$

PROOF. Notice, first, that if $dx > 0$, then
$$\frac{dx}{ds} = \frac{1}{\sqrt{1 + (\frac{dy}{dx})^2}}$$

and if $dx < 0$
$$\frac{dx}{ds} = -\frac{1}{\sqrt{1 + (\frac{dy}{dx})^2}}.$$

Thus, dx/ds is always finite, and if $dx \geq dy$, dx/ds is not infinitesimal. The same is true for dy/ds.

We shall show, first, that (1) is equivalent to (2). Assume (1). We have
$$dz = Dz + \varepsilon_1\, dx + \varepsilon_2\, dy,$$
for certain $\varepsilon_1 \approx 0 \approx \varepsilon_2$. Dividing both sides by ds and rearranging
$$\frac{dz}{ds} - \frac{Dz}{ds} = \varepsilon_1 \frac{dx}{ds} + \varepsilon_2 \frac{dy}{ds} \approx 0.$$

Thus, we have (2).

Assume that (1) is not true for certain infinitesimal dx and dy. Suppose, also, that $dx \geq dy$. (The proof for the case $dx \leq dy$ is similar.) Let $\varepsilon_2 \approx 0$. Then, since (1) does not hold
$$\frac{dz - Dz}{dx} - \varepsilon_2 \frac{dy}{dx}$$
is not infinitesimal. Because $dx \geq dy$, dy/dx is finite and, hence, $\varepsilon_2(dy/dx) \approx 0$. Thus
$$\frac{dz - Dz}{dx}$$
is not infinitesimal. We have
$$\left| \frac{dz - Dz}{ds} \right| = \left| \frac{dz - Dz}{dx} \frac{dx}{ds} \right|.$$

As we saw above, since $dx \geq dy$, dx/ds is not infinitesimal. Therefore
$$\frac{dz - Dz}{ds}$$
is not infinitesimal. Thus we have shown the negation of (2). Hence, we have completed the proof of the equivalence between (1) and (2).

Assume, now, (2) in order to prove (3). Let $\varepsilon > 0$, real. Then the set A of the δ such that
$$\|(\Delta x, \Delta y)\| < \delta \implies \left\| \frac{F(x_0 + \Delta x, y_0 + \Delta y) - F(x_0, y_0) - DF(x_0, y_0)}{\sqrt{\Delta x^2 + \Delta y^2}} \right\| < \varepsilon$$

is standard and contains all positive infinitesimals, hence, by standard overflow, Theorem VIII.7, it contains a certain $\delta > 0$ real. With this we have (3).

Assume (3). The standard set B of the ε such that there is a δ satisfying

$$\|(\Delta x, \Delta y)\| < \delta \implies \left\| \frac{F(x_0 + \Delta x, y_0 + \Delta y) - F(x_0, y_0) - DF(x_0, y_0)}{\sqrt{\Delta x^2 + \Delta y^2}} \right\| < \varepsilon$$

contains all positive reals, and hence, all positive hyperreals. Thus, we have shown (4).

Finally, assume the negation of (2). Let ds be such that

$$\frac{dz - Dz}{ds}$$

is not infinitesimal, with $ds \approx 0$. Thus

$$\left| \frac{dz - Dz}{ds} \right| > \varepsilon$$

for certain $\varepsilon > 0$ real. The standard set

$$C = \{\delta \mid \text{ there is a } \Delta s \text{ with } |\Delta s| < \delta \text{ and } \left| \frac{dz - Dz}{\Delta s} \right| > \varepsilon\}$$

contains all positive reals and, hence, all positive hyperreals. Therefore, we have shown the negation of (4). □

An almost immediate corollary of the definition of differentiability is:

COROLLARY VIII.25. *If F is differentiable at a point $\langle x_0, y_0 \rangle$ in its domain, then F is nearly continuous at $\langle x_0, y_0 \rangle$.*

PROOF. Assume that
$$dz = Dz + \varepsilon \, ds$$
with $\varepsilon \approx 0$. Then
$$F(x_0 + dx, y_0 + dy) = F(x_0, y_0) + Dz + \varepsilon \, ds.$$
But both Dz and $\varepsilon \, ds$ are infinitesimals. Therefore
$$F(x_0 + dx, y_0 + dy) \approx F(x_0, y_0).$$

□

DEFINITION VIII.16 (SMOOTH FUNCTIONS). *We call a standard function F smooth at an interior point $\langle x_0, y_0 \rangle$ of its domain, if both partial derivatives exist and are continuous at $\langle x_0, y_0 \rangle$.*

We shall not prove the Mean Value Theorem for the differential calculus, but we assume it for standard functions.

THEOREM VIII.26. *Assume that $z = F(x, y)$ is smooth (and standard) at an interior point $\langle x_0, y_0 \rangle$ of its domain. Then F is differentiable at $\langle x_0, y_0 \rangle$.*

7. DIFFERENTIALS AND DERIVATIVES

PROOF. We shall show that
$$dz = Dz + \varepsilon_1\, dx + \varepsilon_2\, dy$$
at $\langle x_0, y_0 \rangle$ with ε_1 and ε_2 infinitesimals.

Assume, first, that (x_0, y_0) is real. We have

(1)
$$\begin{aligned}dz &= F(x_0 + dx, y_0 + dy) - F(x_0, y_0) \\ &= F(x_0 + dx, y_0 + dy) - F(x_0 + dx, y_0) + \\ & \quad F(x_0 + dx, y_0) - F(x_0, y_0)\end{aligned}$$

It is clear that

(2)
$$\begin{aligned}F(x_0 + dx, y_0) - F(x_0, y_0) &= \partial_x z \\ &= F_x(x_0, y_0)\, dx + \varepsilon_1\, dx\end{aligned}$$

with $\varepsilon_1 \approx 0$.

We shall show

(3) $$F(x_0 + dx, y_0 + dy) - F(x_0 + dx, y_0) = F_y(x_0, y_0)\, dy + \varepsilon_2\, dy$$

with $\varepsilon_2 \approx 0$.

As $(x_0 + dx, y) \approx (x_0, y)$, $F_y(x_0 + dx, y)$ is continuous at every y between y_0 and $y_0 + dy$, by the Mean Value Theorem
$$F(x_0 + dx, y_o + dy) - F(x_0 + dx, y_0) = F_y(x_0 + dx, y_1)\, dy$$
for a certain y_1 between y_0 and $y_0 + dy$. Since F_y is continuous at $\langle x_0, y_0 \rangle$ and $y_1 \approx y_0$
$$F_y(x_0 + dx, y_1) = F_y(x_0, y_0) + \varepsilon_2$$
with $\varepsilon_2 \approx 0$. Thus
$$F(x_0 + dx, y_o + dy) - F(x_0 + dx, y_0) = (F_y(x_0, y_0) + \varepsilon_2)\, dy,$$
which is equivalent to (3). Replacing (2) and (3) in (1), we obtain the theorem for $\langle x_0, y_0 \rangle$ real. Because the condition of differentiability is standard, we obtain the theorem for all hyperreal vectors. □

The generalization to functions of n variables is straightforward. Let $w = F(x_1, x_2, \ldots, x_n)$ be a function of n variables. Then, the differential dw is
$$dw = F(x_1 + dx_1, x_2 + dx_2, \ldots, x_n + dx_n) - F(x_1, x_2, \ldots, x_n)$$
and the total derivative
$$Dw = \frac{\partial w}{\partial x_1}\, dx_1 + \frac{\partial w}{\partial x_2}\, dx_2 + \cdots + \frac{\partial w}{\partial x_n}\, dx_n,$$
for $dx_1, dx_2, \ldots, dx_n$ infinitesimals. The same theorems are valid with similar proofs.

We also shall have occasion to use the Implicit Function Theorem for standard functions, which we assume proved. We also assume the Chain Rule.

CHAPTER IX

Integration

1. Hiperfinite sums

With the use of hyperfinite sums, we can get approximations to the Riemann integral. The basic nonstandard theorem that justifies the main theorems for the Riemann integral is the following. It is useful for the proof of several theorems such as the Fundamental Theorem of Calculus and the rule for change of variables.

THEOREM IX.1. *Assume that, for $1 \leq i \leq \nu$, where ν is a (finite or infinite) natural number, x_i, z_i, and y_i are internal sequences which satisfy:*

(1) $z_i > 0$, for $1 \leq i \leq \nu$.
(2) $x_i z_i \approx y_i$ (z_i), for $1 \leq i \leq \nu$.
(3) x_i is finite for $1 \leq i \leq \nu$.
(4) $\sum_{i=1}^{\nu} z_i$ is finite.

Then
$$\sum_{i=1}^{\nu} x_i z_i \approx \sum_{i=1}^{\nu} y_i.$$

We postpone the proof. It is clear by external induction that we have

THEOREM IX.2. *If $x_i \approx y_i$ for all $i \leq n$, with n a finite natural number, then*
$$\sum_{i=1}^{n} x_i \approx \sum_{i=1}^{n} y_i.$$

This theorem is not true, in general, if n is infinite. As a simple counterexample, suppose that $\nu \approx \infty$ and take $x_i = 1/\nu$ and $y_i = 0$. Then $x_i \approx y_i$, for $1 \leq i \leq \nu$, but
$$\sum_{i=1}^{\nu} x_i = 1 \quad \text{and} \quad \sum_{i=1}^{\nu} y_i = 0.$$

It is convenient, as in [71, p. 17], to introduce:

DEFINITION IX.1. x *is asymptotically near* y, *in symbols* $x \asymp y$, *if*
$$\frac{x}{y} \approx 1.$$

It is easy to show that $\asymp$ is an equivalence relation.
The following theorem is proved, in part, in [71, pp. 17, 18].

THEOREM IX.3.

(1) *If* $x_1 \asymp y_1$ *and* $x_2 \asymp y_2$, *then* $x_1 x_2 \asymp y_1 y_2$.
(2) *If* x *is finite, then* $x \asymp y$ *implies* $x \approx y$. *If* $x \not\approx 0$, *then* $x \approx y$ *implies* $x \asymp y$.
(3) *Let* $y > 0$, *then* $x \asymp y$ *if and only if for all* $\varepsilon \gg 0$ *we have* $(1-\varepsilon)y \le x \le (1+\varepsilon)y$.
(4) *Let* x_i, y_i, *for* $1 \le i \le \nu$, *where* $\nu \in {}^*\mathbb{N}$ *(ν can be finite or infinite), be internal sequences such that* $x_i > 0$ *(or* $x_i < 0$*) and* $x_i \asymp y_i$ *for every* i *with* $1 \le i \le \nu$. *Then*
$$\sum_{i=1}^{\nu} x_i \asymp \sum_{i=1}^{\nu} y_i.$$

The proof of (1) and (2) are easy.

PROOF OF (3). Suppose that $x \asymp y$. Then
$$\frac{x}{y} \approx 1.$$

That is
$$\left|\frac{x}{y} - 1\right| \approx 0.$$

This means that for every $\varepsilon \gg 0$
$$\left|\frac{x}{y} - 1\right| < \varepsilon.$$

This last statement is equivalent to:
$$(1-\varepsilon)y \le x \le (1+\varepsilon)y.$$

□

PROOF OF (4). Suppose that $x_i \asymp y_i$ and $x_i > 0$, for every $i \le \nu$. Then $y_i > 0$, for every $i \le \nu$. Let $\varepsilon \gg 0$. Then, by (3)
$$(1-\varepsilon)y_i \le x_i \le (1+\varepsilon)y_i,$$
for every $i \le \nu$. Therefore
$$(1-\varepsilon)\sum_{i=1}^{\nu} y_i \le \sum_{i=1}^{\nu} x_i \le (1+\varepsilon)\sum_{i=1}^{\nu} y_i.$$

Using (3) in the other direction, we obtain (4). □

We also need the following lemma for the proof of IX.1 and other later theorems.

LEMMA IX.4. *Let x_i, y_i, z_i, for $1 \leq i \leq \nu$, be internal sequences such that $x_i \approx 0$, $z_i > 0$ and $y_i \approx x_i z_i$ (z_i) for every i with $1 \leq i \leq \nu$. Then*

$$\sum_{i=1}^{\nu} y_i \approx 0 \quad (\sum_{i=1}^{\nu} z_i).$$

In particular, we have

$$\sum_{i=1}^{\nu} x_i z_i \approx 0 \quad (\sum_{i=1}^{\nu} z_i).$$

Therefore, if $\sum_{i=1}^{\nu} z_i$ is finite, we have

$$\sum_{i=1}^{\nu} y_i \approx 0 \quad \text{and} \quad \sum_{i=1}^{\nu} x_i z_i \approx 0.$$

PROOF. Let $\varepsilon \gg 0$. Then, since

$$\frac{y_i}{z_i} \approx x_i \approx 0,$$

we have

$$|y_i| < \varepsilon z_i,$$

for $1 \leq i \leq \nu$. Thus

$$|\sum_{i=1}^{\nu} y_i| \leq \sum_{i=1}^{\nu} |y_i|$$
$$\leq \varepsilon \sum_{i=1}^{\nu} z_i.$$

Since $\sum_{i=1}^{\nu} z_i$ is positive and ε is arbitrary

$$\sum_{i=1}^{\nu} y_i \approx 0 \quad (\sum_{i=1}^{\nu} z_i).$$

The last conclusion is clear from the definitions of approximate equality and relative approximate equality. □

Nelson asserts in [71, p. 18] that IX.3 is what makes the integral work. This is not exactly right, since one also needs Lemma IX.4 and the technical Lemma VIII.10, which are needed for the proof of IX.1. We have not seen Theorem IX.1 or Lemma IX.4 used anywhere. The usual nonstandard proofs for Riemann integrals (see, for instance [52] and [51]) do not use these theorems.

Let $[a, b]$ be an interval. We say that $T = \{x_0, x_1, \ldots, x_\nu\}$ is *a partition of* $[a, b]$, if $x_0 = a$, $x_\nu = b$, and $x_0 < x_1 < \cdots < x_\nu$. For any partition, T, we write

$$dx_i = x_{i+1} - x_i,$$

for $i = 0, 1, \ldots, \nu - 1$, without necessarily assuming that dx_i is infinitesimal. Recall that T is a near interval for $[a, b]$, if $dx_i \approx 0$, for $i = 0, 1, \ldots, \nu - 1$.

We know that if
$$dF(x, dx) \approx f(x)\, dx \quad (dx),$$
then f is a derivative of F. Suppose that we have a finite interval $[a, b]$ and a near interval for $[a, b]$
$$T = \{x_0, x_1, \ldots, x_\nu\}.$$
Suppose that f is an internal derivative of F, and that f is finite on $[a, b]$; we have
$$dF(x_i, dx_i) \approx f(x_i)\, dx_i \quad (dx_i),$$
for $i = 0, 1, \ldots, \nu - 1$. Now, since $dF(x_i, dx_i) = F(x_i + dx_i) - F(x_i) = F(x_{i+1}) - F(x_i)$
$$\sum_{i=0}^{\nu-1} dF(x_i, dx_i) = F(x_\nu) - F(x_0)$$
$$= F(b) - F(a).$$
Since f is finite, if $b - a$ is finite, a consequence of Theorem IX.1 is
$$\sum_{i=0}^{\nu-1} dF(x_i, dx_i) \approx \sum_{i=0}^{\nu-1} f(x_i)\, dx_i.$$
Thus, we can show, from Theorem IX.1, that
$$\sum_{i=0}^{\nu-1} f(x_i)\, dx_i \approx F(b) - F(a),$$
and this is an approximate form of the Fundamental Theorem of Calculus.

PROOF OF THEOREM IX.1. Since x_i is finite, for every i in the domain of the sequence, by Proposition VIII.9, $x_i \leq M$, for $1 \leq i \leq \nu$, where M is a certain finite number. Thus, we have, if the hypotheses are satisfied
$$\left| \sum_{i=1}^{\nu} x_i\, z_i \right| \leq \sum_{i=1}^{\nu} |x_i|\, z_i$$
$$\leq M \sum_{i=1}^{\nu} z_i$$
so that $\sum_{i=1}^{\nu} x_i\, z_i$ or any partial sum are finite.

Let
$$N^+ = \{i \leq \nu \mid x_i \geq 0\},$$
$$N^- = \{i \leq \nu \mid x_i < 0\}.$$

Then N^+ and N^- are internal and, as subsets of a hyperfinite set, hyperfinite. We shall show the approximate equation which we denote (*)

$$\sum_{i \in N^+} y_i \approx \sum_{i \in N^+} x_i z_i.$$

Similarly, it can be shown that

$$\sum_{i \in N^-} y_i \approx \sum_{i \in N^-} x_i z_i$$

and, since

$$\sum_{i=1}^{\nu} y_i = \sum_{i \in N^+} y_i + \sum_{i \in N^-} y_i$$

and

$$\sum_{i=1}^{\nu} x_i z_i = \sum_{i \in N^+} x_i z_i + \sum_{i \in N^-} x_i z_i,$$

we obtain the conclusion of the theorem.

We now show (*). Let

$$N_m^+ = \{i \leq \nu \mid x_i \geq \frac{1}{m}\},$$

for any $m \in {}^*\mathbb{N}$. Assume, now, that m is a finite natural number. For $i \in N_m^+$, $\frac{y_i}{z_i} \approx x_i$. Since $x_i \geq 1/m$, x_i is not infinitesimal. Then, $\frac{y_i}{x_i z_i} \approx 1$, and so $y_i \asymp x_i z_i$. By Theorem IX.3

$$\sum_{i \in N_m^+} y_i \asymp \sum_{i \in N_m^+} x_i z_i.$$

But $\sum_{i \in N_m^+} x_i z_i$ is finite. Therefore, by Theorem IX.3, again

$$\sum_{i \in N_m^+} y_i \approx \sum_{i \in N_m^+} x_i z_i,$$

and, thus

$$\left| \sum_{i \in N_m^+} y_i - \sum_{i \in N_m^+} x_i z_i \right| \approx 0, \tag{1}$$

is true for every finite m. By Robinson's Lemma VIII.10, there is a $\nu \approx \infty$ such that

$$\sum_{i \in N_\nu^+} y_i \approx \sum_{i \in N_\nu^+} x_i z_i$$

with $N_\nu^+ = \{i \mid x_i \geq 1/\nu\}$.

Suppose, now, that $i \in N^+ - N_\nu^+$. Then $0 \le x_i < 1/\nu$. Thus, $x_i \approx 0$. Hence, by Lemma IX.4

$$\sum_{i \in N^+ - N_\nu^+} x_i z_i \approx 0 \quad \text{and} \quad \sum_{i \in N^+ - N_\nu^+} y_i \approx 0.$$

Thus, we have

$$\begin{aligned}
\sum_{i \in N^+} y_i &= \sum_{i \in N_\nu^+} y_i + \sum_{i \in N^+ - N_\nu^+} y_i \\
&\approx \sum_{i \in N_\nu^+} y_i \\
&\approx \sum_{i \in N_\nu^+} x_i z_i \\
&\approx \sum_{i \in N_\nu^+} x_i z_i + \sum_{i \in N^+ - N_\nu^+} x_i z_i \\
&= \sum_{i \in N^+} x_i z_i
\end{aligned}$$

□

We can deduce the following corollary, which was mentioned before the last proof:

COROLLARY IX.5 (APPROXIMATE FUNDAMENTAL THEOREM). *Let f and F be internal functions on the finite interval $[a, b]$. Let $T = \{x_0, x_1, \ldots, x_\nu\}$ be a near interval for $[a, b]$. Assume that f is finite on $[a, b]$, and that*

$$dF(x_i, dx_i) \approx f(x_i)\, dx_i \quad (dx_i)$$

for $i = 0, 1, \ldots, \nu - 1$. Then

$$\sum_{i=0}^{\nu-1} f(x_i)\, dx_i \approx F(b) - F(a).$$

We also have the following useful lemma, which implies an approximate version of the second form of the Fundamental Theorem, which we give below.

LEMMA IX.6. *Let u_i and z_i, for $1 \le i \le \nu$ be internal sequences such that $u_i \approx x$ and $z_i > 0$ for every i with $1 \le i \le \nu$. Then*

$$\sum_{i=1}^\nu u_i z_i \approx x \sum_{i=1}^\nu z_i \quad \left(\sum_{i=1}^\nu z_i\right).$$

PROOF. We have, $u_i = x + \varepsilon_i$ with $\varepsilon_i \approx 0$ for $i = 1, \ldots, \nu$. Then

$$\sum_{i=1}^{\nu} u_i\, z_i = \sum_{i=1}^{\nu} (x + \varepsilon_i)\, z_i$$
$$= x \sum_{i=1}^{\nu} z_i + \sum_{i=1}^{\nu} \varepsilon_i\, z_i.$$

By Lemma IX.4

$$\sum_{i=1}^{\nu} \varepsilon_i\, z_i \approx 0 \quad (\sum_{i=1}^{\nu} z_i).$$

□

COROLLARY IX.7 (APROXIMATE FUNDAMENTAL THEOREM). *Let f be an internal nearly continuous function on the interval $[x, x + dx]$ with $dx \approx 0$, let*

$$T = \{u_0, u_1, \ldots, u_\mu\}$$

be a partition of $[x, x + dx]$. Then

$$\sum_{j=0}^{\mu-1} f(u_j)\, du_j \approx f(x)\, dx \quad (dx).$$

PROOF. This is obtained immediately from Lemma IX.6 by noticing that $f(u_j) \approx f(x)$, for $j = 0, \ldots, \mu$, and

$$\sum_{j=0}^{\mu-1} du_j = dx.$$

□

2. Integrals

By the fundamental theorem of calculus and the approximate fundamental theorem, Corollary IX.5, we get what was announced at the beginning of this chapter, namely, that for f a standard continuous function on the interval $[a, b]$

$$\int_a^b f(t)\, dt \approx \sum_{i=0}^{n-1} f(t_i)\, dt_i,$$

where $T = \{t_0, t_1, \ldots, t_n\}$ is a near interval for $[a, b]$.

In fact:

DEFINITION IX.2 (INTEGRALS). Suppose that a standard function f is defined on a standard interval I. We say that f is *integrable on I* if for every $a < b$ real, $a, b \in I$, and for every near intervals T and T' for $[a, b]$, we have that

$$\sum_{t \in T} f(t)\, dt$$

is finite, and
$$\sum_{t \in T} f(t)\, dt \approx \sum_{t \in T'} f(t)\, dt.$$

We can then define
$$F(a, b) = \int_a^b f(t)\, dt = \operatorname{st} \sum_{t \in T} f(t)\, dt$$

for any near interval T for $[a, b]$, and any $a, b \in I$ real. We also define $F(a, a) = 0$ and $F(b, a) = -F(a, b)$, for $a < b$.

The function F is now defined for every pair of real numbers $a, b \in I$. Thus, $F : {}^2(I \cap \mathbb{R}) \to \mathbb{R}$. Since I is standard, by the properties of standard functions, F can be extended to a standard function ${}^*F$ defined on ${}^2I$. We then write
$$\int_a^b f(t)\, dt = {}^*F(a, b)$$

for any $a, b \in I$.

Hence, the integral of a standard integrable function is standard.
We have the following lemma:

LEMMA IX.8. *Let f be a standard integrable function on the standard interval I. Then, for any $a, b \in I$, $a < b$, and $\varepsilon > 0$ (infinitesimal or not), there is a partition of $[a, b]$, $T = \{x_0, \ldots, x_\nu\}$, such that*
$$\left| \int_a^b f(x)\, dx - \sum_{i=0}^{\nu-1} f(x_i)\, dx_i \right| \leq \varepsilon.$$

PROOF. Suppose first that a and b are standard. Then, since f is integrable, for every near interval, $T = \{x_0, \ldots, x_\nu\}$, for $[a, b]$, we have
$$\int_a^b f(x)\, dx \approx \sum_{i=0}^{\nu-1} f(x_i)\, dx_i.$$

Thus, the set A of triples $\langle a, b, \varepsilon \rangle$ such that f is integrable on I, $a, b \in I$, $a < b$, $\varepsilon > 0$, and there is a partition T for $[a, b]$ satisfying
$$\left| \int_a^b f(x)\, dx - \sum_{i=0}^{\nu-1} f(x_i)\, dx_i \right| \leq \varepsilon,$$

is standard and contains all standard triples. Then, it contains all such internal triples. □

We have the following theorem.

THEOREM IX.9. *Let f be a standard finite function defined on the interval I. Then f is integrable, if there is a function F defined on I such that*

$$dF(x,dx) \approx f(x)\,dx \quad (dx)$$

for every x, $x + dx \in I$ and $dx \approx 0$.
If there is such a function, we have

$$\int_a^b f(t)\,dt = F(b) - F(a)$$

for every $a, b \in I$.

PROOF. Suppose that there is a function F with

$$dF(x,dx) \approx f(x)\,dx \quad (dx)$$

for every $x \in I$ and $dx \approx 0$. Let $a < b$, $a, b \in I$, finite. Let $T = \{t_0, t_1, \ldots, t_\nu\}$ be a near interval for $[a,b]$. By Theorem IX.5

$$F(b) - F(a) \approx \sum_{i=0}^{\nu-1} f(t_i)\,dt_i.$$

Since the approximate equation is true for any near interval T, we have that f is integrable. □

We also have:

THEOREM IX.10. *If f is a standard continuous function on the interval I, then f is integrable on I.*

PROOF. We must show the following. Let

$$T = \{x_0, x_1, \ldots, x_\nu\}$$

and

$$T' = \{u_0, u_1, \ldots, u_\mu\}$$

be near intervals for $[a,b]$. Then

$$\sum_{i=0}^{\nu-1} f(x_i)\,dx_i \approx \sum_{i=0}^{\mu-1} f(u_i)\,du_i.$$

We shall show this first for the case $T \subseteq T'$. That is, the partition of $[a,b]$ induced by T' is finer than that induced by T. We then have

$$dx_i = \sum_{j=k_i}^{n_i} du_j$$

for every i and certain k_i and n_i. By Corollary IX.7

$$\sum_{j=k_i}^{n_i} f(u_j)\,du_j \approx f(x_i)\,dx_i \quad (dx_i).$$

By Theorem IX.1

$$\sum_{i=0}^{\nu-1} f(x_i)\,dx_i \approx \sum_{i=0}^{\nu-1} \sum_{j=k_i}^{n_i} f(u_j)\,du_j$$
$$= \sum_{j=0}^{\mu-1} f(u_j)\,du_j.$$

Consider, now, arbitrary near intervals for $[a,b]$, T' and T''. Let $T = T' \cup T''$. Then the sum determined by T' is approximately equal to the sum determined by T, and, also, the sum determined by T'' is approximately equal to the sum determined by T. Therefore, the sums determined by T' and T'' are approximately equal. □

3. Riemann integrals in higher dimensions

We say that a subset R of $^*\mathbb{R}^n$ is a *rectangle*, if $R = I_1 \times I_2 \times \cdots \times I_n$, where each I_i is an interval. If the intervals I_i have real endpoints, for $i = 1, 2, \ldots, n$, we say that R is a *standard rectangle*. A rectangle, R, is *closed* if it is the product of closed intervals. An *S-bounded set* is a set contained in a standard rectangle. We extend the notions of near interval and partition to any interval, not just closed, in the obvious way. Let R be a rectangle; then the closure of R is the closed rectangle R' obtained from R by adding edges and faces.

A partition of a rectangle, $R = I_1 \times I_2 \times \cdots \times I_n$, is obtained as follows. Obtain partitions $T_1, T_2, \ldots, T_n$ for $I_1, I_2, \ldots, I_n$. Then, we denote: $R^\# = T_1 \times T_2 \times \cdots \times T_n$.

Suppose that we have that $T_1 = \{x_1^0, x_1^1, \ldots, x_1^{\nu_1}\}$, $T_2 = \{x_2^0, x_2^1, \ldots, x_2^{\nu_2}\}$, ..., and $T_n = \{x_n^0, x_n^1, \ldots, x_n^{\nu_n}\}$. We let $dx_j^i = x_j^{i+1} - x_j^i$ and $r'_{i_1 i_2 \ldots i_n}$ be the rectangle $[x_1^{i_1}, x_1^{i_1+1}] \times [x_2^{i_2}, x_2^{i_2+1}] \times \cdots \times [x_n^{i_n}, x_n^{i_n+1}]$. Then the volume of $r'_{i_1 \ldots i_n}$ is $dx_1^{i_1} \cdot dx_2^{i_2} \cdots dx_n^{i_n}$.

Now, a *partition of R associated with $R^{\#1}$* is a collection $\mathcal{P}$ of pairwise disjoint half open rectangles such that their closure is an $r'_{i_1 \ldots i_n}$ and their union is R, perhaps except for a side. That is, $\mathcal{P}$ satisfies

(1) If $r \in \mathcal{P}$ then r is a half open rectangle and its closure is an $r'_{i_1 \ldots i_n}$. We denote this r by $r_{i_1 \ldots i_n}$, and call it a *subrectangle of $R^\#$*. The volume of $r_{i_1 \ldots i_n}$ is also $dx_1^{i_1} \cdot dx_2^{i_2} \cdots dx_n^{i_n}$.
(2) If $r, s \in \mathcal{P}$, then $r \cap s = \emptyset$.
(3) There is a rectangle R', whose closure is the closure of R, such that $\bigcup \mathcal{P} = R'$.

The family of subrectangles of $R^\#$ will also be denoted by $R^\#$.

The set $R^\#$ is called a *hyperfinite approximation of the rectangle R* if $R^\# = T_1 \times T_2 \times \cdots \times T_n$, where $T_1, T_2, \ldots, T_n$ are near intervals associated with the sides of R.

---

[1] Notice that we have called both the set of vertices of the subrectagles and the collection of subrectangles themselves a partition. Which is referred to will be clear from the context.

3. RIEMANN INTEGRALS IN HIGHER DIMENSIONS

We now turn to functions $F: {}^{n*}\mathbb{R} \to {}^*\mathbb{R}$. The *support of* F, in symbols $N(F)$, is
$$N(F) = \{\vec{x} \mid F(\vec{x}) \neq 0\}.$$
We say that an internal function F is *S-bounded*, if $N(F)$ is S-bounded and $F(\vec{x})$ is finite, for every $\vec{x}$. By Proposition VIII.9, if F is S-bounded, then $|F(\vec{x})| < M$, for every $\vec{x}$, where M is a finite number.

A step function F on ${}^{n*}\mathbb{R}$ is an internal function such that

(1) The range of F is a hyperfinite subset of ${}^*\mathbb{R}$.
(2) For each x in the range of F, $F^{-1}(x)$ is a rectangle.
(3) There is a rectangle R such that $F(\vec{x}) = 0$ for all $\vec{x} \notin R$.

The step function F *lives on* a hyperfinite family of rectangles $\mathcal{A}$, if F is constant in every rectangle $r \in \mathcal{A}$ and 0 outside of $\bigcup \mathcal{A}$.

If a step function F lives on the hyperfinite approximation $R^\#$ of a rectangle R, we define
$$I(F) = \sum_{r_{i_1 i_2 \ldots i_n} \in R^\#} F(x_1^{i_1}, \ldots, x_n^{i_n}) \, dx_1^{i_1} \ldots dx_n^{i_n}.$$

It can be shown in the usual standard fashion (see, for instance, [67, p. 52–57]) that $I(F)$ is well defined, that is, that $I(F)$ does not depend on the particular partition $R^\#$.

We have the following properties of $I(F)$.

THEOREM IX.11. *Let F and G be step functions. Then*

(1) $I(F + G) = I(F) + I(G)$.
(2) $I(cF) = cI(F)$, *if* $c \in {}^*\mathbb{R}$.
(3) $I(F) \geq 0$, *if* $F \geq 0$.
(4) *If F is S-bounded and $F(\vec{x}) \approx 0$ for every $\vec{x}$, then $I(F) \approx 0$.*

PROOF. The only property that does not have a simple standard proof is 4. So assume that F is S-bounded, $F(\vec{x}) \approx 0$ for every $\vec{x}$, and let F live on $R^\#$, where R is S-bounded. Then
$$I(F) = \sum_{r_{i_1 i_2 \ldots i_n} \in R^\#} F(x_1^{i_1}, \ldots, x_n^{i_n}) \, dx_1^{i_1} \ldots dx_n^{i_n}.$$

Since
$$\sum_{r_{i_1 i_2 \ldots i_n} \in R^\#} dx_1^{i_1} \ldots dx_n^{i_n}$$
is the volume of R, it is finite. Hence, by Lemma IX.4, $I(F)$ is infinitesimal. □

We say that a subset, S, of ${}^{n*}\mathbb{R}$ is *an infinitesimal subset*, if for every $\vec{x}, \vec{y} \in S$, we have $\vec{x} \approx \vec{y}$. We say that an internal set A is a *null set*, if A can be covered by a hyperfinite family of infinitesimal rectangles, such that the sum of their volumes is infinitesimal. It is easy to show that A is a null set if and only if $A \subseteq B$ where B is such that for any step function F that lives on B, if $|F(\vec{x})|$ is finite for every $\vec{x}$, then $I(F) \approx 0$.

Let G and F be functions on $^{n*}\mathbb{R}$. We say that

$$F \approx G \quad a.e.$$

if $F(\vec{x}) \approx G(\vec{x})$ for every $\vec{x} \notin D$, where D is a null set.

From the previous theorem, it is easy to prove

THEOREM IX.12. *Let F and G be S-bounded step functions. Then if $F \approx G$ a.e., we have that $I(F) \approx I(G)$.*

PROOF. We have that $F = F_1 + F_2$, $G = G_1 + G_2$, with $F_1 \approx G_1$ and $I(F_2) \approx 0 \approx I(G_2)$. Then, $F_1 - G_1 \approx 0$, and, hence, $I(F_1) - I(G_1) = I(F_1 - G_1) \approx 0$ and so $I(F_1) \approx I(G_1)$. Since $I(F) = I(F_1) + I(F_2) \approx I(F_1)$ and $I(G) = I(G_1) + I(G_2) \approx I(G_1)$, we get the theorem. □

We say that F is *Riemann integrable*, if there is an S-bounded step function G such that $F \approx G$ a.e. In this case, $I(G)$ is a *near integral* of F. By Theorem IX.12, all near integrals of F are infinitely close.

We now discuss the Riemann near integrals of continuous functions over regions with Jordan content. Suppose that a subset D of $^{n*}\mathbb{R}$ is contained in the S-bounded rectangle R. We say *D has Jordan content* if for every hyperfinite approximation $R^{\#}$ of R we have

(1) For every $\vec{x} \in D$ there is a $\vec{y} \in R^{\#}$ such that $\vec{x} \approx \vec{y}$.
(2)
$$\sum_{r_{i_1\ldots i_n} \subseteq D} dx_1^{i_1} \ldots dx_n^{i_n} \approx \sum_{r_{i_1\ldots i_n} \cap D \neq \emptyset} dx_1^{i_1} \ldots dx_n^{i_n}.$$

where $r_{i_1\ldots i_n}$ are the subrectangles of $R^{\#}$.

Let $\mathcal{F}_{R^{\#},D}$ be the family of subrectangles $r_{i_1\ldots i_n}$ that intersect D but are not included in D. Then it is clear that D satisfies 2 if and only if

$$\sum_{r_{i_1\ldots i_n} \in \mathcal{F}_{R^{\#},D}} dx_1^{i_1} \ldots dx_n^{i_n} \approx 0.$$

Also, it is not difficult to show that if $R^{\#}$ is a hyperfinite approximation of a rectangle R containing D, and $R'^{\#}$, of the rectangle R' containing D, then

$$\sum_{r_{i_1\ldots i_n} \subseteq D} dx_1^{i_1} \ldots dx_n^{i_n} \approx \sum_{r'_{i_1\ldots i_n} \subseteq D} dx_1^{i_1} \ldots dx_n^{i_n}$$

where $r_{i_1\ldots i_n}$ are the subrectangles of $R^{\#}$ and $r'_{i_1\ldots i_n}$, those of $R'^{\#}$.

In case D has Jordan content, $\sum_{r_{i_1\ldots i_n} \subseteq D} dx_1^{i_1} \ldots dx_n^{i_n}$ is a *near volume of D*, for any hyperfinite approximation $R^{\#}$ of any rectangle R that contains D.

For any function F defined on $^{n*}\mathbb{R}$ we set

$$\overline{\sum\sum}_{R^{\#},D} F = \sum_{r_{i_1\ldots i_n} \cap D \neq \emptyset} F(x_1^{i_1},\ldots x_n^{i_n}) dx_1^{i_1} \ldots, dx_n^{i_n},$$

3. RIEMANN INTEGRALS IN HIGHER DIMENSIONS

$$\overline{\sum\sum}_{R\#,D} F = \sum_{r_{i_1\ldots i_n} \subseteq D} F(x_1^{i_1}, \ldots x_n^{i_n}) \, dx_1^{i_1} \ldots, dx_n^{i_n}.$$

Let F be internal, and G_1 and G_2 be the step functions defined by

$$G_1(x_1^{i_1}, \ldots, x_n^{i_n}) = \begin{cases} F(x_1^{i_1}, \ldots, x_n^{i_n}), & \text{if } \vec{x} \in r_{i_1\ldots i_n} \text{ and } r_{i_1\ldots i_n} \subseteq D, \\ 0, & \text{otherwise}; \end{cases}$$

$$G_2(x_1^{i_1}, \ldots, x_n^{i_n}) = \begin{cases} F(x_1^{i_1}, \ldots, x_n^{i_n}), & \text{if } \vec{x} \in r_{i_1\ldots i_n} \text{ and } r_{i_1\ldots i_n} \cap D \neq \emptyset, \\ 0, & \text{otherwise}. \end{cases}$$

Then

$$\underline{\sum\sum}_{R\#,D} F = I(G_1) \quad \text{and} \quad \overline{\sum\sum}_{R\#,D} F = I(G_2).$$

It is clear that if D has Jordan content and $|F(\vec{x})|$ is finite, then

$$\overline{\sum\sum}_{R\#,D} F \approx \underline{\sum\sum}_{R\#,D} F,$$

and

$$\sum_{r_{i_1\ldots i_n} \in \mathcal{F}_{R\#,D}} F(x_1^{i_1}, \ldots x_n^{i_n}) \, dx_1^{i_1} \ldots dx_n^{i_n} \approx 0.$$

If F is continuous, then it can be checked that

$$F \approx G_1 \quad \text{a.e.} \quad \text{and} \quad F \approx G_2 \quad \text{a.e.}$$

Thus, both $\overline{\sum\sum}_{R\#,D} F$ and $\underline{\sum\sum}_{R\#,D} F$ are near integrals for internal continuous functions on regions with Jordan content.

We complete the section with a definition of the Riemann integral in higher dimensions. Let F be a standard function which is Riemann integrable, and a a near integral of F. Then we define

$$\int F = \text{st } a.$$

The standard class of Riemann integrable functions $\mathcal{R}$ is defined as the standard set that contains all standard Riemann integrable functions. We extend the function $\int$ to the standard function defined on $\mathcal{R}$.

It is not to difficult to show that the standard definition of the Riemann integral is equivalent, for standard functions, to the nonstandard definition that we have given. Also, for internal functions, the standard definition implies the nonstandard one. Thus, one can prove that all functions in $\mathcal{R}$, not only the standard functions, are Riemann integrable in our sense.

We also write, as is usual, the integral of F over the region with Jordan content D as

$$\int\cdots\int_D F$$

or
$$\int_D \cdots \int F(\vec{x})\, d\vec{x}$$
or
$$\int_D \cdots \int F(x_1, \ldots, x_n)\, dx_1 \ldots x_n,$$

for the integral $\int G$ of the function G that is equal to F on D, and zero outside of D.

4. Transformations

We consider transformations of $^{n*}\mathbb{R}$ into itself. For simplicity, we shall consider, first, transformations of $^{2*}\mathbb{R}$. Suppose that the two standard functions

$$x = \phi(u,v) = x(u,v) \quad \text{and} \quad y = \psi(u,v) = y(u,v)$$

are continuously differentiable (i.e., they have continuous derivatives) in a region R of the plane uv. We may consider the system as *a mapping* or *transformation*: to the point P with coordinate $\langle u, v \rangle$ in the plane uv corresponds as image the point Q with coordinates $\langle x, y \rangle$ in the plane xy. We shall write this transformation

$$T(u,v) = \langle \phi(u,v), \psi(u,v) \rangle$$

that is, T is a function from a subset of $^{2*}\mathbb{R}$ to $^{2*}\mathbb{R}$.

The region B is the *image by T of the region R*, if

$$B = T(R) = \{T(u,v) \mid \langle u,v \rangle \in R\}.$$

If T is one-one we can obtain the inverse transformation T^{-1}. That is, we can determine u and v uniquely as functions of x and y, namely

$$u = g(x,y) \quad \text{and} \quad v = h(x,y).$$

Then, we have

$$x = \phi(g(x,y), h(x,y)) \quad \text{and} \quad y = \psi(g(x,y), h(x,y)).$$

Thus
$$T^{-1}(x,y) = \langle g(x,y), h(x,y) \rangle$$
and
$$\langle x, y \rangle = T(T^{-1}(x,y)) \quad \text{and} \quad \langle u, v \rangle = T^{-1}(T(u,v)).$$

In the case of three or more variables the situation is analogous. Thus, a system of n standard functions continuously differentiable

$$x_1 = \phi_1(u_1, u_2, \ldots, u_n),\ x_2 = \phi_2(u_1, u_2, \ldots, u_n),\ \ldots,\ x_n = \phi_n(u_1, u_2, \ldots, u_n)$$

defined on a region S of space $u_1 u_1 \ldots u_n$, can be considered as the mapping T of the region S over a region $B = T(S)$ of the space $x_1 x_2 \ldots x_n$, defined by

$$T(u_1, \ldots, u_n) = \langle \phi_1(u_1, \ldots, u_n), \phi_2(u_1, \ldots, u_n), \ldots, \phi_n(u_1, \ldots, u_n) \rangle.$$

4. TRANSFORMATIONS

If we assume that this mapping of S onto B is one-one, then we have the inverse mapping

$$u_1 = g_1(x_1, x_2, \ldots, x_n), u_2 = g_2(x_1, x_2, \ldots, x_n), \ldots, u_n = g_n(x_1, x_2, \ldots, x_n),$$

with

$$T^{-1}(x_1, x_2, \ldots, x_n) = \langle g_1(x_1, x_2, \ldots, x_n), \ldots, g_n(x_1, x_2, \ldots, x_n)\rangle$$

and

$$x_1 = \phi_1(g_1(x_1, x_2, \ldots, x_n), \ldots, g_n(x_1, x_2, \ldots, x_n)),$$
$$x_2 = \phi_2(g_1(x_1, x_2, \ldots, x_n), \ldots, g_n(x_1, x_2, \ldots, x_n)),$$
$$\vdots$$
$$x_n = \phi_n(g_1(x_1, x_2, \ldots, x_n), \ldots, g_n(x_1, x_2, \ldots, x_n)).$$

Consider the differentiable transformation

$$x_1 = x_1(u_1, \ldots, u_n) = \phi_1(u_1, \ldots, u_n), \ldots, x_n = x_n(u_1, \ldots, u_n) = \phi_n(u_1, \ldots, u_n),$$

which can be written

$$T(u_1, \ldots, u_n) = \langle x_1(u_1, \ldots, u_n), \ldots, x_n(u_1, \ldots, u_n)\rangle.$$

The function

$$J_T = \frac{\partial(x_1, \ldots, x_n)}{\partial(u_1, \ldots, u_n)} = \begin{vmatrix} \frac{\partial x_1}{\partial u_1} & \cdots & \frac{\partial x_1}{\partial u_n} \\ \vdots & \ddots & \vdots \\ \frac{\partial x_n}{\partial u_1} & \cdots & \frac{\partial x_n}{\partial u_n} \end{vmatrix} = \begin{vmatrix} (\phi_1)_{u_1} & \cdots & (\phi_1)_{u_n} \\ \vdots & \ddots & \vdots \\ (\phi_n)_{u_1} & \cdots & (\phi_n)_{u_n} \end{vmatrix}$$

is called the *Jacobian of* $x_1, \ldots, x_n$ (or of the functions $\phi_1, \ldots, \phi_n$) with respect to $u_1, \ldots, u_n$. We assume that the main properties of Jacobians are known. In particular, the Jacobian of the inverse transformation and the Chain Rule for Jacobians. From these rules, we obtain that, if a transformation has an inverse, then its Jacobian is not zero. From the Implicit Function Theorem, we obtain a sort of reciprocal of this assertion, that we assume proved:

THEOREM IX.13. *If on a neighborhood of the real point $\langle u_1, \ldots, u_n\rangle$ the standard functions $\phi_1(u_1, \ldots, u_n), \ldots, \phi_n(u_1, \ldots, u_n)$ are continuously differentiable, T is the transformation given by*

$$T(u_1, \ldots, u_n) = \langle \phi_1(u_1, \ldots, u_n), \ldots, \phi_n(u_1, \ldots, u_n)\rangle,$$

$\vec{x}_0 = T(\vec{u}_0)$ *and, further, the Jacobian of T, J_T, is not zero at a real point $\vec{u}_0$, then on a neighborhood of $\vec{u}_0$ the transformation T has a unique inverse, which is continuously differentiable (and, of course, standard).*

Notice that since $\vec{u}_0$ is real and J_T is standard, $J_T(\vec{u}_0)$ is also real. Hence, since it is different from zero, J_T is noninfinitesimal at $\vec{u}_0$.

5. Parametric representation of curves and surfaces

We shall review curves and sufaces in the plane and in three dimensions, without giving most of the proofs, which can be found in [25]. These notions can be generalized to higher dimensions, but the matter becomes more complicated so that we shall treat them in dimensions two and three, and suppose that the treatment is generalized to higher dimensions.

We represent curves in the plane by two functions

$$x = f(t) = x(t) \quad \text{and} \quad y = g(t) = y(t)$$

where t is in a certain interval I. Curves are, more precisely, functions c from intervals in $^*\mathbb{R}$ into $^{2*}\mathbb{R}$ defined by

$$c(t) = \langle f(t), g(t) \rangle$$

where f and g are standard and continuous.

The notion of a curve can be generalized slightly, by considering a finite number of parametrizations. For instance, we could have

$$x = f_1(t) \quad \text{and} \quad y = g_1(t)$$

with $t \in [a, b]$ and

$$x = f_2(t) \quad \text{and} \quad y = g_2(t)$$

with $t \in [c, d]$, such that $f_1(b) = f_2(c)$ and $g_1(b) = g_2(c)$.

A curve is *simple*, if there are no t_1 and t_2 in the interior of the interval, $t_1 \neq t_2$, such that $x(t_1) = x(t_2)$ and $y(t_1) = y(t_2)$.

A curve defined with a parameter on $[a, b]$ is *closed*, if $x(a) = x(b)$ and $y(a) = y(b)$.

We shall always assume that our curves are defined by standard continuously differentiable functions. The derivative of the curve $c(t) = \langle x(t), y(t) \rangle$ is defined by $c'(t) = \langle x'(t), y'(t) \rangle$.

We also assume, in general, that the curve is *regular*, i.e., except for a finite number of points, $c'(t) \not\approx \vec{0}$. This last formula is equivalent to say that not both $x'(t) \approx 0$ and $y'(t) \approx 0$. The reason for demanding regularity, is that regular curves locally are functions. For instance, if $y'(t) \gg 0$ with y' continuous, the function $y(t)$ is increasing on a neighborhood of t and if $y'(t) \ll 0$, it is decreasing. Therefore, it is one-one on the neighborhood and it can be inverted obtaining $t = g^{-1}(y)$. Hence, $x = f(g^{-1}(y))$, and so the curve can be represented on the neighborhood as a function of y. Intuitively, functions are thin, and curves should also be thin.

We define the family of simple closed regular curves to be the standard set containing all standard simple closed regular curves. Since the definition of a standard simple closed regular curve is standard, all curves in the family satisfy it, and all curves that satisfy the definition are in the family. For instance, boundaries of infinitesimal rectangles are in the family. When speaking of simple closed regular curves, we always mean an element of this family.

We shall consider regions of the plane enclosed by a simple closed regular curve, that we shall call *closed bounded regions*. We assume that a simple closed regular

5. CURVES AND SURFACES

curve divides the plane into an inside region and an outside region (Jordan's Theorem).

We say that a subset A of $^{n*}\mathbb{R}$ is in the *interior* of another subset B if every point in A is an interior point of B.

We shall now study the effect of a transformation on a curve and its interior region. Let T be a transformation of $^{2*}\mathbb{R}$ into $^{2*}\mathbb{R}$ defined by

$$x = x(u,v) \quad \text{and} \quad y = y(u,v)$$

with noninfinitesimal Jacobian on all its domain R_0. Let $u = u(t)$, $v = v(t)$ be a regular curve in the interior of R_0 with domain $[a,b]$. The image of the curve by T is parametrized by the functions

$$x(t) = x(u(t), v(t)) \quad \text{and} \quad y(t) = x(u(t), v(t))$$

with the same domain $[a,b]$. It is not difficult to show that the image curve is also regular.

Let R be a closed bounded region included in the interior of the domain R_0 of T, and B the image of R. Then the image of the inside of R is the inside of B, and the image of the boudary of R is the boundary of B. Since regular curves go to regular curves, if the boundary of R is a regular curve, the boundary of B also is a regular curve. Thus, closed bounded regions go to closed bounded regions.

In dimension three, our solids are limited by surfaces, thus, we should discuss parametric representation of surfaces. We need, in this case, three functions with two parameters

$$x = x(t,s) = f(t,s),\ y = y(t,s) = g(t,s),\ z = z(t,s) = h(t,s)$$

where $\langle t, s \rangle$ is in a closed bounded plane region R. The functions f, g and h should be standard and continuously differentiable. The surface is, then, represented by the function S from $^{2*}\mathbb{R}$ in $^{3*}\mathbb{R}$

$$S(t,s) = \langle x(t,s), y(t,s), z(t,s) \rangle.$$

Just as for curves, surfaces should locally be functions of the form $z = F(x,y)$ or $x = F(y,z)$ or $y = F(x,z)$. In order to insure this, by Theorem IX.13, we demand that the surfaces be *regular*, that is, that the three Jacobians

$$\begin{vmatrix} x_t & x_s \\ y_t & y_s \end{vmatrix}, \quad \begin{vmatrix} y_t & y_s \\ z_t & z_s \end{vmatrix}, \quad \begin{vmatrix} z_t & z_s \\ x_t & x_s \end{vmatrix},$$

not be infinitesimal at the same point.

It is not difficult to define simple closed surfaces. They should be surfaces which do not intersect, except on a curve, and that surround a part of the space. The closed bounded solids are precisely those limited by simple closed surfaces. It can also be proved that a transformation with noninfinitesimal Jacobian in its domain transforms regular surfaces into regular surfaces, and closed bounded solids into closed bounded solids.

214   IX. INTEGRATION

We shall now show that closed bounded regions have Jordan content, and, thus, the integral is defined over them. We shall do it for regions in the plane, but the generalization to higher dimensions is straightforward.

Let R be the region enclosed by the simple regular closed curve

$$x = x(t) = f(t) \quad \text{and} \quad y = y(t) = g(t),$$

with $t \in [a, b]$. We assume that the domain $[a, b]$ of the curve is an interval. It is not difficult to generalize the construction to a finite union of intervals.

The circumscribed rectangle, I, is determined by

$$m_x = \min_{a \leq t \leq b} x(t), \ M_x = \max_{a \leq t \leq b} x(t), \ m_y = \min_{a \leq t \leq b} y(t), \ M_y = \max_{a \leq t \leq b} y(t),$$

with $I = [m_x, M_x] \times [m_y, M_y]$. Let $\mathcal{F}$ be the boundary curve of R, we show

LEMMA IX.14.
$$\sum_{r_{ij} \cap \mathcal{F} \neq \emptyset} dx_i \, dy_j \approx 0.$$

PROOF. Assume that the rectangles r_{ij} have sides $[x_i, x_{i+1}]$ and $[y_j, y_{j+1}]$. Since all subrectangles, r_{ij}, are infinitesimal, the curve $\mathcal{F}$ is a function on them, maybe x is function of y, or y is function of x. Let A be the family of the r_{ij} which intersect the curve where y is function of x and B, the set where x is function of y.

Consider the part of the curve where y is a function of x, say, $y = f(x)$, which we shall denote $\overline{f}$. Let $[x_i, x_{i+1}]$, the horizontal side of the rectangle r_{ij}, be one of the subintervals where y is a function of x, i.e., where f is defined. Since the boundary is standard, f is also standard; then minimum and maximum in any closed intervals exist. Let m_i the minimum of f and M_i the maximum of f on $[x_i, x_{i+1}]$. Since f is continuous, $M_i - m_i \approx 0$. Consider all rectangles with side $[x_i, x_{i+1}]$ which intersect the curve and where y is a function of x, as in Fig. IX.7.

Let $[y^i_{j_1}, y^i_{j_2}], [y^i_{j_2}, y^i_{j_3}], \ldots, [y^i_{j_{\eta-1}}, y^i_{j_\eta}]$, where η is a finite or infinite natural number, be the vertical sides of the rectangles with horizontal side $[x_i, x_{i+1}]$ considered above. Assume that

$$y^i_{j_1} < y^i_{j_2} < \cdots < y^i_{j_\eta}.$$

Then
$$y^i_{j_1} \leq m_i \leq y^i_{j_2}, \qquad y^i_{j_{\eta-1}} \leq M_i \leq y^i_{j_\eta}.$$

Therefore
$$0 \leq \sum_{k=1}^{\eta-1} dy^i_{j_k} = y^i_{j_\eta} - y^i_{j_1} \leq M_i - m_i + dy^i_{j_1} + dy^i_{j_{\eta-1}} \approx 0.$$

Hence
$$0 \leq \sum_{r_{ij} \cap \overline{f} \neq \emptyset} dx_i \, dy_j = \sum_{i=0}^{\nu-1} (dx_i \sum_{k=1}^{\eta-1} dy^i_{j_k}) \leq \varepsilon \sum_{i=0}^{\nu-1} dx_i = \varepsilon(b - a),$$

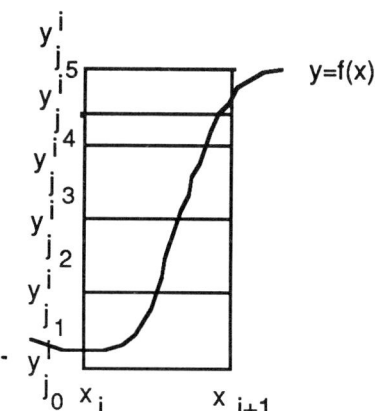

FIGURE IX.7. Intersection of curve with rectangles.

for every $\varepsilon \gg 0$. Thus

$$\sum_{r_{ij} \in A} dx_i dy_j = \sum_{r_{ij} \cap \overline{f} \neq \emptyset} dx_i\, dy_j \approx 0.$$

Similarly, we prove that

$$\sum_{r_{ij} \in B} dx_i dy_j \approx 0.$$

Then

$$\sum_{r_{ij} \cap \mathcal{F} \neq \emptyset} dx_i dy_j = \sum_{r_{ij} \in A} dx_i dy_j + \sum_{r_{ij} \in B} dx_i dy_j$$
$$\approx 0$$

□

We define, now, the volume of a closed bounded region. We, first, divide the rectangle I into infinitesimal subrectangles $r_{i_1 \ldots i_n}$ and define the hyperfinite approximation $R^{\#}$ as before. We also define inferior volume

$$\underline{V_{R^{\#}}(R)} = \sum_{r_{i_1 \ldots i_n} \subseteq R} dx_1^{i_1} \cdots dx_n^{i_n}$$

and the superior volume

$$\overline{V_{R^{\#}}(R)} = \sum_{r_{i_1 \ldots i_n} \cap R \neq \emptyset} dx_1^{i_1} \cdots dx_n^{i_n}.$$

We have the following Corollary:

COROLLARY IX.15.

$$\underline{V_{R^{\#}}(R)} \approx \overline{V_{R^{\#}}(R)}.$$

We now show that for any approximations of R, $R_1^\#$ and $R_2^\#$, the volumes are approxiamtely equal. Let $R_1^\#$ and $R_2^\#$ be two hyperfinite approximations of the closed bounded region R such that $R_1^\#$ is finer that $R_2^\#$. This means that all subrectangles of $R_2^\#$ contained in R is a union of subrectangles of $R_1^\#$. In $R_1^\#$ there could be subrectangles contained in R which are not in subrectangles of $R_2^\#$ contained in R, but in subrectangles of $R_2^\#$ which intersect the boundary. But, since the union of the subrectangles that intersect the boundary has infinitesimal volume
$$V_{R_1^\#}(R) \approx V_{R_2^\#}(R).$$
Now, if we take two arbitrary approximations $R_1^\#$ and $R_2^\#$, their union, $R_3^\#$, is finer than both, and, hence
$$V_{R_1^\#}(R) \approx V_{R_3^\#}(R) \approx V_{R_2^\#}(R).$$
It is also clear, that for a standard closed bounded region, R
$$V_{R^\#}(R) \leq (M_x - m_x)(M_y - m_y)$$
where the number on the right is finite.

Thus, we can define, for R a closed bounded standard region
$$V(R) = \operatorname{st} V_{R^\#}(R)$$
where $R^\#$ is any hyperfinite approximation. We extend the function V to any closed bounded region, standard or not. An infinitesimal closed bounded region has an infinitesimal volume, because it is included in cubes of side ε, for arbitrary $\varepsilon \gg 0$.

6. Change of variables

We shall prove the rule for change of variables for integrals. We have, in this case a function $F(u_1, u_2, \ldots, u_n)$, continuous on a closed bounded region R, and a tranformation T, defined by
$$x_1 = \phi_1(u_1, \ldots, u_n) = x_1(u_1, \ldots, u_n), \ldots, x_n = \phi_n(u_1, \ldots, u_n) = x_n(u_1, \ldots, u_n)$$
on a region whose interior contains R. We also assume that $\phi_1, \phi_2, \ldots,$ and ϕ_n are continuously differentiable with Jacobian
$$J_T \not\approx 0$$
at every point in the region. Then T transforms R in a closed bounded region, B.

The rule that we shall show in this section is
$$\int \cdots \int_B F(x_1, \ldots, x_n)\, dx_1 \ldots, dx_n =$$
$$\int \cdots \int_R F(x_1(u_1, \ldots, u_n), \ldots x_n(u_1, \ldots, u_n))|J_T|\, du_1 \ldots du_n.$$

6. CHANGE OF VARIABLES

We sometimes write

$$\int_R \cdots \int F(x_1, \ldots, x_n) |J_T| \, du_1 \ldots du_n$$

instead of

$$\int_R \cdots \int F(x_1(u_1, \ldots, u_n), \ldots, x_n(u_1, \ldots, u_n)) |J_T| \, du_1 \ldots du_n.$$

It is understood, in this case, that $x_1, x_2, \ldots, x_n$ are expressed in terms of $u_1, u_2, \ldots, u_n$.

We first study the relationship between the volumes of an infinitesimal cube and the region that is obtained by the action of T.

LEMMA IX.16. *Let ΔR be an infinitesimal cube $[u_1, v_1] \times [u_2, v_2] \times \cdots \times [u_n, v_n]$ with $u_i - v_i = u_j - v_j = du \approx 0$, for $1 \leq i, j \leq n$. Let T be a transformation with noninfinitesimal Jacobian, J_T, defined on a region whose interior contains ΔR, and let $T(\Delta R) = B$ be the image of ΔR by T. Then*

$$V(T(\Delta R)) \approx |J_T| \, du^n \quad (du^n).$$

PROOF. We shall prove the theorem for dimension two. The generalization to dimension n is complicated but straightforward. Let T be the trasformation given by

$$x = x(u, v) = \phi(u, v) \quad \text{and} \quad y = y(u, v) = \psi(u, v).$$

Then $J_T(u_1, v_1)$ is

$$\begin{vmatrix} x_u(u_1, v_1) & y_u(u_1, v_1) \\ x_v(u_1, v_1) & y_v(u_1, v_1) \end{vmatrix}.$$

Notice that $|J_T(u_1, v_1)| \, du^2$ is the area (volume) of the parallelogram between the vectors

$$x_u(u_1, v_1) \, du \, \vec{i} + y_u(u_1, v_1) \, du \, \vec{j},$$
$$x_v(u_1, v_1) \, du \, \vec{i} + y_v(u_1, v_1) \, du \, \vec{j}.$$

We denote $x_1 = x(u_1, v_1)$ and $y_1 = y(u_1, v_1)$. The following parallelogram has these vectors as sides. One of the vertices is $\langle x_1, y_1 \rangle$. Its sides are the lines (where the variables are ξ and η, and x_u is $x_u(u_1, v_1)$, x_v is $x_v(u_1, v_1)$, y_u is $y_u(u_1, v_1)$, and y_v is $y_v(u_1, v_1)$)

$$x_v \xi - y_v \eta + y_1 x_v - x_1 y_v = 0, \tag{1}$$
$$y_u \xi - x_u \eta + y_1 x_u - x_1 y_u = 0, \tag{2}$$
$$x_v \xi - y_v \eta + (y_1 + y_u \, du) x_v - (x_1 + x_u \, du) y_v = 0, \tag{3}$$
$$x_u \xi - y_u \eta + (y_1 + y_v \, du) x_u - (x_1 + x_v \, du) y_u = 0. \tag{4}$$

The side $u = u_1$ with $v_1 \leq v \leq v_2$ of the square ΔR is transformed into the parametric curve

$$x = x(u_1, v) \quad \text{and} \quad y = y(u_1, v),$$

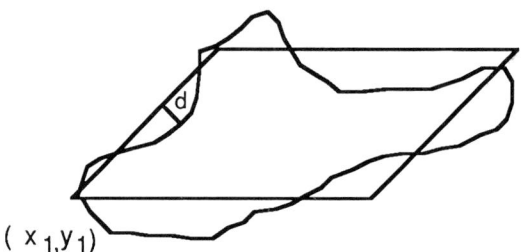

FIGURE IX.8. Transformation of a square.

with $v_1 \leq v \leq v_2$. We shall show that the area between this curve and side (1) of the parallelogram is infinitesimal compared to du^2. The situation is shown in Fig. IX.8.

We know that $x_v^2 + y_v^2$ is finite and noninfinitesimal. We now bound the distance, d, between a point $\langle x(u_1,v), y(u_1,v) \rangle$ of the curve and line (1):

$$d = \frac{|y_v x(u_1,v) - x_v y(u_1,v) + y_1 x_v - x_1 y_v|}{\sqrt{x_v^2 + y_v^2}}$$

$$= \frac{|y_v(x(u_1,v) - x(u_1,v_1)) + x_v(y_1 - y(v_1,v))|}{\sqrt{x_v^2 + y_v^2}}$$

$$= \frac{|(y_v x_v(u_1,v_3) - x_v y_v(u_1,v_4))(v - v_1)|}{\sqrt{x_v^2 + y_v^2}}$$

where v_3 and v_4 are between v_1 and v (by the Mean Value Theorem). We continue with our calculations

$$d = \frac{|y_v(x_v(u_1,v_3) - x_v) + x_v(y_v - y_v(u_1,v_4))(v - v_1)|}{\sqrt{x_v^2 + y_v^2}}$$

$$\leq \frac{|y_v(x_v(u_1,v_3) - x_v) + x_v(y_v - y_v(u_1,v_4)) du|}{\sqrt{x_v^2 + y_v^2}}.$$

We use here the fact that the distance between v_1 and v_2 is du.

On the other hand, the length of the side of the parallelogram is

$$\sqrt{x_v^2 + y_v^2}\, du.$$

Thus, the area between the curve and the side is contained in a rectangle with area

$$2\varepsilon\, du^2$$

for every $\varepsilon \gg 0$. This means that this area is infinitesimal compared with du^2.

Let us now look at the area between the curve

$$x = x(u,v_2) \quad \text{and} \quad y = y(u,v_2)$$

6. CHANGE OF VARIABLES

with $u_1 \leq u \leq u_2$ and side (4). This curve is image of side $y = v_2$ sith $u_1 \leq u \leq u_2$ of the square ΔR. We have that the distance, d, between a point on the curve $(x(u, v_2), y(u, v_2))$ and the line is

$$d = \frac{|y_u x(u, v_2) - x_u y(u, v_2) + x_u(y_1 + y_v\, du) - y_u(x_1 + x_v\, du)|}{\sqrt{x_u^2 + y_u^2}}$$

$$= \frac{|y_u(x(u, v_2) - x_1 - x_v\, du) - x_u(y(u, v_2) - y_1 - y_v\, du)|}{\sqrt{x_u^2 + y_u^2}}.$$

As x and y are differentiable, we have

$$x(u, v_2) - x_1 = x(u, v_2) - x(u_1, v_1)$$
$$= x_u(u_1, v_1)(u - u_1) + x_v(u_1, v_1)\, du + \varepsilon_1(u - u_1) + \varepsilon_2\, du$$
$$= x_u(u - u_1) + x_v\, du + \varepsilon_1(u - u_1) + \varepsilon_2\, du,$$

$$y(u, v_2) - y_1 = y(u, v_2) - y(u_1, v_1)$$
$$= y_u(u_1, v_1)(u - u_1) + y_v(u_1, v_1)\, du + \varepsilon_3(u - u_1) + \varepsilon_4 du$$
$$= y_u(u - u_1) + y_v\, du + \varepsilon_3(u - u_1) + \varepsilon_4\, du,$$

where $\varepsilon_1, \varepsilon_2, \varepsilon_3$ and ε_4 are infinitesimal. Hence

$$d = \frac{|y_u(x_u(u - u_1) + \varepsilon_1(u - u_1) + \varepsilon_2 du) - x_u(y_u(u - u_1) + \varepsilon_3(u - u_1) + \varepsilon_4 du)|}{\sqrt{x_u^2 + y_u^2}}$$

$$= \frac{|y_u(\varepsilon_1(u - u_1) + \varepsilon_2 du) - x_u(\varepsilon_3(u - u_1) + \varepsilon_4 du)|}{\sqrt{x_u^2 + y_u^2}}$$

$$\leq \varepsilon\, du,$$

for every $\varepsilon \gg 0$. The length of the side is

$$\sqrt{x_u^2 + y_u^2}\, du.$$

Thus, the area we are discussing is contained in a rectagle with area less than $2\varepsilon\, du^2$, for every $\varepsilon \gg 0$. Therefore, it is infinitesimal with respect to du^2.

Similarly one can show that the other two areas are infinitesimal compared to du^2. □

We shall define an intermediate sum. Let R be a closed bounded region, let F be a function defined on R, and let $R^\#$ be a hyperfinite approximation of R. Let $\vec{x}_{i_1,\ldots,i_n} = \langle x_1^{i_1}, \ldots, x_n^{i_n} \rangle$. Define

$$\vec{x}'_{i_1,\ldots,i_n} = \begin{cases} \vec{x}_{i_1,\ldots,i_n} & \text{if } \vec{x}_{i_1,\ldots,i_n} \in R, \\ \text{a point in } R \cap r_{i_1\ldots i_n}, & \text{if } \vec{x}_{i_1,\ldots,i_n} \notin R \text{ and } R \cap r_{i_1\ldots i_n} \neq \emptyset. \end{cases}$$

Let $s_{i_1\ldots i_n} = r_{i_1\ldots i_n} \cap R$. It is clear that $s_{i_1\ldots i_n}$ is a closed bounded region so that its volume is defined. We, now define

$$\sum\sum{}_{R\#} F = \sum_{r_{i_1\ldots i_n} \cap R \neq \emptyset} F(\vec{x}'_{i_1\ldots i_n}) V(s_{i_1\ldots i_n}).$$

We also need the following lemma, which is a generalization in higher dimensions of the second form of the approximate fundamental theorem.

LEMMA IX.17. *Let R be an infinitesimal closed bounded region, $R^\#$ a subdivision of R, $\vec{x}$ an element of R, and F a continuous function on R. Then*

$$\sum\sum\nolimits_{R^\#} F \approx F(\vec{x})\, V(R) \quad (V(R)).$$

PROOF. We shall prove the lemma in dimension two. We have, since F is continuous

$$F(x'_i, y'_j) = F(x,y) + \varepsilon_{ij}$$

with $\varepsilon_{ij} \approx 0$. Thus

$$\sum\sum\nolimits_{R^\#} F = \sum_{i=0}^{\nu-1}\sum_{j=0}^{\mu-1} F(x'_i, y'_j)\, V(s_{ij})$$

$$= \sum_{i=0}^{\nu-1}\sum_{j=0}^{\mu-1} (F(x,y) + \varepsilon_{ij})\, V(s_{ij})$$

$$= F(x,y)\, V(R) + \sum_{i=0}^{\nu-1}\sum_{j=0}^{\mu-1} \varepsilon_{ij}\, V(s_{ij})$$

Hence

$$\left|\frac{\sum\sum_{R^\#} F - F(x,y)\,V(R)}{V(R)}\right| \leq \left|\frac{\sum_{i=0}^{\nu-1}\sum_{j=0}^{\mu-1} \varepsilon_{ij}\, V(s_{ij})}{V(R)}\right|$$

$$\leq \frac{\sum_{i=0}^{\nu-1}\sum_{j=0}^{\mu-1} |\varepsilon_{ij}|\, V(s_{ij})}{V(R)}.$$

Thus, for every $\delta \gg 0$

$$\left|\frac{\sum\sum_{R^\#} F - F(x,y)\,V(R)}{V(R)}\right| \leq \frac{\sum_{i=0}^{\nu-1}\sum_{j=0}^{\mu-1} \delta\, V(s_{ij})}{V(R)}$$

$$= \delta$$

Therefore

$$\frac{\sum\sum_{R^\#} F}{V(R)} \approx F(x,y).$$

□

We now show the main theorem.

6. CHANGE OF VARIABLES

THEOREM IX.18. *Let R be a closed bounded region, T a one-one transformation with noninfinitesimal Jacobian, J_T, defined on R, and $T(R)$, the image of R by T. Assume that $F(u_1, \ldots, u_n)$ is a continuous standard function on a region whose interior contains R. Then*

$$\int \cdots \int_{T(R)} F(x_1, \ldots, x_n) \, dx_1 \ldots dx_n = \int \cdots \int_R F(x_1, \ldots, x_n) |J_T| \, du_1 \ldots du_n.$$

PROOF. We prove the theorem for dimension two. Assume, first, that R is standard. Let $R^\#$ be a hyperfinite approximation of R such that $du_i = dv_j = du$. That is, the subrectangles are squares of sides du. Because R is a closed bounded region, we have, by the definition of the integral

$$\iint_R F(x,y)|J_T| \, du \, dv \approx \underline{\sum \sum}_{R^\#} F(x,y)|J_T| \, du \, dv$$

$$= \sum_{r_{ij} \subseteq R} F(x(u_i, v_j), y(u_i, v_j))|J_T(u_i, v_j)| \, du^2$$

By Lemma IX.16

$$V(T(r_{ij})) \approx |J_T| \, du^2 \quad (du^2).$$

Thus, because F is standard continuous, and R is a standard closed bounded region, F is S-bounded on R. Hence, F takes finite values on R, and therefore

$$F(x(u_i, v_j), y(u_i, v_j)) V(T(r_{ij})) \approx F(x(u_i, v_j), y(u_i, v_j))|J_T| \, du^2 \quad (du^2). \tag{1}$$

On the other hand, by Lemma IX.17

$$\iint_{T(r_{ij})} F(x,y) \, dx \, dy \approx F(x,y) V(T(r_{ij})) \quad (V(T(r_{ij}))). \tag{2}$$

From (1) and (2) we obtain

$$\iint_{T(r_{ij})} F(x,y) \, dx \, dy \approx F(x,y)|J_T| \, du^2 \quad (du^2).$$

Thus, by Theorem IX.1

$$\sum_{r_{ij} \subseteq R} \iint_{T(r_{ij})} F(x,y) \, dx \, dy \approx \underline{\sum \sum}_{R^\#} F(x,y)|J_T| \, du \, dv,$$

and

$$\sum_{r_{ij} \cap \mathcal{F} \neq \emptyset} \iint_{T(r_{ij})} F(x,y) \, dx \, dy \approx \sum_{r_{ij} \cap \mathcal{F} \neq \emptyset} F(x,y)|J_T| \, du \, dv \approx 0.$$

Let $s_{ij} = T(r_{ij}) \cap T(R)$. Then, if $r_{ij} \subseteq R$, $s_{ij} = T(r_{ij})$. Since T is one-one, the s_{ij} form a partition of $T(R)$. Hence

$$\iint_{T(R)} F(x,y) \, dx \, dy = \sum_{r_{ij} \subseteq R} \iint_{T(r_{ij})} F(x,y) \, dx \, dy + \sum_{r_{ij} \cap \mathcal{F} \neq \emptyset} \iint_{s_{ij}} F(x,y) \, dx \, dy.$$

But
$$\left| \sum_{r_{ij} \cap \mathcal{F} \neq \emptyset} \iint_{s_{ij}} F(x,y)\,dx\,dy \right| \leq \sum_{r_{ij} \cap \mathcal{F} \neq \emptyset} \iint_{s_{ij}} |F(x,y)|\,dx\,dy$$
$$\leq \sum_{r_{ij} \cap \mathcal{F} \neq \emptyset} \iint_{T(r_{ij})} |F(x,y)|\,dx\,dy$$
$$\approx 0$$

Hence
$$\iint_{T(R)} F(x,y)\,dx\,dy \approx \iint_R F(x,y)|J_T|\,du\,dv.$$

Since both integrals are real numbers, they are equal.

We now extend the theorem to all closed bounded regions, standard or not: The class C of closed bounded regions is standard, because it is defined in standard terms. Since the theorem is standard, the class of closed bounded regions, B, that satisfies the theorem is also standard. The classes B and C are equal, because they contain the same standard objects. □

7. Improper integrals

We need to discuss improper integrals of nonnegative functions. Suppose that F is a standard function on $^{n*}\mathbb{R}$ to $^*\mathbb{R}$, integrable on any closed bounded region, and such that $F(\vec{x}) \geq 0$, for all $\vec{x} \in {}^{n*}\mathbb{R}$ (we say, in this case, that F is *nonnegative*). For $m \in {}^*\mathbb{N}$, let C_m be the cube $^n[-m,m]$. Let R be a standard subset of $^{n*}\mathbb{R}$ such that $C_m \cap R = R_m$ is a closed bounded region, for every $m \in {}^*\mathbb{N}$. We call such regions, *closed regions*. The set $^{n*}\mathbb{R}$ is, of course, a closed region. We have that for a closed region, R, the improper integral

$$\int \cdots \int_R F(\vec{x})\,d\vec{x},$$

which we may write simply as

$$\int \cdots \int_R F,$$

is the limit, if it exists and is finite, when m tends to infinity of the sequence

$$\int \cdots \int_{R_m} F(\vec{x})\,d\vec{x}.$$

Since this sequence is standard, convergence can be taken in any sense, by Theorem VIII.18.

It is clear that if

$$\int \cdots \int_{{}^{n*}\mathbb{R}} F$$

exists (and hence it is finite), then

$$\int\cdots\int_R F$$

also exists.

We can deal similarly with improper integrals over other domains. We say that a standard subset R of $^{n*}\mathbb{R}$ is a *region* if there is a standard sequence of closed bounded regions $\{R_m\}_{m\in {}^*\mathbb{N}}$ such that $R_m \subseteq R_{m+1}$, for all m (i.e., the sequence is *increasing*) and

$$R = \bigcup_{m\in {}^*\mathbb{N}} R_m.$$

When R is a region, we denote, as above, by $\{R_m\}_{m\in {}^*\mathbb{N}}$, an increasing standard sequence of closed bounded regions whose union is R.

Suppose that F is a nonnegative function defined and integrable over the region, R. Then the improper integral over R is the limit, if it exists and is finite, of

$$\int\cdots\int_{R_m} F,$$

when m tends to infinity.

We could develop the theory, without much further cost, of internal regions, instead of just standard regions, but, since we do not need them, we shall not do it.

By the theorems on convergence at infinity and convergence of sequences, and the previous result about integrals, we get the following theorem, which can be generalized to other forms of improper integrals. We shall assume this generalization in some cases.

THEOREM IX.19. *Let F be a standard integrable nonnegative function on a region R. Then*

$$\int\cdots\int_R F$$

exists if and only if there is a hyperfinite approximation, T, for R_ν, where ν is an infinite natural number, such that either of the sequences

$$\overline{\sum\sum}_{T\cap R_m} F,$$

or

$$\underline{\sum\sum}_{T\cap R_m} F,$$

for $m \leq \nu$, nearly converge to a finite number.

In case either of these sequences nearly converge, then

$$\int\cdots\int_R F \approx \overline{\sum\sum}_{T\cap R} F \approx \underline{\sum\sum}_{T\cap R} F.$$

PROOF. Assume that $\int \cdots \int_R F$ exists and is equal to c. Then, the sequence $a_m = \int \cdots \int_{R_m} F$ S-converges to c. Let T' be a hyperfinite approximation for R_ν with $\nu \approx \infty$, and let $T'_m = T' \cap R_m$ for every $m \in {}^*\mathbb{N}$ and $m \leq \nu$. Then T'_m is a hyperfinite approximation for R_m, for any $m \leq \nu$. By the previous characterization of the definite integral, for all finite m

$$\int \cdots \int_{R_m} F \approx \underline{\sum\sum}_{T'_m} F,$$

that is

$$\int \cdots \int_{R_m} F - \underline{\sum\sum}_{T'_m} F \approx 0$$

for all finite $m \in {}^*\mathbb{N}$. Thus, by Robinson's lemma, VIII.10, there is an infinite μ such that

$$\int \cdots \int_{R_\mu} F \approx \underline{\sum\sum}_{T'_\mu} F.$$

Let $T = T'_\mu$. By Theorem VIII.18, for every $\mu \approx \infty$, $\int \cdots \int_{R_\mu} F \approx c$. Thus

$$\underline{\sum\sum}_{T \cap R_\eta} F \approx c = \int \cdots \int_R F,$$

for every $\eta \leq \mu$, $\eta \approx \infty$. The proof for the upper sum is similar.

In order to prove the other direction, suppose that we have a near interval T for R_ν, with $\nu \approx \infty$, and that the sequence

$$\underline{\sum\sum}_{T \cap R_m} F,$$

for $m \leq \nu$, nearly converges to a real number c. For each finite $m \in \mathbb{N}$, we have

$$\int \cdots \int_{R_m} F \approx \underline{\sum\sum}_{T \cap R_m} F.$$

Thus, by Robinson's lemma, there is an infinite $\eta \leq \nu$, such that

$$\int \cdots \int_{R_m} F \approx \underline{\sum\sum}_{T \cap R_m} F$$

for every $m \leq \eta$. Therefore, the sequence $\int \cdots \int_{R_m} F$, $m \leq \eta$, nearly converges to c, and, hence, by Theorem VIII.11, this sequence S-converges to c. Therefore

$$\int \cdots \int_R F = c.$$

Again, the proof for the upper sum is similar. □

We shall need the following theorem, for the one-dimensional case.

7. IMPROPER INTEGRALS

THEOREM IX.20. *Suppose that f is a continuous, nonnegative, and nonincreasing function defined on $^*\mathbb{R}$ for all $x \geq c$, where c is a real number, i.e., $f(x) \geq 0$ and $f(x) \geq f(y)$ for $c \leq x \leq y$, and assume also that*

$$\int_c^\infty f(t)\, dt$$

exists. Then for every near interval, T, for $[c, b]$, where $b \approx \infty$

$$\sum_{t \in T} f(t)\, dt$$

nearly converges to

$$\int_c^\infty f(t)\, dt.$$

PROOF. Let $T = \{t_0, t_1, \ldots, t_\kappa\}$ be a near interval for $[c, b]$. From the proof of the previous theorem in the one-dimensional case, we can get a $\nu \approx \infty$ such that

$$\sum_{t \in T \cap [c, \nu]} f(t)\, dt$$

nearly converges to

$$\int_c^\infty f(t)\, dt.$$

If $\nu \leq b$, we are done. So assume that $b > \nu$. Let

$$T \cap [\nu, b] = \{t_\mu, t_{\mu+1}, \ldots, t_{\mu+\eta}\}.$$

We define

$$a_j = \int_{t_{\mu+j}}^{t_{\mu+j+1}} f(t)\, dt,$$

for $j = 0, 1, \ldots, \eta$. By properties of the integral and the fact that f is nonincreasing, we have

$$a_j \geq f(t_{\mu+j})\, dt_{\mu+j}$$

so that

$$\sum_{j=0}^\xi a_j \geq \sum_{j=0}^\xi f(t_{\mu+j})\, dt_{\mu+j} \geq 0,$$

for any $\xi \leq \eta$. We also have, by the properties of the integral

$$0 \approx \int_\nu^b f(t)\, dt \geq \sum_{j=0}^\xi a_j$$

for every $\xi \leq \eta$. Thus

$$\sum_{j=0}^\xi f(t_{\mu+j})\, dt_{\mu+j} \approx 0$$

for every $\xi \leq \eta$. Therefore
$$\sum_{t \in T} f(t)\, dt$$
nearly converges to
$$\int_c^\infty f(t)\, dt.$$

□

CHAPTER X

Probability distributions

1. Hyperfinite probability spaces

We return, now, to probability theory. In this chapter, we shall consider probability spaces $\langle \Omega, \mathrm{pr} \rangle$, where Ω is a hyperfinite set and pr is a internal function on Ω into $^*\mathbb{R}$. The sum, $\sum_{\omega \in \Omega} \mathrm{pr}(\omega)$, then, is defined, and, as for finite probability spaces, we require

$$\sum_{\omega \in \Omega} \mathrm{pr}(\omega) = 1,$$

and $\mathrm{pr}(\omega) > 0$, for $\omega \in \Omega$.

We only consider as events, in this case, *internal* subsets of Ω. The internal class of all internal subsets of Ω, i.e., the internal power set of Ω, is an algebra of subsets of Ω in the usual sense. Any internal subset A of Ω is also hyperfinite, so we can define its probability as before

$$\Pr A = \sum_{\omega \in A} \mathrm{pr}(\omega).$$

The internal set of all internal subsets of Ω, call it $^*\mathcal{P}\Omega$, is hyperfinite, and, also, Pr is an internal function from $^*\mathcal{P}\Omega$ into $^*\mathbb{R}$, which satisfies Kolmogorov's axioms for finite spaces.

A random variable is an *internal* function from Ω into $^*\mathbb{R}$. The class $^*(^\Omega\mathbb{R})$ of random variables is the internal set of all internal functions from Ω into $^*\mathbb{R}$. It is easy to check that $^*(^\Omega\mathbb{R})$ is an algebra of random variables.

In Appendix B, we shall extend the probability measure, random variables, and expectations, to external objects, via the nonstandard construction of the Daniell integral.

We introduced in Chapter VI, Section 3, the probability distribution $\Pr_X$ of a random variable, and in Chapter VI, Section 10, of a random vector, $\Pr_{\vec{X}}$. Namely, $\Lambda_{\vec{X}}$ is the hyperfinite (and, hence, internal) subset of $^{n*}\mathbb{R}$ of values of $\vec{X}$, and, for any internal subset of $\Lambda_{\vec{X}}$, A

$$\Pr_{\vec{X}} A = \Pr[\vec{X} \in A].$$

We now define probability distributions without an specific reference to a random vector.

DEFINITION X.1 (PROBABILITY DISTRIBUTIONS). A *probability distribution* is a probability, Pr defined from a probability space, $\langle \Omega, \text{pr} \rangle$, where Ω is a hyperfinite subset of $^{n*}\mathbb{R}$, for some n.

As before, we have that for any internal subset, A, of Ω

$$\Pr A = \sum_{\omega \in A} \text{pr}(\omega).$$

The main purpose of this chapter is to approximate probability distributions over hyperfinite spaces by standard functions. The standard functions we need are the usual standard probability distributions, such as normal, Poisson, etc. By these approximations, we can use standard results for our hyperfinite spaces.

Most of the results of the chapter are adaptations of standard results to our hyperfinite probability spaces. One of the reasons for not being able to use the standard results themselves, without adaptation, is that standard distributions, such as the normal, are not probability distributions in our sense, since they are not defined over a hyperfinite space.

Part of the theory of integration that was developed in the last chapter is needed precisely because we must adapt usual results to the hyperfinite case. This is especially the case for the theorem on change of variables. As we shall see, we not only need the standard theorem itself, Theorem IX.18, but also Lemmas IX.16 and IX.17, which are nonstandard in character.

The definition of probability that will be given in Part 3 works directly, in the standard sense, only for finite spaces. We extend then the definition for hyperfinite spaces, and the approximations given in this and the next chapter provide approximations to the usual standard spaces. In Appendix B, a procedure for approximating any probability spaces is sketched.

2. Discrete distributions

We shall discuss especially two types of probability distributions: discrete and continuous. We begin with the definition of standard frequency functions and discrete probability distributions.

DEFINITION X.2 (DISCRETE DISTRIBUTIONS).

(1) A function $p : {}^{n*}\mathbb{R} \to {}^*\mathbb{R}$ is called a *frequency function* if
  (a) p is standard.
  (b) There is a standard sequence $\vec{x}_1, \vec{x}_2, \ldots$ such that
    (i) $p(\vec{x}_i) > 0$, for $i = 1, 2, \ldots$,
    (ii) p vanishes except on $\vec{x}_1, \vec{x}_2, \ldots$ and
    (iii) $\displaystyle\sum_{i \in {}^*\mathbb{N}} p(\vec{x}_i) = 1.$
  The sequence $\{\vec{x}_i\}$ is called the *domain sequence of p*.

2. DISCRETE DISTRIBUTIONS

(2) A probability distribution Pr over the hyperfinite set space $\langle \Omega, \text{pr} \rangle$ is *discrete* if there is a frequency function p with domain sequence $\{\vec{x}_i\}$, such that, there is a $\nu \approx \infty$ and an internal sequence $\{y_i\}_{i=1}^{\nu}$, satisfying $\Omega = \{y_i \mid 1 \leq i \leq \nu\}$, and for any finite i, we have, $\vec{y}_i = \vec{x}_i$ and $\text{pr}(\vec{y}_i) \approx p(\vec{x}_i)$.

It is clear, since $\vec{x}_1, \vec{x}_2, \ldots$ is a standard sequence, that its domain is either a finite number, m, or the set of natural numbers, $^*\mathbb{N}$. That is, we either have, $\vec{x}_1, \ldots, \vec{x}_m$, or $\langle \vec{x}_i \mid i \in {}^*\mathbb{N} \rangle$. In case the sequence is finite, we have that Ω contains all elements of the sequence. It is also clear, that for a finite number, i, $\vec{x}_i$ is a real number, and, hence, finite. We have the following theorem for the discrete case. The main difficulty in the proof is that if two sequences are approximately equal term by term, their sum is not necessarily approximately equal. Thus, we must use results in Chapter IX, Section 1, for hiperfinite sums.

THEOREM X.1. *Let Pr be a probability distribution over* $\langle \Omega, \text{pr} \rangle$. *Then Pr is discrete if and only if there is a frequency function p with domain sequence $\vec{x}_1, \vec{x}_2, \ldots$ such that for any standard subset A of $^{n*}\mathbb{R}$ we have*

$$\Pr(A \cap \Omega) \approx \sum_{\vec{y} \in A \cap \{\vec{x}_1, \vec{x}_2, \ldots\}} p(\vec{y}).$$

PROOF. Suppose, first, that Pr is discrete with frequency function, p. Assume, as a first case, that the domain sequence is finite, thus, we have $\vec{x}_1, \vec{x}_2, \ldots, \vec{x}_m$, where m is finite. Thus, Ω consists of all elements of the domain sequence. Then, since $A \cap \Omega$ is finite, and $\vec{x}_i$ is finite, for every i, we have

$$\Pr(A \cap \Omega) = \sum_{\vec{y} \in A \cap \Omega} \text{pr}(\vec{y})$$
$$\approx \sum_{\vec{y} \in A \cap \Omega} p(\vec{y}).$$

Suppose, now, that the domain sequence is infinite. Let

$$\Omega = \{\vec{y}_1, \vec{y}_2, \ldots, \vec{y}_\nu\},$$

with $\nu \approx \infty$. Since $\sum_{i=1}^{\infty} p(\vec{x}_i)$ converges to one, we have

$$\sum_{i=\nu+1}^{\infty} p(\vec{x}_i) \approx 0.$$

Since $\sum_{i=1}^{m} p(x_i) \approx \sum_{i=1}^{m} \text{pr}(y_i)$, for every finite m, we have, by Robinson's lemma, that there is an infinite $\mu \leq \nu$ such that

$$\sum_{i=1}^{\eta} p(\vec{x}_i) \approx \sum_{i=1}^{\eta} \text{pr}(\vec{y}_i),$$

for every $\eta \leq \mu$. Hence, for such an $\eta \approx \infty$

$$\sum_{i=\eta+1}^{\nu} \mathrm{pr}(\vec{x}_i) = \sum_{i=1}^{\nu} \mathrm{pr}(\vec{x}_i) - \sum_{i=1}^{\eta} \mathrm{pr}(\vec{x}_i)$$

$$\approx 1 - \sum_{i=1}^{\eta} p(\vec{x}_i)$$

$$\approx 0$$

Let A be standard. Then the subsequence, $\{\vec{z}_j\}$, of elements of $A \cap \{\vec{x}_1, \vec{x}_2, \ldots\}$, is standard. Thus, the series $\sum_{j=1}^{\infty} p(\vec{z}_j)$ converges, and so $\sum_{j=\kappa}^{\infty} p(\vec{z}_j) \approx 0$, for every infinite κ. Let $\{\vec{u}_j\}_{j=1}^{\kappa}$ be the correseopondig subsequence of elements of $A \cap \Omega$. Then, $\vec{u}_j = \vec{z}_j$, for finite j. We have, for every finite m

$$\sum_{j=1}^{m} p(\vec{z}_j) \approx \sum_{j=1}^{m} \mathrm{pr}(\vec{u}_j).$$

Hence, by Robinson's lemma, there is an $\eta \approx \infty$, $\eta \leq \mu$ such that

$$\sum_{j-1}^{\eta} p(\vec{y}_j) \approx \sum_{j=1}^{\eta} \mathrm{pr}(\vec{y}_j).$$

We also have

$$\mathrm{Pr}(A \cap \Omega) = \sum_{k=1}^{\kappa} \mathrm{pr}(\vec{u}_k)$$

$$= \sum_{k=1}^{\eta} \mathrm{pr}(\vec{u}_k) + \sum_{k=\eta+1}^{\kappa} \mathrm{pr}(\vec{u}_k)$$

But

$$\sum_{k=\eta+1}^{\kappa} \mathrm{pr}(\vec{u}_k) \leq \sum_{i=\eta+1}^{\nu} \mathrm{pr}(\vec{x}_i) \approx 0.$$

Thus

$$\mathrm{Pr}(A \cap \Omega) \approx \sum_{j=1}^{\eta} \mathrm{pr}(\vec{u}_j)$$

$$\approx \sum_{1=1}^{\eta} p(\vec{z}_j)$$

$$\approx \sum_{j=1}^{\infty} p(\vec{z}_j)$$

$$= \sum_{\vec{y} \in A \cap \{\vec{x}_1, \vec{x}_2, \ldots\}} p(\vec{y}).$$

2. DISCRETE DISTRIBUTIONS

The converse is not difficult to prove: Suppose that Pr has a frequency function, p, such that for every standard A

$$\Pr(A \cap \Omega) = \sum_{\vec{y} \in A \cap \{\vec{x}_1, \vec{x}_2, \ldots\}} p(\vec{y}).$$

Let m be finite. Then $\vec{x}_m$ is real, and, hence, the singleton $\{\vec{x}_m\}$ is standard. Thus

$$\mathrm{pr}(\vec{x}_m) = \Pr(\{\vec{x}_m\} \cap \Omega) \approx p(\vec{x}_m).$$

□

We also have the following theorem, which allows us to compute expectations and variances in some cases:

THEOREM X.2. *Let X be a random variable over $^*\mathbb{R}$ such that $\Pr_X$ is a discrete probability distribution with frequency p and domain sequence $\{x_i\}$. Then:*

(1) *If $\sum_{n \in {}^*\mathbb{N}} |x_n| p(x_n)$ *-converges and $\sum_{x \in \Lambda_X} |x| \mathrm{pr}(x)$ nearly converges, we have*

$$\mathbf{E}\, X \approx \sum_{n \in {}^*\mathbb{N}} x_n\, p(x_n).$$

(2) *If $\sum_{n \in {}^*\mathbb{N}} (x_n - \mathbf{E}\, X)^2 p(x_n)$ *-converges and $\sum_{x \in \Lambda_X} (x - \mathbf{E}\, X)^2 \mathrm{pr}(x)$ nearly converges we have*

$$\mathrm{Var}(X) \approx \sum_{n \in {}^*\mathbb{N}} (x_n - \mathbf{E}\, X)^2 p(x_n).$$

PROOF. We prove (1). Let $z_1, z_2, \ldots, z_\mu, \ldots$ be an enumeration of the positive elements of the domain of p. Since p is standard, the sequence has length m, where m is finite, or it is defined over $^*\mathbb{N}$. The case with length m is simple, so we leave it to the reader. So assume that the sequence is defined over $^*\mathbb{N}$. We have, that z_m, for m finite, is finite. Let $u_1, u_2, \ldots, u_\nu$ be an enumeration of the nonnegative elements of Λ_X.

The hypothesis implies that $u_m = z_m$ and

$$\mathrm{pr}_X(u_m) \approx p(z_m)$$

for every finite m.

We also have

$$\sum_{z \in {}^*\mathbb{R}^+} z\, p(z) \approx \sum_{i=1}^{\nu} z_i\, p(z_i)$$

is finite. That is, the series $\sum_{i=1}^{\nu} z_i\, p(z_i)$ nearly converges to a finite number c.

On the other hand, for every finite m, we have

$$\sum_{i=1}^{m} u_i\, \mathrm{pr}(u_i) \approx \sum_{i=1}^{m} z_i p(z_i).$$

Then, by Robinson's Lemma VIII.10, there is an infinite $\mu \leq \nu$, such that

$$\sum_{i=0}^{m} u_i \operatorname{pr}(u_i) \approx \sum_{i=1}^{m} z_i\, p(z_i)$$

for every $m \leq \mu$. But

$$\sum_{i=1}^{\mu} z_i\, p(z_i) \approx c$$

and

$$\sum_{i=1}^{\nu} u_i \operatorname{pr}_X(u_i) \approx \sum_{i=1}^{\mu} u_i \operatorname{pr}(u_i).$$

Hence

$$\sum_{i=1}^{\nu} u_i \operatorname{pr}_X(u_i) \approx c = \sum_{z \in {}^*\mathbb{R}^+} z\, p(z).$$

We enumerate the negative elements of Λ_X, t_1, t_2, ..., t_η and prove in a similar way that

$$\sum_{i=0}^{\eta} t_i \operatorname{pr}_X(t_i) \approx \sum_{y \in {}^*\mathbb{R}^-} y\, p(y)$$

and thus

$$\sum_{x \in \Lambda_X} x \operatorname{pr}_X(x) = \sum_{i=1}^{\nu} u_i \operatorname{pr}_X(u_i) + \sum_{i=1}^{\eta} t_i \operatorname{pr}_X(t_i) \approx \sum_{y \in {}^*\mathbb{R}} y\, p(y).$$

The proof of (2) is similar. □

We conclude this section with a few examples of discrete distributions.

Bernoulli and binomial distributions. The Bernoulli distribution is very simple. It is determined by the frequency function

$$p(x) = \begin{cases} 1-p, & \text{if } x = 0, \\ p, & \text{if } x = 1, \\ 0, & \text{otherwise.} \end{cases}$$

Thus, the distribution, pr is

$$\operatorname{pr}(x) = \begin{cases} 1-p, & \text{if } x = 0, \\ p, & \text{if } x = 1, \\ 0, & \text{otherwise.} \end{cases}$$

It is clear that if X is a Bernoulli random variable, then pr_X is a Bernoulli frequency function.

A *binomial* $\langle n, p \rangle$ *frequency function*, for n a finite or infinite natural number, is defined by

$$p(i) = \binom{n}{i} p^i (1-p)^{n-i}, \qquad i = 0, 1, \ldots, n$$

and $p(i) = 0$, otherwise. If n is a finite natural number, then we get a standard probability distribution defined over the space $\langle\{0, 1, \ldots, n\}, \text{pr}\rangle$, where:

$$\text{pr}(i) = \binom{n}{i} p^i (1-p)^{n-i}.$$

It is clear that if X is a binomial $\langle n, p\rangle$ random variable, where n is finite and p is real, then pr_X and Pr_X are binomial $\langle n, p\rangle$ frequency function and probability distribution, respectively.

Poisson distribution. We define the standard frequency function

$$p(i) = e^{-\lambda} \frac{\lambda^i}{i!}$$

for $i = 0, 1, \ldots$ This is a frequency function, since

$$\sum_{i=0}^{\infty} p(i) = e^{-\lambda} \sum_{i=0}^{\infty} \frac{\lambda^i}{i!} = e^{-\lambda} e^{\lambda} = 1.$$

By the ratio test, it is easily proved that

$$\sum_{i=0}^{\infty} i p(i) \quad \text{and} \quad \sum_{i=0}^{\infty} (i-\lambda)^2 p(i)$$

converge.

Let X be a binomial $\langle n, p\rangle$ random variable, with $n \approx \infty$, $p \approx 0$, and $\lambda = np$, real. Then, for $i = 0, 1, \ldots, n$

$$\text{pr}_X(i) = \Pr[X = i]$$

$$= \frac{n!}{(n-i)! i!} p^i (1-p)^{n-i}$$

$$= \frac{n!}{(n-i)! i!} \left(\frac{\lambda}{n}\right)^i \left(1-\frac{\lambda}{n}\right)^i$$

$$= \frac{n(n-1)\cdots(n-i+1)}{n^i} \frac{\lambda^i}{i!} \frac{(1-\lambda/n)^n}{(1-\lambda/n)^i}$$

If i is finite, we have

$$\left(1-\frac{\lambda}{n}\right)^n \approx e^{-\lambda}, \quad \frac{n(n-1)\cdots(n-i+1)}{n^i} \approx 1, \quad \left(1-\frac{\lambda}{n}\right)^i \approx 1.$$

Hence, for i finite

$$\text{pr}_X(i) \approx e^{-\lambda} \frac{\lambda^i}{i!} = p(i).$$

We shall now prove that

$$\sum_{i=0}^{n} i \, \text{pr}_X(i) \quad \text{and} \quad \sum_{i=0}^{n} (i-\lambda)^2 \, \text{pr}_X(i)$$

nearly converge. We use the comparison test, Theorem VIII.13. We have, if $i \approx \infty$

$$\left(1 - \frac{\lambda}{n}\right)^i \geq \left(1 - \frac{\lambda}{n}\right)^n,$$

so that

$$\frac{(1 - \lambda/n)^n}{(1 - \lambda/n)^i} \leq 1.$$

Also

$$\frac{n(n-1)\cdots(n-i+1)}{n^i} \leq 1.$$

Thus

$$\mathrm{pr}_X(i) \leq \frac{\lambda^i}{i!}.$$

Hence

$$i \,\mathrm{pr}_X(i) \leq i\frac{\lambda^i}{i!}.$$

The series

$$\sum_{i=0}^{n} \frac{\lambda^i}{i!}$$

obviously nearly converges (it can be proved by the ratio test, for instance). Thus, by Theorem X.2

$$\mathbf{E}\,X \approx \sum_{i=0}^{\infty} ip(i).$$

We have $\mathbf{E}\,X = np = \lambda$, since X is binomial $\langle n, p \rangle$. Thus, since λ is real, we have shown that

$$\sum_{i=0}^{\infty} ip(i) = \lambda.$$

We also have

$$(i - \lambda)^2 \,\mathrm{pr}_X(i) \leq (i - \lambda)^2 \frac{\lambda^i}{i!}.$$

Hence, again by Theorem X.2

$$\mathrm{Var}(x) \approx \sum_{i=0}^{\infty} (i - \lambda)^2 p(i).$$

Now, $\mathrm{Var}(X) = np(1-p) = \lambda - \lambda p \approx \lambda$, so that we also have shown that

$$\sum_{i=0}^{\infty} (i - \lambda)^2 p(i) = \lambda.$$

Here we have a standard result obtained rather simply by nonstandard methods.

3. Continuous distributions

We now introduce densities and continuous probability distributions.

DEFINITION X.3 (CONTINUOUS DISTRIBUTIONS).
(1) A *density* is a standard continuous function, $f : R \to {}^*\mathbb{R}$, where R is a closed region in $^{n*}\mathbb{R}$, such that $f(\vec{x}) \geq 0$, for every $\vec{x} \in R$, and
$$\int \cdots \int_R f = 1.$$

(2) A probability distribution, Pr, defined on the hyperfinite space $\langle \Omega, \mathrm{pr} \rangle$, is *(absolutely) nearly continuous* if there is a density f over a region R, which contains Ω, such that for every standard closed region $A \subseteq R$
$$\Pr(A \cap \Omega) \approx \int \cdots \int_A f.$$

We say that a random vector, $\vec{X}$ is *nearly continuous*, if its probability distribution, $\Pr_{\vec{X}}$ is nearly continuous.

This definition implies that there is at most one density for any probability distribution. We shall not prove this fact here. We shall prove an equivalent nonstandard condition for continuous distributions. We need, for its proof, the following fact:

PROPOSITION X.3. *Let f be a continuous standard function over $^{n*}\mathbb{R}$, and R an infinitesimal closed region. Then*
$$\int \cdots \int_R f \approx f(\vec{x}) V(R) \quad (V(R)),$$
for any finite $\vec{x} \in R$.

PROOF. It is easy to see that
$$m V(R) \leq \int \cdots \int_R f \leq M V(R),$$
where m is the minimum of f in R and M is its maximum. Since R is infinitesimal, for any $\vec{x} \in R$, $f(\vec{x}) \approx m \approx M$. Thus, ths conclusion of the theorem is obtained. □

We need to extend the notion of partition into rectangles to closed bounded regions R. There is always a smallest rectangle, D, containing R, called *the circumscribed rectangle of R*. We obtain a partition into regions $\mathcal{Q}$ of D. Then the set $\mathcal{P}$ of elements of $\mathcal{Q}$ that are contained in R is a partition of R. We say that $\mathcal{P}$ is an *infinitesimal partition* of a closed bounded region R, if $\mathcal{P}$ is a partition of R and the closure of every element of $\mathcal{Q}$ is an infinitesimal closed region.

Since closed bounded regions have Jordan content, we have that the sum of the volume of the elements of $\mathcal{Q}$ that intersect the boudary is zero.

THEOREM X.4. *Let* Pr *be a probability distribution over* $\langle \Omega, \mathrm{pr} \rangle$ *such that* $\Omega \subseteq R$, *where* R *is a region, and let* f *be a density over* R. *Then:*

(1) *If there is an infinite ν and an infinitesimal partition, $\mathcal{P}$ of R_ν such that for every $r \in \mathcal{P}$ and finite $\vec{x} \in R$ we have*
$$\Pr(r \cap \Omega) \approx f(\vec{x})\,V(r) \quad (V(r)),$$
then Pr *is nearly continuous with density f.*

(2) *If* Pr *is nearly continuous with density f over the region R then, there is an infinite $\nu \in {}^*\mathbb{N}$, such that for every infinite $\mu \leq \nu$, if $\mathcal{P}$ is a partition of R_ν into cubes of volume $1/\mu$, then every cube r in $\mathcal{P}$ and every finite $\vec{x} \in r$ satisfies*
$$\Pr(r \cap \Omega) \approx f(\vec{x})\,V(r) \quad (V(r)).$$

PROOF OF (1). Suppose that Pr, the infinitesimal partition, $\mathcal{P}$ of R_ν, and the function f satisfy the hypothesis of (1). Let A be a standard closed bounded region, and let T consist of the elements of $\mathcal{P}$ which intersect A. We have

$$\sum_{\substack{r \subseteq A \\ r \in T}} \Pr(r \cap \Omega) \leq \Pr(A \cap \Omega) \leq \sum_{\substack{r \cap A \neq \emptyset \\ r \in T}} \Pr(r \cap \Omega).$$

We have that $f(\vec{x})$ is bounded for $\vec{x} \in A$, and

$$\sum_{\substack{r \subseteq A \\ r \in T}} V(r)$$

is finite. Thus, by Theorem IX.1

$$\sum_{\substack{r \subseteq A \\ r \in T}} \Pr(r \cap \Omega) \approx \underline{\sum\sum}_T f$$

and

$$\sum_{\substack{r \cap A \neq \emptyset \\ r \in T}} \Pr(r \cap \Omega) \approx \overline{\sum\sum}_T f.$$

By the properties of the integral

$$\underline{\sum\sum}_T f \approx \int \cdots \int_A f \approx \overline{\sum\sum}_T f.$$

Thus
$$\Pr(A \cap \Omega) \approx \int \cdots \int_A f.$$

3. CONTINUOUS DISTRIBUTIONS

Assume, now, that A is a standard unbounded, subregion of R. Let $A_m = R_m \cap A$. Then $\{A_m\}$ is an increasing sequence whose union is A and, for every infinite μ

$$\int \cdots \int_A f \approx \int \cdots \int_{A_\mu} f.$$

For every finite m, we have

$$\Pr(A_m \cap \Omega) \approx \int \cdots \int_{A_m} f.$$

Thus, by Robinson's lemma, VIII.10, there is an infinite η_1 such that for all $\mu \leq \eta_1$

$$\Pr(A_\mu \cap \Omega) \approx \int \cdots \int_{A_\mu} f.$$

We also have that

$$\Pr(R_m \cap \Omega) \approx \int \cdots \int_{R_m} f$$

for every finite m. Thus, again by Robinson's lemma, and the convergence of the integral, there is an infinite η_2 such that for every infinite $\mu \leq \eta_2$

$$\Pr(R_\mu \cap \Omega) \approx \int \cdots \int_{R_\mu} f \approx 1.$$

Hence, for an infinite $\mu \leq \eta_2$

$$\Pr((A - A_\mu) \cap \Omega) \leq \Pr(\Omega - R_\mu) \approx 0.$$

Let $\eta = \min\{\eta_1, \eta_2\}$. Then, for infinite $\mu \leq \eta$

$$\Pr(A \cap \Omega) = \Pr(A_\mu \cap \Omega) + \Pr((A - A_\mu) \cap \Omega)$$
$$\approx \Pr(A_\mu \cap \Omega)$$
$$\approx \int \cdots \int_{A_\mu} f$$
$$\approx \int \cdots \int_A f$$

□

PROOF OF (2). Suppose that $\Pr$ is nearly continuous with density f over the region R. Let $\mathcal{P}_m$ be a partition of R_m into cubes, which are standard, of volume $1/k$, for a certain $k \leq m$, for a finite m. Then, by the near continuity of $\Pr$, for every $r \in \mathcal{P}_m$

$$\Pr(r \cap \Omega) \approx \int \cdots \int_r f.$$

Since $V(r)$ is finite, we have
$$\frac{\Pr(r \cap \Omega)}{V(r)} \approx \frac{\int \cdots \int_r f}{V(r)}.$$

Hence, the set A of natural numbers $m \in {}^*\mathbb{N}$ such that for every $k \leq m$, if $\mathcal{P}_k$ is a partition of R_m into cubes of volume $1/k$, then
$$\left| \frac{\Pr(r \cap \Omega)}{V(r)} - \frac{\int \cdots \int_r f}{V(r)} \right| \leq \frac{1}{k}$$

is internal and contains all finite natural numbers. Hence, by overflow, for a certain $\nu \approx \infty$, it contains all numbers $\mu \leq \nu$. Thus, if $\mathcal{P}_\mu$ is a partition of R_ν into infinitesimal regions of volume $1/\mu$, for μ infinite, $\mu \leq \nu$, then, we have, by the previous proposition, that for every $r \in \mathcal{P}_\mu$ and $\vec{x} \in r$
$$\frac{\Pr(r \cap \Omega)}{V(r)} \approx \frac{\int \cdots \int_r f}{V(r)} \approx f(\vec{x}).$$

□

For nearly continuous random variables, the notion of near independence, which we introduce next, is useful.

DEFINITION X.4 (NEAR INDEPENDENCE). Let m be a finite natural number. The random variables $X_1, X_2, \ldots, X_m$ are *nearly independent* if for every standard intervals, $I_{j_1}, I_{j_2}, \ldots, I_{j_k}$, where $1 \leq j_i \leq m$, and $j_i \neq j_l$, for i and l between 1 and m, we have

$$\Pr[\langle X_{j_1}, \ldots, X_{j_k} \rangle \in I_{j_1} \times \cdots \times I_{j_k}] \approx \Pr[X_{j_1} \in I_{j_1}] \cdot \Pr[X_{j_2} \in I_{j_2}] \cdots \Pr[X_{j_k} \in I_{j_k}].$$

It is clear that independent random variables are nearly independent. We have the following proposition:

PROPOSITION X.5. *Let $X_1, \ldots, X_m$ be nearly continuous random variables, for m finite, over the same probability space, with densities $f_{X_1}, \ldots, f_{X_m}$. Then $X_1, \ldots, X_m$ are nearly independent if and only if the random vector $\vec{X} = \langle X_1, \ldots, X_m \rangle$ is nearly continuous with density*
$$f_{\vec{X}} = f_{X_1} \cdot f_{X_2} \cdots f_{X_m}.$$

PROOF. Suppose, first, that $X_1, \ldots, X_m$ are nearly independent. Let A be a standard rectangle, $A = I_1 \times \cdots \times I_m$, where $I_1, \ldots, I_m$ are intervals. Then
$$\Pr[\vec{X} \in A] \approx \Pr[X_1 \in I_1] \cdots \Pr[X_m \in I_m]$$
$$\approx \int_{I_1} f_{X_1} \cdots \int_{I_m} f_{X_m}$$
$$= \int \cdots \int_A f_{X_1} \cdots f_{X_m}$$

Thus, $f_{\vec{X}} = f_{X_1} \cdots f_{X_m}$ is a density for $\vec{X}$.

3. CONTINUOUS DISTRIBUTIONS

Suppose, now, that A is as above and $f_{\vec{X}} = f_{X_1} \cdots f_{X_m}$ is the density of $f_{\vec{X}}$. Then

$$\Pr[\vec{X} \in A] \approx \int \cdots \int_A f_{X_1} \cdots f_{X_m}$$

$$= \int_{I_1} f_{X_1} \cdots \int_{I_m} f_{X_m}$$

$$\approx \Pr[X_1 \in I_1] \cdots \Pr[X_m \in I_m].$$

□

We also have the following theorem that allows us, in some cases, to compute expectations and variances of nearly continuous distributions.

THEOREM X.6. *Let $X : \Omega \to {}^*\mathbb{R}$ be a random variable with a nearly continuous probability distribution, $\Pr_X$, with density function f. Then:*

(1) *if $\int_{-\infty}^{\infty} |x| f(x)\,dx$ exists and the series $\sum_{x \in \Lambda_X} |x| \operatorname{pr}(x)$ nearly converges, we have*

$$\mathbf{E}\,X \approx \int_{-\infty}^{\infty} x f(x)\,dx.$$

(2) *if $\int_{-\infty}^{\infty} (x - \mathbf{E}\,X)^2 f(x)\,dx$ exists and the series $\sum_{x \in \Lambda_X} (x - \mathbf{E}\,X)^2 \operatorname{pr}(x)$ nearly converges we have*

$$\operatorname{Var}(X) \approx \int_{-\infty}^{\infty} (x - \mathbf{E}\,X)^2 f(x)\,dx.$$

PROOF. We prove (1). Let $x_0, x_1, \ldots, x_\nu$ be an increasing enumeration of Λ_X and let

$$0 < dx < \min\{x_{i+1} - x_i \mid i = 0, 1, \ldots, \nu - 1\}.$$

Then

$$\Pr[x < X < x + dx] = \begin{cases} \operatorname{pr}_X(y), & \text{if } y \in \Lambda_X \cap (x, x + dx], \\ 0, & \text{otherwise.} \end{cases}$$

By the hypothesis and Proposition X.3, we also have

$$\Pr[x < X < x + dx] \approx f(x)\,dx \quad (dx),$$

for every finite $x \in {}^*\mathbb{R}$. Hence

$$\operatorname{pr}_X(x) \approx f(x)\,dx$$

for every finite $x \in \Lambda_X$. By Theorem IX.1, for any $a < 0 < b$ finite

$$\sum_{\substack{x \in \Lambda_X \\ a \leq x \leq b}} x \operatorname{pr}(x) \approx \sum_{\substack{x \in \Lambda_X \\ 0 \leq x \leq b}} f(x)\,dx \approx \int_a^b x f(x)\,dx.$$

By Robinson's Lemma VIII.10, there are infinite $b > 0$ and $a < 0$ such that

$$\sum_{\substack{x \in \Lambda_X \\ a \leq x \leq b}} x\,\mathrm{pr}(x) \approx \int_a^b x f(x)\,dx.$$

Since the series $\sum_{x \in \Lambda_X} |x|\,\mathrm{pr}(x)$ nearly converge, we have that

$$\sum_{x \in \Lambda_X} x\,\mathrm{pr}(x) \approx \sum_{\substack{x \in \Lambda_X \\ a \leq x \leq b}} x\,\mathrm{pr}(x)$$

$$\approx \int_a^b x f(x)\,dx$$

$$\approx \int_{-\infty}^{\infty} x f(x)\,dx$$

The proof of (2) is similar. □

As an example, we shall discuss the uniform distribution:

Uniform distributions. Let a, b be real numbers with $a < b$. Then the standard density function

$$f(x) = \begin{cases} \dfrac{1}{b-a}, & \text{if } a \leq x \leq b, \\ 0, & \text{otherwise.} \end{cases}$$

is called a *uniform* density over $[a, b]$. It is clear that f is a density, since

$$\int_{-\infty}^{\infty} f(x)\,dx = \int_a^b f(x)\,dx = 1.$$

We now show a probability distribution that is approximately uniform on the interval $[0, 1]$, i.e., with a uniform density over $[0, 1]$. We take the interval $[0, 1]$ just for simplicity. Let Ω be an equally spaced hyperfinite approximation of the unit interval $[0,1]$. That is, $\Omega = \{t_0, t_1, \ldots, t_\nu\}$ such that ν is an infinite natural number, $t_0 = 0$, $t_\nu = 1$, and we also asumme that $dt = t_{n+1} - t_n = 1/\nu$ for every $n < \nu$. It is clear that $\#\Omega = \nu + 1$. Thus, we are assuming that $\Omega = \{0, dt, 2\,dt, 3\,dt, \ldots, \nu\,dt\}$. We define the distribution pr on Ω by setting for any $n \leq \nu$

$$\mathrm{pr}(t_n) = \frac{1}{\nu + 1}.$$

Thus, for any internal subset A of Ω, we have the probability distribution

$$\Pr A = \frac{\#A}{\#\Omega}.$$

The probability measure Pr is called the *counting measure on* Ω. It is clear that $\langle \Omega, \mathrm{pr}\rangle$ is an internal probability space. We now show that Pr has a uniform density.

3. CONTINUOUS DISTRIBUTIONS

The partition that we need in order to apply Theorem X.4 is formed by cubes, which in this case are intervals, $[t_0, t_1), \ldots, [t_i, t_{i+1}), \ldots, [t_{\nu-1}, t_\nu)$ of length $dt = 1/\nu$. Then

$$\frac{\Pr([t_i, t_{i+1}) \cap \Omega)}{dt} = \frac{\Pr(\{t_i\})}{dt} = \frac{1}{(\nu+1)dt} \approx 1.$$

Since the uniform density for $[0,1]$ is $f(x) = 1$, for $0 \le x \le 1$, by Theorem X.4, we get that Pr has this uniform density.

Thus, if $c, d \in [0,1]$ with $c < d$, we have

$$\Pr(\Omega \cap [c,d]) \approx \int_c^d dx = d - c.$$

We also can see that, in a sense, Theorem X.4, 2, cannot be improved: it is not true that for any partition with infinitesimal intervals, any interval r in the partition, and $x \in r$, satifies

$$\frac{\Pr(r \cap \Omega)}{V(r)} \approx 1.$$

This formula is true, as it is asserted in the theorem, only for partitions into cubes of volume $\ge 1/\mu$, for $\mu \le \nu$, for a certain fixed ν. In order to find a counterexample, take a partition with subintervals of length $1/2\nu$. Then there are some subintervals which do not contain any point in Ω, and, thus $\Pr(r \cap \Omega) = 0$.

The same construction can be carried out in the unit circle **C**. As in Example VIII.4, let

$$C = \{e^{ik\, d\theta} \mid 0 \le k \le \nu\},$$

where $d\theta = 2\pi/\nu$ and ν is an infinite natural number, be a hyperfinite approximation of **C**. Again, let pr be the counting measure on C. Then, as above, if A is an arc of C, we have that $\Pr(C \cap A) \approx$ the length of A.

We discuss, now, Bertrand's mixture paradox (see page 61) from an infinitesimal standpoint. The uniform distribution of the ratio of water to wine between 1 and 2 is obtained by a near interval, $T_1 = \{t_0, t_1, \ldots, t_\nu\}$, for $[1,2]$. We must assume that $t_{i+1} - t_i = dt = 1/(\nu+1)$. Every point in T_1 is equally probable, that is $\text{pr}_1(t_i) = 1/(\nu+1)$.

On the other hand, the uniform distribution of the ratio of wine to water between $1/2$ and 1, is obtained by an equally spaced near interval for $[1/2, 1]$. If we take the reciprocals of the elements of T_1, we do get a near interval, $T_2 = \{u_0, u_1, \ldots, u_\nu\}$, for $[1/2, 1]$, but its points are not equally spaced. Thus, we do not get in this way a uniform distribution. Suppose that we assign the same probability to each point, that is

$$\text{pr}_2(u_i) = \text{pr}_1(t_i) = \frac{a}{\nu + 1}.$$

We have that

$$du_i = u_{i+1} - u_i = \frac{1}{t_{i+1}} - \frac{1}{t_i} = \frac{t_i - t_{i+1}}{t_i t_{i+1}}.$$

Then
$$\frac{\Pr([u_{i+1}, u_i) \cap \Omega)}{du_i} = \frac{1}{\nu+1} \frac{t_i t_{i+1}}{t_{i+1} - t_i}$$
$$= \frac{\nu}{\nu+1} \cdot t_i t_{i+1}$$
$$\approx t_i^2$$
$$= \frac{1}{u_i^2}.$$

Thus, the density is, in this case, $f(x) = \dfrac{1}{x^2}$, which is clearly not uniform.

We conclude this section with a corollary to the theorem on change of variables (Theorem IX.18). Notice, that if R is a region and T is continuously differentiable transformation defined on R, then $T(R)$ is also a region.

THEOREM X.7. *Let $\vec{X}$ be a continuous random vector with density $f_{\vec{X}}$ over the region R, and let T be a continuously differentiable one-one transformation of R. Then the random vector $\vec{Y} = T(\vec{X})$ is also continuous with density*
$$f_{\vec{Y}}(\vec{y}) = f_{\vec{X}}(T^{-1}(\vec{y}))|J_{T^{-1}}(y)|,$$
for any $\vec{y} \in T(R)$.

PROOF. Let $\vec{X}$ be defined over $\langle \Omega, \text{pr} \rangle$, and let R_ν and $\mathcal{P}$ be as in (2) of Theorem X.7, such that $\mathcal{P}$ is composed of infinitesimal cubes of area $1/\mu$. Thus, we have, for any $r \in \mathcal{P}$ and $\vec{x} \in R$
$$\Pr[\vec{X} \in r \cap \Omega] = \Pr_{\vec{X}}(r \cap \Omega) \approx f(\vec{x})\,V(r) \quad (V(r)).$$

Let $\mathcal{Q}$ be the partition of $T(R)$ obtained by applying T to the elements of $\mathcal{P}$. Then $\mathcal{Q}$ is an infinitesimal partition of $T(R)$, and, by Lemma IX.16
$$\frac{V(T(r))}{V(r)} \approx |J_T(\vec{x})|.$$

We have for any s in $\mathcal{Q}$, since $f(T^{-1}(\vec{x}))$ and $J_T(\vec{x})$ are finite
$$\frac{\Pr_{\vec{Y}}(s \cap \Omega)}{V(s)} = \frac{\Pr[T(\vec{X}) \in s]}{V(s)}$$
$$= \frac{\Pr[\vec{X} \in T^{-1}(s)]}{V(s)}$$
$$= \frac{\Pr[\vec{X} \in T^{-1}(s)]}{V(T^{-1}(s))} \frac{V(T^{-1}(s))}{V(s)}$$
$$\approx f(T^{-1}(\vec{x})) \cdot |J_T^{-1}(\vec{x})|$$

Since $f \circ T^{-1} \cdot |J_{T^{-1}}|$ is continuous and, by Theorem IX.18, integrable over $T(R)$, by Theorem X.7 (1) it is a density of $\vec{Y}$. □

Recall that T is called an *affine transformation* of a region R if there exists an $n \times n$ matrix $\mathbf{A}$ and a vector $\vec{c}$ such that $T(\vec{x}) = \vec{x}\mathbf{A} + \vec{c}$, for $\vec{x} \in R$. If $\vec{c} = \vec{0}$, then T is called *linear*. The function T is one-one, if $\mathbf{A}$ is nonsingular, and then

$$T^{-1}(\vec{y}) = (\vec{u} - \vec{c})\mathbf{A}^{-1},$$

for $\vec{y} \in T(R)$. We have the following corollary.

COROLLARY X.8. *Suppose that $\vec{X}$ is a continuous random vector with density $f_{\vec{X}}$ and that T is a one-one affine transformation, as defined above. Then $\vec{Y} = T(\vec{X})$ is continuous with density*

$$f_{\vec{Y}}(\vec{y}) = |\det \mathbf{A}|^{-1} f_{\vec{X}}((\vec{y} - \vec{c})\mathbf{A}^{-1}),$$

for every $\vec{y} \in {}^{n*}\mathbb{R}$, where $\det \mathbf{A}$ is the determinant of $\mathbf{A}$.

PROOF. The corollary follows from the preceding theorem and

$$J_T(T^{-1}(\vec{y})) = \det \mathbf{A}.$$

□

4. Normal distributions

We discuss, in this section, random variables whose probability functions approximate the normal distribution. We first introduce a definition.

DEFINITION X.5. Let μ be a real number and σ, a positive real number. The random variable X over $\langle \Omega, \mathrm{pr} \rangle$ is said to be *nearly normally distributed with mean μ and variance σ^2* if X is nearly continuous with density

$$f(x) = \frac{1}{\sqrt{2\pi}\sigma} e^{-(x-\mu)^2/2\sigma^2}.$$

We write, in this case, $X \sim \mathcal{N}(\mu, \sigma^2)$.

There is a well-known standard proof for the fact that

$$\frac{1}{\sqrt{2\pi}\sigma} \int_{-\infty}^{\infty} e^{-(x-\mu)^2/2\sigma^2} \, dx = 1.$$

Thus, f is a standard density function.

As an example of the use of Theorem X.7, suppose that $\vec{X} = \langle X_1, X_2 \rangle$ where X_1 and X_2 are nearly independent nearly normally distributed random variables with mean 0 and variances 1 and 4, respectively. We want to calculate the density of the joint distribution of $Y_1 = X_1 + X_2$ and $Y_2 = X_1 - X_2$. Since X_1 is nearly independent of X_2, by Proposition X.5, the joint density is the product of the densities, that is

$$f_{\vec{X}}(x_1, x_2) = \frac{1}{4\pi} e^{-\frac{1}{2}(x_1^2 + \frac{1}{4}x_2^2)}.$$

We put $T(x_1, x_2) = \langle x_1 + x_2, x_1 - x_2 \rangle$. Then $\vec{Y} = T(\vec{X})$. We also have
$$T^{-1}(y_1, y_2) = \langle \frac{1}{2}(y_1 + y_2), \frac{1}{2}(y_1 - y_2) \rangle.$$
Then
$$J_{T^{-1}}(\vec{y}) = \begin{vmatrix} \frac{1}{2} & \frac{1}{2} \\ \frac{1}{2} & -\frac{1}{2} \end{vmatrix} = -\frac{1}{2}.$$
Substituting in Theorem X.7, we obtain
$$f_{\vec{Y}}(y_1, y_2) = \tfrac{1}{2} f_{\vec{X}}(\tfrac{1}{2}(y_1 + y_2), \tfrac{1}{2}(y_1 - y_2))$$
$$= \frac{1}{8\pi} e^{\frac{1}{2}(\frac{1}{4}(y_1+y_2)^2 + \frac{1}{16}(y_1-y_2)^2)}$$
$$= \frac{1}{8\pi} e^{\frac{1}{32}(5y_1^2 + 5y_2^2 + 6y_1 y_2)}.$$

We also have that if $X \sim \mathcal{N}(\mu, \sigma^2)$, then, by Corollary X.8, $Y = \dfrac{X - \mu}{\sigma}$ is $\mathcal{N}(0, 1)$. Thus, we can do all our calculations with standard normal distributions

De Moivre-Laplace Theorem. We, now, discuss an example of a random variable over a hyperfinite probability space, which is nearly normally distributed. That this random variable is nearly normally distributed is, in fact, the de Moivre-Laplace Central Limit Theorem. In the course of explaining the example, we give an elementary proof of this theorem.

We assume without proof the well-known Stirling formula for the factorial. Its nonstandard version is
$$\nu! \asymp \sqrt{2\pi\nu}\left(\frac{\nu}{e}\right)^\nu, \quad \text{for } \nu \approx \infty.$$
Recall that $a \asymp b$ is defined in Definition IX.1.

Let X be a binomial (ν, p) random variable, where $\nu \approx \infty$, $p \not\approx 0$, and $q = 1 - p \not\approx 0$. We have, $\mathbf{E}\, X = \nu p$ and $\mathrm{Var}(X) = \nu p q$. We also have
$$\Pr[X = m] = \binom{\nu}{m} p^m q^{\nu - m}.$$
We define the random variable Y by
$$Y = \frac{X - \nu p}{\sqrt{\nu p q}}.$$
Suppose that v is a number such that $m = \nu p + \sqrt{\nu p q}\, v$ is an integer $m \leq \nu$, and m and $\nu - m$ are infinite. We prove that
$$\Pr[Y = v] \asymp \frac{1}{\sqrt{2\pi \nu p q}} e^{-(1/2)v^2}.$$
We have
$$\Pr[Y = v] = \Pr[X = \nu p + \sqrt{\nu p q}\, v] = \Pr[X = m] = \binom{\nu}{m} p^m q^{\nu-m}.$$

From Stirling's formula we get

4. NORMAL DISTRIBUTIONS

$$\binom{\nu}{m} = \frac{\nu!}{m!(\nu-m)!}$$

$$\asymp \frac{1}{\sqrt{2\pi\nu}}\left(\frac{\nu}{m}\right)^{m+1/2}\left(\frac{\nu}{\nu-m}\right)^{\nu-m-1/2}.$$

Therefore

$$\Pr[Y = v] \asymp \frac{1}{\sqrt{2\pi pq\nu}}\left(\frac{m}{\nu p}\right)^{-m-1/2}\left(\frac{\nu-m}{\nu q}\right)^{-(\nu-m)-1/2}.$$

Thus, the quotient of the left-hand and right-hand side is infinitely close to 1, if ν, m, and $\nu - m$ are infinite. On taking natural logarithms, the difference is infinitesimal:

$$\log \Pr[Y = v] \approx -\log\sqrt{2\pi} - \frac{1}{2}\log \nu pq -$$

$$(\nu p + \sqrt{\nu pq}v + \frac{1}{2})\log(1 + \frac{\sqrt{\nu pq}}{\nu p}v) -$$

$$(\nu q - \sqrt{\nu pq}v + \frac{1}{2})\log(1 - \frac{\sqrt{\nu pq}}{\nu q}v).$$

Using the logarithmic series, we have that for $|x| < 1$

$$\log(1 + x) = x - \frac{x^2}{2} + \varepsilon$$

where $\varepsilon/x^2 \approx 0$. So we have,

$$\log \Pr[Y = v]) \approx -\log\sqrt{2\pi} - \frac{1}{2}\log \nu pq -$$

$$(\nu p + \sqrt{\nu pq}v + \frac{1}{2})(\frac{\sqrt{\nu pq}}{\nu p} - \frac{1}{2}\frac{q}{\nu p}v^2 + \varepsilon_1) -$$

$$(\nu q - \sqrt{\nu pq}v + \frac{1}{2})(-\sqrt{\nu pq}v - \frac{1}{2}\frac{p}{\nu q}v^2 + \varepsilon_2)$$

where $\nu\varepsilon_1 \approx 0$ and $\nu\varepsilon_2 \approx 0$. Multiplying out the parentheses, we get

$$\log \Pr[Y = v] \approx -\log\sqrt{2\pi} - \frac{1}{2}\log \nu pq - \frac{1}{2}v^2.$$

Hence

$$\Pr[Y = v] \asymp \frac{1}{\sqrt{2\pi\nu pq}}e^{-(1/2)v^2}. \tag{1}$$

Let

$$T = \{v \mid \text{for some integer } m, v = \frac{m - \nu p}{\sqrt{\nu pq}}, \frac{-\nu p}{\sqrt{\nu pq}} \leq v \leq \frac{\nu - \nu p}{\sqrt{\nu pq}}\}.$$

We have that T is a hyperfinite set $\{t_0, t_1, \ldots, t_\nu\}$ with $t_0 = -\nu p/\sqrt{\nu pq}$, $t_\nu = (\nu - \nu p)/\sqrt{\nu pq}$ and $dt = 1/\sqrt{\nu pq} \approx 0$. We have that t_0, the first element of

T, is negative infinite and t_ν, the last element of T, is positive infinite. Hence, T is a hyperfinite approximation of the real line.

Let t_m be finite. Then m and $\nu - m$ are infinite. Thus

$$\Pr[Y = t_m] \asymp \frac{1}{\sqrt{2\pi\nu pq}} e^{-t_m^2/2} = \frac{1}{\sqrt{2\pi}} e^{-t_m^2/2} dt.$$

Thus

$$\frac{\Pr[Y = t_m] \cdot \sqrt{2\pi}}{e^{-t_m^2/2} dt} \approx 1.$$

Since, $(1/\sqrt{2\pi})e^{-t_m^2/2} dt$ is finite

$$\Pr[Y = t_m] \approx \frac{1}{\sqrt{2\pi\nu pq}} e^{-t_m^2/2} = \frac{1}{\sqrt{2\pi}} e^{-t_m^2/2} dt \quad (dt).$$

By Theorem X.4, we get that Y is nearly normally distributed.

From these considerations we have proved a version of the de Moivre-Laplace Theorem:

THEOREM X.9 (DE MOIVRE-LAPLACE). *Let* $X_1, X_2, \ldots, X_n, \ldots X_\nu$, *for* $\nu \approx \infty$ *be a sequence of independent Bernoulli random variables with* $\Pr[X_n = 1] = p$, *for a certain* $p \not\approx 0$ *and* $p \not\approx 1$. *Let* $q = 1 - p$ *and*

$$Y_n = \frac{\sum_{i=1}^n X_i - np}{\sqrt{npq}}.$$

Then for any infinite $\mu \leq \nu$, *and any standard closed region*, A, *we have that*

$$\Pr[Y_\mu \in A]) \approx \frac{1}{\sqrt{2\pi}} \int_A e^{-t^2/2} dt.$$

That is, Y_μ *is nearly normally distributed, for any infinite* $\mu \leq \nu$.

It is interesting to note, that the variable Y_n defined above has expectation

$$\mathbf{E}\, Y_n = \frac{\mathbf{E}\, X - np}{\sqrt{npq}} = 0$$

and variance

$$\mathrm{Var}(Y_n) = \frac{\mathrm{Var}(X)}{npq} = 1.$$

We also have

$$\frac{1}{\sqrt{2\pi}} \int_{-\infty}^\infty t e^{-t^2/2}\, dt = 0$$

and

$$\frac{1}{\sqrt{2\pi}} \int_{-\infty}^\infty t^2 e^{-t^2/2}\, dt = 1.$$

Thus

$$\mathbf{E}\, Y_n = \frac{1}{\sqrt{2\pi}} \int_{-\infty}^\infty t e^{-t^2/2}\, dt$$

and
$$\text{Var}(Y_n) = \frac{1}{\sqrt{2\pi}} \int_{-\infty}^{\infty} t^2 e^{-t^2/2} \, dt.$$

5. Distribution theory for samples from a normal population

We, first, introduce two families of distributions. the first family has densities given by
$$g_{p,\lambda}(x) = \frac{\lambda^p x^{p-1} e^{\lambda x}}{\Gamma(p)},$$
for $x > 0$. The parameters p and λ are taken to be positive reals and $\Gamma(p)$ denotes the Euler *gamma function* defined by
$$\Gamma(p) = \int_0^{\infty} t^{p-1} e^{-t} \, dt.$$

It follows by integration by parts that, for all $p > 0$
$$\Gamma(p+1) = p\Gamma(p) \text{ and } \Gamma(k) = (k-1)!$$
for all positive integers k.

The family of distributions with such densities is referred to as the *gamma* family of distributions and we shall write $X \sim \Gamma(p, \lambda)$, if X is nearly continuous with density $g_{p,\lambda}$. The special case $p = 1$ is called the *exponential* family, and is written $\mathcal{E}(\lambda)$. By Corollary X.8, $X \sim \Gamma(p, \lambda)$ if and only if $\lambda X \sim \Gamma(p, 1)$.

Let k be a positive integer. The gamma density with $p = \frac{1}{2}k$ and $\lambda = \frac{1}{2}$ is called the *chi square density with k degrees of freedom* and is denoted by χ_k^2.

The other family is the *beta* family. Its densities are given by
$$b_{r,s} = \frac{x^{r-1}(1-x)^{s-1}}{B(r,s)},$$
for $0 < x < 1$, where r and s are positive real parameters, and $B(r,s) = (\Gamma(r)\Gamma(s))/\Gamma(r+s)$ is the *beta function*.

We have the following theorem:

THEOREM X.10. *If X_1 and X_2 are nearly independent continuous random variables with $\Gamma(p, \lambda)$ and $\Gamma(q, \lambda)$ densities, then $Y_1 = X_1 + X_2$ and $Y_2 = X_1/(X_1 + X_2)$ are nearly independent and have, respectively, $\Gamma(p+q, \lambda)$, and $\beta(p,q)$ densities.*

PROOF. Let $\lambda = 1$. Then, by Proposition X.5, the joint density of X_1 and X_2 is
$$f_{X_1,X_2}(x_1, x_2) = (\Gamma(p)\Gamma(q))^{-1} e^{(x_1+x_2)} x_1^{p-1} x_2^{q-1},$$
for $x_1 > 0$, $x_2 > 0$. Let
$$\langle y_1, y_2 \rangle = T(x_1, x_2) = \langle x_1 + x_2, \frac{x_1}{x_1 + x_2} \rangle.$$

Then T is one-one on $R = \{\langle x_1, x_2\rangle \mid x_1 > 0, x_2 > 0\}$ and its range is $T(R) = \{\langle y_1, y_2\rangle \mid y_1 > 0, 0 < y_2 < 1\}$. We note that on $T(R)$

$$T^{-1}(y_1, y_2) = \langle y_1 y_2, y_1 - y_1 y_2\rangle.$$

Therefore

$$J_{T^{-1}}(y_1, y_2) = -y_1.$$

We now apply the formula of Theorem X.7, and get the joint density of $\vec{Y} = \langle Y_1, Y_2\rangle$:

$$f_{\vec{Y}}(y_1, y_2) = \frac{e^{-y_1}(y_1 y_2)^{p-1}(y_1 - y_1 y_2)^{q-1} y_1}{\Gamma(p)\Gamma(q)},$$

for $y_1 > 0$, $0 < y_2 < 1$. Simplifying

$$f_{\vec{Y}}(y_1, y_2) = g_{p+q,1}(y_1) b_{p,q}(y_2).$$

By Proposition X.5, we obtain the theorem for $\lambda = 1$.

If $\lambda \neq 1$, define $X_1' = \lambda X_1$ and $X_2' = \lambda X_2$. Now X_1' and X_2' are nearly independent $\Gamma(p, 1)$, $\Gamma(q, 1)$ variables respectively. Since $X_1' + X_2' = \lambda(X_1 + X_2)$ and $X_1'(X_1' + X_2')^{-1} = X_1(X_1 + X_2)^{-1}$, the theorem follows. $\square$

By induction, we obtain the following result:

COROLLARY X.11. *If X_1, ..., X_n are nearly independent random variables such that X_i has a $\Gamma(p_i, \lambda)$ density, $i = 1, \ldots, n$, then $\sum_{i=1}^{n} X_i$ has $\Gamma(\sum_{i=1}^{n} p_i, \lambda)$ density.*

Before continuing with distributions derived from the normal, we notice that one-dimensional regions in $^*\mathbb{R}$ are simply intervals. For this case, in order to determine a density we do not need to analyze all intervals. Suppose that X is a random variable. If we know that

$$\Pr[X \leq t] \approx \int_{-\infty}^{t} f(x)\, dx,$$

then f is a density for X. This is so, because the integral over any interval can be obtained from these integrals.

Also, if we determine that

$$\Pr[X \leq t] \approx F(t),$$

for all t, where F is a standard continuous function, and $F'(t) = f(t)$, then f is a density of X, because

$$F(t) = \int_{-\infty}^{t} f(x)\, dx.$$

We can also obtain the density by differentiating functions G and H that satisfy

$$\Pr[0 < X < t] \approx G(t)$$

and

$$\Pr[-t < X < 0] \approx H(-t).$$

5. SAMPLES FROM A NORMAL POPULATION

Similarly, for determining near independence, it is enough to consider standard intervals of the form $(-\infty, t]$.

In the rest of this section, we shall suppose that $\vec{X} = \langle X_1, X_2, \ldots, X_n \rangle$, where the X_i form a sample from a $\mathcal{N}(0, \sigma^2)$ population and n is finite. That is, $X_1, \ldots, X_n$ are nearly independent nearly normal random variables with mean 0 and variance σ^2. We begin by investigating the distribution of $\sum_{i=1}^n X_i^2$, the square distance of $\vec{X}$ from the origin.

THEOREM X.12. *The random variable $V = \sum_{i=1}^n X_i^2/\sigma^2$ is nearly χ_n^2 distributed.*

PROOF. Let $Z_i = X_i/\sigma$. Then $Z_i \sim \mathcal{N}(0,1)$. The Z_i^2 are nearly independent, because

$$\Pr[Z_1^2 \le t_1, \ldots, Z_n^2 \le t_n] = \Pr[-\sqrt{t_1} \le Z_1 \le \sqrt{t_1}, \ldots, \sqrt{t_n} \le Z_n \le \sqrt{t_n}]$$
$$\approx \Pr[-\sqrt{t_1} \le Z_1 \le \sqrt{t_1}] \cdots \Pr[-\sqrt{t_n} \le Z_n \le \sqrt{t_n}]$$
$$= \Pr[Z_1^2 \le t_1] \cdots \Pr[Z_n^2 \le t_n].$$

Hence, it is enough to prove the theorem for $n = 1$ and then apply Corollary X.11. We have

$$\Pr[Z_1^2 \le t] = \Pr[-\sqrt{t} \le Z_1 \le \sqrt{t}]$$
$$\approx F(t),$$

where

$$F(t) = \frac{1}{\sqrt{2\pi}} \int_{-\sqrt{t}}^{\sqrt{t}} e^{-x^2/2} \, dx.$$

Differentiating F, we get the density of Z_1^2:

$$F'(t) = \frac{1}{\sqrt{2\pi}} t^{\frac{1}{2}} e^{-t/2}.$$

Looking at this formula, we see that $F'(t)$ is the same as $g_{\frac{1}{2},\frac{1}{2}}$, except, possible for a numerical constant. Since both are densities, the constants must be equal, and, hence, $F' = g_{\frac{1}{2},\frac{1}{2}}$. □

We, now, introduce two more densities. The first one is the $\mathcal{F}_{k,m}$ density. A random variable with an $\mathcal{F}_{k,m}$ density is said to be nearly distributed with an F distribution with k and m degrees of freedom. The $\mathcal{F}_{k,m}$ density is defined by

$$f(s) = \frac{(k/m)^{\frac{1}{2}k} s^{\frac{1}{2}(k-2)} (1 + (k/m)s)^{\frac{1}{2}(k+m)}}{B(\frac{1}{2}k, \frac{1}{2}m)}$$

for $s > 0$.

The last density we need is that $\mathcal{T}_k$ density. A variable with this density is said to nearly have a t distribution with k degrees of freedom. The $\mathcal{T}_k$ density is defined by
$$f(q) = \frac{\Gamma(\frac{1}{2}(k+1))(1+(q^2/k))^{-\frac{1}{2}(k+1)}}{\sqrt{\pi k}\Gamma(\frac{1}{2}k)}.$$
We have, the following theorem:

THEOREM X.13.

(1) Let V and W be nearly independent and have χ_k^2 and χ_m^2 densities respectively. Then the random variable
$$S = \frac{V/k}{W/m}$$
has an $\mathcal{F}_{k,m}$ density.

(2) Let Z and V be neraly independent random variables such that $Z \sim \mathcal{N}(0,1)$ and $V \sim \chi_k^2$. Then
$$Q = \frac{Z}{\sqrt{V/k}}$$
has a $\mathcal{T}_k$ density.

PROOF OF (1). To derive the density of S note that, if $U = V/(V+W)$, then
$$S = \frac{V/k}{W/m} = \frac{m}{k}\frac{U}{1-U}.$$
Since $V \sim \Gamma(\frac{1}{2}k, \frac{1}{2})$, $W \sim \Gamma(\frac{1}{2}m, \frac{1}{2})$, and V and W are nearly independent, then by Theorem X.10, U has beta density with parameters $\frac{1}{2}k$ and $\frac{1}{2}m$. In order to obtain the density of S we need to apply the change of variable Theorem X.7 to U with
$$T(u) = \frac{m}{k}\frac{u}{1-u}.$$
After some calulations we arrive at the $\mathcal{F}_{k,m}$ density. □

PROOF OF (2). In order to get the density of Q, we argue as follows. Since $-Z$ has nearly the same distribution as Z, we may conclude that Q and $-Q$ are nearly identically distributed. It follows that
$$\Pr[0 < Q < q] \approx \Pr[0 < -Q < q]$$
$$= \Pr[-q < Q < 0]$$
$$= \Pr[0 < Q^2 < q^2]$$
Differentiating the functions
$$\Pr[0 < Q < q] \approx G(q)$$
and
$$\Pr[-q < Q < 0] \approx H(-q),$$

5. SAMPLES FROM A NORMAL POPULATION

we get that the density of Q is

$$f_Q(q) = f_Q(-q) = q f_{Q^2}(q^2),$$

where f_{Q^2} is the density of Q^2, which, by, by 1, is an $\mathcal{F}_{1,k}$ density. After some calculations, we get that f_Q is a $\mathcal{T}_k$ density. □

Finally, we state the following corollary:

COROLLARY X.14. *The random variable*

$$\frac{m}{k} \frac{\sum_{i=1}^{k} X_i^2}{\sum_{i=k+1}^{k+m} X_i^2}$$

has an $\mathcal{F}_{k,m}$ density. The random variable

$$\frac{X_1}{\sqrt{\frac{1}{k}\sum_{i=1}^{k+1} X_i^2}}$$

has a $\mathcal{T}_k$ density.

PROOF. For the first assertion we need only to note that

$$\frac{\sum_{i=1}^{k} X_i^2}{\sum_{i=k+1}^{k+m} X_i^2} = \frac{\frac{1}{\sigma^2}\sum_{i=1}^{k} X_i^2}{\frac{1}{\sigma^2}\sum_{i=k+1}^{k+m} X_i^2}$$

and apply the previous theorem. The second assertion follows the same way. □

CHAPTER XI

Hyperfinite random processes

1. Properties that hold almost everywhere

This chapter continues the study of some of the basic elements of probability theory, especially of stochastic processes, from a nonstandard viewpoint. My development is mostly based on [71], but I shall only deal with the some of the more elementary classical theorems, namely those which will be needed in the sequel.

In this chapter, we shall use the following notation: T will always be a hyperfinite subset of $^*\mathbb{R}$; the first element of T is a, the last is b; for $t \in T$, its successor in T is $t+dt$; and for any $\xi : T \to {}^*\mathbb{R}$ we write $d\xi(t) = \xi(t+dt) - \xi(t)$. Whenever $\xi : T \to {}^*\mathbb{R}$ and we write $\xi(t)$, it is understood that $t \in T$.

Notice that if $\xi : T \to {}^*\mathbb{R}$, then for every $t \in T$

$$\xi(t) = \xi(a) + \sum_{s<t} d\xi(s).$$

We assume that stocastic processes are always internal.

The first part of this chapter deals with processes indexed by the natural numbers. That is, a process of the form $\langle X_n \mid 1 \leq n \leq \nu \rangle$ where ν is a natural number, in general, infinite, or $\langle X_n \mid n \in \mathbb{N} \rangle$ or $\langle X_n \mid n \in {}^*\mathbb{N} \rangle$. In the first case the set T is a discrete hyperfinite set, i.e., $T = \{1, 2, \ldots, \nu\}$, satisfying: $a = 1$, $b = \nu$, and $dt = 1$. Also, in this case, we have

$$dX_n = X_{n+1} - X_n.$$

Unless explicitly excepted, in this chapter, $\langle \Omega, \mathrm{pr} \rangle$ is always an internal hyperfinite probability space. Also, a random variable or stochastic process will always be an internal random variable or internal stochastic process.

We begin with the study of statements that hold almost everywhere.

DEFINITION XI.1 (ALMOST EVERYWHERE). Let $\langle \Omega, \mathrm{pr} \rangle$ be an internal hyperfinite probability space. Then

(1) A subset A of Ω (internal or external) is called a pr–*null set* (or just a *null set*, if there is no danger of confusion) if for every $\varepsilon \gg 0$ there is an internal set $B \supseteq A$ such that $\Pr B < \varepsilon$.
(2) Let $\varphi(\omega)$ be any statement about ω. We say that $\varphi(\omega)$ *holds almost everywhere* (*a.e.*), or *almost surely* (*a.s.*) *on* $\langle \Omega, \text{pr} \rangle$ if the set $[\varphi]^c$ is a null set.

If A is internal, then A is a null set if and only if $\Pr A \approx 0$. Thus, if φ is a standard statement (i.e., with only standard notions), then the set

$$[\varphi] = \{\omega \in \Omega \mid \varphi(\omega)\}$$

is internal and, hence, $\varphi(\omega)$ holds a.s. if and only if $\Pr[\varphi] \approx 1$. But some of the properties that we shall need cannot be expressed by standard notions, so the definition as stated is necessary, at this stage. In Appendix B, however, we indicate how to extend the probability measure to external sets, so that null sets can actually be assigned infinitesimal probability. Then, we can say that $\varphi(\omega)$ holds a.s. if and only if $\Pr[\varphi] \approx 1$, whether φ is internal or not. In any case, the intuitive content of the statement that $\varphi(\omega)$ holds a.s. is near certainty.

In Chapter VIII, Section 3, we discussed convergence of sequences of numbers. We now extend this discussion to random variables:

DEFINITION XI.2. *Let* X_1, ..., X_ν, *where ν is an infinite natural number, be an internal sequence of random variables over the same probability space* $\langle \Omega, \text{pr} \rangle$. *We say that* $X_1, \ldots, X_\nu$ *nearly converges to the random variable X if the sequence of numbers* $X_n(\omega)$ *nearly converges to* $X(\omega)$, *for every* $\omega \in \Omega$.

We also say that the sequence *nearly converges a.s.* (*a.e.*), if the subset of Ω where the sequence does not nearly converge is a null set.

We have:

THEOREM XI.1. *Let X be a random variable. Then the following are equivalent:*

(1) $X \approx 0$ *a.s.*
(2) *For all* $\lambda \gg 0$ *we have* $\Pr[|X| \geq \lambda] \approx 0$.
(3) *There is a* $\lambda \approx 0$ *such that* $\Pr[|X| \geq \lambda] \approx 0$.

PROOF. Suppose (1), and let $\lambda \gg 0$ and $\varepsilon \gg 0$. Then there is an internal subset of Ω, N, with $\Pr N \leq \varepsilon$ such that $X(\omega) \approx 0$ for all $\omega \varepsilon N^c$. Then $[|X| \geq \lambda] \subseteq N$ and thus $\Pr[|X| \geq \lambda] \leq \varepsilon$. Since ε is arbitrary, we have that $\Pr[|X| \geq \lambda] \approx 0$. Thus we have proved that (1) implies (2).

Suppose (2). Then the set of all λ such that $\Pr[|X| \geq \lambda] \leq \lambda$ contains all $\lambda \gg 0$ and so contains, by overflow, some $\lambda \approx 0$. Thus, (2) implies (3).

Suppose (3). Then $\Pr[|X| < \lambda] \approx 1$ for a certain $\lambda \approx 0$. Hence, $X \approx 0$ a.s. So (3) implies (1). □

We now define:

DEFINITION XI.3 (CONVERGENCE IN PROBABILITY).

1. PROPERTIES THAT HOLD ALMOST EVERYWHERE   255

(1) Let $X_1, X_2, \ldots, X_n, \ldots, X_\nu$ be a sequence of random variables with ν infinite. Then $\langle X_n \mid 1 \leq n \leq \nu \rangle$ *nearly converges in probability* to the random variable X if $X_\mu \approx X$ a.s. for all infinite $\mu \leq \nu$.
(2) The sequence $\langle X_n \mid n \in \mathbb{N} \rangle$ *S-converges in probability* to the random variable X if for every $\lambda \gg 0$ and $\varepsilon \gg 0$, there is an $n_0 \in \mathbb{N}$ such that $\Pr[|X_n - X| \geq \lambda] \leq \varepsilon$ for every $n \geq n_0$.

We now prove that Theorem XI.1 implies that near convergence in probability is an internal analogue of the usual convergence in probability:

THEOREM XI.2. *Let $\langle X_n \mid n \in \mathbb{N} \rangle$ be a sequence of random variables. Then the sequence S-converges in probability to X if and only if there is an internal extension $\langle X_n \mid 1 \leq n \leq \nu \rangle$, with $\nu \approx \infty$ which nearly converges in probability to X.*

In fact, if $\langle X_n \mid n \in \mathbb{N} \rangle$ S-converges to X, then for every internal extension $\langle X_n \mid n \leq \mu \rangle$, there is a $\nu \approx \infty$ such that $\langle X_n \mid n \leq \nu \rangle$ nearly converges in probability.

PROOF. We have that if the sequence $\langle X_n \mid n \leq \nu \rangle$ nearly converges in probability to X, then $|X_\mu - X| \approx 0$ a.s., for every infinite $\mu \leq \nu$. Let $\lambda \gg 0$ and $\varepsilon \gg 0$. Then, by Theorem XI.1, $\Pr[|X_\mu - X| \geq \lambda] \ll \varepsilon$, for all infinite $\mu \leq \nu$. Hence, by overflow, there is a finite n_0 such that $\Pr[|X_n - X| \geq \lambda] \ll \varepsilon$, for all finite $n \geq n_0$. Thus, $\langle X_n \mid n \in \mathbb{N} \rangle$ converges in probability to X.

On the other hand, let $\langle X_n \mid n \in \mathbb{N} \rangle$ be a sequence of random variables that S-converges in probability to X. Using denumerable comprehension, extend the sequence to $\langle X_n \mid n \in {}^*\mathbb{N} \rangle$. Let p be a finite natural number. For each finite natural number m, there is a finite number n_m such that $\Pr[|X_n - X| \geq 1/p] \leq 1/m$, for all finite $n > n_m$. Hence, by overflow, there is an infinite μ_m such that $\Pr[|X_n - X| \geq 1/p] \leq 1/m$, for all n with $n_m \leq n \leq \mu_m$. We can assume that the sequence $\langle \mu_m \mid m \in \mathbb{N} \rangle$ is decreasing. Using denumerable comprehension again, extend the sequence $\langle \mu_m \mid m \in \mathbb{N} \rangle$ to $\langle \mu_m \mid m \in {}^*\mathbb{N} \rangle$. The set of m such that $\mu_{n+1} \leq \mu_n$, for all $n \leq m$, contains all finite m, and hence, by overflow, it contains an infinite η. Let $\nu_p = \mu_\eta$. Then $\Pr[|X_n - X| \geq 1/p] \approx 0$, for every infinite $n \leq \nu_p$. We can assume that the sequence $\langle \nu_p \mid p \in \mathbb{N} \rangle$ is nonincreasing. Extend it by denumerable comprehension, and, as above find an infinite κ such that $\nu_{n+1} \leq \nu_n$, for every $n \leq \kappa$. Let $\nu = \nu_\kappa$. Then, the sequence $\langle X_n \mid 1 \leq n \leq \nu \rangle$ nearly converges to X. Because, if $\lambda \gg 0$, then $1/p \leq \lambda$, for some finite natural number p, and, since $\nu \leq \nu_p$, if μ is infinite, $\mu \leq \nu$, $\Pr[|X_\mu - X| \geq 1/p] \approx 0$. Hence, $\Pr[|X_\mu - X| \geq \lambda] \approx 0$. By Theorem XI.1, $X_\mu \approx X$. □

The property of convergence in probability is rather weak. An example discussed in [71, p. 26] shows this. Suppose that the day is divided into ν equal parts of infinitesimal duration $1/\nu$, that we have a device whose malfunction would cause a disaster, that the probability of malfunction in any period is c/ν, $0 \ll c \ll \infty$, and that different periods are independent. If we let X_n be the indicator function of the event of a malfunction in the nth period, then we have

for each n, $X_n \approx 0$ a.s., thus the sequence X_n nearly converges to 0 in probability. But we are really interested in $\max X_n$, the indicator of a disaster sometime during the day. By independence, the probability of no disaster during the day is
$$(1 - \frac{c}{\nu})^\nu \approx e^{-c} \ll 1,$$
so that there is a noninfinitesimal probability of a disaster during the day.

Near convergence a.s. is a stronger property. Combining the definitions of near convergence and a.s., we get that a sequence $\langle X_n \mid 1 \leq n \leq \nu \rangle$, where ν is infinite, nearly converges to X a.s. if and only if for every $\varepsilon \gg 0$ there is an internal set N such that $\Pr N \leq \varepsilon$ and for all $\omega \in N^c$, and all infinite $\mu \leq \nu$, $X_\mu(\omega) \approx X(\omega)$. On the other hand, $\langle X_n \mid 1 \leq n \leq \nu \rangle$ nearly converges in probability to X if and only if for all $\varepsilon \gg 0$ and all infinite $\mu \leq \nu$, there is an internal N with $\Pr N \leq \varepsilon$ such that for all $\omega \in N^c$, $X_\mu(\omega) \approx X(\omega)$. We see that for near convergence a.s. the set N must work for all infinite μ, i.e., N must work uniformly for all infinite μ, while for near convergence in probability, for each μ there may be a different N.

For convergence a.s., we have the following analogue of Theorem XI.2:

THEOREM XI.3. *Let $\langle X_n \mid 1 \leq n \leq \nu \rangle$ be an internal sequence of random variables over $\langle \Omega, \mathrm{pr} \rangle$, where ν is infinite. Then $\langle X_n \mid n \leq \nu \rangle$ nearly converges a.s. if and only if for all $\lambda \gg 0$ and all infinite $\mu \leq \nu$ we have*
$$\Pr[\max\{|X_m - X_\mu| \mid \mu < m \leq \nu\} \geq \lambda] \approx 0.$$

PROOF. Let
$$M(n, \lambda) = [\lambda \leq \max\{|X_m - X_n| \mid n < m \leq \nu\}].$$
Then $M(n, \lambda)$ is internal.

Suppose that $\langle X_n \mid 1 \leq n \leq \nu \rangle$ nearly converges to X a.s., and let $\lambda \gg 0$ and $\varepsilon \gg 0$. There is an internal set N with $\Pr N \leq \varepsilon$ such that $\langle X_n \mid n \leq \nu \rangle$ nearly converges to X on N^c.

If η, μ are infinite, and $\omega \in N^c$, then
$$|X_\mu(\omega) - X_\eta(\omega)| \leq |X_\mu(\omega) - X(\omega)| + |X(\omega) - X_\eta(\omega)| \approx 0.$$
So $|X_\mu(\omega) - X_\eta(\omega)| \leq \lambda$. Then $M(\mu, \lambda) \subseteq N$, if μ is infinite, so that $\Pr M(\mu, \lambda) \leq \varepsilon$. Since $\varepsilon \gg 0$ is arbitrary, $\Pr M(\mu, \lambda) \approx 0$.

Conversely, suppose that $\Pr M(\mu, \lambda) \approx 0$ for $\mu \approx \infty$ and $\lambda \gg 0$, and for $j \neq 0$ in $^*\mathbb{N}$, let n_j be the least natural number such that
$$\Pr M(n_j, \frac{1}{j}) \leq \frac{\varepsilon}{2^j}.$$
Let
$$N = \bigcup \{M(n_j, \frac{1}{j}) \mid j \in {}^*\mathbb{N}\}.$$
Then,
$$\Pr N \leq \sum_{j \in {}^*\mathbb{N}} \frac{\varepsilon}{2^j} = \varepsilon.$$

If j is finite, then n_j is also finite, since otherwise, $n_j - 1$ would be infinite and, because $1/j \gg 0$, $\Pr M(n_{j-1}, 1/j)$ would be infinitesimal by hypothesis, and so $\leq \varepsilon/2^j$, contradicting the definition of n_j.

Consequently, if $\omega \in N^c$, j is finite, and η and μ are infinite, then $n_j \leq \eta, \mu$ and $\omega \notin M(n_j, 1/j)$, and hence

$$|X_\eta(\omega) - X_\mu(\omega)| \leq |X_\eta(\omega) - X_{n_j}(\omega)| + |X_{n_j}(\omega) - X_\mu(\omega)| \leq \frac{2}{j}.$$

So, if η and μ are infinite and $\omega \in N^c$, $|X_\eta(\omega) - X_\mu(\omega)| \approx 0$. That is, in N^c, $\langle X_n \mid n \leq \nu \rangle$ nearly converges to X_μ.

Since $\varepsilon \gg 0$ is arbitrary, this shows that $\langle X_n \mid n \leq \nu \rangle$ nearly converges a.s. to X_μ. □

The following theorem is an immediate consequence of the definitions:

THEOREM XI.4. *Let $\langle X_n \mid 1 \leq n \leq \nu \rangle$ be an internal sequence of random variables with ν an infinite number. Then, if $\langle X_n \mid 1 \leq n \leq \nu \rangle$ nearly converges a.s. and, moreover, it nearly converges in probability to X, then it nearly converges a.s. to X.*

2. The law of large numbers

I shall only give a proof of simplest nonstandard forms of the weak and strong laws of large numbers, and not of their strongest forms. We shall only need, in the sequel, these simplest forms.

We first prove a fact that permits a simplification of some of the proofs. Suppose that $X_1, \ldots, X_n, \ldots$ are independent random variables (i.e., they form a random process indexed by a subset of *$\mathbb{N}$) with mean $\mu \ll \infty$ and variance $\sigma^2 \ll \infty$. The random variable

$$U_i = \frac{X_i - \mu}{\sigma}$$

has mean 0 and variance 1. Consider the random variables

$$Y_n = \frac{\sum_{i=1}^n X_i}{n}$$

and

$$Z_n = \frac{\sum_{i=1}^n U_i}{n} = \frac{Y_n}{\sigma} - \frac{\mu}{\sigma}.$$

We have that Y_n nearly converges in probability to μ if and only if Z_n nearly converges in probability to 0. For we have, for $\eta \approx \infty$, since σ is finite

$$Y_\eta \approx \mu \quad \text{a.s.} \iff \frac{Y_\eta}{\sigma} \approx \frac{\mu}{\sigma} \quad \text{a.s.}$$
$$\iff Z_\eta \approx 0 \quad \text{a.s.}$$

Similarly, we have that Y_n nearly converges a.s. to μ if and only if Z_n nearly converges a.s. to 0. For, by Theorem XI.3, Y_n nearly converges a.s. if and only if for all $\lambda \gg 0$ and all infinite $n \leq \nu$

$$\Pr[\max\{|Y_m - Y_n| \mid n < m \leq \nu\} \geq \lambda] \approx 0.$$

So, if we are given any $\lambda \gg 0$, we have that

$$\lambda \leq \max\{|Y_m - Y_n| \mid n < m \leq \nu\} \iff \frac{\lambda}{\sigma} \leq \max\{|Z_n - Z_m| \mid n < m \leq \nu\}.$$

Thus, the result follows.

Hence, for studying convergence, we can always replace random variables with finite mean an variance with random variables with mean 0 and variance 1.

Weak law of large numbers.

THEOREM XI.5 (WEAK LAW OF LARGE NUMBERS). *Let $\langle X_n \mid 1 \leq n \leq \nu \rangle$ be an internal sequence of independent random variables of mean $\mu \ll \infty$ and variance $\sigma^2 \ll \infty$, where ν is an infinite natural number. Let*

$$Y_n = \frac{X_1 + \cdots + X_n}{n},$$

for $n \leq \nu$. Then the sequence $\langle Y_n \mid 1 \leq n \leq \nu \rangle$ nearly converges in probability to μ.

PROOF. By the previous remarks, we may assume that $\mu = 0$ and $\sigma = 1$. By the definition of convergence in probability, we must prove that $Y_\mu \approx 0$ a.s. for every $\mu \approx \infty$. We have

$$\mathbf{E}|Y_n|^2 = \frac{1}{n}.$$

By Chebishev's inequality, Theorem VI.3, for $\lambda > 0$

$$\Pr[|Y_n| \geq \lambda] \leq \frac{\mathbf{E}|Y_n|^2}{\lambda^2} = \frac{1}{\lambda^2 n}.$$

Hence, if $\lambda \gg 0$ and μ is infinite, $\Pr[|Y_\mu| \geq \lambda] \approx 0$. By Theorem XI.2, if μ is infinite, $Y_\mu \approx 0$ a.e. □

Strong law of large numbers. We turn, now, to a version of the strong law of large numbers. We need the following lemma.

LEMMA XI.6. *Let $\langle Y_n \mid 1 \leq n \leq \nu \rangle$ be an internal sequence of random variables over the same probability space, where ν is an infinite natural number. Then, if*

$$\sum_{n=1}^{\nu} \mathbf{E}|dY_n|$$

nearly converges, we have that $\langle Y_n \mid 1 \leq n \leq \nu \rangle$ nearly converges a.s.

2. THE LAW OF LARGE NUMBERS

PROOF. Since
$$Y_n = Y_1 + \sum_{k=1}^{n-1} dY_k,$$
we only have to show that $\sum_{n=1}^{\nu-1} dY_n$ nearly converges. By Theorem XI.2, we need only to show that for all $\lambda \gg 0$ and infinite $\mu < \nu$, we have
$$\Pr[\sum_{k=\mu}^{\nu-1} dY_k \geq \lambda] \approx 0.$$
By Chebishev's inequality, since $\lambda \gg 0$ and, by assumption
$$\sum_{k=\mu}^{\nu-1} \mathbf{E}\,|dY_k| \approx 0,$$
we have, if μ is infinite
$$\Pr[\sum_{k=\mu}^{\nu-1} |dY_k| \geq \lambda] \leq \frac{\sum_{k=\mu}^{\nu-1} \mathbf{E}\,|dY_k|}{\lambda} \approx 0.$$
□

We now formulate the strong law of large numbers.

THEOREM XI.7 (STRONG LAW OF LARGE NUMBERS). *Let $X_1, X_2, \ldots, X_\nu$ be an internal sequence of independent random variables of mean $\mu \ll \infty$ and variance $\sigma^2 \ll \infty$, where ν is an infinite natural number. Let*
$$Y_n = \frac{X_1 + \cdots + X_n}{n},$$
for $1 \leq n \leq \nu$. Then the sequence $\langle Y_n \mid 1 \leq n \leq \nu \rangle$ nearly converges a.s. to μ.

PROOF. Again, we can assume that $\mu = 0$ and $\sigma = 1$. We have that
$$dY_n = \frac{X_1 + \cdots + X_{n+1}}{n+1} - \frac{X_1 + \cdots + X_n}{n} = \frac{X_{n+1}}{n+1} - \frac{Y_n}{n+1}.$$
Since $Y_n = Y_1 + \sum_{k=1}^{n-1} dY_k$, we have that
$$Y_n = Y_1 + \sum_{k=1}^{n-1} \frac{X_{k+1}}{k+1} - \sum_{k=1}^{n-1} \frac{Y_k}{k+1}.$$
So in order to prove that $\langle Y_n \mid 1 \leq n \leq \nu \rangle$ nearly converges a.s., we just have to prove that
$$\sum_{n=1}^{\nu-1} \frac{Y_n}{n+1}$$
and
$$\sum_{n=1}^{\nu-1} \frac{X_{n+1}}{n+1}$$

nearly converge a.s.

We begin by proving that the first sum nearly converges a.s. Since the X_i are independent

$$\left(\mathbf{E}\left|\frac{Y_n}{n+1}\right|^2\right)^{1/2} = \left(\frac{\sum_{k=1}^n \mathbf{E}\,|X_k|^2}{(n+1)^2}\right)^{1/2} = \frac{1}{n+1}\frac{1}{\sqrt{n}}.$$

By Theorem VI.3

$$\mathbf{E}\,\frac{|Y_n|}{n+1} \leq \left(\mathbf{E}\left|\frac{Y_n}{n+1}\right|^2\right)^{1/2}.$$

Since

$$\sum_{n=1}^{\nu-1} \frac{1}{n+1}\frac{1}{\sqrt{n}}$$

nearly converges, we have by the comparison test, Theorem VIII.13, that

$$\sum_{n=1}^{\nu-1} \mathbf{E}\,\frac{|Y_n|}{n+1}$$

also nearly converges. So, by the previous lemma

$$\sum_{n=1}^{\nu-1} \frac{Y_n}{n+1}$$

nearly converges a.s.

We now prove that

$$\sum_{n=1}^{\nu-1} \frac{X_{n+1}}{n+1}$$

nearly converges a.s. Let

$$Z_n = \sum_{k=1}^{n-1} \frac{X_{k+1}}{k+1}.$$

Thus, we must prove that $\langle Z_n \mid 1 \leq n < \nu \rangle$ nearly converges a.s. We have that

$$dZ_n = \frac{X_{n+1}}{n+1}.$$

By Theorem XI.3, it is enough to prove that for all $\lambda \gg 0$

$$\Pr[\lambda \leq \max\{|Z_m - Z_\mu| \mid \mu < m < \nu\}] \approx 0,$$

for all infinite $\mu < \nu$.

Let $\lambda \gg 0$ and $\mu \approx \infty$, $\mu < \nu$. Let

$$M(\mu, m, \lambda) = [\lambda \leq \max\{|Z_i - Z_\mu| \mid \mu \leq i < m\}].$$

For m, with $\mu \leq m < \nu$ define

$$V_m(\omega) = \begin{cases} 0, & \text{if } \omega \in M(n, m, \lambda), \\ 1, & \text{otherwise.} \end{cases}$$

2. THE LAW OF LARGE NUMBERS

Thus, V_m is independent of dZ_m (because V_m is defined in terms of the Z_i for $i < m$). Define

$$U_m = \sum_{k=\mu}^{m-1} V_k\, dZ_k.$$

We have

$$U_m > \lambda \cdot \chi_{M(n,m,\lambda)} + |Z_m - Z_\mu| \cdot \chi_{M(n,m,\lambda)^c},$$

and hence

$$\mathbf{E}\, U_m > \mathbf{E}(\lambda \cdot \chi_{M(n,m,\lambda)} + (Z_m - Z_\mu) \cdot \chi_{M(n,m,\lambda)^c})$$
$$= \lambda \Pr M(n,m,\lambda) + \mathbf{E}((Z_m - Z_\mu) \cdot \chi_{M(n,m,\lambda)^c}).$$

We can calculate the expectation of U_m:

$$\mathbf{E}\, U_m = \sum_{k=\mu}^{m-1} \mathbf{E}(V_k\, dZ_k).$$

But, since V_k and Z_k are independent

$$\mathbf{E}(V_k\, dZ_k) = \mathbf{E}\, V_k\, \mathbf{E}\, dZ_k = 0.$$

Thus, $\mathbf{E}\, U_m = 0$, and

$$\lambda \Pr M(n,m,\lambda) \leq \mathbf{E}((Z_\mu - Z_m) \cdot \chi_{M(n,m,\lambda)^c}) \leq \mathbf{E}\, |Z_m - Z_\mu|.$$

Hence

$$\Pr M(n,m,\lambda) \leq \frac{1}{\lambda} \mathbf{E}\, |Z_m - Z_\mu|.$$

In particular, we get

$$\Pr M(\mu,\nu,\lambda) \leq \frac{1}{\lambda} \mathbf{E}\, |Z_\nu - Z_\mu|.$$

We also have

$$\mathbf{E}\, |dZ_n|^2 = \mathbf{E}\left|\frac{X_{n+1}}{n+2}\right|^2 = \frac{1}{(n+2)^2},$$

so that

$$\sum_{n=1}^{\nu-1} \mathbf{E}\, |dZ_n|^2$$

nearly converges. That is

$$\sum_{k=\mu}^{\nu-1} \mathbf{E}\, |dZ_k|^2 \approx 0,$$

for all μ infinite, $\mu < \nu$. We have, since the X_i are independent

$$\mathbf{E}\,|Z_m - Z_\mu|^2 = \mathbf{E}(\sum_{k=\mu}^{\nu-1} dZ_k)^2 = \sum_{k=\mu}^{\nu-1} \mathbf{E}\,|dZ_k|^2 \approx 0.$$

But, by Jensen's inequality, Theorem VI.1

$$\mathbf{E}\,|Z_\nu - Z_\mu| \leq (\mathbf{E}\,|Z_\nu - Z_\mu|^2)^{1/2}.$$

Thus, $\Pr M(\mu, \nu, \lambda) \approx 0$ for every μ infinite and $\lambda \gg 0$. Since

$$M(\mu, \nu, \lambda) = [\lambda \leq \max\{|Z_m - Z_\mu| \mid \mu < m < \nu\}],$$

by Theorem XI.3, the sequence $\langle Z_n \mid 1 \leq n < \nu \rangle$ nearly converges a.s.

So we have proved that the sequence $\langle Y_n \mid 1 \leq n < \nu \rangle$ nearly converges a.e. By Theorem XI.5, it converges in probability to 0. Hence, by Theorem XI.4, it converges to 0 a.e. □

3. Brownian motion and Central Limit Theorem

We now turn to a nonstandard analysis of Brownian motion.

DEFINITION XI.4 (WIENER WALK). Let T be an equally spaced (i.e., $dt = ds$ for every $t, s \in T$) near interval for $[a, b]$. Then w is a *Wiener walk* if w is a stochastic process indexed by T such that $dw(t)$ is independent of $dw(s)$ for $s \neq t$, and

$$dw(t) = \begin{cases} \sqrt{dt}, & \text{with probability } 1/2, \\ -\sqrt{dt}, & \text{with probability } 1/2. \end{cases}$$

For a Wiener walk w we have that for $t \in T$

$$w(t) = w(a) + \sum_{s<t} dw(s),$$

$$\mathbf{E}\,w(t) = w(a) + \sum_{s<t} \mathbf{E}\,dw(s) = w(a),$$

$$\text{Var}(w(t)) = \sum_{s<t} \mathbf{E}(dw(s))^2 = t - a.$$

Let w be a Wiener walk indexed by T, and let $\#T = \nu$. For simplicity, we assume that $w(a) = 0$. Let $\Pi = {}^\nu\{0, 1\}$, and for $\pi \in \Pi$, let

$$\text{pr}(\pi) = \frac{1}{2^\nu}.$$

Then w can be considered as a stochastic process over $\langle \Pi, \text{pr} \rangle$ with

$$w(t, \pi) = \sum_{s<t} (-1)^{\pi(s)} \sqrt{dt}.$$

Assume, also, that $c = b - a \ll \infty$, and $\#T = \nu \approx \infty$. Then $dt = c/\nu$ and $dw(t) = \sqrt{c/\nu}$. Let n be infinite with $n\,dw(t)$ finite. Then $n/\sqrt{\nu}$ is finite, and, hence, $\nu - n$ is infinite. We have that $w(b, \pi) = n\,dw(t)$ if and only if π has m values

3. BROWNIAN MOTION AND CENTRAL LIMIT THEOREM

0 and k values 1 with $m+k = \nu$ and $m-k = n$. Then $m = (n+\nu)/2$. Consider a sequence of Bernoulli random variables $\langle X_i \mid i \leq \nu \rangle$ such that $\Pr[X_i = 1] = 1/2$. Then, we have, for v finite with $v = n\, dw(t)$

$$\Pr[w(b) = v] = \Pr[\sum_{i \leq \nu} X_i = m].$$

Since n and $\nu - n$ are infinite, we have that m and $\nu - m$ are also infinite. Let

$$Y_\nu = \frac{\sum_{i \leq \nu} X_i - \nu/2}{\sqrt{\nu}/2}.$$

Then by the considerations leading to Theorem X.9

$$\Pr[w(b) = v] = \Pr[\sum_{i \leq \nu} X_i = m]$$

$$= \Pr[Y_\nu = \frac{2m - \nu}{\sqrt{\nu}}]$$

$$= \Pr[Y_\nu = \frac{n}{\sqrt{c}} dw(t)]$$

$$\asymp \frac{2}{\sqrt{2\pi}} e^{-(1/2)(n/\sqrt{c}\, dw(t))^2} \frac{dw(t)}{\sqrt{c}}$$

$$= \frac{2}{\sqrt{2\pi}} e^{-(1/2)(v^2/c)} \frac{dw(t)}{\sqrt{c}}.$$

Hence, as in Theorem X.9

$$\Pr[w(b) \leq v] \asymp \frac{1}{\sqrt{2\pi c}} \sum_{t \in S, t \leq v} e^{-(1/2)t^2/c}\, dt,$$

for a certain hyperfinite S; and, then, by Theorem IX.20

$$\Pr[w(b) \leq v] \approx \frac{1}{\sqrt{2\pi c}} \int_{-\infty}^{v} e^{-(1/2)t^2/c}\, dt.$$

Thus, $w(b)$ is nearly normally distributed with mean 0 and variance c.

Recall that from Definition VI.20, two stochastic processes ξ and η indexed by the same hyperfinite set T are called equivalent in case $\langle \Lambda_\xi, \mathrm{pr}_\xi \rangle = \langle \Lambda_\eta, \mathrm{pr}_\eta \rangle$; that is, in case they have the same trajectories with the same probabilities. Λ_ξ is the set of trajectories of ξ, i.e., the set of functions from T to $^*\mathbb{R}$, $\xi(\ ,\omega)$ for $\omega \in \Omega$, and

$$\mathrm{pr}_\xi(\lambda) = \Pr[\xi(t) = \lambda(t) \text{ for all } t \in T].$$

Recall, also, that a process ξ may be considered to be defined over the space of its trajectories $\langle \Lambda_\xi, \mathrm{pr}_\xi \rangle$. We now introduce two notions of approximate equivalence.

DEFINITION XI.5. Let ξ and η be two stochastic processes indexed by the same hyperfinite set T but defined over possibly different hyperfinite probability spaces.

(1) We say that ξ and η are *very nearly equivalent* if $\Lambda_\xi = \Lambda_\eta$ and
$$\sum_{\lambda \in \Lambda_\xi} |\mathrm{pr}_\xi(\lambda) - \mathrm{pr}_\eta(\lambda)| \approx 0.$$

(2) We write $\xi \approx \eta$, if ξ and η are defined over the same probability space and $\xi(t) \approx \eta(t)$ a.s., for all $t \in T$.

(3) The processes ξ and η are *nearly equivalent* if there are processes ξ' and η' indexed by T, such that $\xi \approx \xi'$, $\eta \approx \eta'$ and ξ' is very nearly equivalent to η'.[1]

It is clear that very nearly equivalent processes are nearly equivalent, and equivalent processes are very nearly equivalent.

DEFINITION XI.6 (FUNCTIONALS). Let Λ be a finite subset of $^{T*}\mathbb{R}$. Then an internal function $F : \Lambda \to {}^*\mathbb{R}$ is called a *functional defined over* Λ. If F is a functional defined over Λ and $\Lambda_\xi \subseteq \Lambda$, then we write $F(\xi)$ for the random variable whose value at each $\omega \in \Omega$ is $F(\xi(\,,\omega))$. A functional F defined over Λ is *finite* if $F(\lambda)$ is finite for every $\lambda \in \Lambda$. (Since $\max |F|$ exists, this is the same as saying that $\max |F|$ is finite.)

We have the following proposition:

PROPOSITION XI.8. *Let ξ and η be two stochastic processes indexed by the same hyperfinite set T but defined over possibly different probability spaces. Then ξ and η are equivalent if and only if $\mathbf{E} F(\xi) = \mathbf{E} F(\eta)$ for every functional F defined over $\Lambda_\xi \cup \Lambda_\eta$.*

PROOF. In order to prove this proposition, assume that $\mathbf{E} F(\xi) = \mathbf{E} F(\eta)$ for every functional F defined over $\Lambda_\xi \cup \Lambda_\eta$, and define the functional F_λ for each $\lambda \in \Lambda_\xi \cup \Lambda_\eta$ by
$$F_\lambda(\mu) = \begin{cases} 1, & \text{if } \mu = \lambda, \\ 0, & \text{if } \mu \neq \lambda. \end{cases}$$
Then, since for any functional F, $\mathbf{E} F(\xi) = \sum F(\xi(\,,\lambda)) \mathrm{pr}_\xi(\lambda)$, we have that $\mathbf{E} F_\lambda(\xi(\,,\lambda)) = \mathrm{pr}_\xi(\lambda)$ and $\mathbf{E} F_\lambda(\eta(\,,\lambda)) = \mathrm{pr}_\eta(\lambda)$. So $\Lambda_\xi = \Lambda_\eta$ and $\mathrm{pr}_\xi = \mathrm{pr}_\eta$. The implication in the other direction is obvious. □

We also have the following implication, which relates very near equivalence with functionals:

LEMMA XI.9. *Let ξ and ξ' be very nearly equivalent stochastic process. Then, for every finite functional F defined over Λ_ξ, we have*
$$\mathbf{E} F(\xi) \approx \mathbf{E} F(\eta).$$

---

[1] The definition of near equivalence introduced here is apparently stronger that the notion in Nelson's [71, Chapter 17], using functionals (functionals will be discussed below). In the chapter mentioned, Nelson proves that my notion implies his; I do not know whether the converse implication is true or not.

PROOF. Let F be a finite functional. Then

$$|\mathbf{E}\,F(\xi) - \mathbf{E}'\,F(\xi')| \leq \sum_{\lambda \in \Lambda_\xi} |F(\xi(\lambda))|\,|\mathrm{pr}_\xi(\lambda) - \mathrm{pr}_{\xi'}(\lambda)|$$

$$\leq \max |F| \sum_{\lambda \in \Lambda_\xi} |\mathrm{pr}_\xi(\lambda) - \mathrm{pr}_{\xi'}(\lambda)|$$

$$\approx 0.$$

□

It is proved in [71, Chapter 17], but we do not need it here, that if ξ and η are nearly equivalent, then for every finite continuous functional F defined over $\Lambda_\xi \cup \Lambda_\eta$, we have $\mathbf{E}\,F(\xi) \approx \mathbf{E}\,F(\eta)$, where F continuous means that $\mu \approx \lambda$ implies $F(\mu) \approx F(\lambda)$.

We shall now see that very nearly equivalent and nearly equivalent processes have nearly equal distributions, in the sense to be explained below. We consider, first, ξ and η very nearly equivalent processes. Let A be any internal set of hyperreal numbers. Define, for each $t \in T$, the functional $F_{t,A}$ by

$$F_{t,A}(\lambda) = \begin{cases} 1 & \text{if } \lambda(t) \in A, \\ 0, & \text{otherwise.} \end{cases}$$

Then, $\mathbf{E}\,F_{t,A}(\xi) = \Pr[\xi(t) \in A]$ and $\mathbf{E}\,F_{t,A}(\eta) = \Pr[\eta(t) \in A]$. Now, $F_{t,A}$ is clearly finite. Hence, by very near equivalence and Lemma XI.9

$$\mathbf{E}\,F_{t,A}(\xi) \approx \mathbf{E}\,F_{t,A}(\eta).$$

Thus, for any t and A

$$\Pr[\xi(t) \in A] \approx \Pr[\eta(t) \in A].$$

Hence, two very nearly equivalent stochastic processes have nearly equal probability distributions in a very strong sense.

For nearly equivalent processes, the equality of distributions is not so strong. We shall only prove the following result, which we shall need later.

PROPOSITION XI.10. *Let ξ and η be nearly equivalent stochastic processes indexed by T, and suppose that for a certain $t \in T$, η is nearly continuous with density f. Then ξ is also nearly continuous with the same density, f.*

PROOF. By the previous remarks, we only have to prove the proposition for the case that ξ and η are defined over the same probability space and $\xi \approx \eta$. Let

$[a, b]$ be an interval with a, b, real. Then, for every finite natural number n

$$\int_{a+\frac{1}{n}}^{b-\frac{1}{n}} f \approx \Pr[a + \frac{1}{n} \leq \eta(t) \leq b - \frac{1}{n}]$$
$$\leq \Pr[a \leq \xi \leq b]$$
$$\leq \Pr[a - \frac{1}{n} \leq \eta(t) \leq b + \frac{1}{n}]$$
$$\approx \int_{a-\frac{1}{n}}^{b+\frac{1}{n}} f.$$

Since n is arbitrary

$$\Pr[a \leq \xi \leq b] \approx \int_a^b f.$$

By taking limits, we can extend the result to any standard closed region. □

Before proceeding to prove that certain processes are nearly equivalent to Wiener walks, we need to extend the notion of a Wiener walk:

DEFINITION XI.7. Let T be a near interval. Then, a stochastic process ξ indexed by T is called a T-*Wiener walk* if there is an infinite natural number ν and a Wiener walk w such that

(1) If $\varepsilon = (b-a)/\nu$ then w is indexed by $S = \{a, a+\varepsilon, a+2\varepsilon, \ldots, a+\nu\varepsilon\}$.
(2) For $t \in T$, let $\nu(t) =$ the largest $s \in S$ such that $s \leq t$. Then $\xi(t) = w(s)$.

It is clear that $\xi(b) = w(b)$, and hence, the distribution of $\xi(b)$ is the same as the distribution $w(b)$. When there is no danger of confusion we shall identify ξ with w.

Given any near interval T for $[a, b]$, with $b-a$ finite, and any $\nu \approx \infty$, one can construct a T-Wiener walk, ξ, as follows. Let $\varepsilon = (b-a)/\nu$. We first define a Wiener, w, walk indexed by $S = \{a, a+\varepsilon, \ldots, a+\nu\varepsilon\}$. Define the space $\langle \Pi, \mathrm{pr} \rangle$, by taking Π as the set of all function from $\{0, 1, \ldots, \nu-1\}$ to $\{0, 1\}$, and $\mathrm{pr}(\pi) = \frac{1}{2^\nu}$. Let ξ be the process defined over this space and indexed by T

$$\xi(t, \pi) = \sum_{a+n\varepsilon < t} (-1)^{\pi(n)} \sqrt{\varepsilon}.$$

Then ξ is a T-Wiener walk with Wiener walk, w given by

$$w(a+n\varepsilon, \pi) = \sum_{k=0}^{n-1} (-1)^{\pi(k)} \sqrt{\varepsilon}.$$

Before going to the proof of our theorem on processes equivalent to Wiener walks, we need the following lemma.

3. BROWNIAN MOTION AND CENTRAL LIMIT THEOREM

LEMMA XI.11. *Let ξ be a process indexed by a hyperfinite set T, such that the increments $d\xi(t)$ are independent and $\mathbf{E}\,d\xi(t) = 0$. Then for any $\lambda > 0$ we have*

$$\Pr[\max(\xi(t) - \xi(a)) \geq \lambda] \leq \frac{1}{\lambda}\mathbf{E}\,|\xi(b) - \xi(a)|.$$

PROOF. Define the process ζ by

$$\zeta(t) = \sum_{s<t} \eta(s)\,d\xi(s)$$

where

$$\eta(s) = \begin{cases} 1, & \text{if } \xi(r) < \lambda \text{ for all } r \leq s, \\ 0, & \text{otherwise.} \end{cases}$$

$\eta(s)$ depends only on the values of ξ for $r \leq s$, so that $\eta(s)$ is independent of $d\xi(s)$. Let Λ be the event

$$\Lambda = [\max(\xi(t) - \xi(a)) \geq \lambda].$$

We have

$$\zeta(b) \geq \lambda \cdot \chi_\Lambda + (\xi(b) - \xi(a)) \cdot \chi_{\Lambda^c}.$$

Thus

$$\lambda \cdot \chi_\Lambda \leq \zeta(b) + (\xi(a) - \xi(b)) \cdot \chi_{\Lambda^c}.$$

But

$$\mathbf{E}\,\zeta(b) = \sum_{s<b} \mathbf{E}\,\eta(s)\,\mathbf{E}\,d\xi(s) = 0.$$

Hence

$$\lambda\Pr[\max(\xi(t) - \xi(a)) \geq \lambda] \leq \mathbf{E}\,|\xi(b) - \xi(a)| \cdot \chi_{\Lambda^c} \leq \mathbf{E}\,|\xi(b) - \xi(a)|,$$

and, since $\lambda > 0$

$$\Pr[\max(\xi(t) - \xi(a)) \geq \lambda] \leq \frac{1}{\lambda}\mathbf{E}\,|\xi(b) - \xi(a)|.$$

□

We now introduce the definition of a condition for stochastic processes that is similar to the well known Lindeberg condition. The theorem we are about to prove is true for any process satisfying the Lindeberg condition, but we would have to develop too much more of the theory in order to prove the theorem with this condition.[2]

DEFINITION XI.8. Let ξ be a stochastic process indexed by a near interval T. We say that ξ satisfies the *independent increments condition* if we have:
(1) The increments $d\xi(t)$ for $t \in T$ are independent.
(2) For every $t \in T$, $\mathbf{E}\,d\xi(t) = 0$.
(3) For every $t \in T$, $d\xi(t)^2 \approx 0$ and $\sum_{s<t}\mathrm{Var}(d\xi(s)) \approx t - a$.

---

[2] See [71, Chapter 18] for a more general theorem.

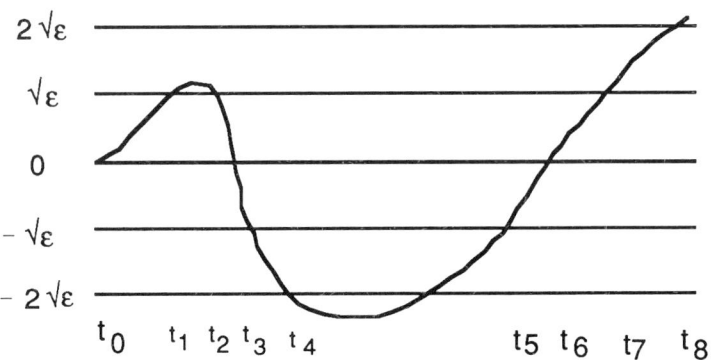

FIGURE XI.9. A T-Wiener walk obtained from the process ξ

It is clear that if w is a T-Wiener walk, then it satisfies the independent increments condition. We are now ready for our theorem.

THEOREM XI.12. *Let ξ be a stochastic process indexed by a near interval T for $[a, b]$, which satisfies the independent increments condition. Then ξ is nearly equivalent to a T-Wiener walk.*

PROOF. The idea of the proof, which has been extracted from [71, Chapter 18], is simple, although the details are complicated: observe the process ξ at the times t_n as in Fig. XI.9, where ε is a "huge infinitesimal". Then almost surely it goes up and down by nearly $\sqrt{\varepsilon}$, and by the independent increments condition these occur with nearly equal probability. The times t_n are random variables, but since the quadratic variation of the process (which is nearly $n\sqrt{\varepsilon}$) is nearly equal to the elapsed time, $t - a$, they behave as if they were spaced ε apart on the average (this is not indicated in Fig. XI.9), and the process looks like a Wiener process.

We may assume, without loss of generality, that $\xi(a) = 0$, and hence

$$\xi(t) = \sum_{s<t} d\xi(s).$$

Hence, $\mathbf{E}\,\xi(t) = 0$ and

$$\mathrm{Var}(\xi(t)) = \mathbf{E}\,\xi(t)^2 = \sum_{s<t} \mathbf{E}\,d\xi(s)^2 = \sum_{s<t} \mathrm{Var}(d\xi(s)) \approx t - a,$$

for every $t \in T$. Assume, also, that ξ is defined over the hyperfinite probability space $\langle \Omega, \mathrm{pr} \rangle$.

First notice that, since T is hyperfinite, for some $\alpha \approx 0$, we have $|d\xi(t)| \leq \alpha$, for every $t \in T$. We begin by constructing a process that looks like a Wiener

3. BROWNIAN MOTION AND CENTRAL LIMIT THEOREM

walk (see Fig. XI.9). We start the construction by extending the process ξ. Let c be a number such that $c - b \approx \infty$ and let $\bar{T}$ be the union of T and the set of all numbers of the form

$$b + k\frac{c-b}{m} \quad \text{for } k = 1, 2, \ldots, m, \text{ where } \frac{c-b}{m} \approx 0.$$

We extend ξ to $\bar{T}$ by setting $\xi(t) = \xi(b) + w'(t)$, where w' is the Wiener walk on $\bar{T} - T$. This extension preserves the properties specified above for ξ. The purpose of this extension is to avoid having to worry about when the times of the new process we are about to define, t_n, become undefined in the original process.

We denote the predecessor of t by $t - d * t$. Let $\omega \in \Omega$. We say that $\xi(\cdot, \omega)$ crosses the number λ at t if and only if $\xi(t - d * t, \omega) < \lambda$ and $\xi(t, \omega) \geq \lambda$, or else $\xi(t - d * t, \omega) > \lambda$ and $\xi(t, \omega) \leq \lambda$. Let $\varepsilon > 0$ with $\sqrt{\varepsilon} > 2\alpha$, and let $\nu = (b-a)/\varepsilon$. For each $\omega \in \Omega$, we define by recursion elements of $\bar{T}$, $t_n(\omega)$, and integers $k_n(\omega)$, for $n = 0, 1, \ldots, \nu$. Let $t_0(\omega) = a$ and $k_0(\omega) = 0$. For $n > 0$, let $t_n(\omega)$ be the first element of t subsequent to $t_{n-1}(\omega)$ at which ξ crosses $(k_{n-1}(\omega)+1)\sqrt{\varepsilon}$ or $(k_{n-1}(\omega)-1)\sqrt{\varepsilon}$ (and if there is no such time, then we define $t_n(\omega)$ to be c). Let $k_n(\omega) = k_{n-1}(\omega) + 1$, if ξ crosses $(k_{n-1}(\omega)+1)\sqrt{\varepsilon}$ at $t_n(\omega)$, and $k_n(\omega) = (k_{n-1}(\omega) - 1)$, if ξ crosses $(k_{n-1}(\omega)-1)\sqrt{\varepsilon}$ at $t_n(\omega)$. (Then $k_n(\omega)$ has the same parity as n.) In Fig. XI.9, we have pictured a trajectory, ω, with $t_1(\omega), t_1(\omega), t_3(\omega), \ldots, t_8(\omega)$. In this case, $k_1 = 1$, $k_2 = 0$, $k_3 = -1$, $k_4 = -2$, $k_5 = -1$, $k_6 = 0$, $k_7 = 1$ and $k_8 = 2$.

Consider $\xi(t_n)$ as a process indexed by ν, i.e., define the process

$$\eta(n, \omega) = \xi(t_n(\omega), \omega)$$

for $n \leq \nu$ and $\omega \in \Omega$. From now on, we write $\xi(t_n)$ for the process $\eta(n)$.

We have that

$$\xi(t_{n+1}) - \xi(t_n) = \sum \mu(s) d\xi(s),$$

where μ is the process defined by

$$\mu(s) = \begin{cases} 1 & \text{if } t_n \leq s < t_{n+1}, \\ 0 & \text{otherwise.} \end{cases}$$

We have that $\mu(s)$ is independent of $d\xi(s)$, since $t_n \leq s < t_{n+1}$ can be defined in terms of t_n and the u with $t_n < u \leq s$. Then

$$\mathbf{E}(\xi(t_{n+1}) - \xi(t_n)) = \sum \mathbf{E}(\mu(s) d\xi(s))$$
$$= \sum \mathbf{E}\mu(s) \mathbf{E} d\xi(s)$$
$$= 0.$$

It is also clear that the increments $\xi(t_{n+1}) - \xi(t_n)$ of the process $\xi(t_n)$ are independent.

Suppose that $0 \ll \varepsilon \ll \infty$. Then $\nu \ll \infty$. We claim that a.s.

$$|\xi(t_{n+1}) - \xi(t_n) - \sqrt{\varepsilon}| \leq 2\alpha \quad \text{or} \quad |\xi(t_{n+1}) - \xi(t_n) + \sqrt{\varepsilon}| \leq 2\alpha \tag{1}$$

for all $n < \nu$. The proof of claim (1) is as follows. The only way (1) could fail for a particular n, with t_n finite, is for t_{n+1} to be c. In fact, if t_{n+1} is finite, then either

$$\sqrt{\varepsilon} \leq \xi(t_{n+1}) - \xi(t_n) \leq \sqrt{\varepsilon} + 2\alpha,$$

or

$$-\sqrt{\varepsilon} - 2\alpha \leq \xi(t_{n+1}) - \xi(t_n) \leq \sqrt{\varepsilon}.$$

With the first alternative we get the first alternative of (1) and with the second, the second alternative. Since ν is finite, we now prove by external induction the t_n is a.s. finite. Assume, as inductive hypothesis, that t_n is a.s. finite. We have added a Wiener walk to our original process with $c - b$ infinite. Since for $t \gg s > b$, $\xi(t) - \xi(s)$ is nearly normally distributed, the probability

$$\Pr[-\sqrt{\varepsilon} < \xi(t) - \xi(s) < \sqrt{\varepsilon}] \ll 1.$$

Partition the interval $[b, c]$ into equally spaced $s_1 \ll s_2 \ll \cdots \ll s_n \ll \cdots$, where $|s_i| \ll \infty$. Then the set that contains the $\omega \in \Omega$ where the difference is less that $\sqrt{\varepsilon}$ in absolute value for all those intervals is a null set. This completes the proof of claim (1).

We have, then

$$\left| \sum_{|\xi(t_{n+1},\omega)-\xi(t_n,\omega)|<\sqrt{\varepsilon}} (\xi(t_{n+1},\omega) - \xi(t_n,\omega)) \operatorname{pr}(\omega) \right| \leq$$

$$\sum_{|\xi(t_{n+1},\omega)-\xi(t_n,\omega)|<\sqrt{\varepsilon}} |\xi(t_{n+1},\omega) - \xi(t_n,\omega)| \operatorname{pr}(\omega).$$

$$\leq \sqrt{\varepsilon} \sum_{|\xi(t_{n+1},\omega)-\xi(t_n,\omega)|<\sqrt{\varepsilon}} \operatorname{pr}(\omega)$$

$$\approx 0.$$

Thus, for every $n < \nu$

$$\sum_{|\xi(t_{n+1},\omega)-\xi(t_n,\omega)|<\sqrt{\varepsilon}} (\xi(t_{n+1},\omega) - \xi(t_n,\omega)) \operatorname{pr}(\omega) \approx 0.$$

Notice also that $|\xi(t_{n+1}) - \xi(t_n)| \leq \sqrt{\varepsilon} + 2\alpha \ll \infty$ everywhere, for all $n < \nu$.

3. BROWNIAN MOTION AND CENTRAL LIMIT THEOREM

Therefore, since $\mathbf{E}(\xi(t_{n+1}) - \xi(t_n)) = 0$, we have, for every $n < \nu$

$$0 \approx \sum_{\sqrt{\varepsilon} < \xi(t_{n+1},\omega) - \xi(t_n,\omega) \leq \sqrt{\varepsilon}+2\alpha} (\xi(t_{n+1},\omega) - \xi(t_n,\omega))\,\mathrm{pr}(\omega) +$$

$$\sum_{-\sqrt{\varepsilon}-2\alpha \leq \xi(t_{n+1},\omega) - \xi(t_n,\omega) < \sqrt{\varepsilon}} (\xi(t_{n+1},\omega) - \xi(t_n,\omega))\,\mathrm{pr}(\omega)$$

$$\leq (\sqrt{\varepsilon} + 2\alpha) \sum_{\sqrt{\varepsilon} < \xi(t_{n+1},\omega) - \xi(t_n,\omega) \leq \sqrt{\varepsilon}+2\alpha} \mathrm{pr}(\omega) -$$

$$(\sqrt{\varepsilon} + 2\alpha) \sum_{-\sqrt{\varepsilon}-2\alpha \leq \xi(t_{n+1},\omega) - \xi(t_n,\omega) < \sqrt{\varepsilon}} \mathrm{pr}(\omega)$$

$$\approx \sqrt{\varepsilon}\Big(\sum_{\sqrt{\varepsilon} < \xi(t_{n+1},\omega) - \xi(t_n,\omega) \leq \sqrt{\varepsilon}+2\alpha} \mathrm{pr}(\omega) -$$

$$\sum_{-\sqrt{\varepsilon}-2\alpha \leq \xi(t_{n+1},\omega) - \xi(t_n,\omega) < \sqrt{\varepsilon}} \mathrm{pr}(\omega) \Big)$$

So that

$$0 \lesssim \sqrt{\varepsilon}\Big(\sum_{\sqrt{\varepsilon} < \xi(t_{n+1},\omega) - \xi(t_n,\omega) \leq \sqrt{\varepsilon}+2\alpha} \mathrm{pr}(\omega) - \sum_{-\sqrt{\varepsilon}-2\alpha \leq \xi(t_{n+1},\omega) - \xi(t_n,\omega) < \sqrt{\varepsilon}} \mathrm{pr}(\omega) \Big)$$

We also have, by a similar argument

$$\sqrt{\varepsilon}\Big(\sum_{\sqrt{\varepsilon} < \xi(t_{n+1},\omega) - \xi(t_n,\omega) \leq \sqrt{\varepsilon}+2\alpha} \mathrm{pr}(\omega) - \sum_{-\sqrt{\varepsilon}-2\alpha \leq \xi(t_{n+1},\omega) - \xi(t_n,\omega) < \sqrt{\varepsilon}} \mathrm{pr}(\omega) \Big) \lesssim 0.$$

Thus

$$\sqrt{\varepsilon}\Big(\sum_{\sqrt{\varepsilon} < \xi(t_{n+1},\omega) - \xi(t_n,\omega) \leq \sqrt{\varepsilon}+2\alpha} \mathrm{pr}(\omega) - \sum_{-\sqrt{\varepsilon}-2\alpha \leq \xi(t_{n+1},\omega) - \xi(t_n,\omega) < \sqrt{\varepsilon}} \mathrm{pr}(\omega) \Big) \approx 0.$$

Since $\sqrt{\varepsilon} \gg 0$

$$\sum_{\sqrt{\varepsilon} < \xi(t_{n+1},\omega) - \xi(t_n,\omega) \leq \sqrt{\varepsilon}+2\alpha} \mathrm{pr}(\omega) \approx \sum_{-\sqrt{\varepsilon}-2\alpha \leq \xi(t_{n+1},\omega) - \xi(t_n,\omega) < \sqrt{\varepsilon}} \mathrm{pr}(\omega),$$

i.e.

$$\Pr[\sqrt{\varepsilon} < \xi(t_{n+1}) - \xi(t_n) \leq \sqrt{\varepsilon} + 2\alpha] \approx \Pr[-\sqrt{\varepsilon} - 2\alpha \leq \xi(t_{n+1}) - \xi(t_n) < \sqrt{\varepsilon}],$$

and, hence

$$\Pr[|\xi(t_{n+1}) - \xi(t_n) - \sqrt{\varepsilon}| \leq 2\alpha] \approx \frac{1}{2},$$

$$\Pr[|\xi(t_{n+1}) - \xi(t_n) + \sqrt{\varepsilon}| \leq 2\alpha] \approx \frac{1}{2}.$$

Since ν is finite for $\varepsilon \gg 0$, this means that

$$\sum_{\pi} |\Pr[|\xi(t_{n+1}) - \xi(t_n) - (-1)^{\pi(n)}\sqrt{\varepsilon}| \leq 2\alpha \text{ for all } n] - \frac{1}{2^\nu}| \approx 0, \qquad (2)$$

where the sum is over all 2^ν mappings π of $\{0,\ldots,\nu-1\}$ into $\{0,1\}$.

Now use overflow. Since the set of all $\varepsilon \gg 0$ for which the left side of (2) (i.e., everything but '≈ 0') is $\leq \varepsilon$, contains all $\varepsilon \gg 0$, it also contains all sufficiently large $\varepsilon \approx 0$. Now fix ε (the "huge infinitesimal") so that $\varepsilon \approx 0$, $\alpha/\varepsilon \approx 0$, and (2) holds.

Let the random variable $\nu(t)$ be such that $\nu(t,\omega)$ is the largest n with $n \leq \nu$ such that $t_n(\omega) \leq t$. For $t \in T$, let

$$\zeta(t) = \sum_{n=0}^{\nu(t)-1} (d\xi(t_n))^2 + (\xi(t) - \xi(t_{\nu(t)}))^2. \tag{3}$$

Then

$$d\zeta(t) = d\xi(t)^2 + 2(\xi(t) - \xi(t_{\nu(t)}))d\xi(t),$$

so that $|d\zeta(t)| \leq \gamma |d\xi(t)|$ for all $t \in T - \{b\}$ and all ω, with $\gamma = \alpha + 2(\sqrt{\varepsilon} + \alpha) \approx 0$. Consequently, since the $d\zeta(t)$ are independent and $\sum \mathbf{E}\, d\xi(t)^2 \approx b - a$ is finite

$$\mathbf{E}\,|\zeta(t) - \mathbf{E}\,\zeta(t)|^2 = \mathbf{E}\,|\sum d\zeta(t) - \sum \mathbf{E}\, d\zeta(t)|^2$$
$$\leq \sum \mathbf{E}\, d\zeta(t)^2$$
$$\leq \gamma^2 \sum \mathbf{E}\, d\xi(t)^2$$
$$\approx 0.$$

By Jensen's inequality, Theorem VI.1

$$\mathbf{E}\,|\zeta(t) - \mathbf{E}\,\zeta(t)| \leq (\mathbf{E}\,|\zeta(t) - \mathbf{E}\,\zeta(t)|^2)^{1/2}.$$

Hence, by the previous lemma, for any $\lambda \gg 0$

$$\Pr[\max_{s \leq t}(\zeta(s) - \mathbf{E}\,\zeta(s)) \geq \lambda] \leq \frac{1}{\lambda}\mathbf{E}(\zeta(t) - \mathbf{E}\,\zeta(t)) \approx 0,$$

and, thus, by Theorem XI.1, a.s.

$$\zeta(t) \approx \mathbf{E}\,\zeta(t).$$

Thus

$$\zeta(t) \approx \sum_{s<t} \mathbf{E}\, d\zeta(s) = \sum_{s<t} \mathbf{E}\, d\xi(s)^2 \approx t - a. \tag{4}$$

The last term in (3), $(\xi(t) - \xi(t_{\nu(t)}))^2$, is infinitesimal. By (1), since $|d\xi(t_n) - \sqrt{\varepsilon}| < 1$ and $|d\xi(t_n) + \sqrt{\varepsilon}| < 1$ we get that

$$|(d\xi(t_n))^2 - \varepsilon| = |d\xi(t_n) - \sqrt{\varepsilon}|\,|d\xi(t_n) + \sqrt{\varepsilon}| \leq 2\alpha,$$

and, hence

$$\left|\frac{d\xi(t_n)^2}{\varepsilon} - 1\right| \leq \frac{2\alpha}{\varepsilon}.$$

Since $\alpha/\varepsilon \approx 0$, for each term in the sum in (3), $d(\xi(t_n))^2$, we have $d(\xi(t_n))^2 \asymp \varepsilon$ a.s. Thus, by Theorem IX.3, a.s.

$$\zeta(t) \approx \sum_{n=1}^{\nu(t)-1} (d\xi(t_n))^2 \approx \nu(t)\varepsilon.$$

Thus, by (4), a.s.

$$\nu(t)\varepsilon \approx t - a \tag{5}$$

for all $t \in T$.

Let Π be the finite probability space of all mappings π of $\{0, 1, \ldots, \nu - 1\}$ into $\{0, 1\}$, with $\mathrm{pr}'(\pi) = 1/2^\nu$ for all π in Π. Let w_ε be the stochastic process indexed by T and defined over $\langle \Pi, \mathrm{pr}' \rangle$ by

$$w_\varepsilon(t) = \sum_{n \leq (t-a)/\varepsilon} (-1)^{\pi(n)} \sqrt{\varepsilon}.$$

w_ε is the T-Wiener walk defined after Definition XI.7. Define pr'' on Π by

$$\mathrm{pr}''(\pi) = \Pr[|d\xi(t_n) - (-1)^{\pi(n)}\sqrt{\varepsilon}| \leq 2\alpha \text{ for all } n].$$

By (2) and Lemma XI.9, w_ε on $\langle \Pi, \mathrm{pr}' \rangle$ is nearly equivalent to w_ε on $\langle \Pi, \mathrm{pr}'' \rangle$. For each $\omega \in \Omega$, define $f(\omega) \in \Pi$ by

$$f(\omega)(n) = \begin{cases} 0, & \text{if } |d\xi(t_n) - \sqrt{\varepsilon}| \leq 2\alpha, \\ 1, & \text{otherwise.} \end{cases}$$

Then, by (1), since $\nu\alpha \approx 0$ (because $\nu = (b-a)/\varepsilon$, $b - a$ is finite, and $\alpha/\varepsilon \approx 0$),

$$\Pr\{\omega \mid |d\xi(t_n, \omega) - (-1)^{f(\omega)(n)}\sqrt{\varepsilon}| \leq 2\alpha \text{ for all } n\} \approx 1. \tag{6}$$

Define the process w on $\langle \Omega, \mathrm{pr} \rangle$ by

$$w(t, \omega) = w_\varepsilon(t, f(\omega)).$$

Let Λ be the set of trajectories of w_ε. Then Λ is also the set of trajectories of w. Let pr''_w be the probability on Λ obtained from $\langle \Pi, \mathrm{pr}'' \rangle$ and pr_w the probability on Λ obtained from $\langle \Omega, \mathrm{pr} \rangle$. Then, w_ε and w can be considered as the same function over Λ, but as processes defined over $\langle \Lambda, \mathrm{pr}''_w \rangle$ and $\langle \Lambda, \mathrm{pr}_w \rangle$, respectively.

Thus, for $\lambda \in \Lambda$

$$\mathrm{pr}''_w(\lambda) = \mathrm{pr}''(\{\pi \mid w_\varepsilon(t, \pi) = \lambda(t) \text{ for all } t\})$$
$$= \Pr(\{\omega \mid |d\xi(t_n, \omega) - (-1)^{f(\omega)(n)}\sqrt{\varepsilon}| \leq 2\alpha \text{ for all } n \text{ and } w_\varepsilon(t, f(\omega)) = \lambda(t) \text{ for all } t\})$$

and

$$\mathrm{pr}_w(\lambda) = \Pr(\{\omega \mid w(t, \omega) = \lambda(t) \text{ for all } t\}).$$

Since $w_\varepsilon(t, f(\omega)) = w(t, \omega)$, we have, by (5)

$$\sum_{\lambda \in \Lambda} |\mathrm{pr}_w(\lambda) - \mathrm{pr}''_w(\lambda)| \leq \Pr(\{\omega \mid |d\xi(t_n, \omega) - (-1)^{f(\omega)(n)}\sqrt{\varepsilon}| \leq 2\alpha \text{ for all } n\}^c)$$
$$\approx 0.$$

Thus, by Lemma XI.9, w_ε and w are very nearly equivalent.

We have for $\omega \in \Omega$ and $t \in T$

$$|\xi(t, \omega) - w(t, \omega)| \lesssim \sum_{n \leq \nu(t, \omega)} |\xi(t_n, \omega) - (-1)^{f(\omega)(n)}\sqrt{\varepsilon}| + |\xi(t, \omega) - \xi(t_{\nu(t,\omega)}, \omega)|.$$

We also have that $|\xi(t, \omega) - \xi(t_{\nu(t,\omega)}, \omega)| \leq \sqrt{\varepsilon}$; so, by (1), a.s.

$$|\xi(t) - w(t)| \leq \nu\alpha + \sqrt{\varepsilon} \approx 0,$$

that is, $\xi \approx w$. Therefore, we have shown that ξ is nearly equivalent to w_ε, a T-Wiener walk. □

We have the following corollaries.

COROLLARY XI.13. *Let ξ be a stochastic process indexed by a near interval T such that for every $t \in T$*

$$d\xi(t) = \begin{cases} \sqrt{dt}, & \text{with probability } 1/2, \\ -\sqrt{dt}, & \text{with probability } 1/2. \end{cases}$$

Suppose also that the increments $d\xi(t)$ are independent for every $t \in T$. (For instance, ξ may be a Wiener walk.) Then if $s, t \in T$ with $0 \ll t - s \ll \infty$, we have that $\xi(t) - \xi(s)$ is nearly normally distributed with mean 0 and variance $t - s$.

PROOF. The process ξ restricted to $\{u \in T \mid s \leq u \leq t\}$ satisfies the hypothesis of Theorem XI.12, and hence it is nearly equivalent to a T-Wiener walk, which is nearly normally distributed. We obtain the corollary by Proposition XI.10. □

Thus, we have proved for Wiener walks all the main properties of Brownian motion.

COROLLARY XI.14 (CENTRAL LIMIT THEOREM). *Let $X_1, X_2, \ldots, X_\nu$, with $\nu \approx \infty$, be independent, identically distributed random variables, with finite values, finite mean μ and finite variance σ^2. Then*

$$Y_\nu = \frac{\sum_{i=1}^\nu X_i - \nu\mu}{\sqrt{\nu}\sigma}$$

is nearly normally distributed with mean 0 and variance 1.

3. BROWNIAN MOTION AND CENTRAL LIMIT THEOREM

PROOF. By taking $X_i - \mu$, we can assume without loss of generality that the random variables have mean 0.

Let T be the near interval $\{t_0, t_1, \ldots, t_\nu\}$, with $t_0 = 0$, $dt_n = 1/\nu$, for $n = 0$, $1, \ldots, \nu - 1$, and, hence, $t_\nu = 1$. Let

$$Y_n = \sum_{i=1}^{n} X_n.$$

Define the process ξ over T by

$$\xi(\frac{n}{\nu}) = \frac{Y_n}{\sqrt{\nu}\sigma},$$

for $n = 0, 1, \ldots, \nu$. Then

$$d\xi(t_n) = \frac{X_{n+1}}{\sqrt{\nu}\sigma},$$

for $n = 0, 1, \ldots, \nu - 1$. Since the X_n have finite values and $\nu \approx \infty$, $d\xi(t_n) \approx 0$. Also

$$\text{Var}(d\xi(t_n)) = \frac{\text{Var}(X_{n+1})}{\nu\sigma^2} = \frac{1}{\nu}.$$

Thus

$$\sum_{s<t} \text{Var}(d\xi(s)) = t.$$

Therefore, the process ξ satisfies the independent increments condition. We apply Theorem XI.12, and obtain that ξ is nearly equivalent to a T-Wiener walk. Hence, by Proposition XI.10

$$\xi(1) = \frac{Y_\nu}{\sqrt{\nu}\sigma}$$

is nearly normally distributed with mean 0 and variance 1. □

COROLLARY XI.15. *Let $X_1, X_2, \ldots, X_\nu$ be a sequence of random variables, with ν infinite, such that for each $n \leq \nu$*

$$X_n = \begin{cases} \varepsilon_n, & \text{with probability } 1/2, \\ -\varepsilon_n, & \text{with probability } 1/2, \end{cases}$$

where $\varepsilon_1, \varepsilon_2, \ldots, \varepsilon_\nu$ is a sequence of infinitesimals such that

$$0 \ll \sum_{n=1}^{\nu} \varepsilon_n^2 = \alpha \ll \infty.$$

Let

$$Y = \sum_{n=1}^{\nu} X_n.$$

Then Y is nearly normally distributed with mean 0 and variance α.

PROOF. Consider the process ξ defined over
$$T = \{0, \varepsilon_1^2, \ldots, \sum_{m=1}^{n} \varepsilon_m^2, \ldots, \alpha\}$$
by
$$d\xi(\sum_{m=1}^{n} \varepsilon_m^2) = \begin{cases} \varepsilon_{n+1}, & \text{with probability } 1/2, \\ -\varepsilon_{n+1}, & \text{with probability } 1/2. \end{cases}$$
Then, by Theorem XI.12, ξ is nearly equivalent to a T–Wiener walk. Thus, by Proposition XI.10, $\xi(\alpha)$ is nearly normally distributed with variance α. But Y has the same distribution as $\xi(\alpha)$. □

Part 3

Probability models

CHAPTER XII

Simple probability structures

In Chapter III, we discussed informally relational systems and probability structures and their dual role. On one hand, they are possible models of the world. On the other hand, they are possible models of sentences of a language. Notice that although the systems or structures remain the same, they are models in different senses. In the first sense, a relational system may be a model of reality, because the properties and relations of the system represent faithfully the real relations which are intended as their interpretations. In the second sense, the sentences are interpreted in the relational system, and the system is a possible model of the sentences. That is, for a sentence φ, the system may be a model of φ (i.e., φ is true in ω) or not (i.e., φ is false in ω).

I shall define formally in Sections 2–4 probability structures that will play this dual role. On one hand, they will be possible models of reality, when reality is a chance setup. On the other hand, probability structures will serve as interpretations of a language, which will be a language of random variables. These interpretations will not determine truth or falsity for sentences, but only their probabilities. Probability structures will be defined as sets of relational systems, and the definition of probability for sentences will be based on the definition of truth for relational systems. Thus, first we have to study relational systems more formally.

The probability structures that are studied in this chapter, the simple probability structures, are a special case of the general structures that will be studied in the next chapters, and the examples of simple structures are rather simple-minded. I could define directly the general structures,[1] but in order to make the subject more understandable, I prefer to treat first these simple structures.

In this and the next two chapters, we shall only use standard notions, unless explicitly mentioned otherwise. These standard notions will have their usual interpretations in the nonstandard universe. Thus, for instance, if we say that r is a real number, we mean a hyperreal number. Similarly, "finite" means "hyperfinite", and so on.

---

[1]This approach was followed in [20].

1. Relational systems

This section contains a brief survey of relational systems with one or more universes. In the next chapters, we shall study systems that incorporate the notion of causal dependence. These more complicated systems will be needed for the general notion of probability structures.

We are especially interested in the semantic aspects of logic; that is, in the notions of interpretation and truth. As was mentioned in the previous section, our languages are interpreted in relational systems. We shall deal with the appropriate languages in Chapter XVII. In this section, we study the relational systems themselves. Before giving the formal definition, we begin with some examples. A relational system, in general, has one or more universes of objects, and also indicates properties of these objects, relations or operations between these objects, and distinguished objects. All these relations and operations are explicitly displayed in the system.

An example of a relational system with one universe is the arithmetical structure of the natural numbers, namely

$$\mathfrak{N} = \langle \mathbb{N}; <, +, \times, 0, 1 \rangle,$$

where $\mathbb{N}$, the set of natural numbers, is the universe of the system. After the semicolon, the relations, operations, and distinguished elements appear, for which I use their usual symbols. As will be explained later, the distinguished elements will be coded as a special kind of operation, zero-ary operations, and the properties, i.e., subsets of the universe, as a special kind of relation, i.e., one-ary relations. So we shall have just relations and operations in our systems.

A vector space is a system with two universes, which can be represented as,

$$\mathfrak{V} = \langle V, F; +, \times, 0, 1, \vec{+}, \vec{\times}, \vec{0} \rangle,$$

where the two universes are V, the set of vectors, and F, the field of scalars; $+$, $\times$, 0, and 1 are the operations of the field, $\vec{+}$ is vector addition, $\vec{\times}$ is scalar product, and $\vec{0}$ is the zero vector.

I shall not give a general definition of relational systems, since we will have almost no use for systems with more than one universe. I shall just define relational systems with one universe and the notions that we need for them. It is easy to generalize these notions to systems with several universes.

A *relational system* is a pair $\omega = \langle A; R \rangle$, where A is a nonempty set, the universe of ω, and $R = \langle R_i \mid i \in I \rangle$ is a system such that for each $i \in I$, R_i is a relation between elements of the universe or an operation that applied to elements of the universe has as value also an element of the universe. To be more precise, there is a natural number n_i such that $R_i \subseteq {}^{n_i}A$ or $R_i : {}^{n_i}A \to A$. We also write ω as $\langle A; R_i \rangle_{i \in I}$. The set A is called the *universe* or *domain of* ω and is designated by $|\omega|$.

The *similarity type*, or simply the *type of* ω is a way of coding the arities of the relations and operations and of indicating which indices correspond to relations or operations. It can be taken to be the function ν with domain I, such that

1. RELATIONAL SYSTEMS

for each $i \in I$, $\nu(i)$ is a pair $\langle m, n \rangle$, where m is 0 or 1 and n a natural number; $\nu(i) = \langle 0, n \rangle$ indicates that R_i is a relation that is a subset of $^n A$, and $\nu(i) = \langle 1, n \rangle$ that R_i is an operation from $^n A$ to A. In general, we write $\nu(i) = \langle \nu(i)_1, \nu(i)_2 \rangle$, and the second component of $\nu(i)$, i.e., $\nu(i)_2$, is the arity of R_i.

As we mentioned in the Preliminaries, Section 1, we identify $^1 A$, the set of one-termed sequences of elements of A, with A itself. Hence, one-ary relations are subsets of A, which represent properties of elements of A. We should also have distinguished elements of A in our systems. For these elements, we can use 0-ary operations. A 0-ary operation is a function $Q : {}^0 A \to A$. But $^0 A = \{\emptyset\}$, since the empty function is the only sequence with empty domain. Hence, $Q = \{\langle \emptyset, a \rangle\}$, for a certain $a \in A$. We identify Q with this a. Thus, Q represents the distinguished element a.

Since an operation can be considered as a relation that is a function, i.e., a functional relation, we could only have relations in our systems. This is a theoretical simplification that will be used in the next chapter. Many situations are simpler to describe, however, with operations. Thus, the restriction to relations will only be used for theoretical purposes. In case we only use relations, a distinguished element a can more simply be treated as the singleton $\{a\}$, which is a subset of the universe, i.e., a 1-ary relation.

We need to define isomorphisms of relational systems. Let

$$\omega = \langle A; \langle R_i \mid i \in I \rangle \rangle$$

be a relational system of type ν, then we define:[2]

(1) Let f be a one-one function such that f has domain containing A. Then:
  (a) If R is an n–ary relation, then $f''R$ is the n–ary relation on $f''A$ defined by
  $$f''R = \{\langle f(a_1), \ldots, f(a_n) \rangle \mid \langle a_1, \ldots, a_n \rangle \in R\}.$$
  (b) If R is an n–ary operation, then $f''R$ is the n–ary operation on $f''A$, such that
  $$(f''R)(f(a_1), \ldots, f(a_n)) = f(R(a_1, \ldots, a_n)).$$
  (c) If ω is a relational system $\langle A, R_i \rangle_{i \in I}$
  $$f''(\omega) = \langle f''A, f''R_i \rangle_{i \in I}.$$
  (d) f is called an isomorphism of ω onto $f''(\omega)$.
(2) ω is isomorphic to ω' if and only if there is a one-one function f on A such that $\omega' = f''(\omega)$.

For instance, if we take $f(n) = 2n$ for all $n \in \mathbb{N}$, $f''(\langle \mathbb{N}; <, + \rangle) = \langle \mathbb{E}; <, + \rangle$, where $\mathbb{E}$ is the set of even numbers, and $<$ and $+$ have their usual meanings, then f is an isomorphism of $\langle \mathbb{N}; <, + \rangle$ onto $\langle \mathbb{E}; <, + \rangle$.

---

[2] As explained in Preliminaries Section 1, if B is a set and g a function, $g''B$ denotes the image of B under g. Conditions (a), (b), and (c) are the obvious generalizations of this notion to relations, operations, and relational systems, respectively.

2. The structure of chance

In this section, I shall begin the study of chance structures of simple setups. The rest of the sections of this chapter contains the definition of the algebra of events and the probability measure for these structures.

Simple setups are those where essentially one random choice is made, for instance the selection of n balls from an urn or the selection of a point from a circle. The simple probability structures which will be defined for these setups cannot model setups with more than one choice, such as the choosing of an urn and then the choosing of balls from this urn. The general notion of probability structure that is able to deal with these cases will be studied in the next chapter.

As was explained in Part 1, a chance setup can be identified with its set of possible outcomes. Each possible outcome is a possible state of the world, which can be represented by a completed world. As we saw in Chapter III, possible states of the world can be partially modeled (or approximated) by relational systems. The relations in the systems should indicate the particular result plus the chance structure of the setup, i.e., its mechanism. In order to include the chance mechanism, the universe of the systems should include all the relevant objects. In general, since there is just one mechanism for each setup, there will be a common universe for all the possible outcomes of a setup. However, we don't have to require it.

DEFINITION XII.1 (SIMPLE CHANCE STRUCTURES). A set $\mathbf{K}$ of relational systems with a common similarity type is called a *simple chance structure*. The union of the universes of the systems in $\mathbf{K}$ is denoted by $|\mathbf{K}|$ and is called the *universe of* $\mathbf{K}$. More precisely, if $\omega \in \mathbf{K}$ is of the form $\omega = \langle A_\omega; R \rangle$, then $|\mathbf{K}| = \bigcup \{A_\omega \mid \omega \in \mathbf{K}\}$.

The universe $|\mathbf{K}|$ represents the objects that are involved in the setup, and the similarity type indicates the types of relations and operations that are essential for the description of the chance character of the setup. Each relational system in $\mathbf{K}$ represents a possible outcome, and it codes the particular result plus the chance mechanism.

Some examples might help to clarify the definitions.[3] Some of these examples, were already discussed in Chapter III.

EXAMPLE XII.1. Choosing at random a sample S of m elements from a finite population A.

The relational systems in the chance structure $\mathbf{K}_0$ that models this setup are very simple. The structure $\mathbf{K}_0$ consists of all systems $\omega_S = \langle A, S \rangle$, where S is a subset of A of m elements. There is no more structure for this setup than to choose the sample, and the set A is just a set with no more structure. Although the result is determined just by S, in order to code the mechanism we must

---

[3]Examples XII.1 and XII.2 were already discussed in [13] and [16]. In these two papers, however, I have simple probability structures for the different statistics for the distribution of r balls into n cells. I have decided that is more natural to have compound structures, to be seen in the next chapters, for these setups. In Chapter XVI, we shall see the general model for the distribution of r balls into n cells, which will be given as a compound structure.

include A in all the outcomes. In this case, $|\mathbf{K}_0| = A$, and the similarity type may be taken as the function ν with domain $\{1\}$, and such that $\nu(1) = \langle 0, 1 \rangle$. The number 0 indicates that S is a relation, and 1, its arity. The index set is $I = \{1\}$.

EXAMPLE XII.2. Placing at random one ball of a population of r balls into one of n cells.

There are at least two ways of setting up chance structures for this case. The first is to construct a structure similar to that of Example XII.1. The relational systems in the chance structure are of the form $\langle B, P_1, P_2, \ldots, P_n \rangle$, where B is the set of balls and P_i is the set of balls that is in cell i in the particular outcome that the system is modeling. Hence, $|\mathbf{K}| = B$, and the similarity type of $\mathbf{K}$ indicates n unary relations. Since we are placing just one ball, all cells are empty, but one. However, this chance structure is not adequate, because in it the cells are not considered as objects but as properties. It seems more appropriate that both the balls and the cells should be objects in the universes.

Thus, structures with two universes are more appropriate. The chance structure $\mathbf{K}_1$ for this setup contains the relational systems with two universes $\omega_O = \langle B, C; O \rangle$, where B is the set of r balls, C is the set of n cells, and O is a singleton that contains the pair $\langle i, m \rangle$, which indicates that ball i was put in cell m. For each such singleton O, there is an outcome in $\mathbf{K}_1$. The set O is a binary relation between an element of B and an element of C.

EXAMPLE XII.3. The choosing of a point at random from a circle.

The systems for this chance structure are similar to those of $\mathbf{K}_0$ with a sample of one element. We have to choose the sample from a circle, however, and not just from a set without structure. Thus, the systems should have relations or operations that insure the structure of a circle. One way of doing this is to consider systems of the form $\omega_S = \langle \mathbf{C}, r, S \rangle$, where $\mathbf{C}$ is the set of points in the circle considered as complex numbers in the unit circle, i.e., $\mathbf{C} = \{e^{i\theta} \mid 0 \leq \theta \leq 2\pi\}$, r is multiplication between complex numbers, and S is a one element subset of $\mathbf{C}$. It is well known that $r(u_1, u_2)$, represents a rotation of u_1 in the arc of length θ, where $u_2 = e^{i\theta}$. If we had just $\langle \mathbf{C}, S \rangle$, we would be choosing an element from the set $\mathbf{C}$, and not from the circle $\langle \mathbf{C}, r \rangle$.

As was developed in the previous chapters, we shall only deal with hyperfinite structures. Thus, instead of dealing with the circle $\langle \mathbf{C}, r \rangle$ we consider the hyperfinite approximation $\langle C, r \rangle$ defined in Example VIII.4. Recall that, in this example, $C = \{e^{ik\,d\theta} \mid 0 \leq k \leq \nu\}$ with $d\theta = 2\pi/\nu$, where ν is an infinite natural number. Notice that C is closed under the operation r, so that $\langle C, r \rangle$ is a hyperfinite approximation of $\langle \mathbf{C}, r \rangle$. In order to preserve r, we are not free to choose any hyperfinite approximation to $\mathbf{C}$, but it must be closed under multiplication.

The chance structure $\mathbf{K}_2$ is then the set of all systems $\omega_S = \langle C, r, S \rangle$, where S is a subset of one element of C. The structure $\mathbf{K}_2$ has universe C and a similarity type with one binary operation and one unary relation. For some purposes, it is better to have a different similarity type for $\mathbf{K}_2$. Namely, the systems $\omega_c = \langle C, r, c \rangle$, where $c \in C$ is a distinguished element of C. It is clearly equivalent to have ω_S or ω_c, if we put $S = \{c\}$.

Any chance structure **K** determines a group of invariance or symmetries $G_\mathbf{K}$.[4] Suppose that we are modeling a certain chance setup by the chance structure **K**. Then **K** represents, in a certain sense, "the laws of the chance setup". The group $G_\mathbf{K}$ is a group under which these laws, i.e., **K** itself, are invariant. It is natural then to take $G_\mathbf{K}$ as the group of automorphisms of **K**. Thus, we must define what is a automorphism of a chance structure. In order to understand the definition, notice that if f is a one-one function on the universe $|\mathbf{K}|$, then f'' is one-one on **K**.

DEFINITION XII.2 (AUTOMORPHISMS). Let **K** and **H** be chance structures, $\mathbf{A} \subseteq \mathbf{K}$, and f a one-one function on the universe $|\mathbf{K}|$. Then
(1) $\mathbf{A}^f = \{f''\omega \mid \omega \in \mathbf{A}\}$.
(2) f is an isomorphism from **K** onto **H**, if $\mathbf{K}^f = \mathbf{H}$.
(3) f is an automorphism of **K**, if f is an isomorphism from **K** onto **K**.
(4) $G_\mathbf{K}$ is the set of all automorphisms of **K**.

The set $\mathbf{A}^f$ is the image of **A** under f'', and thus it may be written $\mathbf{A}^f = f''''(\mathbf{A})$. When there is no danger of confusion, we shall not distinguish between f, f'', and f''''. Thus, $\mathbf{A}^f$ will also be written $f(\mathbf{A})$, and $f''(\omega)$, $f(\omega)$. We have the following easy proposition.

PROPOSITION XII.1. $G_\mathbf{K}$ *is a group of functions, i.e., it is closed under composition and inverse. Also, for* f, $g \in G_\mathbf{K}$, $(f \circ g)'' = f'' \circ g''$ *and* $f^{-1\prime\prime} = f''^{-1}$. *Thus,* $\{f'' \mid f \in G_\mathbf{K}\}$ *is a group of permutations of* **K**.

As it will be explained later, the probability measure should be invariant under $G_\mathbf{K}$, or more properly, under $\{f'' \mid f \in G_\mathbf{K}\}$. That is, **A** and $\mathbf{A}^f$ should have the same measure, for any $f \in G_\mathbf{K}$.

It is easy to determine which are the groups for the examples discussed above. In Example XII.1, $G_{\mathbf{K}_0}$ consists of all permutations of the population, A, of balls. Since the systems in $\mathbf{K}_1$ contain two universes, isomorphisms on these systems are pairs of one-one functions, one for each universe. Thus, $G_{\mathbf{K}_1}$ contains all pairs $f = \langle f_1, f_2 \rangle$, such that f_1 is a permutation of B and f_2 is a permutation of C. Finally, $G_{\mathbf{K}_2}$ contains all automorphism of $\langle C, r \rangle$, i.e., all rotations of the circle. These functions have to be automorphisms of $\langle C, r \rangle$, since they are isomorphisms of systems in $\mathbf{K}_2$ onto systems in $\mathbf{K}_2$, and all systems in $\mathbf{K}_2$ contain $\langle C, r \rangle$.

I now make more precise the discussion of randomness of Chapter III, Section 2. Intuitively, a setup is random, when all outcomes are pairwise indistinguishable from the point of view of the setup. Loosely speaking, a setup is random when it does not contain information that would permit one to discriminate between the different outcomes. The following is a nice formal characterization of randomness.

A simple chance structure **K** is *random*, if for every ω, $\omega' \in \mathbf{K}$, there is an $f \in G_\mathbf{K}$ such that $\omega = f''\omega'$. That is, if every pair of systems in **K** are isomorphic

---
[4]Cf. Chapter III, Section 2. This group of invariance has always been an essential element of my definition; for instance see [13].

by an automorphism of **K**. I shall not adopt this definition as the official definition of randomness, since randomness has been associated with random sequences, whose definition has a very different character. I believe, however, that the traditional notion of randomness is akin to the concept presented here. In the next section, I shall formally introduce this notion, giving it a different name.

Most of the simple chance structures that occur in practice are random in this sense. In particular, $\mathbf{K}_0$, $\mathbf{K}_1$, and $\mathbf{K}_2$ are random. The chance structure for the setup in Fig. III.3 in page 68, which we discuss next, is an example of a simple structure which is not random:

EXAMPLE XII.4. Choosing a point in a line symmetrically around a given point.

The description of an experiment where this situation may arise is given in Fig. III.3. Here, I will give the chance structure for such a setup. The systems of the chance structure, call it $\mathbf{K}_3$, should express the fact that we have a line, whose only symmetry is reflection around a certain fixed point p. Thus, the systems can be of the form $\omega_c = \langle \mathbb{R}; +, ^{-1}, B, c \rangle$, where $\mathbb{R}$ is the set of real numbers, $+$ and $^{-1}$ are the usual addition, and multiplicative inverse, B is a ternary relation such that $B(x, y, z)$, if y is between x and z; and c is the chosen point, which varies from system to system. The group $G_{\mathbf{K}_3}$ consists of the automorphisms of $\langle \mathbb{R}; +, ^{-1}, B \rangle$. It is not difficult to prove that the only nontrivial automorphisms are reflections r around p (which we identify with the number zero) given by $r(x) = -x$. It is clear that ω_c is isomorphic to ω_d by a function in $G_{\mathbf{K}_3}$ if and only if $c = d$ or $c = -d$. Hence not all systems are isomorphic, and $\mathbf{K}_3$ is not a random structure.

It is easy to find a hyperfinite approximation to the structures $\langle \mathbb{R}; +, ^{-1}, B \rangle$. Just take a hyperfinite approximation to $\mathbb{R}$, as it is explained in Example VIII.3, which is closed under the operations $+$ and $^{-1}$.

3. The probability space

The chance structure **K** should determine the family of events and the probability measure. Thus, we should have a way of obtaining this family and the corresponding probability, uniformly, from **K**. In fact, as we have seen, **K** determines the group of invariance $G_\mathbf{K}$. The group $G_\mathbf{K}$, in its turn, determines the strictly positive and refining equivalence relation $\sim_{G_\mathbf{K}}$ over subsets of **K**. Since $G_\mathbf{K}$ is determined by **K**, we shall call $\sim_{G_\mathbf{K}}$, simply, $\sim_\mathbf{K}$. Namely, if **A** and **B** are subsets of **K**, we have

$$\mathbf{A} \sim_\mathbf{K} \mathbf{B} \text{ if and only if } \mathbf{B} = \mathbf{A}^g, \text{ for some } g \in G_\mathbf{K}.$$

All we need for determining the family of events and the probability measure is this equivalence relation. Thus, I shall introduce a special kind of structure in the following definition. In the rest of this section, we can forget about **K** as a set of relational systems and just proceed with these structures.

DEFINITION XII.3 (SIMPLE EQUIPROBABILITY STRUCTURES). A *simple equiprobability structure* is a pair $\langle \mathbf{K}, \sim \rangle$, where **K** is a set and $\sim$ is a refining and strictly positive equivalence relation on subsets of **K**. A refining and strictly positive equivalence relation shall be called an *equiprobability relation*.

If **K** is a simple chance structure, $\langle \mathbf{K}, \sim_K \rangle$ is the simple equiprobability structure determined by **K**. When there is no danger of confusion, we shall not distinguish between a simple chance structure and its simple equiprobability structure.

We turn, now, to the definition of the family of events. We assume that we start with a simple equiprobability structure $\langle \mathbf{K}, \sim \rangle$. Events are, then, the subsets of **K** where the probability measure will be defined. For a subset of **K** to be an event, it should be in some sense identifiable. Since we are just dealing with finite (or hyperfinite) structures, all internal subsets of **K** can be considered to be identifiable, and we shall adopt this standpoint. If we were dealing with infinite structures, we would need to restrict in some way the permissible subsets of **K**, in order not to have the nonmeasurable subsets as events.[5]

We need another requirement, however, for a subset of **K** to be an event. For a subset of **K** to be a basic event, the equiprobability relation $\sim$ should "determine" its probability. In the next few paragraphs, I shall define more precisely what this means.

First, we must discuss the main characteristic of the probability measure. As has been mentioned before, the measure should be invariant under $\sim$. The definition of a $\sim$-invariant measure is given in Definition VII.9, 5. Namely:

A measure Pr defined on subsets of a simple equiprobability structure **K** is $\sim$-invariant, if for any $\mathbf{A}, \mathbf{B} \subseteq \mathbf{K}$ such that $\mathbf{A} \sim \mathbf{B}$, $\Pr \mathbf{B}$ is defined whenever $\Pr \mathbf{A}$ is defined, and $\Pr \mathbf{B} = \Pr \mathbf{A}$.

This principle of symmetry is the formal version of the objective principle discussed informally in Chapter III, Section 2.

Thus, we require a $\sim$-invariant measure. We shall always deal with finite algebras of events, but they may be disjunctive algebras and not algebras. The family of events should at least be a disjunctive algebra of subsets of **K**, $\mathcal{F}$, such that:

(1) If $\mathbf{A} \in \mathcal{F}$ and $\mathbf{B} \sim \mathbf{A}$, then $\mathbf{B} \in \mathcal{F}$.
(2) There is a $\sim$-invariant measure on $\mathcal{F}$.

Recall that if a family of subsets of **K** satisfies (1), we say that it is $\sim$-closed (Definition VII.9, 4).

We now introduce a certain kind of subset of **K** whose measure is, in a certain sense, determined by $\sim$.

DEFINITION XII.4 (MEASURE-DETERMINED). Let $\langle \mathbf{K}, \sim \rangle$ be an equiprobability structure. A subset **H** of **K** is *measure-determined*, if there is a unique real number r such that if Pr is a $\sim$-invariant measure defined on the disjunctive algebra of subsets of **K** generated by $\{\mathbf{A} \subseteq \mathbf{K} \mid \mathbf{H} \sim \mathbf{A}\}$, then $\Pr \mathbf{H} = r$.

For instance, if there is a partition of **K**, **A**, **B**, and **C** such that $\mathbf{H} \sim \mathbf{A} \sim \mathbf{B} \sim \mathbf{C}$, then $\Pr \mathbf{H} = 1/3$, for any $\sim$-invariant measure Pr defined on $\{\mathbf{A} \subseteq \mathbf{K} \mid$

---

[5] See [20] for an attempt at a definition of identifiable subset for standard infinite structures.

3. THE PROBABILITY SPACE

$\mathbf{H} \sim \mathbf{A}\}$. It is clear from the definition that the family of measure-determined subsets of $\mathbf{K}$ is $\sim$-closed.

Now we are ready to define the family of events. We assume, for the rest of this section, that $\langle \mathbf{K}, \sim \rangle$ is an equiprobability structure. We often write only $\mathbf{K}$ for $\langle \mathbf{K}, \sim \rangle$.

DEFINITION XII.5 (FAMILY OF EVENTS).
(1) The family of basic events for $\mathbf{K}$, $\mathcal{F}_{b\mathbf{K}}$, is the least disjunctive algebra of subsets of $\mathbf{K}$ generated by the set of measure-determined subsets of $\mathbf{K}$.
(2) The family of events for $\mathbf{K}$, $\mathcal{F}_{\mathbf{K}}$, is the least algebra generated by $\mathcal{F}_{b\mathbf{K}}$, if there is a unique $\sim$-invariant measure on this algebra, and is equal to $\mathcal{F}_{b\mathbf{K}}$, otherwise.

Since the family of measure-determined subsets of $\mathbf{K}$ is $\sim$-closed, by Theorem VII.14, $\mathcal{F}_{\mathbf{K}}$ is also $\sim$-closed.

One of the reasons for choosing $\mathcal{F}_{\mathbf{K}}$ as the family of events is given in the following theorem, which is an immediate consequence of Theorem VII.4.

THEOREM XII.2. *There is at most one $\sim$-invariant probability measure on $\mathcal{F}_{\mathbf{K}}$.*

By Theorem VII.15 and the previous theorem, we get:

THEOREM XII.3. *Let $\mathbf{K}$ be a simple equiprobability structures such that $\mathbf{K}$ is finite. Then there is always exactly one invariant probability $\text{Pr}_{\mathbf{K}}$ on $\mathcal{F}_{\mathbf{K}}$.*

If we were dealing with infinite $\mathbf{K}$, then the existence of an invariant measure would not be guaranteed.[6]

DEFINITION XII.6 (PROBABILITY SPACE). Let $\langle \mathbf{K}, \sim \rangle$ be a simple equiprobability structure. Then $\langle \mathbf{K}, \mathcal{F}_{\mathbf{K}}, \text{Pr}_{\mathbf{K}} \rangle$ is the *probability space of* $\langle \mathbf{K}, \sim \rangle$.

The notion of random structure discussed at the end of the previous section will be introduced formally in the next definition. I shall call these structures, however, classical equiprobability structures, so as not to confuse them with the current notion of random sequence.[7]

DEFINITION XII.7 (CLASSICAL STRUCTURES). The simple equiprobability structure $\langle \mathbf{K}, \sim \rangle$ is called *classical* if $\langle \mathbf{K}, \sim \rangle$ is a simple equiprobability structure such that for any elements of $\mathbf{K}$, ω and ω', we have $\{\omega\} \sim \{\omega'\}$.

It is clear that if $\langle \mathbf{K}, \sim \rangle$ is a classical structure, then $\mathcal{F}_{\mathbf{K}}$ is the power set of $\mathbf{K}$ and $\text{Pr}_{\mathbf{K}}$ is the counting measure. That is, for any $\mathbf{A} \subseteq \mathbf{K}$

$$\text{Pr}_{\mathbf{K}} \mathbf{A} = \frac{\#\mathbf{A}}{\#\mathbf{K}},$$

---

[6] The problem of the existence of invariant measures has been extensively studied in mathematics. In [15], conditions of an algebraic character for the existence of invariant measures on infinite Boolean algebras are given.

[7] See [65] or [96] for a bibliography and a discussion on this extensively studied subject.

where $\#\mathbf{A}$ is, as was introduced before, the number of elements of $\mathbf{A}$. This is the classical definition of probability, namely, the probability of an event is the number of favorable cases divided by the total number of cases.

We now discuss the probability spaces of the structures in the examples we have discussed in this chapter. We keep the numbers that were assigned to them in our further discussion.

Example XII.1. Choosing of a sample from a population of balls.

Since all outcomes are isomorphic by elements of $G_{\mathbf{K}_0}$, $\langle \mathbf{K}_0, \sim \rangle$ is classical, and, hence, each singleton subset of $\mathbf{K}_0$ is measure-determined. Thus, as we saw above, $\mathcal{F}_{\mathbf{K}_0}$ is the power set of $\mathbf{K}_0$, and the measure $\Pr_{\mathbf{K}_0}$ is the counting measure, i.e., for any subset $\mathbf{A}$ of $\mathbf{K}_0$

$$\Pr_{\mathbf{K}_0} \mathbf{A} = \frac{\#\mathbf{A}}{\#\mathbf{K}_0}.$$

Example XII.2. Placing at random one ball of a population of r balls into one of n cells.

Here the singletons are also measure-determined, since all outcomes are isomorphic by elements of $G_{\mathbf{K}_1}$. Hence, the algebra of events $\mathcal{F}_{\mathbf{K}_1}$ is the power set of $\mathbf{K}_1$, and the measure, $\Pr_{\mathbf{K}_1}$, the counting measure.

Example XII.3. Choosing at random a point from a circle.

For simplicity, we can identify the outcome ω_c with the point c on the circle. Since any point in the hyperfinite approximation to the circle, C, can be rotated by an element of $G_{\mathbf{K}_2}$ to any other point, all singletons are measure-determined and, hence, $\mathcal{F}_{\mathbf{K}_2}$ is the power set of $\mathbf{K}_2$. The measure, $\Pr_{\mathbf{K}_2}$ is the counting measure, because, for each singleton its measure is one over the number of elements of C, ν. As we saw in Chapter X, Section 3, when discussing the uniform distribution, the arcs of length a have measure a. Now, if we pass to the infinite circle $\mathbf{C}$, the measure is Lebesgue measure, which is the only measure invariant under rotations.

Example XII.4. The choosing of a point in a line, symmetrically around a given point p.

We consider, as usual, the hyperfinite approximation. Here, the only measure-determined sets are those which are symmetric around the given point p. Hence, the basic events are the unions of intervals that are symmetric around p. Since these events form a disjunctive algebra, $\mathcal{F}_{b\mathbf{K}_3}$ is the family of these sets. The measure $\Pr_{\mathbf{K}_3}$ assigns $1/2$ to all events in $\mathcal{F}_{b\mathbf{K}_3}$ that are not empty or the whole of $\mathbf{K}_3$. The algebra generated by $\mathcal{F}_{b\mathbf{K}_3}$ is the internal power set of $\mathbf{K}_3$. It is easy to see that there are many extensions of $\Pr_{\mathbf{K}_3}$ to the internal power set that are $G_{\mathbf{K}_3}$-invariant. Hence, $\mathcal{F}_{\mathbf{K}_3}$ is $\mathcal{F}_{b\mathbf{K}_3}$, and it is not an algebra. In this case, $\langle \mathbf{K}_3, \sim \rangle$ is not classical.

4. Simple probability structures

Besides the relations and operations directly connected with the chance structure of the setup, there may be other features, not involved in the chance mechanism, that are needed for the description of the chance setup. For instance, suppose that we are considering the chance setup of choosing at random m balls from an urn that contains a set A of balls (Example XII.1). Then we might be interested in the color of the balls, say black or red. Although the color of the balls is not essential for the description of the chance mechanism of the setup, this property is important for the full description of the setup itself. Properties such as the color of the balls are not essential for the chance mechanism, since with the same setup other properties may be of interest, such us the hardness or smoothness of the balls.

So there may be features, not involved in the mechanism of the setup, that are actually needed for its full description. In probability theory, characteristics of outcomes are usually given by random variables, which correspond to what will be called measurement functions. In the case we are considering, we could add a function Y_{0r} from $\mathbf{K}_0$ to the reals such that $Y_{0r}(\omega_S)$ is the number of red balls in the sample S.

In order to assure that all balls are identifiable, i.e., that each particular ball in the sample is distinguishable from any other, we could assign a number to each ball and define the functions X_{0i} (for ball number i) by $X_{0i}(\omega_S) = 1$, if ball number i is in the sample S, and 0, otherwise.

DEFINITION XII.8 (MEASUREMENT FUNCTIONS). Let $\mathbf{K}$ be a simple chance structure. Then

(1) X is a *measurement system over* $\mathbf{K}$ if X is a system $\langle X_k \mid k \leq n \rangle$, for some natural number n, where each X_k is a function from $\mathbf{K}$ into $\mathbb{R}$ (or $^*\mathbb{R}$). When the system consists of just one function X, we write X instead of $\langle X \rangle$.
(2) A *simple weak probability structure* is a pair $\mathcal{K} = \langle \mathbf{K}, X \rangle$, where $\mathbf{K}$ is a simple chance structure and X is a measurement system over $\mathbf{K}$. X is called the *measurement system of* $\mathcal{K}$. A *simple probability structure* $\mathcal{K} = \langle \mathbf{K}, X \rangle$ is a weak simple probability structure such that the system X is a system of random variables with respect to $\mathcal{F}_\mathbf{K}$.

As we shall see later when analyzing Example XII.4, interesting measurement functions are not always random variables.

Examples of measurement functions and simple probability structures are the following:

Example XII.1. A simple probability structure for the example of the choosing of a sample from a poapulation may be

$$\mathcal{K}_0 = \langle \mathbf{K}_0, \langle X_{01}, \ldots, X_{0p} \rangle \rangle,$$

where the balls are numbered $1, \ldots, p$. Another simple probability structure with the same chance structure $\mathbf{K}_0$ may be $\langle \mathbf{K}_0, Y_{0r} \rangle$, where $Y_{0r}(\omega)$ is the number of red balls in the sample of the outcome ω.

The measurements in X reflect intrinsic properties of the elements of $\mathbf{K}$ and do not have to do with the chance mechanism itself, although they may be essential for the full description of the setup. The same chance structure admits different measurement functions and, thus, gives rise to several probability structures. The purpose of these properties, which are expressed by measurement functions, is to obtain an adequate description of a setup, including, not only the chance mechanism, but also other important properties. I will now complete the description of some probability structures for the examples discussed in this chapter. I shall keep the numbering of these examples.

Example XII.2. Placing at random one ball of a population of r balls into one of n cells.

In this case, the cells as well as the balls are identifiable in principle. We may number the balls and the cells, and introduce a measurement system

$$\langle X_{1ij} \mid 1 \leq i \leq r, 1 \leq j \leq n \rangle$$

such that $X_{1ij}(\omega_O) = 1$, if ball number i is in cell number j (i.e., if $\langle i, j \rangle \in O$), and 0 otherwise. A possible probability structure is

$$\mathcal{K}_1 = \langle \mathbf{K}_1, \langle X_{1ij} \mid 1 \leq i \leq r, 1 \leq j \leq n \rangle \rangle.$$

Just as in the previous example, we may be interested in other properties of the balls or the cells.

Example XII.3. Choosing a point at random from a circle.

We, first, define one measurement function X_2. In order to have its values in the real numbers, we assume that we have associated the real numbers in an interval to the points in the circle, for instance to each point we associate the length of the arc from the point $\langle 1, 0 \rangle$ to the given point. Thus, $X_2(e^{i\theta}) = \theta$. We could also have the values of X_2 in the complex numbers. A probability structure is $\mathcal{K}_2 = \langle \mathbf{K}_2, X_2 \rangle$.

Suppose that we are modeling a roulette over which a fixed force is applied, but starting from a point chosen at random. The point chosen at random is the initial position. Since we are interested in the point that is the final position of the roulette, we could introduce another measurement function Y_2, such that $Y_2(\omega_c)$ is the final position of the roulette with outcome ω_c, i.e., with initial position c. We assume that Y_2 is internal. The function Y_2 depends on the bias of the roulette, and is determined by the equations for the laws of motion. Thus, the function f, defined by $f(c) = c'$ if and only if $Y_2(\omega_c) = c'$, is internal, continuous and one-one. We shall say in cases like this one that Y_2 itself is internal, continuous and one-one. So, we now have a different probability structure $\langle K_2, Y_2 \rangle$ supported by the same chance structure.

The measurement functions may be random variables with respect to the probability space of the chance structure. It is clear that in Examples XII.1, XII.2, and XII.3, this is the case, but, as is seen in the next example, it is not always so.

Example XII.4. The choosing of a point on a line symmetrically around a given point p.

The measurement system may consist of just one function X_3 such that $X_3(\omega_c) = c$, i.e., the point chosen in ω_c. A weak probability structure is $\mathcal{K}_5 = \langle \mathbf{K}_5, X_3 \rangle$. The function X_3 is not a random variable with respect to $\mathcal{F}_{\mathbf{K}_5}$.

4. SIMPLE PROBABILITY STRUCTURES

This is so, because the sets $[X_3 = r]$ are not in $\mathcal{F}_{\mathbf{K}_5}$, for any r. The only sets in $\mathcal{F}_{\mathbf{K}_5}$ are those sets of real numbers which are symmetric around the origin. Hence, $\mathcal{K}_5$ is not a probability structure, but only a weak probability structure.

It may be mathematically convenient to increase the family of events including sets that are not in $\mathcal{F}_{\mathbf{K}}$. For instance, it is easier to work with an algebra and not just a disjunctive algebra; thus, in Example XII.4, so that X_3 be a random variable, we may consider working with the field generated by $\mathcal{F}_{\mathbf{K}_3}$. In case we extend $\mathcal{F}_{\mathbf{K}}$, the only events that we are sure that have their probabilities logically determined are still those in $\mathcal{F}_{\mathbf{K}}$. For the others, as is the case for Example XII.4, there may be a wide range of possible assignment of measures invariant under $G_{\mathbf{K}}$, and, hence, the probabilities that we assign to them may not be very informative, from the point of view of the structure of $\mathbf{K}$.

In Definition VI.1, we introduced the probability distribution of a random variable X, pr_X. Thus, if $\langle \mathbf{K}, X \rangle$ is a simple probability structure, where X is a random variable, we have

$$\text{pr}_X(r) = \text{Pr}_{\mathbf{K}}(\{\omega \in \mathbf{K} \mid X(\omega) = r\}),$$

where r is a real (or hyperreal) number. Very often, what is observable is X, and so, we can estimate directly pr_X and not $\text{Pr}_{\mathbf{K}}$. For instance, in the case of a biased roulette discussed above, in general we can observe the final position and not the initial position. Hence, the observable random variable is Y_2. We can then estimate its probability distribution and only postulate that the underlying probability structure is $\langle \mathbf{K}, Y_2 \rangle$.

When using statistical inference, we usually have an observable measurement and we want to see which hypothesis is the best to explain it. Thus, we must consider several possible interpretations of this measurement, which assign to it different probabilities. This is a situation similar to that in mathematical logic, where a sentence has several possible interpretations each of which assigns to it either truth or falsity. Thus, it is natural to introduce a logical language for treating statistical inference. We have discussed this language informally in Chapter III, Section 5, and we shall introduce it formally in Chapter XVII.

CHAPTER XIII

The structure of chance

Most chance setups have possible results which are compounded of several partial results. For instance, when we toss several coins, the result is the combination of the results of the toss for each coin. In this chapter, we shall study models for this kind of setups.

Recall that in the formalization of an outcome we include, besides the particular result, the mechanism of the setup. Since in case of composite results the mechanism involves the type of connection between the different partial results, we should be able to give a formalization of this connection. There are two main types of connections: of dependence and of independence. In the case of the toss of several coins mentioned above, the results are independent.

On the other hand, when we choose an urn and then a ball from the chosen urn, the second partial results depends on the first one. In order to formalize both dependence and independence, we use certain type of binary relations, which I shall call causal relations. I will only use finite causal relations, i.e., relations with finite fields. Adding some conditions, we could also have infinite causal relations.[1] Using nonstandard analysis in all cases where an infinite causal relation would appear natural, however, the causal relation can be reduced, in a sense that will be clear from some of the examples in Chapter XVI, to a finite, or more properly, hyperfinite relation.

It is not possible to model all setups using the simple probability structures of Chapter XII. Simple examples of setups where simple structures don't work are the urn models discussed in Chapter III, Section 4. We give, here another reason for simple structures not to work, in these cases. Suppose that we choose at random an urn from a set of two urns, and then a ball, also at random, from the urn chosen. Assume that one urn contains two white ball and the other, one black ball. Since the equiprobability relation for simple chance structures is refining, if A and B are events with the same probability, and A can be partitioned into sets A_1 and A_2, then B can also be partitioned into sets B_1 and B_2, such that A_1 has the same measure as B_1 and A_2 has the same measure as B_2. In the example

---

[1] See [20], for a version with infinite causal relations.

we are considering, the event of choosing a white ball can be divided into two events, while the equivalent event of choosing a black ball, cannot. Thus, neither the relation of equiprobability nor its finitely additive closure are refining, and, hence, since relations defined by groups are refining, it is unlikely that either of these two relations comes from a group.

1. Binary relations and causal relations

We now use the notions for binary relations introduced in the Preliminaries, Section 1. We need some more facts about binary relations. Let R be a binary relation. We sometimes write $a\mathrm{R}b\mathrm{R}c$ for the conjunction of $a\mathrm{R}b$ and $b\mathrm{R}c$. The symbol R^{-1} denotes the inverse relation, i.e., $a\mathrm{R}^{-1}b$ if and only if $b\mathrm{R}a$. Similarly as for functions, $\mathrm{R}''B$ is the image of the set B under R, i.e.

$$\mathrm{R}''B = \{a \mid \text{ there is a } b \in B \text{ such that } b\mathrm{R}a\}.$$

The inverse image is

$$\mathrm{R}^{-1}{}''A = \{b \mid \text{ there is an } a \in A \text{ such that } b\mathrm{R}a\}.$$

An especially important image for our purposes is the image and inverse image of a singleton

$$\mathrm{R}''\{t\} = \{a \mid t\mathrm{R}a\},$$

$$\mathrm{R}^{-1}{}''\{t\} = \{a \mid a\mathrm{R}t\}.$$

We shall also use the interval notation. For instance, $(-\infty, t)$ is the set of elements of the field of R that are before t according to R, i.e.

$$(-\infty, t) = \{s \mid s\mathrm{R}t \text{ and } s \neq t\} = (\mathrm{R}^{-1}{}''\{t\}) - \{t\}.$$

We also write

$$(-\infty, t] = \{s \mid s\mathrm{R}t\} = \mathrm{R}^{-1}{}''\{t\}.$$

If R is a binary relation over T, i.e., if $\mathrm{R} \subseteq {}^2T$, we sometimes write $\langle T, \mathrm{R}\rangle$, and call T the *field of* $\langle T, \mathrm{R}\rangle$ (or, of R, if there is no danger of confusion). We say that t is a *last element of* $\langle T, \mathrm{R}\rangle$, if $t \in T$ but there is no $s \neq t$, $s \in T$, such that $t\mathrm{R}s$. On the other hand, t is a *first element of* R, if t is in the field of R but there is no $s \neq t$ such that $s\mathrm{R}t$, i.e, if $(-\infty, t) = \emptyset$. We also extend these last two definitions to nonempty subsets S of the field of R, and talk about first and last elements of S, with the obvious definitions. The element s is an immediate predecessor of t, and t is an immediate successor of s, if $s\mathrm{R}t$, $s \neq t$, and there is no $u \neq s$, $u \neq t$, such that $s\mathrm{R}u\mathrm{R}t$. Finally, we say that s and t are R–incomparable, if not $s\mathrm{R}t$ and not $t\mathrm{R}s$.

We defined reflexive, symmetric, transitive, and equivalence relations in Preliminaries, Section 1. When we say that $\langle T, \mathrm{R}\rangle$ is reflexive, we mean that R is reflexive on T. Similarly for the other notions.

We also defined in Preliminaries, Section 1, the notions of antisymmetric and connected relations, and partial orderings. The equality relation is a partial ordering. I shall write partial orderings over T by $\langle T, \preceq \rangle$, or, simply, when there is no danger of confusion, $\preceq$. A *linear ordering* $\langle T, \preceq \rangle$ is a partial ordering that is *connected*, i.e., such that for every $x, y \in T$ we have $x \preceq y$ or $y \preceq x$.

We could have defined a partial ordering $\langle T, R \rangle$ as a transitive and *asymmetric* relation, i.e, such that $x R y$ implies not $y R x$. I shall call these relations, *nonreflexive partial orderings*. I find it more convenient to have reflexive partial orderings, because, among other reasons, the nonreflexive counterpart of the equality relation is the empty relation, which is not a very natural relation. In any case, it is easy to transform a nonreflexive partial ordering into a reflexive partial ordering and vice-versa: if $\langle T, \preceq \rangle$ is a reflexive partial ordering, then we define $a \prec b$ if and only if $a \preceq b$ and $a \neq b$, and we obtain, thus, a nonreflexive partial ordering $\langle T, \prec \rangle$. On the other hand, if $\langle T, \prec \rangle$ is a nonreflexive partial ordering, then defining $a \preceq b$ to be $a \prec b$ or $a = b$, we obtain a reflexive partial ordering. We shall use both $\preceq$ and $\prec$, when dealing with partial orderings.

A relation $\langle T, R \rangle$ is *well-founded*, if every nonempty set A included in its field T has a first element according to R, i.e., if for every $A \subseteq T$, $A \neq \emptyset$, there is a $t \in A$ such that $(-\infty, t) \cap A = \emptyset$.

A finite partial ordering is always well-founded: Suppose that $\preceq$ is a partial ordering with finite field T. Let A be a nonempty subset of T. If A has no first element, we can define a sequence c_n of elements of A, for $n \in \mathbb{N}$, such that $c_{n+1} \prec c_n$ for every $n \in \mathbb{N}$. Since A is finite, we must have $c_m = c_n$ for certain n, m, with $n < m$. Hence, we have $c_m \prec c_{n-1}$ and $c_{n-1} \preceq c_m$, which is impossible since $\preceq$ is a partial ordering.

The transitive closure of a binary relation $\langle T, R \rangle$, in symbols R^∞, is the least transitive relation containing R. It is clear that R^∞ has the same field as R, since $T \times T$ is transitive and contains R, so that $R^\infty \subseteq T \times T$. The relation R^∞, as defined, is the intersection of all transitive relations which contain R or, equivalently, R^∞ is the relation S, with field T, such that aSb if and only if there is a natural number n and there are $c_1, \ldots, c_n$ in T with $a R c_1 R \cdots R c_n R b$. It is clear that $S \subseteq R^\infty$. On the other hand, it is easy to show that S is transitive and that it contains R. Thus, $R^\infty \subseteq S$.

We define $R^{\infty=}$, the transitive reflexive closure of R, to be $R^\infty \cup \{\langle x, x \rangle \mid x \in T\}$; that is, $R^{\infty=}$ is the least transitive reflexive relation containing R. If R is well-founded, then $R^{\infty=}$ is a partial ordering. The proof of this fact goes as follows. By definition, $R^{\infty=}$ is transitive and reflexive. So we just have to prove that it is antisymmetric. In order to prove this, we first show that R itself cannot have any cycles, i.e., that there are no $c_1, c_2, \ldots, c_n$ with $c_1 R c_2 R \cdots R c_{n-1} R c_n$, $c_i \neq c_1$, for a certain $i = 2, \ldots, n-1$, and $c_1 = c_n$. If there were such a cycle, then the set $\{c_1, c_2, \ldots, c_n\}$ would have no first element, which is impossible because R is well-founded.

We now prove that $R^{\infty=}$ is antisymmetric. Suppose that $aR^{\infty=}b$ and $bR^{\infty=}a$; then there are $c_1, \ldots, c_n$, and $d_1, \ldots, d_m$ such that $a R c_1 R \cdots R c_n R b$ and $b R d_1 R \cdots R d_m R a$. If $a \neq b$, then $a, c_1, \ldots, c_n, b, d_1, \ldots, d_m, a$ would be a cycle. Since this is impossible, $a = b$. It is easy to show that $R^{\infty=}$ is also well-founded, but we will not need this fact, since we shall be mainly dealing with finite partial orderings, which are always well-founded.

If $\langle T, \mathrm{R} \rangle$ is a relation, f is an automorphism of $\langle T, \mathrm{R} \rangle$, if f is a permutation of T, such that for any $a, b \in T$, $a\mathrm{R}b$ if and only if $f(a)\,\mathrm{R}\,f(b)$. If f is a permutation of the field T of a relation R, then we next show that f is an automorphism of R if and only if it is an automorphism of R^∞:

Since $\mathrm{R} \subseteq \mathrm{R}^\infty$, every automorphism of R^∞ is an automorphism of R. On the other hand, suppose that f is an automorphism of R. We have that $a\mathrm{R}^\infty b$ if and only if there are $c_1, \ldots, c_n$ with $a\mathrm{R}c_1\mathrm{R}\cdots\mathrm{R}c_n\mathrm{R}b$; this is true if and only if $f(a)\,\mathrm{R}\,f(c_1)\,\mathrm{R}\cdots\mathrm{R}\,f(c_n)\,\mathrm{R}\,f(b)$; and this last formula is true if and only if $f(a)\,\mathrm{R}^\infty\,f(b)$. Similarly, it can be proved that f is an automorphism of R if and only if f is an automorphism of $\mathrm{R}^{\infty=}$.

The field T of a finite well-founded relation R can be divided recursively into levels T_i for $i < n$, for a certain $n \in \mathbb{N}$. The set T_0 contains exactly the first elements of T. The set T_{i+1} is the class of first elements of $T - \bigcup\{T_j \mid j < i\}$. Since T is finite, $T_n = \emptyset$, for a certain n. The first such n is the *height of T*. The levels of the partial orderings $\langle T, \mathrm{R}^\infty \rangle$ and $\langle T, \mathrm{R}^{\infty=} \rangle$ are the same as those of $\langle T, \mathrm{R} \rangle$.

We, now proceed to causal relations. For simplicity, we require that causal relations be reflexive. This requirement does not imply any loss of generality, since, as we saw above, it is always possible to pass from a reflexive to a nonreflexive relation and vice-versa.

DEFINITION XIII.1 (CAUSAL RELATIONS).

(1) A *causal relation* $\langle T, \mathrm{C} \rangle$ is a finite reflexive relation with field T such that for every $t, s \in T$, $(-\infty, s) = (-\infty, t) \neq \emptyset$, implies $s = t$. Although we do not identify C with the ordering of time, we call the elements of T, the *causal moments of T*. When there is no danger of confusion, we sometimes write T or C for the causal relation $\langle T, \mathrm{C} \rangle$.

(2) A relation R is *forwardly linear* if for every s in its field, there is at most one $t \neq s$ such that t is an immediate successor of s.

(3) A causal relation is a *strict causal relation* if it is forwardly linear.

(4) A causal relation that is a partial ordering is called a *causal ordering*. Similarly, a strict causal relation that is a partial ordering is called a *strict causal ordering*.

The requirement that t be determined in T by $(-\infty, t)$, if t is not a first element, is an embodiment of the idea that a set of causes should not have two effects. In the chance structures that will be introduced later, what happens before a point t determines the possibilities at t. There may be, and usually are, more than one possibility at t, but there is only one set of possibilities. This is the requirement implicit in the condition imposed on causal relations.

We shall work mainly with partial orderings, since they are well-founded and most of our proofs will be by induction on the levels of the relation.

In Fig. XIII.10, a few causal relations are pictured. The direction of the arrow indicates increasing order. In the figure, #1–#4 are well-founded, but #5 is not. Relations #1–#3 are forwardly linear, but #4 and #5 are not.

1. BINARY RELATIONS AND CAUSAL RELATIONS

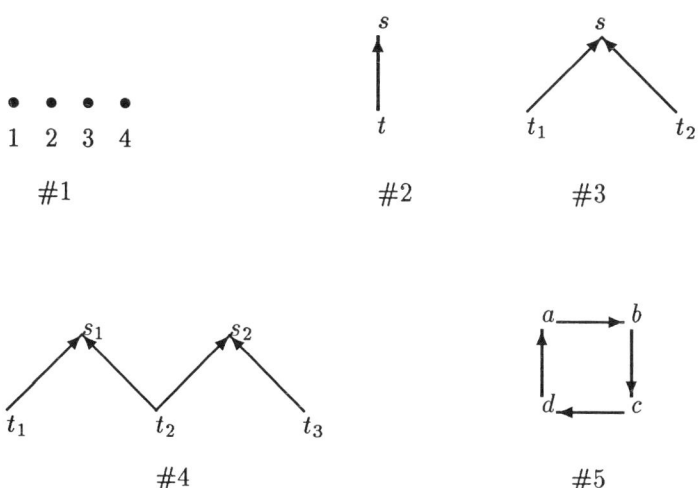

FIGURE XIII.10. Causal relations

Most causal relations that we shall need will be partial orderings. If a causal relation is well-founded, then one can take the reflexive transitive closure of the relation and obtain a partial ordering. We already have shown that every automorphism of R is an automorphism of $R^{\infty =}$ and vice versa.

In Fig. XIII.10, #1 is of height 1, and #2, #3, and #4 are of height 2.

PROPOSITION XIII.1. *If $\langle T, \preceq \rangle$ is a partial ordering, then T is a strict causal ordering, if and only if the set of successors, $\{s \in T \mid t \preceq s\}$, of each element $t \in T$, is linearly ordered.*

PROOF. Suppose that the set of successors of each element of T is linearly ordered. Then, if $(-\infty, t) = (-\infty, s) \neq \emptyset$, t and s are larger than an element u of T. Hence, $t \preceq s$ or $s \preceq t$. If $t \prec s$, then $t \in (-\infty, s)$, but $t \notin (-\infty, t)$, contradicting $(-\infty, t) = (-\infty, s)$. The case $s \prec t$ is symmetric. So we have proved that T is a causal ordering. The forward linearity is obvious. The proof in the other direction is obvious. □

What we have just proved implies that every linear ordering is a strict causal ordering. However, as can be seen in #4 of Fig. XIII.10, there are causal orderings that have elements t such that the set of successors of t are not linearly ordered. On the other hand, we have the following simple theorem, which is easily obtained from the fact the if $\langle T, R \rangle$ is well-founded, then $\langle T, R^{\infty =} \rangle$ is a partial ordering.

PROPOSITION XIII.2. *If $\langle T, R \rangle$ is a well-founded strict causal relation, then*

$$\langle T, R^{\infty =} \rangle$$

is a strict causal ordering.

We shall mainly be concerned with causal orderings and strict causal orderings. Strict causal orderings express more simply the notions of dependence and independence than other causal relations.

For instance, in a causal ordering $\langle T, R \rangle$, we can naturally define two elements s and t of T to be independent, if $(-\infty, t) \cap (-\infty, s) = \emptyset$. This is so, because independence of t and s means that nothing that has happened before s influences what happens at t and vice-versa. For strict causal orderings, we have the following equivalence:

PROPOSITION XIII.3. *Let $\langle T, \preceq \rangle$ be a strict causal ordering. Then two elements of $T - T_0$, s and t, are incomparable in $\preceq$, i.e., not $s \preceq t$ and not $t \preceq s$, if and only if $(-\infty, t) \cap (-\infty, s) = \emptyset$.*

PROOF. Suppose that $(-\infty, t) \cap (-\infty, s) \neq \emptyset$ and let $u \in (-\infty, s) \cap (-\infty, t)$; then $u \prec t$ and $u \prec s$. Hence, by forward linearity, $t \preceq s$ or $s \preceq t$ and, hence s and t are $\preceq$-comparable. On the other hand, if s and t are $\preceq$-comparable, say $s \preceq t$, then $(-\infty, s) \subseteq (-\infty, t)$ and, since $s \notin T_0$, $(-\infty, s) \neq \emptyset$. Thus, $(-\infty, s) \cap (-\infty, t) = (-\infty, s) \neq \emptyset$. □

The definitions of outcome and chance structure that follow in the next section, imply that a well-founded relation can always be replaced by its transitive closure, with a few changes in the structure. Since most cases of interest are well-founded, almost always causal orderings suffice. There seem to be some examples from physics, however, where the causal relations are not well-founded.

2. Relational systems with causal structures

We proceed, now, with the description of an outcome. Although there are important differences, a composite outcome is similar to a trajectory in a stochastic process. A composite outcome λ can be described as a function that associates, with each element of the field T of a causal ordering, a simple outcome. As we saw in the last chapter, simple outcomes can be represented by relational systems. That is, λ can be taken to be a function with domain T and such that $\lambda(t)$ is a relational system, for every $t \in T$. Thus, T represents the set of causal moments and $\lambda(t)$, what happens at t.

For instance, in case we are modeling the toss of four coins, T should contain four elements, and, for each $t \in T$, $\lambda(t)$ should be the relational system modeling the toss of coin t. As was mentioned before, since the tosses are supposed independent, no element of T should be before any other, i.e., $s \preceq t$ if and only if $s = t$, and, hence, the causal relation is $\langle T, = \rangle$, where $=$ is the identity restricted to T. The relation $\langle T, = \rangle$, represented by #1 in Fig. XIII.10, is a strict causal ordering.

If we were modeling the choosing of an urn, and then the choosing of a ball from the urn, then T should be a causal ordering containing two elements, say s and t, with $t \prec s$, as in #2 of Fig. XIII.10. The value of λ at t, $\lambda(t)$, should be

2. RELATIONAL SYSTEMS WITH CAUSAL STRUCTURES

a relational system for the selection of an urn, and $\lambda(s)$ a relational system for a selection of a ball from the urn selected according to $\lambda(t)$.

This representation of outcomes as functions, has the disadvantage of not displaying the causal relation explicitly. We could represent outcomes by the pair $\langle \lambda, C \rangle$, where λ is the function discussed above and C is the causal relation with field T, the domain of λ. It seems more convenient, however, to code the outcomes as relational systems with several universes, one of the universes being T. We must be able to recover the function from the relational system in a uniform way. For simplicity, I will have only one universe of objects, i.e., only one universe besides T.

Suppose that we start from a function λ with domain T whose values are relational systems with one universe, which we may assume to be of the same similarity type. The universe of $\lambda(t)$ is $|\lambda(t)|$, call it A_t. We can code all the universes for all t's in T by the binary relation

$$R = \bigcup \{\{t\} \times A_t \mid t \in T\}$$
$$= \{\langle t, a \rangle \mid t \in T \text{ and } a \in A_t\}.$$

This relation, R, is determined by λ and T. On the other hand, we can recover A_t from R and t by

$$A_t = R''\{t\} = \{a \mid \langle t, a \rangle \in R\},$$

the image of the singleton $\{t\}$ by R.

In a similar way, if we have an n–ary relation S_t in each system $\lambda(t)$, we can code them all by the $n+1$–ary relation

$$R_1 = \bigcup \{\{t\} \times S_t \mid t \in T\}$$
$$= \{\langle t, a_1, \ldots, a_n \rangle \mid t \in T \text{ and } \langle a_1, \ldots, a_n \rangle \in S_t\}.$$

Again

$$S_t = R_1''\{t\} = \{\langle a_1, \ldots, a_n \rangle \mid \langle t, a_1, \ldots, a_n \rangle \in R_1\}.$$

Thus, we can code the universes and relations occurring in λ. We could also devise ways of coding operations; for simplicity, however, we restrict our relational systems to relations. As it was explained in Chapter XII, Section 1, this restriction implies no loss of generality.

The restriction to a common similarity type for the $\lambda(t)$'s implies no loss of generality either, because we can always add empty relations to our systems. Since we here have only relations, the similarity type needs just to code the arities of the relations. Thus, the similarity type of a relational system $\langle A, R_i \rangle_{i \in I}$ is a function ν, with domain I, such that for every $i \in I$, $\nu(i)$ is the arity of R_i.

In the following definition, I proceed in the opposite direction. First, the relational systems with two universes are introduced, and then the functions are obtained from them.

DEFINITION XIII.2 (OUTCOMES).

(1) An *outcome* is a relational system

$$\lambda = \langle T, U; \preceq, R_i \rangle_{i \in I \cup \{0\}}$$

(we assume that $0 \notin I$), where:
  (a) $\langle T, \preceq \rangle$ is a causal ordering. It is called the *causal relation* or *causal ordering of* λ. T is called the *causal universe of* λ.
  (b) $R_0 \subseteq T \times U$ and $U = R_0'' T$ (i.e.,

$$U = \{a \mid \langle t, a \rangle \in R_0, \text{ for some } t \in T\}).$$

U is called the *object universe of* λ and is denoted by $|\lambda|$.
  (c) For each $i \in I$, there is a natural number n, such that $R_i \subseteq T \times {}^n U$.
  (d) For each $t \in T$, $\langle R_0''\{t\}; R_i''\{t\} \rangle_{i \in I}$ is a relational system.
When we want to stress that R_i is a relation of the outcome λ, we write R_i^λ.

(2) The *similarity type of* λ is the triple $\langle T, \preceq, \nu \rangle$, where ν is a function from I to the natural numbers, such that $\nu(i)$ is the arity of $R_i''\{t\}$, for any $t \in T$ and $i \in I$.

(3) If λ is an outcome, as in (1), we use the symbol $\hat\lambda$ for the function with domain T and value

$$\hat\lambda(t) = \langle R_0''\{t\}; R_i''\{t\}\rangle_{i \in I},$$

for each $t \in T$.

The only restriction imposed by (1d) is that the relations $R_i''\{t\}$ are relations over $R_0''\{t\}$, i.e., $R_i''\{t\} \subseteq {}^n R_0''\{t\}$, for a certain $n \in \mathbb{N}$.

Notice that we include in the similarity type the causal relation of λ.

It is clear that from the outcomes as relational systems λ we can recover the outcomes as functions, $\hat\lambda$, and from the outcomes as functions plus the causal ordering $\preceq$, we can recover the outcomes as relational systems. The next definition introduces some notation.

DEFINITION XIII.3 (RESTRICTION). Let $\lambda = \langle T, U; \preceq, R_i \rangle_{i \in I \cup \{0\}}$ be an outcome, $S \subseteq T$, and $t \in T$. Then

$$\lambda \upharpoonright S = \langle S, U_S; \preceq \upharpoonright S, R_i \upharpoonright S \rangle_{i \in I \cup \{0\}},$$

where

$$U_S = R_0'' S,$$

$$\preceq \upharpoonright S = \{\langle t, s \rangle \mid t, s \in S \text{ and } t \preceq s\},$$

and $R_i \upharpoonright S$ is R_i restricted to S, i.e.

$$R_i \upharpoonright S = \bigcup \{\{t\} \times R_i''\{t\} \mid t \in S\}.$$

2. RELATIONAL SYSTEMS WITH CAUSAL STRUCTURES

We have that the restriction of an outcome λ to S, considered as a function, is the restriction of the function $\hat{\lambda}$ to S, i.e.

$$\widehat{\lambda \restriction S} = \hat{\lambda} \restriction S.$$

We write

$$\lambda_t = \begin{cases} \lambda \restriction (-\infty, t), & \text{if } (-\infty, t) \text{ is not empty,} \\ t, & \text{otherwise,} \end{cases}$$

$$\lambda_{t]} = \lambda \restriction (-\infty, t],$$

and

$$\lambda_n = \lambda \restriction \bigcup \{T_i \mid i \leq n\}.$$

The idea is that $\lambda(t)$ should depend on λ_t, i.e., on what has happened before t, if something has happened, or on t itself, otherwise. So we have that always $\lambda(t)$ depends of λ_t.

We next define isomorphisms of outcomes. Recall that in outcome $\lambda = \langle T, U; \preceq, R_i \rangle_{i \in I \cup \{0\}}$, the domain of R_0 is T and, for each $t \in T$, $R_0''\{t\}$ is the universe of $\lambda(t)$.

DEFINITION XIII.4 (ISOMORPHISMS). Let

$$\lambda = \langle T, U; \preceq, R_i \rangle_{i \in I \cup \{0\}}$$

be an outcome. Then:

(1) An *isomorphism of* λ is function g with domain containing R_0 and values ordered pairs such that:
   (a) If $t \in T$, $x, y \in R_0''\{t\}$, $g(\langle t, x \rangle) = \langle u, z \rangle$, and $g(\langle t, y \rangle) = \langle v, w \rangle$, then $u = v$.
   (b) If $s, t \in T$, $x \in R_0''\{s\}$, $y \in R_0''\{t\}$, $g(\langle s, x \rangle) = \langle u, z \rangle$, and $g(\langle t, y \rangle) = \langle u, w \rangle$, then $t = s$.

(2) If g is an isomorphism of λ, we define:
   (a) g_0 is the one-one function with domain T given, for $t \in T$, by $g_0(t) = u$ if and only if $g(\langle t, x \rangle) = \langle u, z \rangle$, for some $x \in R_0''\{t\}$ and some z.
   (b) If $t \in T$ define the function g_t with domain $R_0''\{t\}$ by $g_t(x) = z$ if and only if $g(\langle t, x \rangle) = \langle g_0(t), z \rangle$, for $x \in R_0''\{t\}$.
   (c) We define the image of λ by g

$$g''\lambda = \langle g''T, g''U; g'' \preceq, g'' R_i \rangle_{i \in I \cup \{0\}},$$

with

$$g''T = g_0''T,$$
$$g''R_0 = \{\langle g_0(t), g_t(x)\rangle \mid \langle t, x\rangle \in R_0\},$$
$$g''U = (g''R_0)''(g_0''T),$$

$$g'' \preceq = \{\langle g_0(t), g_0(s)\rangle \mid t \preceq s\},$$
$$g'' R_i = \{\langle g_0(t), g_t(a_1), \ldots, g_t(a_n)\rangle \mid \langle t, a_1, \ldots, a_n\rangle \in R_i\},$$

for $i \in I$.

(3) If μ is an outcome then μ is isomorphic to λ, if there is an isomorphism of λ, g, such that $\mu = g''\lambda$.

The definition of isomorphism implies that g_0 is well defined and is a one-one function with domain T. Also, for $t \in T$, g_t is a one-one function on $R''\{t\} = |\hat{\lambda}(t)|$. Thus, g_t is an ismorphism of $\hat{\lambda}(t)$. It is also clear that g determines the functions g_0 and g_t, for $t \in T$.

We also have, as an immediate consequence of the definition, that if g and f are isomorphisms of the outcome λ such that $f_0 = g_0$ and f_t restricted to $|\hat{\lambda}(t)|$ is the same as g_t restricted to $|\hat{\lambda}(t)|$, for every $t \in T$, then $f''\lambda = g''\lambda$.

We also have that g'' is the obvious generalization for relations, relational systems, and outcomes, of the image function of g. The following theorem, stated for further reference, is an immediate consequence of the definitions, using also the definition of isomorphism of relational systems in Chapter XII, Section 1, for $g''_t \lambda(t)$.

PROPOSITION XIII.4. *Let λ be an outcome with causal universe T and g an isomorphism of λ. Then:*

(1) *$g''\lambda$ is an outcome.*
(2) *g_t is an isomorphism of the relational system $\lambda(t)$ onto the relational system $(g''\lambda)(g_0(t))$ (according to the definition in Chapter XII, Section 1).*
(3) *For each $t \in T$*
$$g''\lambda_t = (g''\lambda)_{g_0(t)}.$$

Condition (1) is clear form the definitions.

PROOF OF (2). Since g is a isomorphism of λ, g is one-one, and hence, g_t is a one-one function with domain $|\hat{\lambda}(t)|$, the universe of the relational system $\hat{\lambda}(t)$. We also have the following chain of equivalences:

$$x \in (g''R_0)''\{g_0(t)\} \iff \langle g_0(t), x\rangle \in g''R_0,$$
$$\iff \langle g_0(t), x\rangle = \langle g_0(t), g_t(y)\rangle$$

for some y with $\langle t, y\rangle \in R_0$,

$$\iff x = g_t(y)$$

for some $y \in R''_0\{t\}$,

$$\iff x \in g''_t(R''_0\{t\}).$$

So $g''_t(R''_0\{t\}) = (g''R_0)''\{g_0(t)\}$. Similarly we can prove that $g''_t(R''_i\{t\}) = (g''R_i)''\{g_0(t)\}$, for $i \in I$. □

PROOF OF (3). Condition (3) is obtained from (2), noticing that the causal universe of λ_t is $(-\infty, t)$, and that $g_0''(-\infty, t) = (-\infty, g_0(t))$. □

The reason for choosing this definition of isomorphism of composite outcomes is to combine an isomorphisms of T with isomorphisms that move independently each $\lambda(t)$. In fact, we have:

PROPOSITION XIII.5. *Let λ be an outcome with causal relation $\langle T, R \rangle$ and μ an outcome with causal relation $\langle S, C \rangle$. Then λ is isomorphic to μ if and only if there is an isomorphism g_0 of $\langle T, R \rangle$ onto $\langle S, C \rangle$, and for each $t \in T$ there are isomorphisms g_t of $\hat{\lambda}(t)$ onto $\hat{\mu}(g_0(t))$.*

PROOF. By the previous proposition, if λ is isomorphic to μ, then the conclusion follows. So assume that there are an isomorphism g_0 of $\langle T, R \rangle$ onto $\langle S, C \rangle$ and isomorphisms g_t of $\hat{\lambda}(t)$ onto $\hat{\mu}(g_0(t))$. Then if we define $g(\langle t, x \rangle) = \langle g_0(t), g_t(x) \rangle$, g is an isomorphism of λ onto μ. □

Thus, we can say that an isomorphism of λ preserves the causal relation and the function $\hat{\lambda}$ with domain T. The last proposition states that the process can be reversed, i.e., from mappings that preserve the causal relation and the function $\hat{\lambda}$, we can obtain an isomorphism of the outcome λ. From now on, we shall not distinguish between the outcome λ and the corresponding function $\hat{\lambda}$, writing both as λ.

A relational system with one universe $\langle A; S_i \rangle_{i \in I}$ can be identified with the outcome $\lambda = \langle T, U; \preceq, R_i \rangle_{i \in I \cup \{0\}}$, where T contains just one element, say $T = \{t\}$, $U = A$, $\preceq$ is the identity, $R_0 = \{t\} \times A$, and $R_i = \{t\} \times S_i$, for $i \in I$. We could also consider outcomes with several object universes, complicating a little the definitions. Thus, all notions of the previous chapter are particular cases, with appropriate restrictions, of the notions that occur in this chapter.

3. Chance structures

The definition of a chance structure is similar to that of simple chance structures, but we must add a condition that insures that all possibilities are present.

DEFINITION XIII.5 (CHANCE STRUCTURES). A chance structure is a set **K** of outcomes of the same similarity type, say $\langle T, \preceq, \nu \rangle$, such that

(*): for every $\lambda_1, \ldots, \lambda_n \in \mathbf{K}$, and every set of pairwise incomparable elements of T, $\{s_1, \ldots, s_n\}$, if there is a μ with domain T such that $\mu_{s_i]} = (\lambda_i)_{s_i]}$ for $i = 1, 2, \ldots, n$, then there is a $\mu \in \mathbf{K}$ with $\mu_{s_i]} = (\lambda_i)_{s_i]}$ for $i = 1, 2, \ldots, n$.

The common causal ordering $\langle T, \preceq \rangle$ of the outcomes in **K** is called the *causal relation of* **K**. The field T of this relation is called the *causal universe of* **K**. The union of the object universes of the outcomes in **K** is called the *object universe of* **K** and is denoted by $|\mathbf{K}|$.

If $\langle T, \preceq \rangle$ is a strict causal ordering we call **K** a *strict chance structure*.

Condition (*) asserts that if $(\lambda_i)_{s_i]}$ for $i = 1, \ldots, n$ are compatible as functions (that is, if in the domain that they may have in common, they are equal), then there is an extension, μ of all of them in **K**. We shall be mainly concerned with strict chance structures, and in this case the $(\lambda_i)_{s_i]}$ for $i = 1, \ldots, n$, are always compatible as functions, since the sets $(-\infty, s_i]$ are disjoint, when the s_i are incomparable.

We now introduce some notation. If $\lambda \in \mathbf{K}$, $t \in T$, the causal universe of **K**, and $\mathbf{A} \subseteq \mathbf{K}$, we define $\mathbf{A}_{\lambda_t}$ to be

$$\mathbf{A}_{\lambda_t} = \{\mu(t) \mid \mu \in \mathbf{A} \text{ and } \lambda_t = \mu_t\}.$$

If $S \subseteq T$, we define

$$\mathbf{A} \upharpoonright S = \{\lambda \upharpoonright S \mid \lambda \in \mathbf{A}\}.$$

In particular, we write $\mathbf{A}(t)$ for $\mathbf{A} \upharpoonright \{t\}$, $\mathbf{A}_t$ for $\mathbf{A} \upharpoonright (-\infty, t)$, $\mathbf{A}_{t]}$ for $\mathbf{A} \upharpoonright (-\infty, t]$, and $\mathbf{A}_n$ for $\mathbf{A} \upharpoonright \bigcup\{T_i \mid i \leq n\}$, i.e.[2]

$$\mathbf{A}(t) = \{\lambda(t) \mid \lambda \in \mathbf{A}\},$$
$$\mathbf{A}_t = \begin{cases} \{\lambda_t \mid \lambda \in \mathbf{A}\}, & \text{if } (-\infty, t) \neq \emptyset, \\ \emptyset & \text{otherwise,} \end{cases}$$
$$\mathbf{A}_{t]} = \{\lambda_{t]} \mid \lambda \in \mathbf{A}\},$$
$$\mathbf{A}_n = \{\lambda_n \mid \lambda \in \mathbf{A}\}.$$

The structures $\mathbf{K}_{\lambda_t}$, $\mathbf{K}_t$, $\mathbf{K}_{t]}$, and $\mathbf{K}_n$ are called the *basic blocks of* **K**. In this chapter, we shall just use this notation with $\mathbf{A} = \mathbf{K}$, but in the next chapter we shall need it for other subsets of **K** as well. Notice that

$$\mathbf{A}(t) = \bigcup\{\mathbf{A}_{\lambda_t} \mid \lambda_t \in \mathbf{A}_t\} = \bigcup\{\mathbf{A}_{\lambda_t} \mid \lambda \in \mathbf{A}\}.$$

That all outcomes in a chance structure have a common similarity type, means that they all have the same causal relation and relations of corresponding arities. Also, all relational system in $\mathbf{K}_{\lambda_t}$ and $\mathbf{K}(t)$ are similar (i.e., have the same type), and hence these sets of relational systems are simple chance structures. If $\langle T, \preceq, \nu \rangle$ is the similarity type of **K**, then $\mathbf{K}_t$ and $\mathbf{K}_{t]}$ are chance structures of type $\langle (-\infty, t), \preceq, \nu \rangle$ and $\langle (-\infty, t], \preceq, \nu \rangle$, respectively, if $(-\infty, t) \neq \emptyset$. If t is a first element of T, then $\mathbf{K}_t$ is empty, and $\mathbf{K}_{t]} = \mathbf{K}_{\lambda_t} = \mathbf{K}(t)$, for every $\lambda \in \mathbf{K}$. The structure $\mathbf{K}_n$ is of type $\langle \bigcup\{T_i \mid i \leq n\}, \preceq, \nu \rangle$.

When there is no danger of confusion, we shall use the following notation for concatenation of functions. Suppose that μ is a function with domain $(-\infty, t) \neq \emptyset$. Then we write $\mu \frown \omega$ for the extension of μ to $(-\infty, t]$ such that $\mu(t) = \omega$. That is,

$$\mu \frown \omega = \mu \cup \{\langle t, \omega \rangle\}.$$

---

[2] To be more precise, $\mathbf{A} \upharpoonright \emptyset = \{\emptyset\}$, but we shall take $\mathbf{A}_t = \emptyset$, when t is a first element.

3. CHANCE STRUCTURES

Since $(-\infty, t)$ determines t in the causal ordering T, there is little danger of confusion. If $\mu = t$ and t is a first element of T (i.e., $t \in T_0$), then

$$\mu \frown \omega = \{\langle t, \omega \rangle\},$$

that is, the function with domain $\{t\}$ such that $\mu(t) = \omega$.

We now analyze the set theoretical structure of a chance structure $\mathbf{K}$. For this analysis, we consider $\mathbf{K}$ as a set of functions with domain T. The general idea behind a chance structure $\mathbf{K}$ is that what has happened before t determines the possibilities at t. The set $\mathbf{K}_{\lambda_t}$ is the set of possibilities at t when λ_t, which represents what has happened before t, has happened. This is expressed in the condition that for $t \notin T_0$, $\mathbf{K}_{t]}$ is obtained by concatenating the functions $\lambda_t \in \mathbf{K}_t$ with the possibilities in $\mathbf{K}_{\lambda_t}$. That is

$$\mathbf{K}_{t]} = \{\lambda \frown \omega \mid \lambda \in \mathbf{K}_t \text{ and } \omega \in \mathbf{K}_\lambda\}.$$

Hence, $\mathbf{K}_{t]}$ can be identified with the set of pairs $\{\langle \lambda_t, \lambda(t)\rangle \mid \lambda \in \mathbf{K}_{t]}\}$, by the function $f(\lambda) = \langle \lambda_t, \lambda(t)\rangle$. We shall use this identification when convenient. That is, we shall consider $\mathbf{K}_{t]}$ to be identified with

$$\bigcup \{\{\mu\} \times \mathbf{K}_\mu \mid \mu \in \mathbf{K}_t\}.$$

So, for instance, when we talk about the average measure on $\mathbf{K}_{t]}$ we shall mean the corresponding average measure on

$$\bigcup \{\{\mu\} \times \mathbf{K}_\mu \mid \mu \in \mathbf{K}_t\}.$$

For $t \notin T_0$, $\mathbf{K}_t$ can be considered as a subset of a Cartesian product. We have

$$\lambda \in \mathbf{K}_t \iff \lambda = \bigcup \{\lambda_{s]} \mid s \text{ an immediate predecessor of } t\}.$$

From now on, I shall abbreviate "s is an immediate predecessor of t" by "s i.p. t". Thus, the function $h(\lambda) = \langle \lambda_{s]} \mid s \text{ i.p. } t\rangle$ is a one to one function from $\mathbf{K}_t$ into $\prod \langle \mathbf{K}_{s]} \mid s \text{ i.p. } t\rangle$. Condition (*) in Definition XIII.5, insures that all the sequence that are theoretically possible in $\mathbf{K}_t$ are present in $\mathbf{K}$, i.e., they are really possible.

We shall use the identification h of $\mathbf{K}_t$ with the product, whenever convenient. In particular, we shall talk of the product measure on $\mathbf{K}_t$, meaning, the product measure in

$$\prod \langle \mathbf{K}_{s]} \mid s \text{ i.p. } t\rangle.$$

This identification is especially useful for strict causal structures. If $\mathbf{K}$ is a strict chance structure then the function h defined above goes onto the product

$$\prod \langle \mathbf{K}_{s]} \mid s \text{ i.p. } t\rangle.$$

That is, for any

$$j \in \prod \langle \mathbf{K}_{s]} \mid s \text{ i.p. } t\rangle,$$

there is a $\lambda \in \mathbf{K}_t$ such that

$$\lambda = \bigcup \{j(s) \mid s \text{ i.p. } t\}.$$

In order to show this, by (*), we just must prove that the $j(s)$, for s immediate predecessors of t are compatible functions. But, for s and u immediate predecessors of t with $s \neq u$, s and u are incomparable. Since T is a strict causal ordering, by Proposition XIII.3, $(-\infty, s]$ and $(-\infty, u]$ are disjoint. But this means that the domains of $j(s)$ and $j(u)$ are disjoint. Hence, $j(s)$ and $j(u)$ are compatible. By Condition (*), there is a $\lambda \in \mathbf{K}_t$ such that $\lambda_{s]} = j(s)$ for every s that is an immediate predecessor of t. Therefore, for $\mathbf{K}$ a strict chance structure, we identify $\mathbf{K}_t$, for $t \notin T_0$, with the product

$$\prod \langle \mathbf{K}_{s]} \mid s \text{ i.p. } t \rangle.$$

Similarly, since any pair of last elements of $\bigcup \{T_i \mid i \leq n\}$ is incomparable, $\mathbf{K}_n$ can be identified with the product

$$\prod \langle \mathbf{K}_{t]} \mid t \text{ is a last element of } \bigcup \{T_i \mid i \leq n\} \rangle.$$

If $\mathbf{K}$ is a strict chance structure, we can construct $\mathbf{K}$ inductively on the basis of the sets $\mathbf{K}_{\lambda_t}$ as follows. For t a first element, i.e., $t \in T_0$, we begin with $\mathbf{K}(t)$, which is $\mathbf{K}_{\lambda_t}$ and also $\mathbf{K}_{t]}$. Suppose, now, that we have constructed $\mathbf{K}_{s]}$ for all $s \in T_i$ and that $t \in T_{i+1}$. Let $S = \{s_1, \ldots, s_n\}$ be the set of immediate predecessors of t. Then, take

$$\mathbf{K}_t = \{ \bigcup \{j(s) \mid s \text{ i.p. } t\} \mid j \in \prod \langle \mathbf{K}_{s]} \mid s \text{ i.p. } t \rangle \},$$

i.e., the product of all $\mathbf{K}_{s]}$ for s an immediate predecessor of t. The set $\mathbf{K}_{t]}$, in its turn, can be obtained by taking each $\lambda \in \mathbf{K}_t$ and concatenating all the possibilities in $\mathbf{K}_\lambda$; that is, we take

$$\mathbf{K}_{t]} = \{\lambda \frown \omega \mid \omega \in \mathbf{K}_\lambda \text{ and } \lambda \in \mathbf{K}_t\}.$$

Similarly, when we have $\mathbf{K}_{t]}$ for all t that are last elements of T, we construct $\mathbf{K}$ from the product of these sets, as was done for $\mathbf{K}_t$. As we can see, only basic blocks of $\mathbf{K}$ are used in this construction. This is one of the reasons for calling them basic.

Some examples might help to clarify the definitions.

EXAMPLE XIII.1. Several independent choices from a circle.

The causal ordering is a set of n elements with an ordering in which all elements are incomparable, i.e., the causal relation is the identity, since all causal moments are independent. For $n = 4$, this ordering is pictured in #1 of Fig. XIII.10.

From now on, we shall call this ordering I_n, identifying the ordering $\langle I_n, = \rangle$ with the set I_n. The ordering I_n is of height one. At each point in I_n we put a system for the choice of a point of a circle, i.e., an element of the simple chance structure $\mathbf{K}_2$ of Example XII.3. Thus, this chance structure, call it $\mathbf{IC}$, consists of all outcomes λ with causal ordering I_n, and such that $\lambda(t) \in \mathbf{K}_2$, for every $t \in I_n$. Since all elements of I_n are first elements, $\lambda_t = t$, for every $\lambda \in \mathbf{IC}$ and $t \in I_n$. Hence, $\mathbf{IC}_{\lambda_t} = \mathbf{IC}(t)$; also, $\mathbf{IC}(t)$, the set of $\lambda(t)$ for $\lambda \in \mathbf{IC}$, is $\mathbf{K}_2$. Thus, $\mathbf{IC}$, considering the outcomes as functions, can be identified with the power $^{I_n}\mathbf{K}_2$ of n copies of $\mathbf{K}_2$. Here, all elements of I_n are both first and last elements.

3. CHANCE STRUCTURES

EXAMPLE XIII.2. Choose at random a point from a circle and then choose again a point at random from the same circle.

We call the chance structure for this setup **DC**. The causal ordering is the linearly ordered set, which we shall call L_2, of two elements, say t and s, with $t \prec s$, which is pictured as #2 in Fig. XIII.10. The ordering L_2 is of height two, with its level 0, $L_{20} = \{t\}$, and its level 1, $L_{21} = \{s\}$. At causal moments t and s we choose a point of the circle; thus, an outcome $\lambda \in \mathbf{DC}$ is a function with domain L_2, and such that $\lambda(t) \in \mathbf{K}_2$ and $\lambda(s) \in \mathbf{K}_2$.

Let us see how the construction indicated above proceeds for this case. We have that $\mathbf{DC}_{\lambda_t} = \mathbf{DC}(t) = \mathbf{K}_2$. Now, $\mathbf{DC}_s = \mathbf{DC}(t) = \mathbf{K}_2$ and $\mathbf{DC}_{s]} = \mathbf{DC}$. Since $\mathbf{DC}_{\lambda_s}$, i.e., the set of choices at s, is also $\mathbf{K}_2$, we obtain **DC** by concatenating an element of $\mathbf{K}_2$ (considered as $\mathbf{DC}_s$) with another element of $\mathbf{K}_2$ (considered now as $\mathbf{DC}_{\lambda_s}$).

EXAMPLE XIII.3. Choose an urn at random from a set of urns and then a ball at random from the chosen urn.

In this example, we have the same causal ordering, L_2, as for Example XIII.2, but the set of possibilities at s may be different for different choices at t. The chance structure for this example will be called **UB**. Choices at t, i.e., $\mathbf{UB}(t)$ (which is equal to $\mathbf{UB}_{\lambda_t}$), can be represented by systems $\mathfrak{U}_S = \langle U, S \rangle$, where U is the set of urns and S is a subset of U of one element containing the urn selected. Thus, $\mathbf{UB}(t)$ is the simple chance structure for the selection of a sample of one element form the set of urns. For $\lambda \in \mathbf{UB}$, if $\lambda(t) = \mathfrak{U}_S$ then $\mathbf{UB}_{\lambda_s}$ represents the selection of a sample of one ball from the population of balls of the urn in S, i.e., the set of systems $\mathfrak{B}(S)_V = \langle B_S, V \rangle$, where B_S contains the balls in the urn in S and V is a one element subset of this set, containing the ball selected. Thus, we have that if $\lambda(t) = \mathfrak{U}_S$, then $\lambda(s) = \mathfrak{B}(S)_V$, for a certain V.

EXAMPLE XIII.4. Selecting an urn from each of two sets of urns, mixing the balls in the urns, and then choosing a ball from the mixture.

Suppose that we have two set of urns U_1 and U_2, containing balls. We choose independently at random an urn from each set, mix thoroughly the balls from the two chosen urns, and finally choose at random a ball from the mixture. Here, the causal ordering T, pictured as # 3 in Fig. XIII.10, contains three causal moments, say t_1, t_2, and s. The moments t_1 and t_2 are incomparable, but both are before s. At t_1 we choose an urn from U_1, independently at t_2 we choose another urn from U_2, then mix the balls, and at s we choose a ball from the mixture.

Let **D** be the model for this chance setup. The moments t_1 and t_2 are first elements, so we begin to construct **D** by $\mathbf{D}(t_1)$ and $\mathbf{D}(t_2)$. This two sets are similar to $\mathbf{UB}(t)$ of the previous example. They contain systems $\langle U_i, S_i \rangle$ for $i = 1, 2$. Now, $(-\infty, s) = \{t_1, t_2\}$ and $(-\infty, s]$ is the whole causal ordering. Thus, $\mathbf{D}_s$ is the product of $\mathbf{D}(t_1)$ and $\mathbf{D}(t_2)$. If $\lambda(t_1) = \langle U_1, S_1 \rangle$ and $\lambda(t_2) = \langle U_2, S_2 \rangle$, then $\mathbf{D}_{\lambda_s}$ consists of systems $\langle B_{S_1 S_2}, V \rangle$ where the universe contains the mixture of the balls in the urns in S_1 and S_2, and V is a singleton. Hence, as usual, the outcomes λ are obtained by concatenating each element of $\mathbf{D}_s$, λ_s, with those of $\mathbf{D}_{\lambda_\bullet}$. Also, the set $\mathbf{D}_s$ can be identified with the product $\mathbf{D}_{t_1]} \times \mathbf{D}_{t_2]}$ by the function h such that, for any element $j \in \mathbf{D}_{t_1]} \times \mathbf{D}_{t_2]}$, we have

$$h(j) = \bigcup j.$$

The next two examples will not be used in the sequel and will only be described briefly. They are just mentioned here in order to make clear certain possible features of causal relations.

EXAMPLE XIII.5. This is an example that uses as causal relation # 4 of Fig. XIII.10. Thus, this is a chance structure that is not a strict chance structure, as the previous examples. Suppose that at t_1, t_2, and t_3 we choose independently an urn from a three sets of urns containing balls. Then we mix the balls in the urns chosen at t_1 and t_2 and, at s_1, choose a ball from the mixture. Similarly, we mix the balls of the urns chosen at t_2 and t_3, and, at s_2, choose a ball from this mixture. We can see that, although s_1 and s_2 are incomparable, the choices at these causal moments are not independent, since the mixtures at both s_1 and s_2 contain balls from the same urn chosen at t_2. We see from this example, that in order to equate incomparability with independence we must have a forwardly linear causal relation, which is not the case for # 4.

EXAMPLE XIII.6. This is a rather artificial example that has a causal relation that is not well-founded, and hence, strictly speaking, it does not satisfy Definition XIII.5 of chance structures. I give it here only to show the possibility of its existence. It seems that some nonstandard constructions of measures in quantum field theory provide more natural examples, but they are too complicated to include here.[3] I will not pursue this kind of example much further. The causal relation is that depicted in # 5 of Fig. XIII.10.

Suppose that at each point a, b, c, and d, there are four balls 1, 2, 3, and 4 in two urns u_1 and u_2, such that $u_1 = \{1, 3\}$ and $u_2 = \{2, 4\}$. If 1 or 3 are selected at a and c, then at the adjacent point b, u_1 is selected and then a ball from u_1, i.e., 1 or 3. If 2 or 4 are selected at a and c, the u_2 is selected at b and a ball from u_2, i.e., 2 or 4. We proceed similarly with other triples of adjacent points.

We could also give a more complicated causal relation that separates the selection of the urn from that of the balls, but this relation would also be non well-founded.

4. The group of internal invariance

I shall now introduce the group of transformations under which the probability measure should be invariant. Here, the situation is more complicated than for the simple structures that were discussed in the last chapter. The functions of the group are also automorphisms of the chance structure $\mathbf{K}$, but in the composite case, the automorphisms have to take into account the relations of dependence and independence involved in the causal relation of $\mathbf{K}$. The following considerations will indicate how to formalize this requirement.

---

[3]See [1, Chapter 7].

4. THE GROUP OF INTERNAL INVARIANCE

We now work with a fixed chance structure $\mathbf{K}$ with causal universe T. In order to define the automorphisms, it is necessary to take into account the following. In the first place, the causal relation of $\mathbf{K}$ is not the subject of chance, i.e., there is only one possibility for it. In the second place, the possibilities at a certain causal moment t are not determined by t itself, but by what has happened before t, i.e., by λ_t. For each λ_t, there is a different set of possibilities $\mathbf{K}_{\lambda_t}$. Thus, the universe of $\mathbf{K}_{\lambda_t}$ should be moved independently of the universe of $\mathbf{K}_{\mu_t}$, when λ and μ are different outcomes in $\mathbf{K}$, although these universes may contain elements in common or, even, they may be the same.

An isomorphism g of $\mathbf{K}$ can be obtained as follows from isomorphisms g_{λ_t} of the different $\mathbf{K}_{\lambda_t}$, for $\lambda \in \mathbf{K}$ and $t \in T$, plus an isomorphism $\bar{g}$ of the causal ordering T. We define

$$g(\langle \lambda_t, x \rangle) = \langle \bar{g}(t), g_{\lambda_t}(x) \rangle,$$

for $\lambda \in \mathbf{K}$ and $t \in T$. So we need the set of parts of elements λ of $\mathbf{K}$, λ_t. This is the reason for introducing in the definition of the isomorphisms, the set of parts λ_t of outcomes λ in $\mathbf{K}$, $P_\mathbf{K}$, which is defined below. We see that an isomorphism g is in fact encoding an isomorphism $\bar{g}$ of T and independent isomorphisms g_{λ_t} for each $\lambda \in \mathbf{K}$ and $t \in T$.

DEFINITION XIII.6 (ISOMORPHISMS). Let $\mathbf{K}$ and $\mathbf{H}$ be chance structures; $\mathbf{K}$ with causal universe T and object universe U. Then
  (1) $P_\mathbf{K} = \{\lambda_t \mid \lambda \in \mathbf{K} \text{ and } t \in T\}$.
  (2) g is an *isomorphism of* $\mathbf{K}$, if g is a function with domain containing $\bigcup \{\{\lambda_t\} \times |\mathbf{K}_{\lambda_t}| \mid \lambda_t \in P_\mathbf{K}\}$ and values ordered pairs, satisfying the following conditions:
     (a) If $t \in T$, $\lambda, \mu \in \mathbf{K}$, $x \in |\mathbf{K}_{\lambda_t}|$, $y \in |\mathbf{K}_{\mu_t}|$, $g(\langle \lambda_t, x \rangle) = \langle u, z \rangle$ and $g(\langle \mu_t, y \rangle) = \langle v, w \rangle$, then $u = v$.
     (b) If $t, s \in T$, $\lambda, \mu \in \mathbf{K}$, $x \in |\mathbf{K}_{\lambda_t}|$, $y \in |\mathbf{K}_{\mu_s}|$, $g(\langle \lambda_t, x \rangle) = \langle u, z \rangle$ and $g(\langle \mu_s, y \rangle) = \langle u, w \rangle$, then $t = s$.
  (3) Let g be an isomorphism of $\mathbf{K}$, then define
     (a) The one-one function $\bar{g}$ with domain T, given for $t \in T$ by $\bar{g}(t) = u$ if and only if $g(\langle \lambda_t, x \rangle) = \langle u, z \rangle$ for some $\lambda \in \mathbf{K}$, $x \in |\mathbf{K}_{\lambda_t}|$, and z.
     (b) If $\lambda \in \mathbf{K}$, define the isomorphism[4] of λ, g_λ, given by

$$g_\lambda(\langle t, x \rangle) = g(\langle \lambda_t, x \rangle)$$

for $\langle t, x \rangle \in R_0^\lambda$. We have that g_λ is an isomorphism of λ with $(g_\lambda)_0 = \bar{g}$. We also write

$$g_{\lambda_t}(x) = g(\langle \lambda_t, x \rangle).$$

  (4) If g is an isomorphism of $\mathbf{K}$ and $\lambda \in \mathbf{K}$, we define $g''\lambda = g''_\lambda \lambda$.
  (5) If g is an isomorphism of $\mathbf{K}$ and $\mathbf{A} \subseteq \mathbf{K}$, then we write

$$\mathbf{A}^g = \{g''\lambda \mid \lambda \in \mathbf{A}\}.$$

(In fact $\mathbf{A}^g = g''''\mathbf{A}$.)

---

[4]Definition XIII.4 gives the definition of isomorphism of λ, say g_λ, and of $(g_\lambda)_0$ and $g''_\lambda \lambda$.

(6) g is an *isomorphism from* **K** *onto* **H**, if g is an isomorphism of **K** and **H** = **K**^{g} (i.e., **H** = $\{g''_\lambda \lambda \mid \lambda \in \mathbf{K}\}$). g is an *automorphism of* **K**, if g is an isomorphism of **K** onto **K**.

(7) $G_\mathbf{K}$, *the group of invariance of* **K**, is the set of all automorphisms of **K**.

(8) For subsets **A**, **B** $\subseteq$ **K**, we write **A** $\sim_\mathbf{K}$ **B** for "there is a $g \in G_\mathbf{K}$ such that $\mathbf{A}^g = \mathbf{B}$".

We shall prove later (Theorem XIII.8) that $G_\mathbf{K}$ is really a group. As for simple structures, we will require that the probability measure be invariant under this group.

In an isomorphism of **K**, we code an isomorphism for each of the λ's in **K**. The function g_λ is well-defined, since λ_t determines t. This is clear, if t is a first element, since then $\lambda_t = t$. If t is not a first element, then the causal relation of λ_t is $(-\infty, t)$, and by the definition of causal relation, Definition XIII.1, $(-\infty, t)$ determines t in T. Notice that all g_λ's have a common $(g_\lambda)_0$, namely $\bar{g}$. Thus the causal relation of **K** is transformed in the same way by g for all λ's in **K**.

Since $g_{\lambda_t}(x) = g(\langle\lambda, x\rangle)$, we can easily see that $g_{\lambda_t} = (g_\lambda)_t$, where $(g_\lambda)_t$ is as in Definition XIII.4.

We see that the universes of the different $\mathbf{K}_{\lambda_t}$ are moved independently by each g_{λ_t}. Thus, if the same x belongs to two different universes $|\mathbf{K}_{\lambda_t}|$ and $|\mathbf{K}_{\mu_s}|$, then $g_{\lambda_t}(x)$ may be different from $g_{\mu_s}(x)$.

Recalling the formulas in Proposition XIII.4, we have, for g an isomorphism of **K**, $\lambda \in \mathbf{K}$, and $t \in T$

$$(g''\lambda)(\bar{g}(t)) = g''_{\lambda_t}\lambda(t),$$

and

$$g''\lambda_t = (g''\lambda)_{\bar{g}(t)}.$$

In order to understand better the notion of isomorphic chance structures, we shall state the next characterization of isomorphism between two chance structures:

PROPOSITION XIII.6. *Let* **K** *and* **H** *be chance structures with causal ordering* $\langle T, R\rangle$ *and* $\langle S, C\rangle$ *respectively, and let g be an isomorphism from* **K** *onto* **H**. *Then*

(1) $\bar{g}$ *is an isomorphism from* $\langle T, R\rangle$ *onto* $\langle S, C\rangle$.

(2) g'' *is a one to one function from* **K** *onto* **H** *such that for any $t \in T$, and λ, $\mu \in \mathbf{K}$, $\lambda_t = \mu_t$ implies $(g''\lambda)_{\bar{g}(t)} = (g''\mu)_{\bar{g}(t)}$, and $\lambda_{t]} = \mu_{t]}$ implies $(g''\lambda)_{\bar{g}(t)]} = (g''\mu)_{\bar{g}(t)]}$.*

(3) *For $\lambda \in \mathbf{K}$ and $t \in T$, g_{λ_t} is an isomorphism of the simple chance structure* $\mathbf{K}_{\lambda_t}$ *onto the simple chance structure* $\mathbf{H}_{g''(\lambda)_{\bar{g}(t)}}$, *satisfying If $\omega \in \mathbf{K}_{\lambda_t}$, then $g''_{\lambda_t}\omega = (g''\mu)(\bar{g}(t))$ for some $\mu \in \mathbf{K}$ with $\mu_t = \lambda_t$ and $\mu(t) = \omega$.*

(4) *If we define the functions g_t on* $\mathbf{K}_t$, *and $g_{t]}$ on* $\mathbf{K}_{t]}$ *by $g_t(\lambda_t) = (g''\lambda)_{\bar{g}(t)}$ and $g_{t]}(\lambda_{t]}) = (g''\lambda)_{\bar{g}(t)]}$ then*

$$g_{t]}(\lambda_{t]}) = g_t(\lambda_t) \frown g''_{\lambda_t}(\lambda(t))$$

$$g_t(\lambda_t) = \bigcup\{g_{s]}(\lambda_{s]}) \mid s \text{ i.p. } t\}$$

4. THE GROUP OF INTERNAL INVARIANCE   311

$$g''\lambda = \bigcup \{g_{t]}(\lambda_{t]}) \mid t \text{ a last element of } T\}.$$

The proof is easily obtained from the definitions.

THEOREM XIII.7. *Let* **K** *and* **H** *be chance structures with causal ordering* $\langle T, R \rangle$ *and* $\langle S, C \rangle$, *respectively. Then if there are functions h and f satisfying the conditions:*

(1) *h is an isomorphism of* $\langle T, R \rangle$ *onto* $\langle S, C \rangle$.
(2) *f is a one-one function from* **K** *onto* **H** *such that for* $\lambda, \mu \in \mathbf{K}$ *and* $t \in T$, $\lambda_t = \mu_t$ *implies* $f(\lambda)_{h(t)} = f(\mu)_{h(t)}$, *and* $\lambda_{t]} = \mu_{t]}$ *implies* $f(\lambda)_{h(t)]} = f(\mu)_{h(t)]}$.
(3) *If we define the function* f_{λ_t} *for* $\omega \in \mathbf{K}_{\lambda_t}$ *by* $f_{\lambda_t}(\omega) = f(\mu)(h(t))$, *where* $\mu \in \mathbf{K}$, *with* $\lambda_t = \mu_t$ *and* $\mu(t) = \omega$, *then* $f_{\lambda_t} = j''_{\lambda_t}$ *for some* j_{λ_t} *which is an isomorphism of* $\mathbf{K}_{\lambda_t}$ *onto* $\mathbf{H}_{f(\lambda)_{h(t)}}$ (*as simple chance structures*).

Then the function g defined by $g(\langle \lambda_t, x \rangle) = \langle h(t), f_{\lambda_t}(x) \rangle$, *for* $\lambda_t \in P_\mathbf{K}$ *and* $x \in |\mathbf{K}_{\lambda_t}|$, *is an isomorphism from* **K** *onto* **H** *such that* $f(\lambda) = g''\lambda$ *for every* $\lambda \in \mathbf{K}$.

PROOF. We show by induction on T, that for $\lambda \in \mathbf{K}$ and $t \in T$, $g''\lambda_t = f(\lambda)_{h(t)} \in P_\mathbf{H}$, and that g'' is onto **H**. Assume that $g''\lambda_s \in P_\mathbf{H}$ for s i.p. t. Then, by the definition of g, $g''\lambda_{s]} = f(\lambda)_{h(s)]}$, and hence, $g''\lambda_t = f(\lambda)_{h(t)}$. On the other hand, if $\mu_{h(t)} \in P_\mathbf{H}$, then $\mu_{h(t)} = f(\lambda)_{h(t)}$, for some $\lambda \in \mathbf{K}$, and hence, $\mu_{h(t)} = g''\lambda_t$. To complete the proof that $g''\lambda \in \mathbf{H}$ and g'' is onto **H**, we proceed in a similar way. □

For the automorphisms g in $G_\mathbf{K}$, these two theorems are valid with $\mathbf{H} = \mathbf{K}$. Thus, we see that we can construct g from $\bar{g}$ and the g_{λ_t}'s.

We now prove that $G_\mathbf{K}$ is a group.

THEOREM XIII.8. *Let* **K** *be a chance structure. Then the set* $\{g'' \mid g \in G_\mathbf{K}\}$ *is a group of permutations of* **K**.

Here g'' is really g'''' the image function of g'', i.e., we are talking of the functions such that for each subset **A** of **K**, $g''\mathbf{A} = \{g''\lambda \mid \lambda \in \mathbf{A}\} = \mathbf{A}^g$. We shall use the same notation, namely $G_\mathbf{K}$, for the original set given in Definition XIII.6 and for $\{g'' \mid g \in G_\mathbf{K}\}$.

PROOF. Let $g, h \in G_\mathbf{K}$. Assume that $\langle T, \preceq \rangle$ is the causal ordering of **K**. Define $g \circ h$ by

$$g \circ h(\langle \lambda_t, x \rangle) = \langle \bar{g}(h_0(t)), g_{h''\lambda_{h_0(t)}}(x) \rangle.$$

It is easy to see that $g \circ h \in G_\mathbf{K}$, and $g'' \circ h'' = (g \circ h)''$.

Let, now, $g \in G_\mathbf{K}$. Then g is onto **K**. Hence, if $\lambda \in \mathbf{K}$, then there is a $\mu \in \mathbf{K}$ such that $g''\mu = \lambda$. Also, $\bar{g}^{-1}$ is a permutation of T. Hence, if $a \preceq b$, then $a = \bar{g}(a')$ and $b = \bar{g}(b')$ for certain $a', b' \in T$ such that $a' \preceq b'$. Thus, $\bar{g}^{-1}$ is an automorphism of $\preceq$. Let g^{-1} be defined by

$$g^{-1}(\langle \lambda_t, x \rangle) = \langle s, g_{\mu_s}(x) \rangle,$$

where $s = \bar{g}^{-1}(t)$. It is easy to prove that $g^{-1} \in G_\mathbf{K}$ and $g''^{-1} = (g^{-1})''$. □

The definition of classical chance structure is exactly the same as Definition XII.7 for simple chance structures, namely:

DEFINITION XIII.7 (CLASSICAL STRUCTURES). A chance structure **K** is *classical*, if for every $\lambda, \mu \in \mathbf{K}$, $\{\lambda\} \sim_\mathbf{K} \{\mu\}$.

An analysis of the groups for the examples discussed in the last section might help to get a hold of the group of invariance:

Example XIII.1. Several (say n) independent choices from a circle.

Let $g \in G_{\mathbf{IC}}$. Any permutation of I_n is an automorphism of the causal ordering $\langle I_n, = \rangle$. Thus, we can take as $\bar{g}$ any permutation of I_n. Since **IC** ($= \mathbf{IC}_{\lambda_t}$) and $\mathbf{IC}(\bar{g}(t))$ are copies of the simple chance structure of choosing a point from a circle, $\mathbf{K}_2$, g_t (which is also $(g_\lambda)_t$) must be an automorphism of $\mathbf{K}_2$, i.e., a rotation of the circle (or, what is the same, of the hyperfinite approximation of the circle). For $t \neq s$, g_t may be a different rotation from g_s. A pair $\langle t, x \rangle \in I_n \times C$ is mapped by g into the pair $\langle \bar{g}(t), g_t(x) \rangle$ (i.e., $g(\langle t, x \rangle) = \langle \bar{g}(t), g_t(x) \rangle$). Hence a choice at t, i.e., x, is transformed into a choice at $\bar{g}(t)$ rotated by g_t.

We can see that **IC** is classical as follows. Let $x_1, \ldots, x_n$, and $y_1, \ldots, y_n$ be the points in the circle chosen at $1, \ldots, n$ in outcomes λ and μ, respectively. Take $\bar{g}$ equal to the identity and g_i, for $i = 1, \ldots, n$, to be the rotation from x_i to y_i. Then we have that $g''\lambda = \mu$.

Example XIII.2. Choose at random a point from a circle and then choose again a point at random for the same circle.

Consider a function g of the group of invariance $G_{\mathbf{DC}}$. Since L_2 is a linear ordering, its only automorphism is the trivial one, the identity. Thus, $\bar{g}$ is the identity on L_2. For t the first element of L_2, g_t, which is g_{λ_t}, has to be a isomorphism of $\mathbf{DC}(t)$ onto $\mathbf{DC}(\bar{g}(t)) = \mathbf{DC}(t)$, i.e., a rotation of the circle. The function g_{λ_s}, in its turn, is an isomorphism of $\mathbf{DC}_{\lambda_s}$, which is $\mathbf{K}_2$, onto $\mathbf{DC}_{(g''\lambda)_s}$, which is also $\mathbf{K}_2$. Hence, g_{λ_s} is also a rotation of the circle, which may be different from g_t, and also different for different λ_s's, i.e., for different choices at t. Notice that λ_s is determined by the choice of the point of the circle at t, and can be identified with $\lambda(t)$. Therefore, a pair $\langle t, x \rangle$ is mapped into a point of the same copy of the circle rotated by g_t, i.e., to $\langle t, g_t(x) \rangle$. A pair $\langle s, y \rangle$ is mapped into $\langle s, g_{\lambda_s}(y) \rangle$. The rotation g_{λ_s} depends on the choice at t, i.e., if $\lambda(t) \neq \mu(t)$, then $g_{\lambda_s}(y)$ may be different from $g_{\mu_s}(y)$.

The structure **DC** is also classical: let x_1, x_2, and y_1, y_2 be the points of the circle chosen at t and s in outcomes λ and μ. Let $\bar{g}$ be the identity, g_t, the rotation from x_1 to y_1, and g_{λ_s}, the rotation from x_2 to y_2. Then $g''\lambda = \mu$.

Example XIII.3. Choose an urn at random from a set of urns and then a ball at random from the chosen urn.

We have the same causal ordering L_2 as in the previous example. Hence, if g is an automorphism of **UB**, $\bar{g}$ is the identity. The function g_t should be an automorphism of the simple chance structure $\mathbf{UB}(t)$, i.e., g_t is a permutation of the set of urns, U. Suppose that $\lambda(t) = \mathfrak{U}_S$ and $g''\lambda(t) = \mathfrak{U}_{S_1}$. Then $S = \{u\}$ and $S_1 = \{g_t(u)\}$, where u is an urn, and, hence, $g_t(u)$ is also an urn. Now, g_{λ_s} should be an isomorphism of $\mathbf{UB}_{\lambda_s}$ onto $\mathbf{UB}_{g(\lambda)_s}$. The set $\mathbf{UB}_{\lambda_s}$ contains the

4. THE GROUP OF INTERNAL INVARIANCE

selections from the urn u and $\mathbf{UB}_{g(\lambda)_s}$, the selections from the urn $g_t(u)$. There is such isomorphism if and only if these two urns contain the same number of balls. Thus, u and $g_t(u)$ must have the same number of balls. Therefore, the family $G_{\mathbf{UB}}$ consists of functions g such that g_t sends urns into urns with the same number of balls. We see that if there are no two urns with the same number of balls, then g_t must also be the identity, and g_{λ_s}, a permutation of the balls in the urn selected at t. So for this example, we cannot recover the family $G_{\mathbf{UB}(t)}$ from $G_{\mathbf{UB}}$. Since the probability measure should reflect the fact that the urns are chosen at random, i.e., that the choosing of each urn is transformable into the choosing of any other urn, and this transformation might not be possible by a member of $G_{\mathbf{UB}}$, we need also to take into account the family $G_{\mathbf{UB}(t)}$ and require that the measure, when restricted to the choosing of an urn, be invariant under $G_{\mathbf{U}(t)}$.

From the last considerations, it is clear that $\mathbf{UB}$, possibly, is not classical. For instance, if $\mathbf{UB}(t)$ represents the choosing from a set of two urns with different numbers of balls, then an outcome where one urn is chosen can never be transformed into an outcome where the other urn is chosen. An example is Hosiasson's urn model of Fig. III.5.

Example XIII.4. Selecting an urn from each of two sets of urns, mixing the balls in the urns, and then choosing a ball from the mixture.

Besides the trivial automorphism, the causal ordering for $\mathbf{D}$ admits the automorphism that sends t_1 into t_2 and t_2 into t_1. The causal moment s has to remain fixed. Suppose that g is a function of our group with $\bar{g}(t_1) = t_2$. Now, g_{t_1} has to be an isomorphism of $\mathbf{D}(t_1)$ onto $\mathbf{D}(t_2)$. Such an isomorphism exists if and only if the universe of $\mathbf{D}(t_1)$ contains the same number of urns as the universe of $\mathbf{D}(t_2)$. If the number of urns is different, then there is no function g in $G_{\mathbf{D}}$ with $\bar{g}(t_1) = t_2$. That is, in the case indicated, the only function allowable as $\bar{g}$ is the identity function. Hence, just as for the previous example, it is possible that $\mathbf{D}$ is not classical.

Example XIII.6. Only a brief discussion of the group of invariance for this example. The automorphisms of the causal relation are the cyclic permutations of a, b, c, and d. It is not difficult to see that this chance structure is classical.

We return, now, to the general discussion. Besides the family $G_{\mathbf{K}}$, we need, for our measure, the families $G_{\mathbf{K}_t}$, $G_{\mathbf{K}_{t]}}$, and $G_{\mathbf{K}_{\lambda_t}}$ for $t \in T$ and $\lambda \in \mathbf{K}$. Notice that $\mathbf{K}_{\lambda_t}$ is a simple chance structure and the corresponding group $G_{\mathbf{K}_{\lambda_t}}$ is defined according to Definition XII.2, (4). As it is apparent from certain of the examples, $G_{\mathbf{K}_{\lambda_t}}$ and the groups for the other basic blocks, $\mathbf{K}_t$ and $\mathbf{K}_{t]}$ are not, in general, recoverable from $G_{\mathbf{K}}$, and we should require our measure restricted to the basic blocks to be invariant under their groups of invariance.

In fact, the only element that we need to obtain from the groups of invariance are the corresponding equivalence relation $\sim_{\mathbf{J}}$, for $\mathbf{J}$ a basic block of $\mathbf{K}$. I shall show in the sequel that, for this purpose, the groups $G_{\mathbf{K}_n}$ suffice, that is, the equivalence relations $\sim_{\mathbf{K}_n}$ suffice, for $n <$ height of T. We first introduce the following notation:

DEFINITION XIII.8. Let $\mathbf{A} \subseteq \mathbf{K} \upharpoonright S$, with $S \subseteq \bigcup \{T_i \mid i \leq n\}$. Then

$$\mathbf{A}_n^\circ = \{\lambda \in \mathbf{K}_n \mid \lambda \upharpoonright S \in \mathbf{A}\}.$$

When n is the height of T, we simply write $\mathbf{A}^\circ$ for $\mathbf{A}_n^\circ$.

We have the following theorem, which shows that the groups of automorphisms of $\mathbf{K}_m$, for m less than the height of T, are enough to determine the groups of the other basic blocks.

THEOREM XIII.9. *Let $\mathbf{K}$ be a strict chance structure, and $\mathbf{A}, \mathbf{B} \subseteq \mathbf{K} \upharpoonright S$, where $S \subseteq \bigcup\{T_i \mid i \leq n\}$, and $\mathbf{K} \upharpoonright S = \mathbf{K}_t$, $\mathbf{K} \upharpoonright S = \mathbf{K}_{t]}$, or $\mathbf{K} \upharpoonright S = \mathbf{K}_m$, with $m < n$. Then*

(1) *If $\mathbf{A}_n^\circ \sim_{\mathbf{K}_n} \mathbf{B}_n^\circ$, then $\mathbf{A} \sim_{\mathbf{K} \upharpoonright S} \mathbf{B}$.*
(2) *If t is a last element of $\bigcup\{T_i \mid i \leq n\}$ and $\mathbf{A}, \mathbf{B} \subseteq \mathbf{K}_{t]}$, then $\mathbf{A} \sim_{\mathbf{K}_{t]}} \mathbf{B}$ implies $\mathbf{A}_n^\circ \sim_{\mathbf{K}_n} \mathbf{B}_n^\circ$.*
(3) *If t is a last element of $\bigcup\{T_i \mid i \leq n+1\}$ and $\mathbf{A}, \mathbf{B} \subseteq \mathbf{K}_t$, then $\mathbf{A} \sim_{\mathbf{K}_t} \mathbf{B}$ implies $\mathbf{A}_n^\circ \sim_{\mathbf{K}_n} \mathbf{B}_n^\circ$.*

PROOF OF (1). The proof is the similar for $\mathbf{K}_t$, $\mathbf{K}_{t]}$, or $\mathbf{K}_m$, so we assume that $\mathbf{K} \upharpoonright S = \mathbf{K}_{t]}$. Suppose, first, that $\mathbf{A}_n^\circ \sim_{\mathbf{K}_n} \mathbf{B}_n^\circ$. Then there is a $g \in G_{\mathbf{K}_n}$ such that $(\mathbf{A}_n^\circ)^g = \mathbf{B}_n^\circ$. We consider two cases:

(1) $g(t) \neq t$. Let $\bar{g}(t) = s$ with $s \neq t$. Since s and t are at the same level of T, the causal moment s must be incomparable with t, and, hence, $(-\infty, t] \cap (-\infty, s] = \emptyset$ and $\bar{g}$ is an isomorphisms of $(-\infty, t]$ onto $(-\infty, s]$. Thus, we have, $(\mathbf{K}_{t]})^g = \mathbf{K}_{s]}$. We claim, first, that $\mathbf{A}^g \subseteq \mathbf{K}_{s]}$. The proof of this first claim is as follows: let $\mu \in \mathbf{A}^g$. Hence, $\mu = g(\nu)$ for a $\nu \in \mathbf{A}$. Thus, $\nu \in \mathbf{K}_{t]}$, and, therefore, $\mu = g(\nu) \in \mathbf{K}_{s]}$.
The second claim is $(\mathbf{B}_n^\circ)_{s]} = \mathbf{K}_{s]}$, whose proof is as follows. It is clear that $(\mathbf{B}_n^\circ)_{s]} \subseteq \mathbf{K}_{s]}$. On the other hand, let $\lambda \in \mathbf{K}_{s]}$. Choose any $\mu \in \mathbf{B}$. Then, since the domains of λ and μ are disjoint, $\lambda \cup \mu \subseteq \nu \in \mathbf{K}_n$, and $\nu_{s]} = \lambda$. Thus, $\lambda \in (\mathbf{B}_n^\circ)_{s]}$.
Since g is onto $\mathbf{B}_n^\circ$, we also claim that $\mathbf{A}^g = \mathbf{K}_{s]}$. This third claim has the following proof: suppose that $\lambda \in \mathbf{K}_{s]}$. Then, by the second claim, $\lambda \in (\mathbf{B}_n^\circ)_{s]}$, that is, there is a $\mu \in \mathbf{B}_n^\circ$ such that $\mu_{s]} = \lambda$. Thus, $\mu = g(\nu)$, with $\nu \in \mathbf{A}_n^\circ$. But, $g(\nu_{t]}) = \mu_{s]} = \lambda$. Therefore, $\lambda \in \mathbf{A}^g$.
From the third claim we obtain: $\mathbf{A} = \mathbf{K}_{t]}$. Similarly, taking g^{-1}, we have that $g^{-1}(t) \neq t$, and so by a symmetric argument, $\mathbf{B} = \mathbf{K}_{t]}$. Thus, $\mathbf{A}$ and $\mathbf{B}$ are equal, and, hence, $\mathbf{A} \sim_{\mathbf{K}_{t]}} \mathbf{B}$.
(2) $g(t) = t$. The proof of $\mathbf{A}^g = \mathbf{B}$ is easy in this case, and we leave the details to the reader.

□

PROOF OF (2). We again consider the case $\mathbf{K} \upharpoonright S = \mathbf{K}_{t]}$, the other case being similar. Suppose that $h \in G_{\mathbf{K}_{t]}}$ with $\mathbf{A}^h = \mathbf{B}$. Define $g \in G_{\mathbf{K}_n}$ as follows:

$$\bar{g}(s) = \begin{cases} \bar{h}(s), & \text{if } s \preceq t, \\ s, & \text{otherwise}, \end{cases}$$

$$g(\langle \lambda_s, x \rangle) = \begin{cases} \langle \bar{h}(s), h_{\lambda_s}(x) \rangle, & \text{if } s \preceq t, \\ \langle s, x \rangle & \text{otherwise}. \end{cases}$$

We now prove that $g \in G_{\mathbf{K}_n}$. It is clear that $\bar{g}$ is an automorphism of $\bigcup \{T_i \mid i \leq n\}$. We also have that $g_{\lambda_s} = h_{\lambda_s}$, if $s \preceq t$, and g_{λ_s} is the identity, otherwise. We only need to prove that $g''\lambda \in \mathbf{K}_n$ for every $\lambda \in \mathbf{K}_n$, and that, if $\mu \in \mathbf{K}_n$, then there is a $\lambda \in \mathbf{K}_n$ such that $\mu = g''\lambda$.

Suppose that $\lambda \in \mathbf{K}_n$. We have that $g''\lambda_{t]} = h''\lambda_{t]} \in \mathbf{K}_{t]}$. If s is a last element of $\bigcup \{T_i \mid i \leq n\}$, $t \neq s$, then t is incomparable with s (because t is also a last element). Hence, $g''\lambda_{s]} = \lambda_{s]} \in \mathbf{K}_{s]}$. Then there is a $\mu \in \mathbf{K}$ such that $\mu_{s]} = g''\lambda_{s]}$ for every last element of $\bigcup \{T_i \mid i \leq n\}$, s. Thus, $\mu_n \in \mathbf{K}_n$ and $g''\lambda = \mu_n$.

Suppose, now, that $\mu \in \mathbf{K}_n$. Define $\lambda_{t]} = h^{-1''}\mu_{t]}$ and $\lambda_{s]} = \mu_{s]}$, for s a last element of $\bigcup \{T_i \mid i \leq n\}$ different from t. Then $\lambda_{s]} \in \mathbf{K}_{s]}$ for every last element s. As above, there is a $\lambda \in \mathbf{K}$ that extends the $\lambda_{s]}$, and, hence, $g''\lambda_n = \mu$.

The proof of (3) is similar. □

We also have that, if $\lambda \in \mathbf{K}$, then $\sim_{\mathbf{K}_{\lambda_t}}$ also can be obtained from $\sim_{\mathbf{K}_n}$:

THEOREM XIII.10. *Let $\lambda \in \mathbf{K}$ and $t \in T$, and let $\mathbf{A}, \mathbf{B} \subseteq \mathbf{K}_{\lambda_t}$. Then $\mathbf{A} \sim_{\mathbf{K}_{\lambda_t}} \mathbf{B}$ if and only if $\{\lambda_t\} \times \mathbf{A} \sim_{\mathbf{K}_{t]}} \{\lambda_t\} \times \mathbf{B}$.*

PROOF. Suppose that $(\{\lambda_t\} \times \mathbf{A})^g = \{\lambda_t\} \times \mathbf{B}$ with $g \in G_{\mathbf{K}_{t]}}$. Then, if $h = g_{\lambda_t}$, $\mathbf{A}^h = \mathbf{B}$ and $h \in G_{\mathbf{K}_{\lambda_t}}$.

Suppose, now, that $\mathbf{A}^h = \mathbf{B}$ with $h \in G_{\mathbf{K}_{\lambda_t}}$. Then if we define g by taking $\bar{g}$ to be the identity, and

$$g_{\mu_s} = \begin{cases} h_{\lambda_t}, & \text{if } \mu_s = \lambda_t, \\ \mu_s, & \text{otherwise}, \end{cases}$$

then $g \in G_{\mathbf{K}_{t]}}$ and $(\{\lambda_t\} \times \mathbf{A})^g = \{\lambda_t\} \times \mathbf{B}$. □

5. Classical chance structures

In order to obtain the probability measure and field of events determined by chance structures, we must study more carefully some special types of structures. Although we shall study in the next chapter the measure starting from equiprobability structures, just as was done for simple structures, it is interesting to analyze classical structure in the setting of groups of transformations.

We already introduced in the previous section classical chance structures. Recall that they are chance structures $\mathbf{K}$ where outcomes can be transformed into each other by functions of the group of invariance $G_\mathbf{K}$. We shall require that the probability measure be invariant under $G_\mathbf{K}$. This means that for classical structures all outcomes will be equiprobable.

The next theorem gives conditions which imply that structures are classical.

THEOREM XIII.11. *Let $\mathbf{K}$ be a chance structure. Then any of the two following equivalent conditions imply that $\mathbf{K}$ is classical :*

(1) *For every λ, $\mu \in \mathbf{K}$ and $t \in T$, $\mathbf{K}_{\lambda_t}$ is classical and $\mathbf{K}_{\lambda_t}$ is isomorphic to $\mathbf{K}_{\mu_t}$.*
(2) *For every λ, $\mu \in \mathbf{K}$ and $t \in T$ there is an isomorphism f from $\mathbf{K}_{\lambda_t}$ onto $\mathbf{K}_{\mu_t}$ such that $f''\lambda(t) = \mu(t)$.*

PROOF. We first prove that (1) implies (2). Assume (1), and let $h : \mathbf{K}_{\lambda_t} \to \mathbf{K}_{\mu_t}$ and $\mu(t) = h(\omega)$. Since $\mathbf{K}_{\lambda_t}$ is classical, there is a $g \in G_{\mathbf{K}_{\lambda_t}}$ such that $g(\lambda(t)) = \omega$. We take $f = h \circ g$, and obtain (2).

In order to complete the proof, we assume (2) and prove that $\mathbf{K}$ is classical. Let T and U be the causal and object universes of $\mathbf{K}$, and let λ, $\mu \in \mathbf{K}$. For each $t \in T$, let f_t be an isomorphism of $\mathbf{K}_{\lambda_t}$ onto $\mathbf{K}_{\mu_t}$ such that $f_t''\lambda(t) = \mu(t)$. We define a function g on $P_{\mathbf{K}} \times U$, for which we shall prove that it is in $G_{\mathbf{K}}$ and that $g''\lambda = \mu$. Let $\bar{g}$ be the identity on T. For $\eta \in \mathbf{K}$, $t \in T$ (i.e., $\eta_t \in P_{\mathbf{K}}$), and $x \in U$, define $g(\langle\eta_t, x\rangle)$, by recursion on t. First set

$$g(\langle\eta_t, x\rangle) = \begin{cases} \langle t, f_t(x)\rangle, & \text{if } \eta_t = \lambda_t \text{ and } x \in |\mathbf{K}_{\lambda_t}|, \\ \langle t, f_t^{-1}(x)\rangle, & \text{if } \eta_t = \mu_t \text{ and } x \in |\mathbf{K}_{\mu_t}|. \end{cases}$$

For the case when $\eta_t \neq \lambda_t, \mu_t$, we proceed in the following way. Suppose that $g(\langle\eta_s, x\rangle)$ has been defined for every s immediate predecessor of t. Let $g_{\eta_t}(\langle s, x\rangle) = g(\langle\eta_s, x\rangle)$ for $s \preceq t$ (where $\preceq$ is the causal ordering). Assume that $g''_{\eta_t}\eta_t = \xi_t \in P_{\mathbf{K}}$. (In case $\eta_t = \lambda_t$ or μ_t this is obviously true, in the other cases, we shall continue the proof by induction.) By assumption, there is an isomorphism h from $\mathbf{K}_{\eta_t}$ onto $\mathbf{K}_{\xi_t}$. Define

$$g(\langle\eta_t, x\rangle) = \langle t, h(x)\rangle,$$

for $x \in \mathbf{K}_{\eta_t}$. If u is an immediate successor of t, then

$$(g_{\eta_u})''_t\eta(t) \in \mathbf{K}_{\xi_t},$$

and

$$g''_{\eta_u}\eta_t = g''_{\eta_t}\eta_t = \xi_t,$$

so

$$g''_{\eta_u}\eta_t \cup ((g_{\eta_u})_t)''\eta(t) \in P_{\mathbf{K}}.$$

By Condition (*) of Definition XIII.5 of chance structure

$$(g_{\eta_u})''_t\eta_u = \bigcup\{g''_{\eta_u}\eta_t \cup (g''_{\eta_u})''_t\eta(t) \mid t \text{ i.p. } u\} \in P_{\mathbf{K}}.$$

The same argument proves that $g''_\eta\eta \in \mathbf{K}$ and $g''_\lambda\lambda = \mu$. The proof that g is onto $\mathbf{K}$ is similar. □

On the other hand, we also have the following:

THEOREM XIII.12. *Let $\mathbf{K}$ be a classical chance structure with causal universe T. Then for any $\lambda \in \mathbf{K}$ and $t \in T$, $\mathbf{K}_{\lambda_t}$ is classical (as a simple chance structure).*

5. CLASSICAL STRUCTURES

PROOF. Let $\omega, \omega' \in \mathbf{K}_{\lambda_t}$. Then there are α and $\beta \in \mathbf{K}$ such that $\alpha_t = \beta_t = \lambda_t$, $\alpha(t) = \omega$ and $\beta(t) = \omega'$. So there is a $g \in G_\mathbf{K}$ such that $g''\alpha = \beta$. We have that g_{λ_t} is an automorphism of $\mathbf{K}_{\lambda_t}$, that is, $g_{\lambda_t} \in G_{\mathbf{K}_{\lambda_t}}$. Also $g''_{\lambda_t}\omega = \omega'$. So $\mathbf{K}_{\lambda_t}$ is classical. □

From these theorems, we obtain that if at each level of T we put the same simple chance structure, the chance structure thus obtained is classical. That is, if $\mathbf{K}$ is such that $\mathbf{K}_{\lambda_t} = \mathbf{K}_{\mu_s}$, if t and s have the same level, then $\mathbf{K}$ is classical. This type of classical structures, which will formally introduced in the next chapter with the name super classical, plays an important technical role in the construction of the measure that will be accomplished in the next chapter.

CHAPTER XIV

Equiprobability structures

As for simple structures, we can define the probability measure and the family of events determined by equiprobability structures.[1] For the definition of the measure and the family of events, we do not need the relational system structure in the simple structures $\mathbf{K}_{\lambda_t}$. These relational systems were only useful to determine the groups of invariance of the basic blocks.

The situation for the general case is, however, more complicated than for simple chance structures. Although we do not need the relational systems attached to each node of the causal relation, we do need the causal relation itself. We also require not just one equiprobability relation determined by a group of permutations, but a family of groups of permutations. We actually need a family that corresponds to the family $G_{\mathbf{K}_n}$, for n less than the height of T. We shall also restrict ourselves to strict causal orderings. Many of the theorems could be proved, with suitable modifications, for general causal orderings, but their statements and proofs would be considerably more complicated.

This chapter, although complicated, is rather elementary, since we prove all theorems for finite sets. As in the previous chapters, everything proved here should be extended to hyperfinite sets, with the restriction that sets and functions should always be internal sets or functions. In order to make the chapter easier to understand, I shall interweave a discussion of Hosiasson's examples which is discussed in page 71. I shall explain, in each case, how the definitions and theorems apply to this example, especially to the structures in Figs. III.5 and III.6.

The structure of the chapter is the following: in the first section, we define equiprobability structures, simplifying the notion of chance structure by considering that the possible simple partial outcomes are just points, and not relational systems. Since the groups of automorphisms are determined by the relational systems, and these are now ommitted, we must add explicitly these groups in the description of the equiprobability structures. The groups should determine the probability measures, which should be invariant under these groups.

---

[1] The construction and theorems of this section were originally published in [21].

The second section introduces, for technical reasons, a special type of equiprobability structures, which are called super classical structures. For these structures, there is a unique invariant measure, which is the counting measure, and is defined over all subsets of the structures. Several combinatorial (or better, counting) lemmas are proved for super classical structures, which culminate in the proof of Theorem XIV.9, that shows that the invariant measure has the right properties: namely, it is the product measure for independent causal moments, and the average measure, for dependent causal moments.

In the third section, the notion of homomorphism of equiprobability structures is introduced. Homomorphisms are functions between structures that preserve the functional structure and the equiprobability relations determined by the groups. It is shown, Theorem XIV.11, that for every equiprobability structure, **H**, there is a super classical structure, **K**, and a homomorphism from **K** onto **H**.

The fourth section deals with measures induced from one structure to another via a homomorphism. The induced measure of a subset **A** of **H** is the measure in **K** of the inverse image of **A**. We show that the induced measures preserve the "structure" of the equiprobability structures.

In the last section, the External Invariance Principle is discussed. As was mentioned in Chapter III, Section 4, this principle asserts that homorphic structures should assign the same measure to corresponding sets, i.e., if **B** is the inverse image of **A** by a homomorphism, then the probabilities of **B** and **A** should be the same. With this principle, one is able to assign the right probabilities in all equiprobability structures, as is shown by the main theorem of this chapter, Theorem XIV.15, which asserts, as it was announced in Chapter III, Section 4, that in case of independence, the probability measure that is determined by equiprobability structures is the product measure and, in case of dependence, it is the average measure. Thus, we obtain what is expected.

In the next chapter, we return to chance structures and apply the results of this chapter to them.

1. Equiprobability structures

We first recall the definitions of the last chapter, but now they are given for arbitrary functions and sets of functions with domain T. In order to obtain this functional structures, we consider chance structures as sets of functions with domain T, and eliminate the relational systems of the simple chance structures $\mathbf{K}_{\lambda_t}$, considering, thus, this set as just an arbitrary set.

DEFINITION XIV.1 (FUNCTIONAL STRUCTURES).

(1) A pair $\langle \mathbf{K}, \preceq_T \rangle$ is called a *functional structure* if $\langle T, \preceq_T \rangle$ is a strict causal ordering and **K** is a set of functions with domain T.

(2) Let $\langle \mathbf{K}, \preceq_T \rangle$ be a functional structure and $\lambda \in \mathbf{K}$. Let $S \subseteq T$ and $t \in T$. Then $\lambda \upharpoonright S$ is λ with its domain restricted to S, and

$$\lambda_t = \begin{cases} t, & \text{if } t \in T_0, \\ \lambda \upharpoonright \{s \in T \mid s \prec t\}, & \text{otherwise,} \end{cases}$$

1. EQUIPROBABILITY STRUCTURES

$$\lambda_{t]} = \lambda \upharpoonright \{s \in T \mid s \preceq t\},$$
$$\lambda_n = \lambda \upharpoonright \bigcup \{T_i \mid i \leq n\},$$
$$\mathbf{K} \upharpoonright S = \{\lambda \upharpoonright S \mid \lambda \in \mathbf{K}\},$$
$$\mathbf{K}_t = \{\lambda_t \mid \lambda \in \mathbf{K}\},$$
$$\mathbf{K}_{t]} = \{\lambda_{t]} \mid \lambda \in \mathbf{K}\},$$
$$\mathbf{K}_n = \{\lambda_n \mid \lambda \in \mathbf{K}\}.$$

If $\lambda \in \mathbf{K}_t$, then

$$\mathbf{K}_\lambda = \{\omega \mid \mu(t) = \omega \text{ for some } \mu \in \mathbf{K} \text{ with } \mu_t = \lambda_t\},$$

$$\lambda \frown \omega = \begin{cases} \lambda \cup \{\langle t,\omega\rangle\} & \text{if } \lambda \in \mathbf{K}_t \text{ and } t \notin T_0, \\ \{\langle t,\omega\rangle\} & \text{if } \lambda = t \in T_0. \end{cases}$$

Also, if $\mathbf{A} \subseteq \mathbf{K} \upharpoonright S$ and $S \subseteq \bigcup\{T_i \mid i \leq n\}$, then

$$\mathbf{A}_n^\circ = \{\lambda \in \mathbf{K}_n \mid \lambda \upharpoonright S \in \mathbf{A}\},$$
$$\mathbf{A}^\circ = \{\lambda \in \mathbf{K} \mid \lambda \upharpoonright S \in \mathbf{A}\}.$$

(3) A functional structure $\langle \mathbf{K}, \preceq_T \rangle$ is called *full* if we have
(*): if $\{s_1, s_2, \ldots, s_p\}$ is a set of pairwise incomparable elements of T, and $\lambda_1, \lambda_2, \ldots, \lambda_p \in \mathbf{K}$, then there is a $\mu \in \mathbf{K}$ such that $\mu_{s_i]} = (\lambda_i)_{s_i]}$ for $i = 1, \ldots, p$.

The functional structure, $\mathbf{Ho}_2$, for Hosiasson's example in Fig. III.5, page 72, is the following. The causal structure, T, which is the same for the structures of the three Hosiasson games, is L_2, i.e., it consists of two elements, say t and s, with $t \prec s$. The elements of $\mathbf{Ho}_2$ are, then, functions with domain T. Let $\lambda \in \mathbf{Ho}_2$. Then $\lambda(t)$ has values c_1, c_2, or c_3, the three counters. Let the balls in urn I be b_1, b_2, and b_3, and the balls in urn II, d_1 and d_2. We assume that b_1, b_2 and d_1 are white, and the b_3 and d_2 are black. Then, the functions in $\mathbf{Ho}_2$ are:

$$\lambda_1(u) = \begin{cases} c_1, & \text{if } u = t, \\ b_1, & \text{if } u = s, \end{cases}$$

$$\lambda_2(u) = \begin{cases} c_1, & \text{if } u = t, \\ b_2, & \text{if } u = s, \end{cases}$$

$$\lambda_3(u) = \begin{cases} c_1, & \text{if } u = t, \\ b_3, & \text{if } u = s, \end{cases}$$

$$\lambda_4(u) = \begin{cases} c_2, & \text{if } u = t, \\ b_1, & \text{if } u = s, \end{cases}$$

$$\lambda_5(u) = \begin{cases} c_2, & \text{if } u = t, \\ b_2, & \text{if } u = s, \end{cases}$$

$$\lambda_6(u) = \begin{cases} c_2, & \text{if } u = t, \\ b_3, & \text{if } u = s, \end{cases}$$

$$\lambda_7(u) = \begin{cases} c_3, & \text{if } u = t, \\ d_1, & \text{if } u = s, \end{cases}$$

$$\lambda_8(u) = \begin{cases} c_3, & \text{if } u = t, \\ d_2, & \text{if } u = s. \end{cases}$$

It is easy to check that $\mathbf{Ho_2}$ is full. Some examples of the different symbols that were defined are the following. The function $\lambda_{1s} = \lambda_{1t]} = \{\langle t, c_1 \rangle\}$. The set $(\mathbf{Ho_2})_0$ is the set of functions with domain $\{t\}$ and vaues in $\{c_1, c_2, c_3\}$, i.e.

$$(\mathbf{Ho_2})_0 = \{\{\langle t, c_1 \rangle\}, \{\langle t, c_2 \rangle\}, \{\langle t, c_3 \rangle\}\}.$$

The set $(\mathbf{Ho_2})_1 = \mathbf{Ho_2}$. Suppose that $\mathbf{A} \subseteq (\mathbf{Ho_2})_0$ is the set of functions, μ, with domain $\{t\}$ such that urn I is selected, that is $\mu(t) = c_1$ or c_2. Thus

$$\mathbf{A} = \{\{\langle t, c_1 \rangle\}, \{\langle t, c_2 \rangle\}\}.$$

Then $\mathbf{A}_0^\circ = \mathbf{A}$ and

$$\mathbf{A}^\circ = \mathbf{A}_1^\circ = \{\lambda_1, \lambda_2, \lambda_3, \lambda_4, \lambda_5, \lambda_6\}.$$

Finally, let μ be the function with domain $\{t\}$ and such that $\mu(t) = c_1$. Then $\mu \frown b_1 = \lambda_1$.

The chance structure corresponding to $\mathbf{Ho_2}$ is obtained from $\mathbf{Ho_2}$ by replacing c_i by the system $\langle \{c_1, c_2, c_3\}, \{c_i\} \rangle$, b_i, by the system $\langle \{b_1, b_2, b_3\}, \{b_i\} \rangle$, and d_j, by $\langle \{d_1, d_2\}, \{d_j\} \rangle$, for $i = 1, 2, 3$, and $j = 1, 2$.

We also need an extended notion of isomorphism.

DEFINITION XIV.2 (ISOMORPHISMS). Let $\langle \mathbf{K}, \preceq_T \rangle$ and $\langle \mathbf{H}, \preceq_S \rangle$ be functional structures. Let f be a function from $\mathbf{K}$ into $\mathbf{H}$.

(1) f is a *homomorphism (of functional structure) from* $\langle \mathbf{K}, \preceq_T \rangle$ *onto* $\langle \mathbf{H}, \preceq_S \rangle$ if f is onto $\mathbf{H}$ and there is an isomorphism $\bar{f}$ from $\langle T, \preceq_T \rangle$ onto $\langle S, \preceq_S \rangle$ such that

(•): if $\lambda, \mu \in \mathbf{K}$ and $t \in T$, then $\lambda_{t]} = \mu_{t]}$ implies $f(\lambda)_{\bar{f}(t)]} = f(\mu)_{\bar{f}(t)]}$, and $\lambda_t = \mu_t$ implies $f(\lambda)_{\bar{f}(t)} = f(\mu)_{\bar{f}(t)}$.

(2) f is an *isomorphism from* $\langle \mathbf{K}, \preceq_T \rangle$ *onto* $\langle \mathbf{H}, \preceq_S \rangle$ if f is a homomorphism from $\langle \mathbf{K}, \preceq_T \rangle$ onto $\langle \mathbf{H}, \preceq_S \rangle$, f is one-one, and f^{-1} is a homomorphism from $\langle \mathbf{H}, \preceq_S \rangle$ onto $\langle \mathbf{K}, \preceq_T \rangle$.

(3) f is an *automorphism of* $\langle \mathbf{K}, \preceq_T \rangle$ if f is an isomorphism from $\langle \mathbf{K}, \preceq_T \rangle$ onto $\langle \mathbf{K}, \preceq_T \rangle$.

1. EQUIPROBABILITY STRUCTURES

(4) Let f be a homomorphism from $\langle \mathbf{K}, \preceq_T \rangle$ onto $\langle \mathbf{H}, \preceq_S \rangle$. For $t \in T$, $\lambda \in \mathbf{K}$, and $m <$ height of T, we define the functions f_t on $\mathbf{K}_t$, $f_{t]}$ on $\mathbf{K}_{t]}$, f_{λ_t} on $\mathbf{K}_{\lambda_t}$, and f_m on $\mathbf{K}_m$, by

$$f_t(\lambda_t) = f(\lambda)_{\bar{f}(t)},$$
$$f_{t]}(\lambda_{t]}) = f(\lambda)_{\bar{f}(t)]},$$
$$f_{\lambda_t}(\mu(t)) = f(\mu)(\bar{f}(t)), \text{ where } \mu_t = \lambda_t$$
$$f_m(\lambda_m) = f(\lambda)_m.$$

(These functions are well defined, by (•), and are onto the structures $\mathbf{H}_{\bar{f}(t)}$, $\mathbf{H}_{\bar{f}(t)]}$, $\mathbf{H}_{f(\lambda)_{\bar{f}(t)}}$, and $\mathbf{H}_m$, respectively.)

When there is no danger of confusion, we say that f is a homomorphism or isomorphism (of functional structures) from $\mathbf{K}$ onto $\mathbf{H}$, meaning from $\langle \mathbf{K}, \preceq_T \rangle$ onto $\langle \mathbf{H}, \preceq_S \rangle$.

It is clear that f is an isomorphism from $\mathbf{K}$ onto $\mathbf{H}$ if and only if f^{-1} is an isomorphism from $\mathbf{H}$ onto $\mathbf{K}$.

As an instance of an automorphism of $\mathbf{Ho}_2$, let f be the function from $\mathbf{Ho}_2$ onto $\mathbf{Ho}_2$ defined as follows:

$$f(\lambda_1) = \lambda_4$$
$$f(\lambda_2) = \lambda_6$$
$$f(\lambda_3) = \lambda_5$$
$$f(\lambda_4) = \lambda_1$$
$$f(\lambda_5) = \lambda_3$$
$$f(\lambda_6) = \lambda_2$$
$$f(\lambda_7) = \lambda_7$$
$$f(\lambda_8) = \lambda_8.$$

In this case, $\bar{f}$ is the identity restricted to T. We shall check that it is a homomorphism, for a few cases. We have that $\lambda_{1s} = \lambda_{1t]} = \lambda_{2t]} = \lambda_{2s}$, $f(\lambda_1) = \lambda_4$, and $f(\lambda_2) = \lambda_5$. But, $\lambda_{4s} = \lambda_{4t]} = \lambda_{5t]} = \lambda_{5s}$, so that the conditions for homormophisms are satisfied. All the other cases are similar. The function f_t is the identity restricted to T. The function f_s is the function that sends $\{\langle t, c_1 \rangle\}$ to $\{\langle t, c_2 \rangle\}$ and vice-versa, and keeps $\{\langle t, c_3 \rangle\}$ fixed. We have that $\lambda_{1s} = \{\langle t, c_1 \rangle\}$; then $f_{\lambda_{1s}}$ maps $\{b_1, b_2, b_3\}$ into itself, and, for instance

$$f_{\lambda_{1s}}(b_2) = (f(\lambda_2))_t = b_3.$$

We return to the general case. It is easy to show that if f is a homomorphism from $\mathbf{K}$ onto $\mathbf{H}$, $\lambda \in \mathbf{K}$, $t \in T$, and m less than the height of T, then

$$f_t(\lambda_t) = \bigcup \{f_{s]}(\lambda_s) \mid s \text{ i.p } t\},$$
$$f_{t]}(\lambda_{t]}) = f_t(\lambda_t) \frown f_{\lambda_t}(\lambda(t)),$$
$$f_m(\lambda_m) = \bigcup \{f_{s]}(\lambda_s) \mid s \text{ a last element of } \bigcup \{T_i \mid i \leq m\}\}.$$

The same as for chance structures, if $\mathbf{K}$ is a full functional structure and t is not a first element of T, then $\mathbf{K}_{t]}$ is

$$\{\lambda \frown \omega \mid \lambda \in \mathbf{K}_t \text{ and } \omega \in \mathbf{K}_\lambda\},$$

and $\mathbf{K}_t$ can be identified with the product

$$\prod \langle \mathbf{K}_{s]} \mid s \text{ i.p. } t \rangle.$$

If $m <$ height of T, then $\mathbf{K}_m$ can also be identified with the product

$$\prod \langle \mathbf{K}_{s]} \mid s \text{ l.e. } \bigcup\{T_i \mid i \leq m\}\rangle.$$

Here, and in the sequel, we abbreviate "s is a last element of R" by "s l.e. R". For an example of these identifications, see Example XIII.4.

We have the following simple proposition, which will be needed later:

PROPOSITION XIV.1. *Let $\langle \mathbf{K}, \preceq_T \rangle$ and $\langle \mathbf{H}, \preceq_S \rangle$ be full functional structures, and let f be a homomorphism from $\mathbf{K}$ onto $\mathbf{H}$. Let m be less than the height of T (which is the same as the height of S), $s \in S$, and $t = \bar{f}^{-1}(s)$. Then:*

(1) *If $\lambda \in \mathbf{H}_m$, then $\nu \in f_m^{-1}(\lambda)$ if and only if*

$$\nu = \bigcup\{h(s) \mid s \text{ l.e. } \bigcup\{S_i \mid i \leq m\}\},$$

where $h(s) \in f_{t]}^{-1}(\lambda_s])$ for $s \in \bigcup\{S_i \mid i \leq m\}$ and $t = \bar{f}^{-1}(s)$; in symbols, $f_m^{-1}(\lambda) =$

$$\{\bigcup\{h(s) \mid s \text{ l.e. } \bigcup\{S_i \mid i \leq m\}\} \mid$$
$$h \in \prod \langle f_{t]}^{-1}(\lambda_s]) \mid s \text{ l.e. } \bigcup\{S_i \mid i \leq m\} \text{ and } t = \bar{f}^{-1}(s)\rangle\}.$$

and $t = \bar{f}^{-1}(s)$.

(2) *If $\lambda \in \mathbf{H}_{s]}$, then $\mu \in f_{t]}^{-1}(\lambda_s]$ if and only if $\mu = \nu \frown \omega$, where $\nu \in f_t^{-1}(\lambda_s)$ and $\omega \in f_\nu^{-1}(\lambda(s))$; that is*

$$f_{t]}^{-1}(\lambda_s]) = \bigcup\{\{\nu\} \times f_\nu^{-1}(\lambda(s)) \mid \nu \in f_t^{-1}(\lambda_s)\}$$
$$= \{\nu \frown \omega \mid \nu \in f_t^{-1}(\lambda_s) \text{ and } \omega \in f_\nu^{-1}(\lambda(s))\}.$$

PROOF OF (1). Let A be the set

$$\{\bigcup\{h(s) \mid s \text{ l.e. } \bigcup\{S_i \mid i \leq m\}\} \mid$$
$$h \in \prod \langle f_{t]}^{-1}(\lambda_s]) \mid s \text{ l.e. } \bigcup\{S_i \mid i \leq m\} \text{ and } t = \bar{f}^{-1}(s)\rangle\}.$$

Assume, first, that $\nu \in f_m^{-1}(\lambda)$. Then $f_m(\nu) = \lambda$. But

$$\lambda = \bigcup\{\lambda_s] \mid s \text{ l.e. } \bigcup\{S_i \mid i \leq m\}\}.$$

Now

$$\lambda_s] = f_m(\nu)_s] = f_{t]}(\nu_{t]}),$$

where $t = \bar{f}^{-1}(s)$; so
$$\nu_{t]} \in f_{t]}^{-1}(\lambda_{s]}).$$
Since
$$\nu = \bigcup \{\nu_{t]} \mid t \text{ l.e. } \bigcup \{T_i \mid i \leq m\}\},$$
we have that $\nu \in A$.

Assume, now, that $\nu \in A$. That is, $\nu_{t]} \in f_{t]}^{-1}(\lambda_{s]})$ for s l.e. of $\bigcup \{S_i \mid i \leq m\}$ and $t = \bar{f}^{-1}(s)$. Since $\bar{f}$ is an isomorphism, this set of t's is the set of last elements of $\bigcup \{T_i \mid i \leq m\}$. Since **K** is full, there is a $\mu \in \mathbf{K}_m$ such that $\mu_{t]} = \nu_{t]}$. Thus, $f_m(\mu) = \lambda$, and so $\mu \in f_m^{-1}(\lambda)$. But
$$\mu = \bigcup \{\nu_{t]} \mid t \text{ l.e. } \bigcup \{T_i \mid i \leq m\}\} = \nu.$$
□

PROOF OF (2). We again have that $\nu \in f_{t]}^{-1}(\lambda)$ if and only if $f_{t]}(\nu) = \lambda$. But λ can be identified with $\langle \lambda_s, \lambda(s) \rangle$, so that $f_{t]}(\nu)$ is $\langle f_{t]}(\nu)_s, f_{t]}(\nu)(s) \rangle$. That is, $f_{t]}(\nu) = \langle f_t(\nu_t), f_\nu(\nu(t)) \rangle$. Thus, if $\nu \in f_{t]}^{-1}(\lambda)$, then $\nu_t \in f_t^{-1}(\lambda_s)$ and $\nu(t) \in f_\nu^{-1}(\lambda(s))$.

On the other hand, if $\mu \in f_t^{-1}(\lambda_s)$ and $\omega \in f_\mu^{-1}(\lambda(s))$, then it is clear that $\mu \frown \omega \in f_{t]}^{-1}(\lambda)$. □

We have the following simple properties of homomorphisms.

PROPOSITION XIV.2. *Let $\langle \mathbf{K}, \preceq_T \rangle$ and $\langle \mathbf{H}, \preceq_T \rangle$ be full functional structures (with the same domain T), $t \in \bigcup \{T_i \mid i \leq m\}$, and let f be a homomorphism of $\mathbf{K}$ onto $\mathbf{H}$. Then*
  (1) *If $\mathbf{A} \subseteq \mathbf{H}_t$ and $\mathbf{B} = f_t^{-1}(\mathbf{A})$, then $\mathbf{B}_m^\circ = f_m^{-1}(\mathbf{A}_m^\circ)$.*
  (2) *If $\mathbf{A} \subseteq \mathbf{H}_{t]}$ and $\mathbf{B} = f_{t]}^{-1}(\mathbf{A})$, then $\mathbf{B}_m^\circ = f_m^{-1}(\mathbf{A}_m^\circ)$.*
  (3) *If $\mathbf{A} \subseteq \mathbf{H}_t$, $\mathbf{B} = f_t^{-1}(\mathbf{A})$, and*
$$\mathbf{A} = \{\bigcup\{j(s) \mid s \text{ i.p. } t\} \mid j \in \prod \langle \mathbf{A}_{s]} \mid s \text{ i.p. } t \rangle\},$$
then
$$\mathbf{B} = \{\bigcup\{h(s) \mid s \text{ i.p. } t\} \mid h \in \prod \langle f_{s]}^{-1}(\mathbf{A}_{s]}) \mid s \text{ i.p. } t \rangle\}.$$

PROOF OF (1). We have the following chain of equivalences
$$\lambda \in \mathbf{B}_m^\circ \iff \lambda_t \in \mathbf{B},$$
$$\iff f_t(\lambda_t) \in \mathbf{A},$$
$$\iff f(\lambda)_t \in \mathbf{A},$$
$$\iff f(\lambda) \in \mathbf{A}_m^\circ,$$
$$\iff \lambda \in f_m^{-1}(\mathbf{A}_m^\circ).$$

The proof of 2 is similar. □

PROOF OF (3). We have

$$\lambda \in \mathbf{B} \iff \lambda = f_t(\bigcup\{\alpha_s \mid s \text{ i.p. } t\}) \text{ for some } \alpha \in \prod \langle \mathbf{A}_{s]} \mid s \text{ i.p. } t\rangle,$$
$$\iff \lambda = \bigcup\{f_{s]}^{-1}(\alpha_s) \mid s \text{ i.p. } t\} \text{ for some } \alpha \in \prod \langle \mathbf{A}_{s]} \mid s \text{ i.p. } t\rangle,$$
$$\iff \lambda = \bigcup\{h(s) \mid s \text{ i.p. } t\} \text{ for some } h \in \prod \langle f_{s]}^{-1}(\mathbf{A}_{s]}) \mid s \text{ i.p. } t\rangle.$$

□

Functional structures do not determine, as chance structures do, the automorphism groups, $G_\mathbf{J}$, of their basic blocks, $\mathbf{J}$. Thus, we must add them to the description of the structure. This is done in the next definition. By Theorem XIII.9, it is enough to include the groups G_m of $\mathbf{K}_m$, for m less than the height of T. The other equivalence relations of the groups of other of basic blocks are dertermined by these.

DEFINITION XIV.3 (EQUIPROBABLITY STRUCTURES).
(1) A triple
$$\langle \mathbf{K}, \preceq_T, G \rangle,$$
is called an *equiprobability structure* if:
  (a) $\langle \mathbf{K}, \preceq_T \rangle$ is a full functional structure.
  (b) $G = \langle G_i \mid i < n \rangle$, where n is the height of T, is a system such that G_i is a group of automorphisms of $\mathbf{K}_i$, with the property that
    (**): if $g \in G_m$, $i < m$, and $g \upharpoonright \mathbf{K}_i$ is an automorphism of $\mathbf{K}_i$, then $g \upharpoonright \mathbf{K}_i \in G_i$.
(2) If $\mathbf{K}$ is a chance structure, then $\langle \mathbf{K}, \preceq, \langle G_{\mathbf{K}_i} \mid i < n \rangle \rangle$ is called the *equiprobability structure associated with* $\mathbf{K}$.

We shall also use the notation $\sim_\mathbf{K}$ for $\sim_{G_n}$, where $n+1$ is the height of T, and $\sim_i$ for $\sim_{G_i}$.

When it is clear from the context, an equiprobability structure $\langle \mathbf{K}, \preceq, G \rangle$, will be designated just by $\mathbf{K}$.

It is clear, by Theorem XIII.11, that if $\mathbf{K}$ is a chance structure then

$$\langle \mathbf{K}, \preceq, \langle G_{\mathbf{K}_i} \mid i < n \rangle \rangle$$

is an equiprobability structure.

As it was mentioned above, one of the differences between chance structures and equiprobability structures is that the relational systems in a chance structure $\mathbf{K}$ determine the groups of automorphisms $G_{\mathbf{K}_m}$, i.e., $G_{\mathbf{K}_m}$ is the set of *all* automorphisms of $\mathbf{K}_m$, while in an equiprobability structure $\langle \mathbf{H}, \preceq, G \rangle$, we have to make explicit G_m, since it is not determined by $\mathbf{H}_m$. In general, G_m cannot the group of all automorphisms of $\mathbf{H}_m$ (considered as an equiprobability structure), but only a subgroup. There is not enough structure in $\mathbf{H}$ to determine G_m. For the definition of the measure and the family of events, the groups of automorphisms, G_m, suffice, so that once they are given, we can forget about the structure that determines them. This is the reason for using equiprobability structures instead of chance structures.

1. EQUIPROBABILITY STRUCTURES

We give an example of an equiprobability structure defined on Hosiasson's example, $\mathbf{Ho_2}$, Fig. III.5. We can take G_0 as the set of the automorphisms of $\{c_1, c_2, c_3\}$ which contains the identity and

$$g(u) = \begin{cases} c_2, & \text{if } u = c_1, \\ c_1, & \text{if } u = c_2, \\ c_3, & \text{if } u = c_3. \end{cases}$$

Let H_1 be the group of all permutations of $\{b_1, b_2, b_3\}$ and H_2 the group of all permutations of $\{d_1, d_2\}$. A function f in the group G_1 is obtained from each function $h \in G_0$ and permutations $h_1 \in H_1$ and $h_2 \in H_2$, for each $\lambda \in \mathbf{Ho_2}$, as follows

$$f(\lambda)(t) = h(\lambda(t)),$$

$$f(\lambda)(s) = \begin{cases} h_1(\lambda(s)), & \text{if } \lambda(t) = c_1 \text{ or } c_2, \\ h_2(\lambda(s)), & \text{if } \lambda(t) = c_3. \end{cases}$$

Thus, $\langle \mathbf{Ho_2}, \preceq_T, \langle G_0, G_1 \rangle \rangle$ is an equiprobability structure. This is the structure that is obtained, if $\mathbf{Ho_2}$ is considered as a chance structure with relational systems instead of points.

We can notice that in the case of $\mathbf{Ho_2}$ that we have have been considering, G_1 contains all automorphisms of $\mathbf{Ho_2}$ as a functional structure, and, thus, for this case, the groups of automorphisms as functional structure coincide with the coresponding groups as chance structure. This is true, whenever the simple structures in the chance structure, which are replaced by sets to form the functional structure, are classical. For instance, in the case of $\mathbf{Ho_2}$, the simple probability structure

$$\{\langle\{b_1, b_2, b_3\}, \{b_1\}\rangle, \langle\{b_1, b_2, b_3\}, \{b_3\}\rangle\langle\{b_1, b_2, b_3\}, \{b_3\}\rangle\},$$

which is classical, is replaced by $\{b_1, b_2, b_3\}$, and all the other simple chance structures of $\mathbf{Ho_2}$ are classical.

When these simple structures are not classical, then the groups cannot be the groups of all automorphisms of the functional structure. For instance, if instead of the simple structure just mentioned we had the structure discussed as Example XII.4, then the group of invariance determined by the chance structure would not be the group of all automorphisms of the functional structure.

We have to introduce some more notation.

DEFINITION XIV.4 (EQUIPROBABILITY RELATIONS). Let $\langle \mathbf{K}, \preceq, G \rangle$ be an equiprobability structure with causal relation $\langle T, \preceq \rangle$. For $t \in T$ and λ in $\mathbf{K}$, we define the relations $\sim_t$, $\sim_{t]}$, and $\sim_{\lambda_t}$, by

(1) $\mathbf{A} \sim_{t]} \mathbf{B}$, if $\mathbf{A}, \mathbf{B} \subseteq \mathbf{K}_{t]}$ and $\mathbf{A}_m^\circ \sim_m \mathbf{B}_m^\circ$, where t is a last element of $\bigcup \{T_i \mid i \leq m\}$.

(2) $\mathbf{A} \sim_t \mathbf{B}$, if $\mathbf{A}, \mathbf{B} \subseteq \mathbf{K}_t$ and $\mathbf{A}_m^\circ \sim_m \mathbf{B}_m^\circ$, where t is a last element of $\bigcup \{T_i \mid i \leq m+1\}$.

(3) $\mathbf{A} \sim_{\lambda_t} \mathbf{B}$, if $\mathbf{A}, \mathbf{B} \subseteq \mathbf{K}_{\lambda_t}$ and $\{\lambda_t\} \times \mathbf{A} \sim_{t]} \{\lambda_t\} \times \mathbf{B}$.

For instance, in $\mathbf{Ho_2}$, let $\mathbf{B} = \{\langle t, c_1\rangle\}$ and $\mathbf{C} = \{\langle t, c_1\rangle\}$. Then these two sets are subset of $(\mathbf{Ho_2})_s$. We have that $\mathbf{B}^\circ = \{\lambda_1, \lambda_2, \lambda_3\}$ and $\mathbf{C}^\circ = \{\lambda_4, \lambda_5, \lambda_6\}$. It is clear that $\mathbf{B}^\circ \sim_1 \mathbf{C}^\circ$, and, hence, $\mathbf{B}_{s]} \sim_{s]} \mathbf{C}_{s]}$.

From Theorems XIII.11 and XIII.12, we see that if $\mathbf{K}$ is a chance structure and $G = \langle G_{\mathbf{K}_m} \mid m < \text{height of } T\rangle$, then $\sim_t = \sim_{\mathbf{K}_t}$, $\sim_{t]} = \sim_{\mathbf{K}_{t]}}$, and $\sim_{\lambda_t} = \sim_{\mathbf{K}_{\lambda_t}}$. When $\langle \mathbf{K}, \preceq, G\rangle$ is an arbitrary equiprobability structure, we shall also use the notations $\sim_{\mathbf{K}_t}$ for $\sim_t$, $\sim_{\mathbf{K}_{t]}}$ for $\sim_{t]}$, and $\sim_{\mathbf{K}_{\lambda_t}}$ for $\sim_{\lambda_t}$.

2. Measures on super classical structures

We begin by the study of the probability space determined by an equiprobability structure. We assume, as usual, that our equiprobability structures $\mathbf{K}$ are finite (or hyperfinite), that is $\mathbf{K}$ is finite and T, the causal universe of $\mathbf{K}$ is finite. When using "finite" we also include "hyperfinite".

We begin by discussing the definition of the probability measure for super classical equiprobability structures, whose definition follows, and which are only introduced for technical reasons. Many of the results of this section are also true for any classical structures. In order to simplify the proofs, however, we shall prove them first for super classical structures and generalize them later for other classical structures.

DEFINITION XIV.5 (CLASSICAL STRUCTURES). An equiprobability structure $\mathbf{K}$ is *classical*, if $\{\lambda\} \sim_{\mathbf{K}} \{\mu\}$ for any $\lambda, \mu \in \mathbf{K}$.

An equiprobability structure $\mathbf{K}$ is *super classical*, if $\mathbf{K}$ is classical and $\mathbf{K}_{\lambda_t} = \mathbf{K}_{\mu_s}$ for any $\lambda, \mu \in \mathbf{K}$ and $t, s \in T_m$, for some m less that the height of T.

The structure $\langle \mathbf{Ho_2}, \preceq_T, \langle G_0, G_1\rangle\rangle$, which we have been considering as an example, is not a classical equiprobability structure, because, for instance, $\{\lambda_1\}$ is not equiprobable with $\{\lambda_7\}$. For two outcomes to be equiprobable, they have to choose the same urn.

Hosiasson's example in Fig. III.6 of page 72 can be constructed as a super classical structure, $\mathbf{Ho_3}$, as follows. Consider the set of six balls

$$\{e_1, e_2, e_3, e_4, e_5, e_6\}.$$

Then the elements of $\mathbf{Ho_3}$ are the functions μ_{ij} with domain T, such that

$$\mu_{ij}(t) = c_i$$
$$\mu_{ij}(s) = e_j,$$

for $i = 1, 2, 3$, and $j = 1, \ldots, 6$. The group G_0 is, now, the group of all permutations of $\{c_1, c_2, c_3\}$. Let H be the group of permutations of $\{e_1, \ldots, e_6\}$. Then the group G_1 is the set of functions f on $\mathbf{Ho_3}$ such that there are $g \in G_0$ and $h_1, h_2, h_3 \in H$ satisfying

$$f(\mu)(t) = g(\mu(t))$$

2. MEASURES ON SUPER CLASSICAL STRUCTURES

$$f(\mu)(s) = \begin{cases} h_1(\mu(s)), & \text{if } g(\mu(t)) = c_1, \\ h_2(\mu(s)), & \text{if } g(\mu(t)) = c_2, \\ h_3(\mu(s)), & \text{if } g(\mu(t)) = c_3, \end{cases}$$

Then, μ_{ij} is sent to μ_{kl} if we take the function f defined from g, h_1, h_2, and h_3 with

$$g(c_i) = c_k \quad \text{and} \quad h_k(e_j) = e_l.$$

We return, now, to the general case. It is clear from Definitions XIV.3 and XIV.5, that if **K** is classical, then so are $\mathbf{K}_m$, $\mathbf{K}_t$, $\mathbf{K}_{t]}$, and $\mathbf{K}_{\lambda_t}$. The same is true for super classical.

From the remarks at the end of the previous chapter, we see that we can always obtain super classical chance structures, and, hence, super classical equiprobability structures.

The first requirements for the measure and the field of events are the same as for the simple structures of Chapter XII, namely, the measure for an equiprobability structure **K** should be invariant under the relation $\sim_\mathbf{K}$, and the field should contain the disjunctive field generated by the measure-determined subsets of **K**. This is what I have called the Internal Invariance Principle. We repeat here the definitions.

DEFINITION XIV.6 (INVARIANT MEASURES). Let the equiprobability structure **K** be given. A probability measure Pr defined on subsets of **K** is $\sim_\mathbf{K}$-*invariant* (or, simply, *invariant*), if for any **A**, **B** $\subseteq$ **K** with **A** $\sim_\mathbf{K}$ **B**, Pr **B** is defined whenever Pr **A** is defined, and Pr **B** = Pr **A**.

DEFINITION XIV.7 (MEASURE-DETERMINED). A subset **A** of an equiprobability structure **K** is **K**–*measure-determined*, if there is a unique real number r such that if Pr is a $\sim_\mathbf{K}$-invariant probability measure defined on the disjunctive algebra of subsets of **K** generated by $\{\mathbf{B} \mid \mathbf{B} \sim_\mathbf{K} \mathbf{A}\}$, then Pr **A** = r.

In the case of classical structures, the situation is very simple. Singletons are measure-determined, so the algebra of events (which should be the disjunctive algebra generated by the measure-determined sets) is the power set. The unique invariant measure is the counting measure.

DEFINITION XIV.8 (PROBABILITY SPACE). Let **K** be a classical equiprobability structure. The *probability space of* **K** is the triple $\langle \mathbf{K}, \mathcal{P}(\mathbf{K}), \mathrm{Pr}_\mathbf{K} \rangle$, where for any **A** $\subseteq$ **K**

$$\mathrm{Pr}_\mathbf{K}(\mathbf{A}) = \frac{\#\mathbf{A}}{\#\mathbf{K}}.$$

We have the obvious theorem:

THEOREM XIV.3. *Let* **K** *be a classical equiprobability structure. Then the unique* **K**-*invariant probability measure on* $\mathcal{P}(\mathbf{K})$ *is* $\mathrm{Pr}_\mathbf{K}$.

We need to introduce the following notation:

DEFINITION XIV.9. Let **K** be an equiprobability structure with causal universe T, let $S \subseteq T$, and $\mathcal{A} \subseteq \mathbf{K} \restriction S$, and let $\langle \mathbf{K}, \mathcal{A}, \Pr \rangle$ be a disjunctive probability space. Then
  (1) $\mathcal{A} \restriction S$ is the set of all $\mathbf{B} \subseteq \mathbf{K} \restriction S$ such that $\mathbf{B}^\circ \in \mathcal{A}$.
  (2) For $\mathbf{B} \in \mathcal{A} \restriction S$, define $(\Pr \restriction S)(\mathbf{B}) = \Pr \mathbf{B}^\circ$. We write $\Pr_t$ for $\Pr \restriction (-\infty, t)$, $\Pr_{t]}$ for $\Pr \restriction (-\infty, t]$, and $\Pr_m$ for $\Pr \restriction \bigcup \{T_i \mid i \leq m\}$.
  (3) If $\lambda \in \mathbf{K}$ and $t \in T$, we write $\Pr_{\lambda_t}$ for $\Pr_{\mathbf{K}_{\lambda_t}}$, the probability on the simple equiprobability structure $\mathbf{K}_{\lambda_t}$.

When **K** is classical we omit $\mathcal{A}$, since $\mathcal{A}$, for this case, is $\mathcal{P}(\mathbf{K})$.

We begin by proving certain properties of $\Pr_\mathbf{K}$ for super classical equiprobability structures. Lemmas XIV.4–XIV.8 are combinatorial in nature and enable us to calculate the number of elements of certain sets. Since, as in Definition XIV.9, the invariant probability in a super classical structure is determined by the number of elements of the set, these calculations permit us to show properties of this measure. For the rest of this section, we assume that **K** is a super classical structure with causal universe T.

LEMMA XIV.4. *Let* $\lambda, \mu \in \mathbf{K}$, $t \in T$. *Then*

$$\#(\{\lambda_t\}^\circ) = \#(\{\mu_t\}^\circ)$$

and

$$\#(\{\lambda_{t]}\}^\circ) = \#(\{\mu_{t]}\}^\circ).$$

PROOF. Define the function $f : \{\lambda_t\}^\circ \to \{\mu_t\}^\circ$, for $\alpha \in \{\lambda_t\}^\circ$, by

$$f(\alpha)(s) = \begin{cases} \mu(s), & \text{if } s \prec t, \\ \alpha(s), & \text{otherwise.} \end{cases}$$

where $s \in T$.

We, first, prove that $f(\alpha) \in \{\mu_t\}^\circ$ for every $\alpha \in \{\lambda_t\}^\circ$. By induction on T, we begin proving that $f(\alpha)_s \in (\{\mu_t\}^\circ)_s$. If $s \prec t$, then $f(\alpha)_s = \mu_s \in (\{\mu_t\}^\circ)_s$.

Suppose, now, that A is the set of immediate predecessors of s, with $s \not\prec t$, and that if $u \in A$, then $f(\alpha)_u \in (\{\mu_t\}^\circ)_u$. By Condition (a) on equiprobability structures, $f(\alpha)_u \in \Pr_\mathbf{K}$. For $u \in A$, we have $f(\alpha)(u) = \alpha(u) \in \mathbf{K}_{\alpha_u}$. Since **K** is super classical, $\mathbf{K}_{\alpha_u} = \mathbf{K}_{f(\alpha)_u}$. Thus, $f(\alpha)_{u]} \in (\{\mu_t\}^\circ)_{u]}$. Hence, by Condition (a) on equiprobability structures again, there is a $\beta \in \mathbf{K}$ such that $f(\alpha)_{u]} = \beta_{u]}$, for every $u \in A$. Therefore, $f(\alpha)_s = \beta_s \in (\{\mu_t\}^\circ)_s$. To prove that $f(\alpha) \in \{\mu_t\}^\circ$ we proceed as above starting with $f(\alpha)_s$, for s a last element of T.

It is clear that f is one-one. To prove that f is onto, let $\beta \in \{\mu_t\}^\circ$. Then we define

$$\alpha(s) = \begin{cases} \lambda(s), & \text{if } s \prec t, \\ \beta(s), & \text{otherwise,} \end{cases}$$

We have that $f(\alpha) = \beta$.

The proof for the second conclusion is similar. □

2. MEASURES ON SUPER CLASSICAL STRUCTURES

From the preceding lemma, if we take $k_t = \#(\{\mu_t\}^\circ)$ and $j_t = \#(\{\mu_{t]}\}^\circ)$, we immediately obtain the following corollary.

COROLLARY XIV.5. *There is a number k_t such that for every $\mathbf{A} \subseteq \mathbf{K}_t$ we have that*
$$\#(\mathbf{A}^\circ) = k_t \cdot \#\mathbf{A}.$$
Similarly, there is a number j_t such that for every $\mathbf{A} \subseteq \mathbf{K}_{t]}$
$$\#(\mathbf{A}) = j_t \cdot \#\mathbf{A}.$$

Hence, we have, that for $\mathbf{A} \subseteq \mathbf{K}_t$ and $\mathbf{B} \subseteq \mathbf{K}_{t]}$, $\Pr_{\mathbf{K}_t}(\mathbf{A}) = \Pr_{\mathbf{K}}(\mathbf{A}^\circ)$ and $\Pr_{\mathbf{K}_{t]}}(\mathbf{B}) = \Pr_{\mathbf{K}}(\mathbf{B}^\circ)$.

Notice that $\Pr_{\mathbf{K}_t}$ is the probability in the classical equiprobability structure $\mathbf{K}_t$, that is
$$\Pr_{\mathbf{K}_t}(\mathbf{A}) = \frac{\#\mathbf{A}}{\#\mathbf{K}_t},$$
for $\mathbf{A} \subseteq \mathbf{K}_t$. Similarly
$$\Pr_{\mathbf{K}_{t]}}(\mathbf{A}) = \frac{\#\mathbf{A}}{\#\mathbf{K}_{t]}},$$
for $\mathbf{A} \subseteq \mathbf{K}_{t]}$.

On the other hand, if $\mathbf{A} \subseteq \mathbf{K}_t$, then $\Pr_t(\mathbf{A}) = \Pr_{\mathbf{K}}(\mathbf{A}^\circ)$, and if $\mathbf{A} \subseteq \mathbf{K}_{t]}$, then $\Pr_{t]}(\mathbf{A}) = \Pr_{\mathbf{K}}(\mathbf{A}^\circ)$. So the corollary asserts that for super classical structures, $\Pr_{\mathbf{K}_t} = \Pr_t$ and $\Pr_{\mathbf{K}_{t]}} = \Pr_{t]}$.

Recall that if $t \in T - T_0$ and $m <$ height of T, then the following identifications can be made:

- $\mathbf{K}_m$, with $\prod \langle \mathbf{K}_{s]} \mid s \text{ l.e. } \bigcup \{T_i \mid i \leq m\} \rangle$,
- $\mathbf{K}_t$, with $\prod \langle \mathbf{K}_{s]} \mid s \text{ i.p. } t \rangle$, and
- $\mathbf{K}_{t]}$, with $\bigcup \{\{\lambda_t\} \times \mathbf{K}_{\lambda_t} \mid \lambda_t \in \mathbf{K}_t\}$.

Our next aim is to prove for super classical structures, that:

- $\langle \mathbf{K}_t, \Pr_t \rangle$ is the product space (Definition VI.18) of
$$\langle \langle \mathbf{K}_{s]}, \Pr_{s]} \rangle \mid s \text{ i.p. } t \rangle,$$

- $\langle \mathbf{K}_m, \Pr_m \rangle$ is the product space of
$$\langle \langle \mathbf{K}_{s]}, \Pr_{s]} \rangle \mid s \text{ l.e. } \bigcup\{T_i \mid i \leq m\} \rangle,$$

- $\langle \mathbf{K}_{t]}, \Pr_{t]} \rangle$ is the average probability space (Definition VI.17) of
$$\langle \langle \mathbf{K}_{\lambda_t}, \Pr_{\lambda_t} \rangle \mid \lambda_t \in \mathbf{K}_t \rangle$$
with respect to $\langle \mathbf{K}_t, \Pr_t \rangle$.

These results are also true for arbitrary classical structures, but we shall prove them first for super classical structures. The proof will be divided into several lemmas. For all these lemmas, we assume that $\mathbf{K}$ is a super classical structure with strict causal ordering T.

LEMMA XIV.6. *Let* $s_1, s_2, \ldots, s_p \in T_n$ *for some* n, $\lambda_1 \in \mathbf{K}_{s_1]}, \ldots, \lambda_p \in \mathbf{K}_{s_p]}$, *and* $\mathbf{A} = \{\alpha \in \mathbf{K} \mid \alpha_{s_i]} = \lambda_i \text{ for } i = 1, \ldots, p\}$. *Let* $m = \#\mathbf{A}$ *and* $k_i = \#\mathbf{K}_{s_i]}$ *for* $i = 1, \ldots, p$. *Then*

$$\#\mathbf{K} = m \cdot k_1 \cdot \ldots \cdot k_p.$$

PROOF. Let

$$\mathbf{B} = \mathbf{A} \times \mathbf{K}_{s_1]} \times \cdots \times \mathbf{K}_{s_p]}.$$

We shall prove that $\mathbf{B}$ has the same number of elements as $\mathbf{K}$. Define the function f, for $a = \langle \gamma, \beta_1, \ldots, \beta_p \rangle \in \mathbf{B}$ (and $u \in T$), by

$$f(a)(u) = \begin{cases} \beta_i(u), & \text{if } u \preceq s_i, \\ \gamma(u), & \text{otherwise.} \end{cases}$$

The function $f(a)$ is well defined, because T is a strict causal ordering and, hence, the sets $(-\infty, s_i]$ are pairwise disjoint, for $1 \leq i \leq p$.

We prove by induction on T, that $f(a)_t \in \mathbf{K}_t$, for every $t \in T$. It $t \preceq s_i$ for some $i = 1, \ldots, p$, this is clear, since $f(a)_t = (\beta_i)_t$. Suppose that $t \not\preceq s_i$ for any $i = 1, \ldots, p$. (Then t is either incomparable with all s_i or larger than some of them.) Let A be the set of immediate predecessors of t.

Assume as inductive hypothesis that $f(a)_u \in \mathbf{P_K}$ for $u \in A$. Then by Condition (*) of Definition XIV.3, because $\mathbf{K}$ is super classical, $f(a)(u) \in \mathbf{K}_{f(a)_u}$ for $u \in A$. Hence, by Condition (*), $f(a)_t \in \mathbf{K}_t$.

Taking the set of last elements of T, we prove, similarly, that $f(a) \in \mathbf{K}$.

The function f is clearly one-one. To prove that f is onto, take $\mu \in \mathbf{K}$. Define for $u \in T$

$$\gamma(u) = \begin{cases} \lambda_i(u), & \text{if } u \preceq s_i, \\ \mu(u), & \text{otherwise,} \end{cases}$$

and

$$\beta_i = \mu_{s_i]},$$

for $i = 1, \ldots, p$. Then it is easy to show that

$$f(\langle \gamma, \beta_1, \ldots, \beta_p \rangle) = \mu.$$

□

LEMMA XIV.7. *Let* $\{s_1, \ldots, s_n\}$ *be the set of immediate predecessors of* $t \in T$, *and* $\lambda \in \mathbf{K}$. *Let* $m = \#\{\lambda_t\}^\circ$, $k_i = \#\mathbf{K}_{s_i]}$, *and* $j_i = \#\{\lambda_{s_i]}\}^\circ$, *for* $i = 1, \ldots, n$. *Then*

$$j_i = m \cdot \prod \langle k_p \mid 1 \leq p \leq n \text{ and } p \neq i \rangle,$$

for $i = 1, \ldots, n$.

2. MEASURES ON SUPER CLASSICAL STRUCTURES

PROOF. We define the function f for
$$a = \langle \gamma, \langle \alpha_p \mid 1 \leq p \leq n \text{ and } i \neq p \rangle \rangle \in \{\lambda_t\}^\circ \times \mathbf{K}_{s_1]} \times \cdots \times \mathbf{K}_{s_n]}$$
by
$$f(a)(u) = \begin{cases} \lambda(u), & \text{if } u \preceq s_i, \\ \alpha_p(u), & \text{if } u \preceq s_p, \text{ for } p = 1, \ldots, n \text{ with } p \neq i, \\ \gamma(u), & \text{otherwise.} \end{cases}$$

Similarly as for the previous lemmas, we prove that f is one-one and onto $\{\lambda_{s_i]}\}^\circ$. Notice that $\lambda(u) = \gamma(u)$, if $u \preceq s_i$. □

LEMMA XIV.8. *Let* $t \in T$, $\lambda \in \mathbf{K}$. *Then*
$$\#\{\lambda_{t]}\}^\circ = \frac{\#\{\lambda_t\}^\circ}{\#\mathbf{K}_{\lambda_t}}.$$

PROOF. From the definition of $\mathbf{K}_{\lambda_t}$ we immediately see that
$$\#\{\lambda_t\}^\circ = \#\{\lambda_{t]}\}^\circ \cdot \#\mathbf{K}_{\lambda_t}.$$

□

We now prove the main theorem of this section, which gives a characterization of the invariant probability measure in a super classical structure. This characterization will be extended later to other equiprobability structures.

THEOREM XIV.9. *Let* $\mathbf{K}$ *be a super classical equiprobability structure with causal relation* T *and* $t \in T$. *Then:*

(1) $\langle \mathbf{K}_t, \Pr_t \rangle$ *is the product space of* $\langle \langle \mathbf{K}_{s]}, \Pr_{s]} \rangle \mid s \text{ i.p. } t \rangle$;
(2) $\langle \mathbf{K}_m, \Pr_m \rangle$ *is the product space of* $\langle \langle \mathbf{K}_{t]}, \Pr_{t]} \rangle \mid t \text{ i.e. } \bigcup \{T_i \mid i \leq m\}\rangle$, *and*
(3) $\langle \mathbf{K}_{t]}, \Pr_{t]} \rangle$ *is the average probability space of* $\langle \langle \mathbf{K}_{\lambda_t}, \Pr_{\lambda_t} \rangle \mid \lambda_t \in \mathbf{K}_t \rangle$ *with respect to* $\langle \mathbf{K}_t, \Pr_t \rangle$.

PROOF OF (1). Let $\lambda \in \mathbf{K}_t$ and let $\Pr = \Pr_{\mathbf{K}}$. We have that
$$\Pr_t(\lambda) = \Pr\{\lambda\}^\circ = \frac{\#\{\lambda\}^\circ}{\#\mathbf{K}}.$$

Let the set of immediate predecessors of t be $\{s_1, \ldots, s_n\}$. The function λ can be identified with the system $\langle \lambda_{s_i]} \mid i \leq n \rangle$. We have that
$$\Pr_{s]}(\lambda_{s]}) = \frac{\#\{\lambda_{s]}\}^\circ}{\#\mathbf{K}},$$
so it is enough to prove
$$\frac{\#\{\lambda\}^\circ}{\#\mathbf{K}} = \frac{\prod \langle \#\{\lambda_{s_i]}\}^\circ \mid i < n \rangle}{(\#\mathbf{K})^n}.$$

Let $m = \#\{\lambda\}^\circ$ and $k_i = \#\mathbf{K}_{s_i]}$ for $i = 1, \ldots, n$. We have, by Lemma XIV.6
$$\#\mathbf{K} = m \cdot k_1 \cdots \cdot k_n.$$

Hence

$$\frac{\#\{\lambda\}^\circ}{\#\mathbf{K}} = \frac{m}{m \cdot k_1 \cdots k_n}$$

$$= \frac{m^n \cdot (k_1 \cdots k_n)^{n-1}}{m^n \cdot (k_1 \cdots k_n)^n}$$

$$= \frac{\prod \langle \#\{\lambda_{s,i}\}^\circ \mid i < n\rangle}{(\#\mathbf{K})^n}.$$

The proof of 2 is similar. □

PROOF OF (3). Let $\mu \in \mathbf{K}_{t]}$. We have that μ can be identified with $\langle \mu_t, \mu(t) \rangle$. It is enough to prove that $\Pr\{\mu\}^\circ = \Pr_t(\mu_t) \cdot \Pr_{\mu_t}(\mu(t))$. But

$$\Pr\{\mu\}^\circ = \frac{\#\{\mu\}^\circ}{\#\mathbf{K}}.$$

By Lemma XIV.8

$$\Pr\{\mu\}^\circ = \frac{\#\{\mu_t\}^\circ}{\#\mathbf{K} \cdot \#\mathbf{K}_{\mu_t}}.$$

Since $\mathbf{K}_{\mu_t}$ is classical

$$\Pr_{\mu_t}(\mu(t)) = \frac{1}{\#\mathbf{K}_{\mu_t}}.$$

Thus, the statement is proved. □

3. Homomorphisms between equiprobability structures

We have seen that not all chance structures are classical. For instance, if one selects at random one from two urns, the first with one black ball and the second with two white balls, the corresponding structure is not classical (this is an instance of the structure in Example XIII.3, call it $\mathbf{UB}_1$). We already anticipated at the beginning of Chapter XIII that for nonclassical structures such as this, a measure invariant under its equiprobability relation would not be sufficient. In fact, as we saw in the previous chapter, the group of automorphisms $G_{\mathbf{UB}_1}$ is the trivial group which only contains the identity. Hence, there are no measure-determined sets, except for the empty set and the set $\mathbf{UB}_1$, and there are many invariant measures. On the other hand, the natural measure is perfectly clear. You just must assign 1/2 to each urn, 1/4 to each white ball, and 1/2 to the black ball. We also see that if we divide the black ball into two balls, we do get a random structure, say $\mathbf{UB}_2$, with a measure defined over all subsets of $\mathbf{UB}_2$. There is a natural way of transporting the measure on $\mathbf{UB}_2$ over to $\mathbf{UB}_1$ and we obtain the measure described above for $\mathbf{UB}_1$. This method can be used in general and most of the rest of this chapter is devoted to this generalization.

Before going into the construction of the measure in general, I shall discuss the procedure outlined above for Hosiasson's equipobability structure $\mathbf{Ho}_2$, Fig. III.5, which as we have seen is not classical and is a little more complicated than $\mathbf{UB}_1$. We first construct a super classical structure, $\mathbf{K}$, which will turn to be $\mathbf{Ho}_3$, Fig. III.6, and a homomorphism f from $\mathbf{K}$ onto $\mathbf{Ho}_2$. We have that the only first element of T is t. We define $\mathbf{K}_0 = \mathbf{K}_{t]} = (\mathbf{Ho}_2)_{t]}$, i.e., $\mathbf{K}_{t]}$ is the set of functions with domain $\{t\}$ and into $\{c_1, c_2, c_3\}$. The homomorphism f at t, $f_{t]}$ is defined to be the identity restricted to $\mathbf{K}_{t]}$. At level one in $\mathbf{Ho}_2$, i.e., at s, we have two sets, namely, $\{b_1, b_2, b_3\}$ and $\{d_1, d_2\}$. We take the minimum common multiple of the number of elements of these two sets, which is six. We consider a set of six elements, for instance, $\{e_1, \ldots, e_6\}$, and $\mathbf{K}$ as $\mathbf{Ho}_3$. Let g be the function from $\{e_1, \ldots, e_6\}$ onto $\{b_1, b_2, b_3\}$ such that e_1 and e_2 are sent to b_1, e_3 and e_4, into b_2, and e_5 and e_6, into b_3. Let h be the function from $\{e_1, \ldots, e_6\}$ onto $\{d_1, d_2\}$ such that e_1, e_2 and e_3 are sent into d_1, and e_4, e_5, and e_6, into d_2. Then, the homomorphism f is defined to be

$$f_{\mu_s}(u) = \begin{cases} g(u), & \text{if } \mu(t) = c_1 \text{ or } c_2, \\ h(u), & \text{if } \mu(t) = c_3, \end{cases}$$

for $\mu \in \mathbf{K}$.

It is easy to see that any pair of equiprobable events, $\mathbf{B}$ and $\mathbf{C}$, in $\mathbf{Ho}_2$ have the same number of elements. Then, one can check that their inverse images, $f^{-1}(\mathbf{B})$ and $f^{-1}(\mathbf{C})$, also have the same number of elements. Since $\mathbf{Ho}_3$ is a super classical structure, they are equiprobable in $\mathbf{Ho}_3$. The same is true for the equiprobability relations of the basic blocks. Thus, f preserves, what may be called, the "structure of equiprobability".

We shall now introduce the formal definition, in general, of a notion of homomorphism between equiprobability structures, which are functions between these structures that preserve the "structure of equiprobability", and also generalize the example of $\mathbf{Ho}_2$ and $\mathbf{Ho}_3$, and show that every equiprobability structure is the homomorphic image of a super classical structure. A homomorphism must preserve the relations of dependence and independence, i.e., must be a homomorphism of functional structures, and the equiprobability relations of the basic blocks. Thus, we introduce the following definition. I shall abbreviate, as earlier, s is an immediate predecessor of t by s i.p. t. Recall for this definition that if $t \in T_n$ with $n > 0$, then

$$\mathbf{K}_t = \{\bigcup\{h(s) \mid s \text{ i.p. } t\} \mid h \in \prod \langle \mathbf{K}_{s]} \mid s \text{ i.p. } t\rangle\}.$$

DEFINITION XIV.10 (HOMOMORPHISMS). Let

$$\langle \mathbf{K}, \preceq_T, G\rangle \text{ and } \langle \mathbf{H}, \preceq_T, G'\rangle$$

be equiprobability structures with the same causal ordering $\langle T, \preceq\rangle$. The function f is a *homomorphism of equiprobability structures from* $\mathbf{K}$ *onto* $\mathbf{H}$, if f is a homomorphism of functional structure (according to Definition XIV.2) from $\langle \mathbf{K}, \preceq_T\rangle$ onto $\langle \mathbf{H}, \preceq_T\rangle$, such that $\bar{f}$ is the identity on T, and

(**): if $\mathbf{A}, \mathbf{B} \subseteq \mathbf{H}_m$, then $\mathbf{A} \sim_{\mathbf{H}_m} \mathbf{B}$ implies $f_m^{-1}(\mathbf{A}) \sim_{\mathbf{K}_m} f_m^{-1}(\mathbf{B})$, for $m <$ height of T.

Notice that we have defined homomorphism of functional structures, Definition XIV.2, and the stronger notion of homomorphism of equiprobability structure, Definition XIV.10. From now on, when we talk about homomorphisms, unless it is expressly mentioned otherwise, we shall mean homomorphisms of equiprobability structures. As was noted before, we have that for $\lambda \in \mathbf{K}$,

$$f_t(\lambda_t) = \bigcup \{f_s](\lambda_s]) \mid s \text{ i.p. } t\}$$

and

$$f_{t]}(\lambda_{t]}) = f_t(\lambda_t) \frown f_{\lambda_t}(\lambda(t)).$$

It is also easy to see that if f is a homomorphism from $\mathbf{K}$ onto $\mathbf{J}$ and h is a homomorphism from $\mathbf{J}$ onto $\mathbf{H}$, then $h \circ f$ is a homomorphism from $\mathbf{K}$ onto $\mathbf{H}$.

An inessential restriction of this definition is that for homomorphisms of equiprobability structures we require that $\mathbf{K}$ and $\mathbf{H}$ have the same causal ordering, instead of isomorphic orderings. We could widen the definition of homomorphism to include the case of different, but isomorphic causal orderings. Since we do not need the extension, for simplicity, we shall not do it.

Notice that we do not require the homomorphisms to be one-one, as isomorphisms are. Instead, we require that an homomorphism preserve the invariance structure of $\mathbf{H}$.

From Proposition XIV.2, we get:

COROLLARY XIV.10. *Let $\mathbf{K}$ and $\mathbf{H}$ be equiprobability structures with causal universe T, $t \in T$, and let f be a homomorphism from $\mathbf{K}$ onto $\mathbf{H}$. Then*

(1) *If $\mathbf{A}, \mathbf{B} \subseteq \mathbf{H}_t$ and $\mathbf{A} \sim_{\mathbf{H}_t} \mathbf{B}$ then $f_t^{-1}(\mathbf{A}) \sim_{\mathbf{K}_t} f_t^{-1}(\mathbf{B})$.*
(2) *If $\mathbf{A}, \mathbf{B} \subseteq \mathbf{H}_{t]}$ and $\mathbf{A} \sim_{\mathbf{H}_{t]}} \mathbf{B}$ then $f_{t]}^{-1}(\mathbf{A}) \sim_{\mathbf{K}_{t]}} f_{t]}^{-1}(\mathbf{B})$.*
(3) *If $\mathbf{A}, \mathbf{B} \subseteq \mathbf{H}_{f(\lambda)_t}$ and $\mathbf{A} \sim_{\mathbf{H}_{f(\lambda)_t}} \mathbf{B}$, then $f_{\lambda_t}^{-1}(\mathbf{A}) \sim_{\mathbf{K}_{\lambda_t}} f_{\lambda_t}^{-1}(\mathbf{B})$.*

The following theorem is a generalization of the process described above for $\mathbf{UB}_1$ and $\mathbf{UB}_2$, and Hosiasson's examples, $\mathbf{Ho}_2$ and $\mathbf{Ho}_3$. These examples show, for structures that are not classical, a super classical structure that is homomorphic to it. We assume, as earlier, that all our structures are finite. This theorem, however, can be generalized to infinite structures, but the definitions and proof become more involved.

Before going into the theorem, we note that for any natural number ν (finite or infinite), there is a classical simple chance structure $\mathbf{K}_\nu$ that has ν elements. Possibly the simplest such structure is constructed in the following way. Let $\mathbf{K}_\nu$ be the set of relational systems $\omega_a = \langle A, a \rangle$ which have a fixed universe A of ν elements, and a is any element of A. This is the model for the choosing of one ball of a set of ν balls.

It is clear that $\mathbf{K}_\nu$ has ν elements. The group of invariance, $G_{\mathbf{K}_\nu}$ consists of all permutations of A. Thus, any two elements ω_a, ω_b are equivalent under the group. It is enough to take the permutation g of A such that $g(a) = b$, and, then, $g''\omega_a = \omega_b$. Therefore, $\mathbf{K}_\nu$ is classical.

Suppose we construct the (compound) chance structure $\mathbf{K}$, with causal universe T, in the following manner: For each $m <$ height of T we fix a number ν (which depends on m). We place at each $t \in T_m$, the simple structure $\mathbf{K}_\nu$. Then by Theorem XIII.11, $\mathbf{K}$ results classical. The equiprobability structure associated with $\mathbf{K}$ is then super classical. The chance structure $\mathbf{K}$ in the proof of the theorem is of the form just described.

3. HOMOMORPHISMS

THEOREM XIV.11. *For each finite equiprobability structure* **H** *there is a classical chance structure* **K**, *associated with a super classical equiprobability structure, and a homomorphism (of equiprobability structures)* f *of* **K** *onto* **H**.

PROOF. Let T be the causal ordering of **H**. We define $\mathbf{K}_t$, $\mathbf{K}_{\lambda_t}$, $\mathbf{K}_{t]}$, $\mathbf{K}_m$, f_t, f_{λ_t}, $f_{t]}$, and f_m by induction on T. We start with T_0. Notice that for $t \in T_0$, $\lambda_t = t$. Let k be the minimum common multiple of $\#\mathbf{H}_t$, for $t \in T_0$. Put as $\mathbf{K}_t$ the simple chance structure $\mathbf{K}_k$ described above, for every $t \in T_0$. Suppose that $\#\mathbf{H}_t = n$, for a certain $t \in T_0$. Then n divides k. Partition $\mathbf{K}_t$ into n parts of n/k elements each. Define f_t (which is also f_{λ_t}) from $\mathbf{K}_t$ onto $\mathbf{H}_t$ (recall that $\mathbf{K}_t = \mathbf{K}_{\lambda_t}$ and $\mathbf{H}_t = \mathbf{H}_{\lambda_t}$) such that the elements of each class in the partition are sent to a distinct element of $\mathbf{H}_t$.

Suppose that f_t and $\mathbf{K}_t$ are defined for $t \in T_n$. Similarly as above, let k be the minimum common multiple of $\#\mathbf{H}_{\mu_t}$ for $t \in T_n$ and $\mu \in \mathbf{H}$, and, for $\lambda \in \mathbf{K}_t$, let $\mathbf{K}_\lambda = \mathbf{K}_k$. We assume that for each $\lambda \in \mathbf{K}_t$, $f_t(\lambda) \in \mathbf{H}_t$. We define f_λ, for $\lambda \in \mathbf{K}_t$, a function from $\mathbf{K}_\lambda$ (which is $= \mathbf{K}_k$) to $\mathbf{H}_{f_t(\lambda)}$ as f_t was defined for the case $n = 0$.

For each $t \in T_n$, define

$$\mathbf{K}_{t]} = \{\lambda \frown \omega \mid \lambda \in \mathbf{K}_t \text{ and } \omega \in \mathbf{K}_\lambda\}$$

and

$$f_{t]}(\lambda \frown \omega) = f_t(\lambda) \frown f_\lambda(\omega).$$

For $s \in T_{n+1}$ define $\mathbf{K}_s$ by

$$\mathbf{K}_s = \{\bigcup\{h(t) \mid t \text{ i.p. } s\} \mid h \in \prod \langle \mathbf{K}_{t]} \mid t \text{ i.p. } s\rangle\}$$

and f_s, by

$$f_s(\alpha) = \bigcup\{f_t(\alpha_t) \frown f_{f_t(\alpha_t)}(\alpha(t)) \mid t \text{ i.p. } s\}.$$

Finally, define

$$\mathbf{K}_n = \{\bigcup\{h(t) \mid t \text{ l.e. } \bigcup\{T_i \mid i \leq n\}\} \mid h \in \prod \langle \mathbf{K}_{t]} \mid t \text{ l.e. } \bigcup\{T_i \mid i \leq n\}\rangle\}$$

and

$$f_n(\lambda) = \bigcup\{f_{t]}(\lambda_{t]}) \mid t \text{ l.e. } \bigcup\{T_i \mid i \leq n\}\}.$$

Thus, we have defined **K** as a chance structure. As we mentioned before the statement of the theorem, by Theorem XIII.11, **K** is a classical chance structure, and, hence, since it has the same sets at each level, the associated equiprobability structure is super classical.

The function f is clearly a homomorphism of functional structures from $\langle \mathbf{K}, \preceq \rangle$ onto $\langle \mathbf{H}, \preceq \rangle$. We now prove that f is also a homomorphism of equiprobability structures. In order to do this, we prove by induction on T

$$\mathbf{A}, \mathbf{B} \subseteq \mathbf{K}_{t]} \text{ and } \mathbf{A} \sim_{\mathbf{H}_{t]}} \mathbf{B} \implies \#(f_{t]}^{-1}(\mathbf{A})) = \#(f_{t]}^{-1}(\mathbf{B})), \tag{1}$$

$$\mathbf{A}, \mathbf{B} \subseteq \mathbf{K}_t \text{ and } \mathbf{A} \sim_{\mathbf{H}_t} \mathbf{B} \implies \#(f_t^{-1}(\mathbf{A})) = \#(f_t^{-1}(\mathbf{B})), \tag{2}$$

$$\mathbf{A}, \mathbf{B} \subseteq \mathbf{K}_m \text{ and } \mathbf{A} \sim_{\mathbf{H}_m} \mathbf{B} \implies \#(f_m^{-1}(\mathbf{A})) = \#(f_m^{-1}(\mathbf{B})). \qquad (3)$$

Since $\mathbf{K}$ is super classical, (1), (2) and (3) complete the proof of the theorem.

We, now, proceed to prove (1), (2) and (3), by induction on T. It is clear that if $\mathbf{A}, \mathbf{B} \subseteq \mathbf{H}_{\mu_t}$ and $\mathbf{A} \sim_{\mathbf{H}_{\mu_t}} \mathbf{B}$, then $\#\mathbf{A} = \#\mathbf{B}$. Hence, by the definition of $\mathbf{K}$ and f, their inverse images under f also have the same number of elements. Suppose, now, as inductive hypothesis, that (2) is true, and we prove (1). Let $\mathbf{A}, \mathbf{B} \subseteq \mathbf{K}_{t]}$ and $\mathbf{A} \sim_{\mathbf{H}_{t]}} \mathbf{B}$, i.e., $\mathbf{B} = \mathbf{A}^g$ for some isomorphism g. We have

$$\mathbf{A} = \bigcup \{\{\lambda\} \times \mathbf{A}_\lambda \mid \lambda \in \mathbf{A}_t\},$$

$$\mathbf{B} = \bigcup \{\{\lambda\} \times \mathbf{B}_\lambda \mid \lambda \in \mathbf{B}_t\}.$$

By the definitions of isomorphism and homomorphism, $\mathbf{B}_t = (\mathbf{A}_t)^g$, $\mathbf{B}_{g(\lambda)} = (\mathbf{A}_\lambda)^g$, and

$$f^{-1}(\mathbf{A}) = \bigcup \{f^{-1}\{\lambda\} \times f^{-1}(\mathbf{A}_\lambda) \mid \lambda \in \mathbf{A}_t\}$$

and, similarly

$$f^{-1}(\mathbf{B}) = \bigcup \{f^{-1}\{\lambda\} \times f^{-1}(\mathbf{B}_\lambda) \mid \lambda \in \mathbf{B}_t\}.$$

By the induction hypothesis, $\#(f^{-1}(\mathbf{A}_t)) = \#(f^{-1}(\mathbf{B}_t))$. Now, since $\mathbf{B}_{g(\lambda)} = (\mathbf{A}_\lambda)^g$, we have that $\mathbf{H}_{g(\lambda)} = (\mathbf{H}_\lambda)^g$, and, hence, $\#\mathbf{H}_{g(\lambda)} = \#\mathbf{H}_\lambda$, and also $\#\mathbf{B}_{g(\lambda)} = \#\mathbf{A}_\lambda$. With these two facts, we conclude, using the definition of $\mathbf{K}$ and f, that

$$\#(f^{-1}(\mathbf{B}_{g(\lambda)})) = \#(f^{-1}(\mathbf{A}_\lambda)).$$

This proves (1).

Suppose, now, (1) as inductive hypothesis, for all s i.p. t, and let $\mathbf{A}, \mathbf{B} \subseteq \mathbf{H}_t$ with $\mathbf{B} = \mathbf{A}^g$ for some isomorphism g. We have

$$\mathbf{A} = \{\bigcup \{h(s) \mid s \text{ i.p. } t\} \mid h \in \prod \langle \mathbf{A}_{s]} \mid s \text{ i.p. } t \rangle\},$$
$$\mathbf{B} = \{\bigcup \{h(s) \mid s \text{ i.p. } t\} \mid h \in \prod \langle (\mathbf{A}_{s]})^g \mid s \text{ i.p. } t \rangle\}.$$

Also

$$f^{-1}(\mathbf{A}) = \{\bigcup \{h(s) \mid s \text{ i.p. } t\} \mid h \in \prod \langle f^{-1}(\mathbf{A}_{s]}) \mid s \text{ i.p. } t \rangle\},$$
$$f^{-1}(\mathbf{B}) = \{\bigcup \{h(s) \mid s \text{ i.p. } t\} \mid h \in \prod \langle f^{-1}((\mathbf{A}_{s]})^g) \mid s \text{ i.p. } t \rangle\}.$$

By the induction hypothesis

$$\#f^{-1}(\mathbf{A}_{s]}) = \#f^{-1}((\mathbf{A}_{s]})^g).$$

Hence, we obtain (2).

Assertion (3) is obtained similarly. $\square$

4. Induced measures

In this section, we shall study measure transported by a homomorphism from one equiprobability structure to another. We define a provisional measure $\widetilde{\Pr}_K$, which for super classical structures coincides with the definitive measure $\Pr_K$. The probability measure $\widetilde{\Pr}_K$ is the invariant measure defined on the disjunctive field of sets generated by the measure-determined subsets of **K**.

DEFINITION XIV.11. Let **K** be an equiprobability structure. Then $\mathcal{A}_K$ is the disjunctive algebra generated by the measure-determined subsets of **K**, and $\widetilde{\Pr}_K$ is the invariant measure defined on $\mathcal{A}_K$. The *provisional probability space of* **K** is $\langle K, \mathcal{A}_K, \widetilde{\Pr}_K \rangle$.

For a classical structure **K**, we know that $\mathcal{A}_K = \mathcal{P}(K)$ and $\widetilde{\Pr}_K = \Pr_K$.

DEFINITION XIV.12 (INDUCED MEASURES). Let **H** and **K** be equiprobability structures with causal universe T, f a homomorphism from **K** onto **H**, and the space $\langle K, \mathcal{A}_K, \widetilde{\Pr}_K \rangle$, the provisional probability space of **K**. We define

$$\widetilde{\Pr}_{KH}(A) = \widetilde{\Pr}_K(f^{-1}(A)),$$

for any $A \subseteq H$ with $f^{-1}(A) \in \mathcal{A}_K$, and, for **K** super classical

$$\widetilde{\Pr}_{f(\lambda_t)}(A) = \Pr_{K_{\lambda_t}}(f^{-1}_{f(\lambda_t)}(A)),$$

for $A \subseteq H_{f(\lambda_t)}$.

The measure $\widetilde{\Pr}_{f(\lambda_t)}$ is well defined because, since **K** is super classical, $K_{\lambda_t} = K_{\mu_t}$, for any $\lambda, \mu \in K$ and $f(\lambda_t) \in K_t$. Notice that $\widetilde{\Pr}_{KH}$ depends also on f.

As an example, let **D** be the event of a white ball in Hosiasson's example Ho_2, and let f be the homomorphism defined in the last section between Ho_3 and Ho_2. Then $f^{-1}(D)$ consists of the outcomes $\mu_{11}, \mu_{12}, \mu_{13}, \mu_{14}, \mu_{21}, \mu_{22}, \mu_{23}, \mu_{24}, \mu_{31}, \mu_{32},$ and μ_{33}. Thus

$$\Pr_{Ho_3} f^{-1}(D) = \frac{11}{18}.$$

Hence

$$\widetilde{\Pr}_{Ho_3 Ho_2} D = \frac{11}{18}.$$

We shall define

$$\Pr_{Ho_2} D = \widetilde{\Pr}_{Ho_3 Ho_2} D$$

and show that it is well defined.

If **K** is super classical, then $\widetilde{\Pr}_{KH}$ is defined on all subsets of **H**. We now prove that the main properties proved for the measure $\Pr_K$, for **K** super classical, are transferred to $\widetilde{\Pr}_{KH}$, when **K** is super classical. So for the next lemmas, we assume that **K** is super classical and that f is a homomorphism from **K** onto **H**.

LEMMA XIV.12. *If* $\lambda \in H$ *and* $t \in T_n$ *with* $n > 0$, *then*

$$\widetilde{\Pr}_{KH}(\{\lambda_t\}^\circ) = \prod \langle \widetilde{\Pr}_{KH}(\{\lambda_s]\}^\circ) \mid s \text{ i.p. } t \rangle.$$

Thus, $\widetilde{\text{Pr}}_{\mathbf{KH}}$ is the product measure when restricted to $\mathbf{H}_t$.

PROOF. We have, by Corollary XIV.10

$$f^{-1}(\{\lambda_t\}^\circ) = \bigcup \{f^{-1}(\{\lambda_s]\}^\circ) \mid s \text{ i.p. } t\}.$$

Hence, by Theorem XIV.9

$$\text{Pr}_{\mathbf{K}}(f^{-1}(\{\lambda_t\}^\circ)) = \prod \langle \text{Pr}_{\mathbf{K}}(f^{-1}(\{\lambda_s]\}^\circ)) \mid s \text{ i.p. } t \rangle.$$

By the definition of $\widetilde{\text{Pr}}_{\mathbf{KH}}$ we obtain the conclusion of the theorem. □

LEMMA XIV.13. *If $\lambda \in \mathbf{H}$ and $t \in T$, then*

$$\widetilde{\text{Pr}}_{\mathbf{KH}}(\{\lambda_{t]}\}^\circ) = \widetilde{\text{Pr}}_{\mathbf{KH}}(\{\lambda_t\}^\circ) \cdot \widetilde{\text{Pr}}_{\lambda_t}(\{\lambda(t)\}).$$

Thus, $\widetilde{\text{Pr}}_{\mathbf{KH}}$ on $\mathbf{H}_{t]}$ is the average measure.

PROOF. By Corollary XIV.10

$$f^{-1}(\{\lambda_{t]}\}^\circ) = (f_{t]}^{-1}(\{\lambda_{t]}\}))^\circ.$$

By Proposition XIV.1

$$f_{t]}^{-1}\{\lambda_{t]}\} = \bigcup \{\beta_t \frown \omega \mid \beta_t \in f_t^{-1}\{\lambda_t\} \text{ and } \omega \in f_{\beta_t}^{-1}\{\lambda(t)\}\}.$$

Hence, by Theorem XIV.9

$$\text{Pr}_{\mathbf{K}}(f^{-1}\{\lambda_{t]}\}^\circ) = \sum \langle \text{Pr}_{\mathbf{K}}(\{\beta_t\}^\circ) \cdot \text{Pr}_{\mathbf{K}_{\beta_t}}(\omega) \mid \beta_t \in f_t^{-1}\{\lambda_t\} \text{ and}$$
$$\omega \in f_{\beta_t}^{-1}\{\lambda(t)\}\rangle$$
$$= \text{Pr}_{\mathbf{K}}(f_t^{-1}\{\lambda_t\}) \cdot \text{Pr}_{\mathbf{K}_{\beta_t}}(f_{\beta_t}^{-1}\{\lambda(t)\}).$$

Thus, the conclusion of the theorem is obtained using the definition of $\widetilde{\text{Pr}}_{\mathbf{KH}}$ and $\widetilde{\text{Pr}}_{\lambda_t}$. □

5. The external invariance principle

We are now ready to introduce the final definition of the disjunctive field of events and the measure. The basic idea is that two homomorphic equiprobability structures should determine the same probability measure. This is what I call the *External Invariance Principle*.

DEFINITION XIV.13. Let $\mathbf{H}$ be an equiprobability structure with causal universe T.

(1) A subset $\mathbf{A}$ of $\mathbf{H}$ is $\mathbf{H}$-*weakly measure-determined* if for every equiprobability structures $\mathbf{K}$ and $\mathbf{K}'$ and every homomorphism f from $\mathbf{K}$ onto $\mathbf{H}$ and f' from $\mathbf{K}'$ onto $\mathbf{H}$ such that $\widetilde{\text{Pr}}_{\mathbf{KH}}(\mathbf{A})$ and $\widetilde{\text{Pr}}_{\mathbf{K}'\mathbf{H}}(\mathbf{A})$ are defined, we have that $\widetilde{\text{Pr}}_{\mathbf{KH}}(\mathbf{A}) = \widetilde{\text{Pr}}_{\mathbf{K}'\mathbf{H}}(\mathbf{A})$.
(2) The family of events of $\mathbf{H}$, $\mathcal{F}_{\mathbf{H}}$, is the disjunctive algebra of subsets of $\mathbf{H}$ generated by the family of weakly measure-determined sets.

5. THE EXTERNAL INVARIANCE PRINCIPLE

(3) $\mathrm{Pr}_{\mathbf{H}}$ is the measure defined for $\mathbf{A} \in \mathcal{F}_{\mathbf{H}}$ by

$$\mathrm{Pr}_{\mathbf{H}} \mathbf{A} = \widetilde{\mathrm{Pr}}_{\mathbf{KH}}(\mathbf{A}),$$

for any $\mathbf{K}$ homomorphic to $\mathbf{H}$.

First we prove that it is enough to consider in the definition of weakly measure-determined sets, super classical equiprobability structures $\mathbf{K}$ and $\mathbf{K}'$.

THEOREM XIV.14. *Let $\mathbf{H}$ be an equiprobability structure and $\mathbf{A} \subseteq \mathbf{H}$. Then $\mathbf{A}$ is weakly measure-determined if and only if for every pair of super classical structures $\mathbf{K}$ and $\mathbf{K}'$ homomorphic to $\mathbf{H}$, $\widetilde{\mathrm{Pr}}_{\mathbf{KH}}(\mathbf{A}) = \widetilde{\mathrm{Pr}}_{\mathbf{K'H}}(\mathbf{A})$.*

PROOF. Let $\mathbf{J}$ be any equiprobability structure such that h is a homomorphism from $\mathbf{J}$ onto $\mathbf{H}$ and $\widetilde{\mathrm{Pr}}_{\mathbf{JH}}(\mathbf{A})$ is defined. Then there is a unique $\mathbf{J}$–invariant measure on $\mathcal{A}_{\mathbf{J}}$ and we have $h^{-1}(\mathbf{A}) \in \mathcal{A}_{\mathbf{J}}$. By Theorem XIV.11, we can find a super classical structure $\mathbf{K}$ and a homomorphism f from $\mathbf{K}$ onto $\mathbf{J}$. Thus, $g = h \circ f$ is a homomorphism from $\mathbf{K}$ onto $\mathbf{H}$. Since $\mathbf{K}$ is super classical $\widetilde{\mathrm{Pr}}_{\mathbf{KJ}}(\mathbf{B})$ is defined for every $\mathbf{B} \in \mathcal{A}_{\mathbf{J}}$. Since $\widetilde{\mathrm{Pr}}_{\mathbf{KJ}}$ is a $\mathbf{J}$–invariant measure, we have that $\widetilde{\mathrm{Pr}}_{\mathbf{KJ}}$ coincides with $\mathrm{Pr}_{\mathbf{J}}$ on $\mathcal{A}_{\mathbf{J}}$. But $h^{-1}(\mathbf{A}) \in \mathcal{A}_{\mathbf{J}}$; then

$$\widetilde{\mathrm{Pr}}_{\mathbf{KJ}}(h^{-1}(\mathbf{A})) = \mathrm{Pr}_{\mathbf{J}}(h^{-1}(\mathbf{A})).$$

We have

$$\widetilde{\mathrm{Pr}}_{\mathbf{KJ}}(h^{-1}(\mathbf{A})) = \mathrm{Pr}_{\mathbf{K}}(g^{-1}(\mathbf{A})) = \widetilde{\mathrm{Pr}}_{\mathbf{KH}}(\mathbf{A}).$$

Thus

$$\widetilde{\mathrm{Pr}}_{\mathbf{JH}}(\mathbf{A}) = \widetilde{\mathrm{Pr}}_{\mathbf{KH}}(\mathbf{A}).$$

□

By Theorem XIV.11, and Lemmas XIV.12, and XIV.13, we obtain the main theorem of this chapter, which asserts that the measure we have defined behaves as it should:

THEOREM XIV.15. *Let $\mathbf{K}$ be an equiprobability structure with causal universe T, $t \in T$, and m less than the height of T. Then*

(1) $\langle \mathbf{K}_t, \mathcal{A}_t, \mathrm{Pr}_t \rangle$ *is the product space of*

$$\langle \langle \mathbf{K}_{s]}, \mathcal{F}_{s]}, \mathrm{Pr}_{s]} \rangle \mid s \text{ i.p. } t \rangle,$$

(2) $\langle \mathbf{K}_m, \mathcal{F}_m, \mathrm{Pr}_m \rangle$ *is the product space of*

$$\langle \langle \mathbf{K}_{s]}, \mathcal{F}_{s]}, \mathrm{Pr}_{s]} \rangle \mid s \text{ l.e. } \bigcup \{T_i \mid i \leq m\} \rangle,$$

(3) $\langle \mathbf{K}_{t]}, \mathcal{F}_{t]}, \mathrm{Pr}_{t]} \rangle$ *is the average probability space of*

$$\langle \langle \mathbf{K}_{\lambda_t}, \mathcal{F}_{\lambda_t}, \mathrm{Pr}_{\lambda_t} \rangle \mid \lambda_t \in \mathbf{K}_t \rangle$$

with respect to $\langle \mathbf{K}_t, \mathcal{F}_t, \mathrm{Pr}_t \rangle$.

In particular, for classical structures the theorem is true with the power set as field of events and the counting measure as probability measure.

CHAPTER XV

Probability structures

The purpose of this chapter is to define probability structures and prove a representation theorem for stochastic processes. The culmination of the chapter is in Section 4, where we prove that, given any stochastic process ξ, there is a very nearly equivalent (according to Definition XI.5) process η defined over a chance structure. As a corollary to this representation theorem, we shall obtain that, given any stochastic process ξ, there is a very nearly equivalent process, η, which is defined over a probability space, $\langle \Omega, \mathrm{pr} \rangle$, where all elementary events are equiprobable, that is, such that $\mathrm{pr}(\omega) = \mathrm{pr}(\omega')$, for all ω, $\omega' \in \Omega$. This corollary, which may be proved independently of the representation theorem mentioned above, shows that the classical definition of probability suffices, in the sense, that any stochastic process can be approximated by a process defined over a probability space where the classical definition works.

Before going into the matter proper to this chapter, I shall review the definitions and theorems we have discussed this far for compound chance setups. In Chapter XIII, chance structures, Definition XIII.5, for these compound setups were introduced. These chance structures are sets of outcomes (Definition XIII.2), which are modeled by a causal structure T and a function λ with domain T and values relational systems. In a chance structure $\mathbf{K}$, for each $t \in T$ and each element $\lambda \in \mathbf{K}$, there is a simple chance structure, $\mathbf{K}_{\lambda_t}$, which represents the possibilities when λ has happened up to t, i.e., when λ_t has happened.

We defined, in a natural way, taking into consideration both the causal structure and the relational systems, the notion of isomorphism between outcomes (Definition XIII.4). Now, an isomorphism between chance structures (Definition XIII.6) codes and isomorphism between the respective causal relations plus isomorphisms between all the simple structures on each of the nodes of the causal relation. An automorphism of a chance structure $\mathbf{K}$ is an isomorphism of $\mathbf{K}$ onto $\mathbf{K}$.

For each chance structure, $\mathbf{K}$, we define the partial structures which are obtained by taking a subrelation of T. The basic blocks of $\mathbf{K}$ are those partial structures which are obtained by natural basic blocks. Since these basic blocks are also chance structures, each has a group of automorphisms, which is determined by the basic block itself. In particular, there is a group of automorphism of **K** itself. Each of these groups defines an equivalence relation between subsets of the corresponding basic block.

The probability measure we want should be invariant under the group of automorphisms of the chance structure, and should, loosely speaking, "respect" the equivalence relations of the basic blocks. We proved in Theorem XIII.9 that it is enough to consider the equivalence relation on $\mathbf{K}_n$, for n less than the height of T, the basic block determined by $\bigcup \{T_i \mid i \leq n\}$.

In Chapter XIV, in order to simplify the statement of theorems and proofs, we forget about relational systems at the nodes of outcomes and treat them only as points. We need, then, to introduce explicitly the groups, since, now, they are not, in general, the group of all automorphism of the structure, but may be only a subgroup. In fact, as it was stated above, we only need the equivalence relations determined by the groups for $\mathbf{K}_n$. Thus, we consider what we call equiprobability structures with an equiprobability relation $\sim_n$ for each $\mathbf{K}_n$, with n less than the height of T.

We then consider classical and super classical structures. Classical structures are those where all outcomes are equiprobable. The notion of super classical structures, which is only introduced for technical reasons, is stronger: super classical structures are classical, but also have the same sets in nodes of equal height. We noticed that super classical equiprobability structures can always be obtained from chance structures. Several combinatorial lemmas are proved for super classical structures, which culminate in the statement that for independent moments the invariant measure is the product measure, and for dependent moments, the average measure.

Then a notion of homomorphism between equiprobability structures is introduced. It is proved that for every equiprobability structure **H** there is a super classical structure **K** and a homomorphism from **K** onto **H**. As noticed above, this super classical equiprobability strucutre **K** can be taken to be a chance structure.

In a super classical structure, **K**, there is a unique invariant probability measure (Internal Invariance Principle), which is the counting measure. This measure is transported in a unique way (External Inveriance Principle) to the homomorphic image **H**, at least to certain subsets of **H**, which are called weakly measure-determined. This transportation preserves product and average measures. Thus, we also obtain that in case of independence we get the product measure, and of dependence, the average measure.

In this chapter, we return to chance structures, that is, to structures with relational systems. For these structures, the group of invariance of the basic blocks are determined by them: they are the group of all automorphism of the corresponding block.

Since chance structures are equiprobability structures, all results of Chapter XIV apply to chance structures. Thus, the probability measure obeying both the Internal and External Invariance Principles, is the product measure, in case of independence, and the average measure, in case of dependence.

1. Existence of classical structures

As a preliminary for the representation theorem, we need to show that there are enough classical structures. As it was mentioned in page 336, for any natural number ν (finite or infinite), there is a classical simple structure, $\mathbf{K}_\nu$, that has ν elements. The construction of such a simple chance structure, which was sketched in page 336, is the following: Let $\mathbf{K}_\nu$ be the set of relational systems $\omega_a = \langle A, a \rangle$ which have a fixed universe A of ν elements, and a is any element of A. This is the model for the choosing of one ball of a set of ν balls.

It is clear that $\mathbf{K}_\nu$ has ν elements. The group of invariance, $G_{\mathbf{K}_\nu}$ consists of all permutations of A. Thus, any two elements ω_a, ω_b are equivalent under the group. It is enough to take the permutation g of A such that $g(a) = b$, and, then, $g''\omega_a = \omega_b$. Therefore, $\mathbf{K}_\nu$ is classical. Hence, the algebra of events is the internal power set of $\mathbf{K}_\nu$ and

$$\Pr_{\mathbf{K}_\nu}\{\omega_a\} = \frac{1}{\nu}$$

for any $a \in A$.

A similar fact is true for compound chance structures:

THEOREM XV.1. *Let $\langle T, \preceq \rangle$ be an internal strict causal ordering, such that T has μ elements. Then for any natural number $\nu \in {}^*\mathbb{N}$ there is an internal classical chance structure $\mathbf{H}$ with causal relation $\preceq$ such that $\mathbf{H}$ has ν^μ elements.*

PROOF. The construction of $\mathbf{H}$ proceeds by induction on the levels of T. For $t \in T_0$, define $\mathbf{H}_{t]} = \mathbf{K}_\nu$, where $\mathbf{K}_\nu$ is the structure defined above. Suppose that $\mathbf{H}_t$ is defined for $t \in T_m$. For each $\lambda \in \mathbf{H}_t$, define $\mathbf{H}_\lambda = \mathbf{K}_\nu$. By Theorem XIII.11, $\mathbf{H}$ is a classical chance structure. It is clear that $\mathbf{H}$ has ν^μ elements. □

As we know, every chance structure is an equiprobability structure. Thus, we can apply all notions defined for the latter to the former. We now apply the notion of homomorphism. We restate Theorem XIV.11 in the present setting:

THEOREM XV.2. *For each hyperfinite strict chance structure $\mathbf{H}$ there is an internal classical chance structure $\mathbf{K}$ (associated with a super classical equiprobability structure), and an internal homomorphism of equiprobability structure, f, of $\mathbf{K}$ onto $\mathbf{H}$.*

2. Measure-determined sets

We now prove that the notion of weakly measure determined, as applied to chance structures, extends the old notion of measure-determined.

THEOREM XV.3. *Let $\mathbf{H}$ be an internal strict chance structure and $\mathbf{A} \subseteq \mathbf{J}$, $\mathbf{A}$ internal, where $\mathbf{J}$ is a basic block of $\mathbf{H}$. Then if $\mathbf{A}$ is $\mathbf{J}$-measure-determined, we have that $\mathbf{A}^\circ$ is $\mathbf{H}$-weakly measure-determined.*

PROOF. We assume that $\mathbf{J} = \mathbf{H}_t$. The proof for the other basic blocks is similar. Let f and h be homomorphisms from the super classical structures $\mathbf{K}$ and $\mathbf{L}$ onto $\mathbf{H}$. Then f_t and h_t are homomorphisms of $\mathbf{K}_t$ and $\mathbf{L}_t$ onto $\mathbf{J}$. Assume that $\mathbf{A}$ is internal $\mathbf{J}$-measure-determined. It is clear that $\widetilde{\Pr}_{\mathbf{K},\mathbf{J}}(\mathbf{A}) = \widetilde{\Pr}_{\mathbf{L},\mathbf{J}}(\mathbf{A})$. So

$$\Pr_{\mathbf{K}_t}(f_t^{-1}(\mathbf{A})) = \Pr_{\mathbf{L}_t}(h_t^{-1}(\mathbf{A})).$$

But, by Corollary XIV.5

$$\Pr_{\mathbf{K}}((f_t^{-1}(\mathbf{A}))^\circ) = \Pr_{\mathbf{K}_t}(f_t^{-1}(\mathbf{A}))$$

and

$$\Pr_{\mathbf{L}}((h_t^{-1}(\mathbf{A}))^\circ) = \Pr_{\mathbf{L}_t}(h_t^{-1}(\mathbf{A})).$$

By Proposition XIV.2

$$(f_t^{-1}(\mathbf{A}))^\circ = f^{-1}(\mathbf{A}^\circ) \text{ and } (h_t^{-1}(\mathbf{A}))^\circ = h^{-1}(\mathbf{A}^\circ).$$

Hence

$$\Pr_{\mathbf{K}}(f^{-1}(\mathbf{A}^\circ)) = \Pr_{\mathbf{L}}(h^{-1}(\mathbf{A}^\circ)),$$

and, thus

$$\widetilde{\Pr}_{\mathbf{KH}}(\mathbf{A}^\circ) = \widetilde{\Pr}_{\mathbf{LH}}(\mathbf{A}^\circ).$$

□

Since every chance structure is an equiprobability structure, we can define the probability space of chance structures as the probability space of the chance structure as an equiprobability structure. We can now prove that the algebra of events thus defined contains the right sets.

We first define a disjunctive field that we shall prove that consists of weakly measure-determined sets. Recall that for $\mathbf{K}$ a simple chance structure, $\mathcal{F}_{\mathbf{K}}$ is the disjunctive field generated by the measure-determined sets.

DEFINITION XV.1. Let $\mathbf{H}$ be a chance structure with causal universe T. We define by induction on T the families $\mathcal{F}_{b\mathbf{H},t}$, $\mathcal{F}_{b\mathbf{H},m}$, and $\mathcal{F}_{b\mathbf{H},t]}$.

(1) If $t \in T_0$, then

$$\mathcal{F}_{b\mathbf{H},t} = \mathcal{F}_{b\mathbf{H},t]} = \{\mathbf{A}^\circ \mid \mathbf{A} \in \mathcal{F}_{\mathbf{H}_t}\}.$$

(2) For $t \in T_n$ with $0 < n$

$\mathcal{F}_{b\mathbf{H},t} = \{\mathbf{A}^\circ \mid \mathbf{A}$ is in the product disjunctive field of $\mathcal{F}_{b\mathbf{H},s]}$
   for s i.p. $t\}$.

$\mathcal{F}_{b\mathbf{H},m} = \{\mathbf{A}^\circ \mid \mathbf{A}$ is in the product disjunctive field of $\mathcal{F}_{b\mathbf{H},s]}$ for s l.e.
   $\bigcup\{T_i \mid i \leq m\}\}$.

$\mathcal{F}_{b\mathbf{H},t]} = \{\mathbf{A}^\circ \mid \mathbf{A} \subseteq \mathbf{H}(-\infty, t], \{\lambda_t \mid \Pr_{\lambda_t}(\mathbf{A}_{\lambda_t}) = x\} \in \mathcal{F}_{b\mathbf{H},t},$
   for every hyperreal x, and $\mathbf{A}_{\lambda_t} \in \mathcal{F}_{\mathbf{H}_{\lambda_t}}$, for every $\lambda \in \mathbf{H}\}$.

(3) The algebra $\mathcal{F}_{b\mathbf{H}}$ is defined by

$$\mathcal{F}_{b\mathbf{H}} = \bigcup\{\mathcal{F}_{b\mathbf{H},m} \mid m \text{ less than the height of } T\} \cup$$
$$\bigcup\{\mathcal{F}_{b\mathbf{H},t} \mid t \in T\} \cup \bigcup\{\mathcal{F}_{b\mathbf{H},t]} \mid t \in T\}.$$

We have the following theorem.

THEOREM XV.4. *Let* $\mathbf{H}$ *be a strict chance structure. Then* $\mathcal{F}_{b\mathbf{H}} \subseteq \mathcal{F}_{\mathbf{H}}$.

I do not know whether $\mathcal{F}_{b\mathbf{H}} = \mathcal{F}_{\mathbf{H}}$.

PROOF. It is clear that for any $\mathbf{A} \in \mathcal{F}_{\mathbf{H}_{\lambda_t}}$ and $\mathbf{K}$ super classical, $\widetilde{\Pr}_{\mathbf{KH}}(\mathbf{A}^\circ)$ is unique, because there is a unique measure on $\mathcal{F}_{\mathbf{H}_{\lambda_t}}$. Using Lemmas XIV.12 and XIV.13, we can easily prove by induction, that for $\mathbf{A} \in \mathcal{F}_{b\mathbf{H},t}$ and $\mathbf{A} \in \mathcal{F}_{b\mathbf{H},t]}$, and $\mathbf{K}$ super classical, $\widetilde{\Pr}_{\mathbf{KH}}(\mathbf{A})$ is unique. Then we obtain the result for all $\mathbf{A} \in \mathcal{F}_{b\mathbf{H}}$. By Theorem XIV.14, $\widetilde{\Pr}_{\mathbf{KH}}$ is unique for these $\mathbf{A}$'s and any equiprobability structure $\mathbf{K}$ where $\widetilde{\Pr}_{\mathbf{KH}}(\mathbf{A})$ is defined. □

We next analyze the measures obtained for the examples of the previous chapter. We shall consider more complicated examples in Chapter XVI. The structures **IC** and **DC** of Examples XIII.1 and XIII.2 are classical, so their measure is the counting measure for the hyperfinite approximation. The measure for **IC** results, as it should, the product of Lebesgue measure for each circle, and in case of **DC**, the corresponding average measure.

Consider **UB** of Example XIII.3. The structure **UB** is, in general, not classical. For instance, if there are two urns, one with 2 ball, one white and the other black, and the other urn with three balls, 2 black and 1 white, then **UB** is not classical. The super classical structure homomorphic to this **UB** can be obtained by taking the model of the choosing of two urns, each with six balls. It can be easily seen that with this structure the obvious measure is obtained.

The theorems of this and the previous chapters do not apply to Examples XIII.4, XIII.5 and XIII.6. But the external invariance principle and the definition of homomorphism applies to the case of Example XIII.4. With the same techniques as used here, one could obtain an appropriate measure for this case.

3. Probability structures

As in the case of simple structures, we use functions to the real numbers for describing properties of outcomes. But in this case, we need functions similar to stochastic processes. So we call them measurement processes.

DEFINITION XV.2 (PROBABILITY STRUCTURES). Let $\mathbf{K}$ be an internal chance structure with causal universe containing T. Then

(1) An m-*ary* T-*measurement process over* $\mathbf{K}$ is an internal function $\xi : {}^m T \times \mathbf{K} \to {}^*\mathbb{R}$, where m is a natural number. We sometimes write $\xi_{t_1,\ldots,t_m}(\lambda)$ for $\xi(t_1,\ldots,t_m,\lambda)$.

(2) A *weak probability structure of type* $\langle T, d \rangle$ is an internal pair $\mathcal{K} = \langle \mathbf{K}, \xi \rangle$, where $\mathbf{K}$ is a chance structure with causal universe T, and $\xi = \langle \xi_k \mid k \in K \rangle$ is a system of T-measurement system over $\mathbf{K}$. A *probability structure of type* $\langle T, d \rangle$ is a weak probability structure $\mathcal{K} = \langle \mathbf{K}, \xi \rangle$, with $\xi = \langle \xi_k \mid k \in K \rangle$ such that if ξ_k is m-ary, then $[\xi_k(t_1, \ldots, t_m) = r] \in \mathcal{F}_{\mathbf{K}}$ for every $t_1, \ldots, t_m \in T$ and $r \in {}^*\mathbb{R}$.

We can also introduce a natural "filtration" on the probability space of a chance structure.

DEFINITION XV.3 (FILTRATION). Let $\mathbf{K}$ be an internal chance structure with causal universe T and $\langle \mathbf{K}, \mathcal{F}_{\mathbf{K}}, \mathrm{Pr}_{\mathbf{K}} \rangle$, the probability space of $\mathbf{K}$. The *filtration of* $\mathbf{K}$, $\mathcal{F}_{\mathbf{K},t}$ is defined by

$$\mathcal{F}_{\mathbf{K},t} = \{ \mathbf{A} \in \mathcal{F}_{\mathbf{K}} \mid \mathbf{A} = (\mathbf{A}_t)^\circ \}.$$

It is clear that if $s \preceq t$, then $\mathcal{F}_{\mathbf{K},s} \subseteq \mathcal{F}_{\mathbf{K},t}$. We also know, by an obvious extension of Theorem XV.3, that $\mathcal{F}_{b\mathbf{K},t} \subseteq \mathcal{F}_{\mathbf{K},t}$.

DEFINITION XV.4 (STOCHASTIC STRUCTURES). Let $\mathbf{K}$ be an internal chance structure with causal universe T, and let ξ be an internal one-ary measurement process over $\mathbf{K}$. Then ξ is a *stochastic process adapted to* $\mathbf{K}$, if for every $t \in T$, ξ_t is a random variable with respect to $\mathcal{F}_{\mathbf{K},t}$.[1]

If ξ is a stochastic process adapted to $\mathbf{K}$ then $\langle \mathbf{K}, \xi \rangle$ is called a *stochastic structure*.

All definitions and theorems for stochastic processes obtained in Chapters VI to XI apply to our case.

As an example, consider Example XIII.1. We could introduce the binary measurement process ξ_0 defined for t, $s \in I_n$ and $\lambda \in \mathbf{IC}$ by $\xi_0(t, s, \lambda) = $ the distance between the points of the circle selected at t and s in outcome λ. Then, $\langle \mathbf{IC}, \xi_0 \rangle$ is a probability structure.

We shall consider further examples in the next chapter.

4. Representation theorem

For the exact statement of our representation theorem, we need a few notions discussed in Chapter XI, Section 3. I shall repeat some of these results, since they are the only notions that we need of the difficult Chapter XI. We assume that all our processes are indexed by the same hyperfinite set T, but possibly defined over different probability spaces. We begin with the notion of equivalent processes, which is actually defined in Definition VI.20: two processes ξ and η are equivalent if $\Lambda_\xi = \Lambda_\eta$ and $\mathrm{pr}_\xi = \mathrm{pr}_\eta$. In Proposition XI.8, it was proved that two processes ξ and η are equivalent if and only if $\mathbf{E}\, F(\xi) = \mathbf{E}\, F(\eta)$ for every functional F defined over $\Lambda_\xi \cup \Lambda_\eta$. Recall that a functional is just an internal function from a hyperfinite set, in this case $\Lambda_\xi \cup \Lambda_\eta$, to ${}^*\mathbb{R}$.

---

[1] That is, $\{\lambda \in \mathbf{K} \mid \xi_t(\lambda) = r\} \in \mathcal{F}_{\mathbf{K},t}$ for every $t \in T$ and $r \in {}^*\mathbb{R}$.

4. REPRESENTATION THEOREM

The second notion is that of very nearly equivalent processes introduced in Definition XI.5. The processes ξ and η are very nearly equivalent if $\Lambda_\xi = \Lambda_\eta$ and

$$\sum_{\lambda \in \Lambda_\xi} |\operatorname{pr}_\xi(\lambda) - \operatorname{pr}_\eta(\lambda)| \approx 0.$$

It is clear that equivalent processes are very nearly equivalent.

We showed in Lemma XI.9 that, if ξ and η are very nearly equivalent, then for every finite functional F defined over Λ_ξ, we have

$$\mathbf{E}\, F(\xi) \approx \mathbf{E}\, F(\eta).$$

This implies, as it was proved after the proof of Lemma XI.9, that very nearly equivalent processes have approximately equal probability distributions.

We are now ready to state and prove our representation theorem:

THEOREM XV.5. *Let ξ be an internal stochastic process* (according to Definition VI.19) *indexed by a hyperfinite set T. Then there is a very nearly equivalent internal process η indexed by T, defined over the internal probability space $\langle \mathbf{H}, \mathcal{F}_\mathbf{H}, \operatorname{Pr}_\mathbf{H} \rangle$ of a (classical) chance structure $\mathbf{H}$. (That is, there is a probability structure $\langle \mathbf{H}, \eta \rangle$ such that η is very nearly equivalent to ξ.)*

PROOF. We can consider that ξ is defined over the hyperfinite space of trajectories, $\langle \Lambda_\xi, \operatorname{pr}_\xi \rangle$. Suppose that the number of elements of Λ_ξ is κ and of T, μ. Let ν be an infinite natural number $\geq \kappa$. Consider the classical chance structure, $\mathbf{H}$, of $\nu^{2\mu}$ elements, with causal universe T, constructed in Theorem XV.1. Enumerate the elements of Λ_ξ by

$$\Lambda_\xi = \{\lambda_1, \lambda_2, \ldots, \lambda_\kappa\}.$$

We can partition $\mathbf{H}$ into κ pairwise disjoint classes $\mathbf{A}_1, \mathbf{A}_2, \ldots, \mathbf{A}_\kappa$, such that

$$|\operatorname{Pr}_\mathbf{H}(\mathbf{A}_n) - \operatorname{pr}_\xi(\lambda_n)| \leq \frac{1}{\nu^{2\mu}},$$

for $n = 1, 2, \ldots, \kappa$, as follows:

Define by induction natural numbers k_n, for $n \leq \kappa$, using the following procedure: k_n is the largest natural number such that

$$\sum_{i=1}^n \operatorname{pr}_\xi(\lambda_i) \geq \sum_{i=1}^n \frac{k_i}{\nu^{2\mu}}.$$

Partition $\mathbf{H}$ in such a way that k_n elements go into $\mathbf{A}_n$, for $n = 1, \ldots, \kappa$. This is possible because k_κ is defined as the largest natural number such that

$$\sum_{i=1}^\kappa \frac{k_i}{\nu^{2\mu}} \leq \sum_{i=1}^\kappa \operatorname{pr}_\xi(\lambda_i) = 1.$$

and, thus, $\sum_{i=1}^\kappa k_i = \nu^{2\mu}$. We have

$$\operatorname{Pr}_\mathbf{H}(\mathbf{A}_n) = \frac{k_n}{\nu^{2\mu}}, \tag{1}$$

for $n = 1, 2, \ldots, \kappa$.
Also
$$\text{pr}_\xi(\lambda_n) = \sum_{i=1}^{n} \text{pr}_\xi(\lambda_i) - \sum_{i=1}^{n-1} \text{pr}_\xi(\lambda_i). \tag{2}$$

It is clear, from the definition of k_n, that
$$\sum_{i=1}^{n} \frac{k_i}{\nu^{2\mu}} \leq \sum_{i=1}^{n} \text{pr}_\xi(\lambda_i) \leq \sum_{i=1}^{n} \frac{k_i}{\nu^{2\mu}} + \frac{1}{\nu^{2\mu}}.$$

Thus
$$\sum_{i=1}^{n} \frac{k_i}{\nu^{2\mu}} - (\sum_{i=1}^{n} \frac{k_i}{\nu^{2\mu}} + \frac{1}{\nu^{2\mu}}) \leq \sum_{i=1}^{n} \text{pr}_\xi(\lambda_i) - \sum_{i=1}^{n-1} \text{pr}_\xi(\lambda_i) \leq \sum_{i=1}^{n} \frac{k_i}{\nu^{2\mu}} + \frac{1}{\nu^{2\mu}} - \sum_{i=1}^{n-1} \frac{k_i}{\nu^{2\mu}}.$$

Therfore, by (2)
$$\frac{k_n}{\nu^{2\mu}} - \frac{1}{\nu^{2\mu}} \leq \text{pr}_\xi(\lambda_i) \leq \frac{k_n}{\nu^{2\mu}} + \frac{1}{\nu^{2\mu}}$$

and, hence
$$|\text{pr}_\xi(\lambda_i) - \frac{k_n}{\nu^{2\mu}}| \leq \frac{1}{\nu^{2\mu}}, \tag{3}$$

for $n = 1, \ldots, \kappa$.

Define a new probability distribution pr' over Λ_ξ by
$$\text{pr}'(\lambda_n) = \text{Pr}_\mathbf{H}(\mathbf{A}_n),$$

for $n = 1, 2, \ldots, \kappa$. By (1)
$$\text{pr}'(\lambda_n) = \frac{k_n}{\nu^{2\mu}},$$

for $n = 1, 2, \ldots, \kappa$. By (3)
$$\sum_{\lambda \in \Lambda_\xi} |\text{pr}_\xi(\lambda) - \text{pr}'(\lambda)| \leq \frac{\kappa}{\nu^{2\mu}} \leq \frac{\nu}{\nu^{2\mu}} = \frac{1}{\nu^\mu} \approx 0.$$

Thus, by Definition XI.5, ξ considered as a process over $\langle \Lambda_\xi, \text{pr}_\xi \rangle$ and ξ', which is the same function as ξ but over $\langle \Lambda_\xi, \text{pr}' \rangle$, are very nearly equivalent.

Define, next, the process η over $\mathbf{H}$ by
$$\eta(t, \omega) = \xi(t, \lambda_n),$$

for $t \in T$ and $\omega \in \mathbf{A}_n$. Then $\Lambda_\eta = \Lambda_\xi$ and $\text{pr}_\eta = \text{pr}'$. Thus, η is equivalent (Definition VI.20) to ξ' defined over $\langle \Lambda_\xi, \text{pr}' \rangle$. Therefore, η and ξ' defined over $\langle \Lambda_\xi, \text{pr}' \rangle$ are very nearly equivalent. Hence, η and ξ defined over $\langle \Lambda_\xi, \text{pr}_\xi \rangle$ are very nearly equivalent. □

As an immediate corollary of this theorem, we get:

4. REPRESENTATION THEOREM

COROLLARY XV.6. *Let ξ be an internal stochastic process indexed by a hyperfinite set T. Then there is a very nearly equivalent internal process η indexed by T, defined over the internal probability space $\langle \Omega, \mathrm{pr} \rangle$ satisfying*

$$\mathrm{pr}(\omega) = \mathrm{pr}(\omega')$$

for all $\omega, \omega' \in \Omega$.

This corollary has a direct proof, which is similar to the proof of Theorem XV.5. The statement of the corollary and its proof are independent of my definition of probability structures. So this corollary is a general theorem on stochastic processes.

As it is proved in [71, Appendix], every standard stochastic process can be approximated by a hyperfinite process. Thus, by the representation theorem XV.5, every standard process can be approximated by a process defined over a chance structure, in the sense introduced in this book. By the Corollary, on the other hand, every standard process can be approximated by a process defined over a space with equiprobable elementary events.

Although I believe that the general representation theorem, Theorem XV.5, is mathematically important, its philosophical importance consists only in the fact that it shows that given any chance setup that can be modeled by a stochastic process, it is always possible to construct a chance structure that models the setup. We need, however, not just any chance structure, but one that is reasonable and natural according to the physical situation involved. In the next chapter, we shall discuss a few examples of natural chance structures for certain phenomena.

Corollary XV.6 shows that the classical definition of probability is able to account, in an approximate manner, for all stochastic processes, and, hence, for all random variables. Thus, the criticism to the classical definition that it is very restrictive, in the sense that there are many probability distributions which cannot be obtained using this definition, does not seem to be correct: any probability distribution can be approximated in a very precise and close sense by a probability distribution defined in terms of equiprobability.

CHAPTER XVI

Examples of probability structures

1. Distributions of r balls into n cells

In this chapter, we shall discuss several probability models each one adequate for a different chance setup. This first section is devoted to the study of the different possible distributions of r balls into n cells, which are used in physics and other areas. The general situation is that there are r balls and n cells, and that the r balls are put into the n cells. There are several possible distributions for the placing of the balls. There are three main types, called statistics, which go under the names of Maxwell-Boltzmann, Bose-Einstein, and Fermi-Dirac.[1] I will now give the probability structures for the three types of chance setups.

EXAMPLE XVI.1 (MAXWELL-BOLTZMANN STATISTICS.). For this statistics, we think of the balls and the cells as identifiable in principle. The intuitive idea behind this setup is that each ball is placed independently in a cell, and that each cell can contain any number of balls. All distributions of the particular balls in the cells should result equiprobable.

Form these considerations, we see that we can assume that, for each ball, a cell is chosen independently. So we can represent the balls by a set of r independent causal moments B, and at each one of these moments a cell from a set of n cells C is chosen at random. The set B is then I_r, the causal relation with r incomparable moments. For $b \in B$, the outcome $\lambda(b)$ is of the form $\langle C, \{c\} \rangle$, where $c \in C$. Thus, the chance structure **MB** consists of all outcomes λ with domain B, and such that $\lambda(b) = \langle C, \{c\} \rangle$, for some $c \in C$. Put in the form of relational systems, λ looks as follows:

$$\lambda = \langle B, C; =, R_0, R_1 \rangle,$$

where the relation $=$ restricted to B is the causal ordering, $R_0 = B \times C$, and R_1 is a subset of R_0 that is a function with domain B. That is, for each $b \in B$, R_1

---

[1]In [18], simple probability structures were proposed for these statistics. It seems to me now, however, that they are not quite adequate, and that compound probability structures presented here are better.

selects the cell in which b is placed by outcome λ.

We have that, for $b \in B$, $\mathbf{MB}(b)$ is the set of all structures of the form $\langle C, \{c\}\rangle$, i.e., the possibilities for the placing of each ball are all the cells, and these possibilities are independent for the different balls.

Notice that $P_{\mathbf{MB}}$ is B, because no element of B has predecessors. Thus, the group of invariance $G_{\mathbf{MB}}$ consists of the functions g with domain $B \times C$, such that there is a permutation $\bar{g}$ of B and, for each $b \in B$, a permutation g_b of C, with $g(\langle b, c\rangle) = \langle \bar{g}(b), g_b(c)\rangle$, with the additional restriction that for each $\lambda \in \mathbf{MB}$, $g''\lambda \in \mathbf{MB}$. If $\lambda = \langle B, C; =, R_0, R_1\rangle$, this means that $g''(B \times C) = B \times C$ and that $g''R_1$ is a function with domain B and range included in C. That is, the following two conditions are satisfied:

(1) If $\bar{g}(b) \neq \bar{g}(b')$, i.e., $b \neq b'$, we must have that $g_b(c) \neq g_{b'}(c)$, for every $c \in C$; and

(2) for every $\langle b, c\rangle \in B \times C$ there must be b' and c' such that $b = \bar{g}(b')$ and $c = g_{b'}(c')$.

It is easy to see that $\mathbf{MB}$ is classical, i.e., for any two outcomes λ and μ, there is a $g \in G_{\mathbf{MB}}$ such that $\mu = g''\lambda$.

Thus, all singletons are measure-determined. Hence, the algebra of events is the power set of $\mathbf{MB}$, and the probability measure is the counting measure. That is, for any $\mathbf{A} \subseteq \mathbf{MB}$, the measure of $\mathbf{A}$ is the number of elements of $\mathbf{A}$ divided by the total number of elements of $\mathbf{MB}$.

EXAMPLE XVI.2 (BÖSE-EINSTEIN STATISTICS.). This statistics is characterized by the fact that the balls are not identifiable. It is really more a matter of positions in each cell being occupied or not. There may be any number of balls in each cell. In this case, the distributions of numbers of balls in each cell, and not of the balls themselves, are equiprobable.

There are at least two different ways of modeling this setup. The first models the following experimental situation, which I shall give first in a continuous version. Suppose that we have n cells and that we have a fixed amount of water. We pour water into the cells up to a certain point, but the stopping point is randomly chosen. Water is poured into the cells in a certain order. So we must have the cells numbered consecutively. If we begin by cell i, we stop and continue with a cell with a higher number, and so on, until the water is exhausted. We never go back to a previous cell. Water can be replaced by r balls, and then we get the proper model.

The chance structure for this setup has as causal relation the n cells linearly ordered, i.e., L_n, which we can identify with the natural numbers $1, \ldots, n$ in their natural order. For cell 1, the possibilities are to choose any number of balls from 1 to r. Once we have made the choices for all cells less than i, the possibilities for cell i are to choose any number from 1 to r minus the sum of the numbers chosen before i. It is easy to prove that this structure is random and that the measure is the counting measure, which gives the statistics we are seeking.

Although this chance structure seems natural, I prefer another one, which is more similar to $\mathbf{MB}$. We take as the causal structure the set of r balls B, but here linearly ordered. We identify B with L_r, which we may take to be the natural numbers $1, \ldots, r$ in their natural order. We select a cell for each ball, but now the different selections are not independent. Let us also number the cells, say $c_1, \ldots, c_n$. For ball 1, the possibilities are any of the cells. If for ball i we have selected cell c_j, then the possibilities for ball $i+1$ are only cells c_k with $j \leq k \leq n$. Thus, the chance structure, call it **BE**, consists of outcomes λ with causal ordering L_r, such that $\lambda(1) = \langle C, \{c\} \rangle$ where $c \in C$, and $\lambda(i+1) = \langle C, \{c_k\} \rangle$, satisfying the condition that: if $\lambda(i) = \langle C, \{c_j\} \rangle$, then $j \leq k \leq n$. Thus, $\mathbf{BE}(1) = \{\langle C, \{c\}\rangle \mid c \in C\}$, but $\mathbf{BE}(\lambda_{i+1}) =$

$$\{\langle C, \{c\}\rangle \mid c \in C, \text{ and } (\lambda(i) = \langle C, \{c_j\}\rangle) \implies c = c_k, \text{ for } k \text{ with } j \leq k \leq n)\}.$$

It is easy to transform this function version to a relational system version. From now on, I will not do this transformation always explicitly, but only in some cases which will provide good examples.

Since B is linearly ordered, it has no nontrivial automorphisms. The group of invariance $G_{\mathbf{BE}}$ consists of all functions g whose domain is $P_{\mathbf{BE}} \times C$, such that for each λ_i there is a permutation of C, g_{λ_i}, with $g''\langle \lambda_i, c\rangle = \langle i, g_{\lambda_i}(c)\rangle$. the function g has to satisfy the additional property that g (or, more properly, g'') is a function from **BE** onto **BE**. Thus, if $\lambda(i) = \langle C, \{c\}\rangle$, $g_{\lambda_i}(c) = c_j$ and $\lambda(i+1) = \langle C, \{d\}\rangle$, then $g_{\lambda_{i+1}}(d) = c_k$ with $j \leq k \leq n$.

The structure **BE** is clearly classical. Thus, singletons are measure-determined, and, hence, we have that the algebra of events is the power set of **BE** and the measure is the counting measure. But in **BE**, we just have one outcome with each distribution, so the statistics is Böse-Einstein.

EXAMPLE XVI.3 (FERMI-DIRAC STATISTICS.). This statistics is quite similar to Böse-Einstein's, but with the restriction that only one ball may occupy each cell. We could just model this case, as the simple chance structure of the choosing of a sample of r elements from the population of n cells. But, in order to be consistent with the previous statistics, we prefer a chance structure **FD** similar to **BE**, but defining $\mathbf{FD}(\lambda_{i+1}) =$

$$\{\langle C, \{c\}\rangle \mid c \in C, \text{ and } (\lambda(i) = \langle C, \{c_j\}\rangle) \implies c = c_k \text{ for } k \text{ with } j < k \leq n)\}.$$

The rest of the notions for **FD** can be easily obtained from the corresponding notions for **BE**. Again the algebra of events is the power set and the measure, the counting measure.

We proceed now to the measurement processes appropriate for these structures and its possible probability models. Here, we can have a system $Y = \langle Y_j \mid 1 \leq j \leq n\rangle$ of 0-ary measurement functions such that $Y_j(\lambda)$ is the number of balls in cell c_j. The pairs $\langle \mathbf{MB}, Y\rangle$, $\langle \mathbf{BE}, Y\rangle$, and $\langle \mathbf{FD}, Y\rangle$ are probability structures.

In the next chapter, we shall introduce formally languages to deal with these structures. These language were already discussed informally in Chapter III, Section 5. We have that for a sentence φ of this language we can compute the possibly different probabilities $\Pr_{\langle \mathbf{MB}, Y\rangle}(\varphi)$, $\Pr_{\langle \mathbf{BE}, Y\rangle}(\varphi)$, and $\Pr_{\langle \mathbf{FD}, Y\rangle}(\varphi)$. The differences in probabilities gives us a criterion for deciding which is the model acceptable for a particular situation. This criterion was already discussed in general in Chapter IV. It will be further discussed in Part 4.

2. Selection of a chord at random

Next, we discuss how to model through probability structures the three different solutions to Bertrand's chord paradox proposed in Chapter III, Section 2. As in the previous section, I first describe the chance structure for each chance setup, and then the probability structures. Let us recall how Bertrand's paradox arises. The general situation is that of choosing a chord at random from a circle of unit radius. There are several ways of choosing such a chord.

(1) Because any chord of the circle intersects the circle in two points, we may suppose these two points to be independently and uniformly distributed on the circumference of the circle. A possible experiment which gives rise to this situation is the following. Along a circle of radius one a needle circles, spinning also around the needle's center, which lies on the circumference of the circle. Where the needle stops, it determines a chord on the circle. Since the setup is just the independent choosing of two points in the circle, the essential structure of this setup is captured by the probability structure **IC** of Example XIII.1, with causal relation I_2, i.e., two incomparable causal moments.

(2) The length of the chord is determined by its distance from the center of the circle, and is independent of its direction. Thus, we may suppose that the chord has a fixed direction perpendicular to a given diameter of the circle, with distance along the diameter measuring its distance from the center. We may then suppose that its point of intersection with this diameter has a uniform distribution. This situation may arise with the following experiment. Draw parallel lines on the floor at a distance of two (the diameter of the circle). Throw a circle of radius one on top of them. One of the lines has to intersect the circle, and, thus, it determines a chord.

The essential elements for the chance mechanism for this case is the following. We first select a point on the circle and then a point on the radius through the first point. The chord chosen is the perpendicular to the radius passing through the point chosen in the radius. Here we also need a compound structure but now with causal relation L_2, which consists of two elements t and s with $t \prec s$. If we call the chance structure $\mathbf{H}_2$, then $\mathbf{H}_2(t) = \mathbf{K}_2$, the simple structure for the choosing of a point from a circle. If $\lambda(t) = \lambda_c$, the choosing of the point c, then $\mathbf{H}_{\lambda_s}$ consists of the systems $\lambda(s)$ of the form $\langle R, +, \{r\}\rangle$, where R is the radius through c, $+$ represents addition modulo the radius (i.e., if we take c as 0, $a + b$ is the sum of a and b, minus the length of R if $a + b > R$), and r is the point selected in R. The system $\langle R, +\rangle$ gives us the linear structure of the radius, and is necessary in order to select a point from a line segment and not just from a set.

As usual, we consider a hyperfinite approximation of these sets. The measure restricted to $\mathbf{H}_{\lambda_s}$ has to be invariant under translations, and, hence, is Lebesgue measure. With this indications, it is easy to construct a algebra of events and a measure for $\mathbf{H}_2$.

(3) Any chord is uniquely determined by the point of intersection of the chord and a radius perpendicular to the chord (i.e., by the middle point of the chord). So one natural assumption is that this point of intersection is distributed uniformly over the interior of the circle. Then the probability of its lying in any region of area A is A/π, since the total area of the circle is π.

The following experiment is represented by this chance setup. Trace a horizontal line on a wall, and insert a spikes in the wall on the line. Throw a disc of radius one so that it is implanted on the spike. When the disc reaches its equilibrium position by the action of gravity, the line will intersect the circle on a chord, with the spike on the middle point of the chord.

For this case a simple probability structure, call it $\mathbf{H}_3$, is enough. The chance setup is essentially the selection of a point at random from the interior of the circle. Thus, the relational systems in $\mathbf{H}_3$ are of the form $\langle A, t, \{p\}\rangle$, where A is the interior of the circle, $t(a,b)$ is the addition of a and b as vectors modulo the circle, and p is the point chosen. By using hyperfinite approximations, it is easy to see that we obtain Lebesgue measure on the plane.

For the definition of the probability structures, we now introduce a 0-ary measurement process Y, such that $Y(\lambda)$ is the length of the chord chosen when the outcome is λ.

We discussed in Chapter III, Section 5, a language whose sentences are assigned probabilities by these probability structures. We can compute, for a sentence φ of the corresponding language, $\Pr_{\langle \mathbf{IC}, Y\rangle}(\varphi)$, $\Pr_{\langle \mathbf{H}_2, Y\rangle}(\varphi)$, and $\Pr_{\langle \mathbf{H}_3, Y\rangle}(\varphi)$. Again, by applying criteria which were discussed in Chapter IV, we may be able to decide which is the appropriate probability model for the setup at hand.

3. The theory of errors

In this section, I present a theory of random errors.[2] In this theory, a quantity is supposed to have a theoretical value, but the procedure of measurement introduces a random error that comes from a combination of a large number of independent causes, each one producing a very small error in the positive or negative direction. The total error for a particular measurement is obtained by adding up the errors produced by the different causes. For simplicity, we assume that the theoretical value of the measurement is zero.

We can idealize a large number of causes by an infinite number of causes. It is not very difficult to modify our causal relations in order to accommodate and infinite number of independent causal moments. Most theorems are easier to prove and more intuitive to formulate, however, with an infinite but hyperfinite

---

[2] This model was developed by Bertossi, and it appeared in essentially the same form in [23].

number of causes, i.e., the internal cardinality of the set of causes is an infinite nonstandard natural number.

For the particular chance structure at hand we take as the field
$$I_\nu = \{1, 2, \ldots, \nu\}$$
of our causal relation the numbers less than or equal a number $\nu \in {}^*\mathbb{N} - \mathbb{N}$. Since the causes are independent, all elements of I_ν are incomparable in the causal relation. With each cause κ we associate a positive infinitesimal number ε_κ, and κ may cause the error ε_κ or $-\varepsilon_\kappa$ with equal probability. In order to keep the total error within bounds, we assume that
$$\sum_{\kappa=1}^{\nu} \varepsilon_\kappa^2 = \varepsilon$$
is finite.

The chance structure will be denoted by **Er**. The outcomes in **Er** are the functions λ with domain I_ν, such that for each $\kappa \in I_\nu$, $\lambda(\kappa)$ is $\langle\{\varepsilon_\kappa, -\varepsilon_\kappa\}, \{\varepsilon_\kappa\}\rangle$ or $\langle\{\varepsilon_\kappa, -\varepsilon_\kappa\}, \{-\varepsilon_\kappa\}\rangle$. Hence
$$\mathbf{Er}(\kappa) = \{\lambda(\kappa) \mid \lambda \in \mathbf{Er}\} = \{\langle\{\varepsilon_\kappa, -\varepsilon_\kappa\}, \{\varepsilon_\kappa\}\rangle, \langle\{\varepsilon_\kappa, -\varepsilon_\kappa\}, \{-\varepsilon_\kappa\}\rangle\}.$$

As relational systems, these outcomes are of the form
$$\lambda = \langle I_\nu, E; =, R_0, R_1\rangle$$
where
$$E = \{\varepsilon_\kappa \mid \kappa \in I_\nu\} \cup \{-\varepsilon_\kappa \mid \kappa \in I_\nu\},$$
$$R_0 = \{\langle\kappa, \varepsilon_\kappa\rangle \mid \kappa \in I_\nu\} \cup \{\langle\kappa, -\varepsilon_\kappa\rangle \mid \kappa \in I_\nu\}$$
$$R_1 = \{\langle\kappa, \beta_\kappa\rangle \mid \kappa \in I_\nu\}, \text{ with } \beta_\kappa = \varepsilon_\kappa \text{ or } \beta_\kappa = -\varepsilon_\kappa.$$

With this definition we recapture $\lambda(\kappa) = \langle R_0''\{\kappa\}, R_1''\{\kappa\}\rangle$.

Here, $P_{\mathbf{Er}} = I_\nu$, because all $\kappa \in I_\nu$ are first elements and incomparable. The group of invariance $G_{\mathbf{Er}}$ is constituted by all functions g such that the domain of g is $I_\nu \times E$, there is a permutation $\bar{g}$ of I_ν, and for each $\kappa \in I_\nu$, there is a one-one mapping g_κ from the universe of $\mathbf{Er}(\kappa)$ onto the universe of $\mathbf{Er}(\bar{g}(\kappa))$ with $g(\langle\kappa, \beta\rangle) = \langle\bar{g}(\kappa), g_\kappa(\beta)\rangle$, where $\beta = \varepsilon_\kappa$ or $\beta = -\varepsilon_\kappa$.

Since **Er** is classical, the probability space is constituted by the power set and the counting measure.

The measurement process for the probability structure is obtained as follows. First define the nonstandard 0-ary internal function $\xi : \mathbf{Er} \to {}^*\mathbb{R}$, by
$$\xi(\lambda) = \sum_{\kappa=1}^{\nu} X(\kappa, \lambda),$$
where $X(\kappa, \lambda) = \varepsilon_\kappa$ or $-\varepsilon_\kappa$ according to $\lambda(\kappa) = \langle\{\varepsilon_\kappa, \varepsilon_\kappa\}, \{\varepsilon_\kappa\}\rangle$ or $\lambda(\kappa) = \langle\{\varepsilon_\kappa, \varepsilon_\kappa\}, \{-\varepsilon_\kappa\}\rangle$. The quantity $\xi(\lambda)$ measures the total (nonstandard) error for outcome λ. By Corollary XI.15, ξ is nearly normally distributed with mean zero and variance ε.

The probability structure for the theory of errors with theoretical value 0 and standard deviation ε is $\langle \mathbf{Er}, \xi \rangle$.

We could define the standard measurement function $°\xi : \mathbf{Er} \to \mathbb{R}$ by

$$°\xi(\lambda) = \operatorname{st} \xi(\lambda),$$

where st is the standard part function.

It is easy to modify the construction so that the mean (i.e., the theoretical value of the measurement) is any real number r. Thus, in order to obtain a model for the measurement of a certain quantity, we must be given two parameters: the theoretical measurement, r, and the standard error ε. With these two numbers, we construct a probability structure as above with its total error measurement function ξ. Then ξ, which gives the actual value of the measurement, will be normally distributed with mean r and standard deviation ε. This model is compared with the actual data by the methods discussed in Chapter IV, which will be further analyzed in Part 4.

4. Brownian motion

Just as for the theory of errors, I use a nonstandard characterization of Brownian motion. This characterization was obtained by Anderson, [2]. However, I shall follow the modification of Anderson's model proposed by Keisler, [53].

The causal ordering for this structure is the hyperfinite time-line, i.e., the near interval for $[0, 1]$

$$T = \{0, dt, 2\, dt, \ldots, 1\},$$

with $dt = 1/\nu$, where ν is an infinite natural number. The ordering is the natural ordering of the hyperreal numbers. Thus, T, as a causal relation is the same as L_ν. In the terminology used in Chapter VIII, Example VIII.2, T is a near interval for $[0, 1]$.

An outcome λ (considered as a function) has domain T and

$$\lambda(t) = \langle \{\sqrt{dt}, -\sqrt{dt}\}, \{\delta\} \rangle,$$

where $\delta = \sqrt{dt}$ or $\delta = -\sqrt{dt}$. The chance structure **BM** is the set of such functions.

For any $\lambda \in \mathbf{BM}$ and any $t \in T$, $\mathbf{BM}_{\lambda_t}$ has the two systems obtained with $\delta = \sqrt{dt}$ and $\delta = -\sqrt{dt}$. This is a model for a hyperfinite random walk, what we called in Definition XI.4, a Wiener walk. Thus, between times t_0 and $t_0 + dt$ the "particle" moves a distance $\sqrt{dt}$ either to the left or to the right, independently and with equal probability. This equality of probability is determined in our case by the fact that the group of invariance of $\mathbf{BM}_{\lambda_t}$ is the group of all permutations of its universe $\{\sqrt{dt}, -\sqrt{dt}\}$.

The outcomes as relational systems are of the form

$$\lambda = \langle T, U; \leq, R_0, R_1 \rangle,$$

where $\langle T, \leq \rangle$ is the hyperfinite near interval, called a time-line, with its natural ordering,

$$U = \{\sqrt{dt}, -\sqrt{dt}\},$$
$$R_0 = \{\langle t, \sqrt{dt}\rangle \mid t \in T\} \cup \{\langle t, -\sqrt{dt}\rangle \mid t \in T\},$$
$$R_1 = \{\langle t, \alpha_t\rangle \mid t \in T\},$$

where α_t is $\sqrt{dt}$ or $-\sqrt{dt}$.

As it should be, $\lambda(t) = \langle R_0''\{t\}, R_1''\{t\}\rangle$, and $\mathbf{BM}_{\lambda_t} = \{\lambda(t) \mid \lambda \in \mathbf{BM}\}$. We also have, $\lambda_t = 0$, if $t = 0$, and $\lambda_t = \lambda \upharpoonright \{0, \ldots, (k-1)\,dt\}$, if $t = k\,dt$; $\mathrm{P}_{\mathbf{BM}}$ is, as usual, the set of λ_t for $\lambda \in \mathbf{BM}$ and $t \in T$.

Since T is linearly ordered, it has no automorphisms different from the identity. Hence, the group of invariance, $G_{\mathbf{BM}}$ is the set of functions g such that:

(1) The domain of g is $\mathrm{P}_{\mathbf{BM}} \times U$.
(2) For every $\lambda_t \in \mathrm{P}_{\mathbf{BM}}$ there is a permutation g_{λ_t} of U such that

$$g(\langle \lambda_t, \beta\rangle) = \langle t, g_{\lambda_t}(\beta)\rangle.$$

Then, if $\lambda \in \mathbf{BM}$ with $\lambda(t) = \langle U, \{\beta\}\rangle$, $g''\lambda$ is obtained by setting

$$g''\lambda(t) = \langle U, \{g_{\lambda_t}(\beta)\}\rangle.$$

If λ, $\mu \in \mathbf{BM}$, then there is a $g \in G_{\mathbf{BM}}$ such that $\mu = g''\lambda$. Hence, $\mathbf{BM}$ is random and, thus, all singletons of $\mathbf{BM}$ are measure-determined. Therefore, the internal probability space is constituted by the internal power set and the counting measure.

We introduce the following nonstandard internal measurement process B, which is a stochastic process adapted to $\mathbf{BM}$ in the internal probability space $\langle \mathbf{BM}, \mathcal{F}_{\mathbf{BM}}, \mathrm{Pr}_{\mathbf{BM}}\rangle$. First, let the internal function X be defined by, $X : T \times \mathbf{BM} \to {}^*\mathbb{R}$ and $X(t, \lambda) = \beta$, if $\lambda(t) = \langle U, \{\beta\}\rangle$. Then, $B : T \times \mathbf{BM} \to {}^*\mathbb{R}$, with

$$B(t, \lambda) = \sum_{s=0}^{t} X(s, \lambda).$$

The probability structure $\langle \mathbf{BM}, B\rangle$ is of type $\langle T, d\rangle$ with domain of $d = \{0\}$ and $d(0) = 1$. The function B is a stochastic process adapted to $\mathbf{BM}$. It is clear that B is a Wiener walk, according to Definition XI.4. In Corollary XI.13, we obtained a characterization of this process, B, and its probability distribution, pr_B. We know from Corollary XI.13 that its increments are independent, and that if $t - s$ is finite and noninfinitesimal, then $B(t) - B(s)$ is nearly normally distributed with mean zero and variance $t - s$.

We can be content with this nonstandard process. In order to be complete a brief account of how the usual standard space is obtained follows, although it is not necessary for our purposes and, so the rest of this section may be omitted.[3]

---

[3] In Appendix B, a sketch is given of the construction, via the Daniell integral, of the external space considered below. For details of a different construction, see [26] or [1].

We can define first the external process $b : [0,1] \times \mathbf{BM} \to \mathbb{R}$, given by $b(\operatorname{st} t, \lambda) = \operatorname{st}(B(t, \lambda))$. This is a stochastic process in the Loeb space obtained from the internal probability space mentioned above. The Loeb space is the standard measure space $\langle \mathbf{BM}, \mathcal{L}(\mathcal{F}_{\mathbf{BM}}), \mathcal{L}(\Pr_{\mathbf{BM}}) \rangle$, where $\mathcal{L}(\mathcal{F}_{\mathbf{BM}})$ is the σ-algebra generated by $\mathcal{F}_{\mathbf{BM}}$ and $\mathcal{L}(\Pr_{\mathbf{BM}})$ is the extension of $\Pr_{\mathbf{BM}}$ to this σ-algebra.

We can then construct the standard process containing the trajectories $\mu_\lambda : [0,1] \to \mathbb{R}$ defined for $\lambda \in \mathbf{BM}$ by $\mu_\lambda(t) = b(t, \lambda)$, for each $t \in [0,1]$. The set of these trajectories we call W. First, define the space formed by $C[0,1]$, the continuous functions on $[0,1]$, the completion of the family of sets $A \cap W = \{\mu_\lambda \mid \lambda \in \mathbf{A}\}$, for an $\mathbf{A} \in \mathcal{L}(\mathcal{F}_{\mathbf{BM}})$, with respect to the measure $\Pr A = \mathcal{L}(\Pr_{\mathbf{BM}})(\mathbf{A})$, and the measure $\Pr$ itself. Then $\Pr$ is the usual Wiener measure. We can, then define the process $Z : [0,1] \times C[0,1] \to \mathbb{R}$, by $Z(t, \mu_\lambda) = \mu_\lambda(t)$. This process is the usual standard Brownian motion.

I believe that the nonstandard Brownian motion B on the internal space $\langle \mathbf{BM}, \mathcal{F}_{\mathbf{BM}}, \Pr_{\mathbf{BM}} \rangle$ is quite adequate for most purposes. If one wants a space in the usual sense of the word, however, the standard process, b on the Loeb space should be enough. The Loeb space is a space in the usual sence, i.e., it has a σ-algebra and a σ-additive probability measure. Only the base set, $\mathbf{BM}$, is nonstandard. I have only introduced the usual process on $C[0,1]$ in order to show how from the nonstandard processes we can obtain the usual standard ones. We shall not discuss other stochastic processes in this book. But I think that all processes can be treated this way.

5. Models for inference and decision

In this last section of the chapter, I will indicate how to construct precisely the models which will be used in the analysis of statistical inference and decision theory of Part 4.

We may need to compound, here, probability structure that are already composite. That is, our outcomes λ will be such that $\lambda(t)$ is already a compound outcome, i.e., a relational system with a causal relation, for each t in a new causal relation. Since the new causal relation is very simple, either I_n, for a certain n, or L_2, the construction is also simple, and I leave the details to the reader.

Models for inference. We first discuss how to model a number of independent repetitions of an experiment. Suppose that we have a chance structure $\mathbf{K}$ modeling a certain chance setup. We want now to model the setup consisting of n independent repetitions of the original setup, call it $^n\mathbf{K}$. The set $^n\mathbf{K}$ has a structure similar to $\mathbf{IC}$. The outcomes of the composite setup $^n\mathbf{K}$ have causal relation I_n, with n incomparable elements. Each outcome $\lambda \in {}^n\mathbf{K}$ is defined by: $\lambda(i) \in \mathbf{K}$, for every $i \in I_n$. Now, as I said above, $\mathbf{K}$ itself may be a compound structure with a certain causal relation T. Hence, $\lambda(i)$ may be an outcome and not just a one-sorted relational system. We would have to extend our definitions to allow for compound compound structures, and so on, but the procedure is straightforward, so I shall leave the details to the reader.

If we have an algebra of events $\mathcal{F}_\mathbf{K}$ and a measure $\Pr_\mathbf{K}$, then $\mathcal{F}_{^n\mathbf{K}}$ and $\Pr_{^n\mathbf{K}}$ are the product algebra and the product measure, as was determined in Chapter XIV.

Suppose, now, that Y is a 0-ary measurement function defined on elements of $\mathbf{K}$. For $^n\mathbf{K}$, we define the 1-ary measurement process given by

$$Y(i, \lambda) = Y(\lambda(i)).$$

We use the same symbol Y, since there is no danger of confusion: $Y(i, \lambda)$ is usually written $Y_i(\lambda)$. In the language appropriate for this process, discussed informally in Chapter III, Section 5, and, more formally, in the next chapter, we can talk about compositions of the process with continuous functions. In particular, we can have

$$\frac{1}{n} \sum_{i=1}^{n} Y_i(\lambda),$$

the mean or average of $Y_1, Y_2, \ldots, Y_n$. If the values of the original Y are just 0 or 1, then $\sum_{i=1}^n Y_i(\lambda)$ is the number of ones in the outcome, i.e., the frequency, and $(1/n) \sum_{i=1}^n Y_i(\lambda)$, the relative frequency of ones in the outcome. For this case, we use the notation

$$^nY(k, \lambda) = \sum_{i=1}^{k} Y_i(\lambda).$$

If the appropriate language contains a symbol for Y, then it also can refer to ^nY and $(1/n)^nY$. In the next paragraphs, we consider only Y's with values 0 or 1.

For many purposes in probability theory and statistical inference, we need a common chance structure where all functions ^nY and Y_i, for all natural numbers n and i, are defined. We could introduce for this purpose, the structure $^\mathbb{N}\mathbf{K}$ consisting of all infinite sequences of elements of $\mathbf{K}$. This is not very convenient, since we would have to accept infinite causal relations, at least the causal relation $\langle \mathbb{N}, = \rangle$. Although it is not difficult to generalize our definitions and theorems to causal relation that admit an infinite number of incomparable elements, in order to maintain the limitation to finite causal relations, and also because it seems simpler, we adopt another course, which involves nonstandard analysis.

We take an infinite natural number $\nu \in {^*\mathbb{N}} - \mathbb{N}$, and construct the probability structure $^\nu \mathbf{K}$ with causal relation I_ν. The algebra of events is the internal product of the algebras $\mathcal{F}_\mathbf{K}$, and the measure is the internal product measure. Then we consider the 1-ary measurement process $^\nu Y$, defined by

$$^\nu Y(n, \lambda) = \sum_{i=1}^{n} Y_i(\lambda).$$

Thus, for every standard natural number n, $^\nu Y(n, \lambda)$ is the frequency of ones in the first n terms of the sequence λ. Similarly, we can use

$$\mathrm{Fr}_Y(n, \lambda) = \frac{1}{n} {^\nu Y(n, \lambda)},$$

5. MODELS FOR INFERENCE AND DECISION

the relative frequency of ones in the first n terms of λ. We write the function $\mathrm{Fr}_Y(n,\ )$, $\mathrm{Fr}_{Y,n}$. Since these functions are obtained by the use of continuous functions, they are in any language that includes the one-ary function Y.

We briefly discuss the passage to standard functions, which can be ommitted. For n a standard natural number, $\mathrm{Fr}_Y(n,\lambda)$ is standard, but for infinite n, this may not be the case. We can now introduce a new measurement function, which has standard values.

$$^\circ\mathrm{Fr}_Y(j,\lambda) = \begin{cases} \mathrm{st}(\mathrm{Fr}_Y(j,\lambda)), & \text{if } \mathrm{Fr}_Y(j,\lambda) \text{ is finite,} \\ \infty, & \text{otherwise.} \end{cases}$$

If for j and i infinite, $\mathrm{Fr}_Y(i,\lambda)$ and $\mathrm{Fr}_Y(j,\lambda)$ are finite and infinitely close, then we define the standard 0-ary measurement function

$$\lim \mathrm{Fr}_Y(\lambda) = {}^\circ\mathrm{Fr}_Y(j,\lambda),$$

for any infinite j. With these functions it is possible to state and prove in the language which contains the Fr measurement functions the usual laws of large numbers and Chebychev's inequality, which are the standard counterparts of Theorems XI.5, XI.7, and VI.3.

Models for decision. The second model that we have to discuss briefly is similar to the urn model of Example XIII.3. We shall denote the chance structure by $\mathbf{B}$. The causal relation of $\mathbf{B}$ is $L_2 = \{t, s\}$. At t we have a chance structure $\mathbf{K} = \mathbf{B}(t)$ and for each $\omega \in \mathbf{K}$ we have at s a chance structure $\mathbf{B}(\omega)$. Thus, the compound outcomes in the chance structure $\mathbf{B}$ are the functions λ with domain L_2 and such that $\lambda(t) \in \mathbf{B}(t)$ and $\lambda(s) \in \mathbf{B}(\lambda(t))$. Here, since $(-\infty, s) = \{t\}$, λ_s can be identified with $\lambda(t)$, and, hence, $\mathbf{B}_{\lambda_s} = \mathbf{B}(\lambda(t))$. Just as in the example for repetitions of an experiment, the chance structures $\mathbf{B}(t)$ and $\mathbf{B}(\lambda(t))$ may themselves be compound, i.e., they may contain outcomes with causal relations.

We suppose that we have probability spaces

$$\langle \mathbf{B}(t), \mathcal{F}_{\mathbf{B}(t)}, \mathrm{Pr}_{\mathbf{B}(t)} \rangle \text{ and } \langle \mathbf{B}(\lambda(t)), \mathcal{F}_{\mathbf{B}(\lambda(t))}, \mathrm{Pr}_{\mathbf{B}(\lambda(t))} \rangle,$$

for each $\lambda \in \mathbf{B}$. The algebra $\mathcal{F}_\mathbf{B}$ is the family of subsets $\mathbf{A}$ of $\mathbf{B}$ such that the function

$$f_\mathbf{A}(\lambda(t)) = \mathrm{Pr}_{\mathbf{B}(\lambda(t))}(\mathbf{A}(\lambda(t)))$$

is $\mathrm{Pr}_{\mathbf{B}(t)}$-measurable. The measure $\mathrm{Pr}_\mathbf{B}$, for such an $\mathbf{A}$, is the average measure which may be written

$$\mathrm{Pr}_\mathbf{B} \mathbf{A} = \sum_{\lambda(t) \in \mathbf{A}(t)} f_\mathbf{A}(\lambda(t)) \cdot \mathrm{pr}_{\mathbf{B}(t)}(f_\mathbf{A}(\lambda(t)))$$
$$= \mathbf{E}_{\mathbf{B}(t)} f_\mathbf{A} \cdot \chi_{\mathbf{A}(t)},$$

where $\mathbf{E}_{\mathbf{B}(t)}$ is the expectation with respect to the probability $\mathrm{Pr}_{\mathbf{B}(t)}$. Using integrals, we may also write

$$\mathrm{Pr}_\mathbf{B} \mathbf{A} = \int_{\mathbf{A}(t)} f_\mathbf{A}(\lambda(t)) \, d\,\mathrm{Pr}_{\mathbf{B}(t)}.$$

A particularly interesting example is the following. Let $\mathbf{B}$ be obtained as follows: let $\lambda \in \mathbf{B}$, be such that $\lambda(t) = \langle\{1,\ldots,n\},\{j\}\rangle$, for some j with $1 \le j \le n$, and $\lambda(s) \in \mathbf{H}_j$ (i.e., $\mathbf{B}_{\lambda_s} = \mathbf{H}_j$), where $\mathbf{H}_1, \ldots, \mathbf{H}_m$ are chance structures. That is, at t we have the structure $\mathbf{B}(t)$ of the systems $\langle\{1,\ldots,n\},\{j\}\rangle$, for $1 \le j \le n$, and $\mathbf{B}_j = \mathbf{H}_j$, for $1 \le j \le n$.

The algebra $\mathcal{F}_{\mathbf{B}(t)}$ is the power set of $\mathbf{B}(t)$, and the measure $\Pr_{\mathbf{B}(t)}$ is the counting measure. The structure $\mathbf{B}(t)$ is actually the structure for the choosing of a sample of one element from a set of n elements. Let Y a the measurement process with $Y(t,\lambda) = i$, if $\lambda(s) \in \mathbf{H}_i$, that is, if $\lambda(t) = \langle\{1,\ldots,n\},\{i\}\rangle$. Then $\Pr_{\mathbf{B}}[Y_t = i]$ represents the probability of the structure $\mathbf{H}_i$, which can be written $\Pr_{\mathbf{B}}(\mathbf{H}_i)$, and $\Pr_{\mathbf{B}}([Y_s \in A] | [Y_t = i])$ represents the probability that the event $A = [Y_s \in A]$ was obtained when the probability distribution was given by the structure $\mathbf{H}_i$, which can be written as $\Pr_{\mathbf{B}}(A|\mathbf{H}_i)$. Similarly, we can write $\Pr_{\mathbf{B}}([Y_t = i] | [Y_s \in A])$ as $\Pr_{\mathbf{B}}(\mathbf{H}_i|A)$. Then Bayes formula has the normal appearance

$$\Pr_{\mathbf{B}}(\mathbf{H}_i|A) = \frac{\Pr_{\mathbf{B}}(\mathbf{H}_i) \cdot \Pr_{\mathbf{B}}(A|\mathbf{H}_i)}{\sum_{j=1}^m \Pr_{\mathbf{B}}(\mathbf{H}_j) \cdot \Pr_{\mathbf{B}}(A|\mathbf{H}_j)}.$$

Notice that we can assign in this way any prior probability, which is a rational number, to each hypothesis $\mathbf{H}_i$, by using adequate number of elements. For instance, suppose that we would like to assign k_i/n to $\mathbf{H}_i$, $i = 1, \ldots, m$. Then, we must have

$$\sum_{i=1}^m \frac{k_i}{n} = 1,$$

and hence

$$\sum_{i=1}^m k_i = n.$$

Thus, we take, as before

$$\lambda(t) = \langle\{1,\ldots,n\},\{j\}\rangle.$$

We define, if $\lambda(t)$ is as above and

$$\sum_{i=1}^{j-1} k_i + 1 \le j \le \sum_{i=1}^{j} k_i,$$

then $\lambda(s) \in \mathbf{H}_j$. Then the a priori probability of $\mathbf{H}_j$ is k_j/n.

In case we would like to assign an irrational number, r_j to $\mathbf{H}_j$, we have to consider an infinite number ν, and the universe of our systems in $\mathbf{B}(t)$ to be $\{1,\ldots,\nu\}$. We take a k_j, such that $(k_j/\nu) \approx r_i$. Then we can get, as above, a structure with probabilities approximately equal to what we want.

CHAPTER XVII

Logical probability

1. Languages for simple structures

In this section, I shall discuss the logical notion of probability, which I call degree of support, as it is derived from the objective notion of chance defined for simple probability structures in Chapter XII. This logical notion, as has been explained in Part 1, especially in Chapter III, Section 5, is a formalization of the notion of "degree of support of φ given $\langle \mathbf{K}, Y \rangle$", where φ is a sentence of the language appropriate for a simple probability structure $\langle \mathbf{K}, Y \rangle$. That is, $\mathbf{K}$ is a chance structure (in this case, a simple chance structure), and Y is a measurement system. Intuitively, according to my interpretation, this notion can be paraphrased as the "degree of possibility of truth (or partial truth) of φ under the interpretation $\langle \mathbf{K}, Y \rangle$".

The chance structures $\mathbf{K}$ are here the simple chance structures introduced in Chapter XII. The language where φ comes from is a measurement language adequate for the measurement system Y.

The language appropriate for the measurement system $Y = \langle Y_k \mid k \in K \rangle$ consists of sentences of the form $[g(\bar{Y}_{k_1}, \bar{Y}_{k_2}, \ldots, \bar{Y}_{k_n}) = r]$ where r is a real (or hyperreal) number, g is an internal continuous function from $^{n*}\mathbb{R}$ into $^*\mathbb{R}$, and k_1, k_2, ..., $k_n \in K$. It would be useful for some purposes to have terms of the form $\mathbf{E}(g(\bar{Y}_{k_1}, \bar{Y}_{k_2}, \ldots, \bar{Y}_{k_n}))$, which stand for the expectation of $g(\bar{Y}_{k_1}, \bar{Y}_{k_2}, \ldots, \bar{Y}_{k_n})$, and sentences of the form $[\mathbf{E}(g(\bar{Y}_{k_1}, \bar{Y}_{k_2}, \ldots, \bar{Y}_{k_n})) \leq r]$, but we shall not need them in this book.[1]

It is clear that when $\mathbf{K}$ is hyperfinite and τ is a composition

$$\tau = g(Y_{k_1}, Y_{k_2}, \ldots, Y_{k_n})$$

of the measurements Y and an internal continuous function, g, the set of possible values Λ_τ of τ is hyperfinite. Thus, if we fix the set of possible values of each Y_k, for $k \in K$, as

$$\Lambda_Y = \langle \Lambda_{Y_k} \mid k \in K \rangle$$

---

[1] See [54] for a development of this language.

then any sentence
$$[g(\bar{Y}_{k_1}, \bar{Y}_{k2}, \ldots, \bar{Y}_{k_n}) \leq r]$$
can be replaced by the disjunction of the sentences
$$[g(\bar{Y}_{k_1}, \bar{Y}_{k2}, \ldots, \bar{Y}_{k_n}) = s]$$
for $s \leq r$ and $s \in \Lambda_{g(Y_{k_1}, Y_{k2}, \ldots, Y_{k_n})}$. This disjunction is the disjunction of a hyperfinite set of sentences. Since we do allow negations, and finite and hyperfinite conjunctions and disjunctions of formulas, the only type of basic sentences we need is the one mentioned with equality.

For instance, for Example XII.1 in Chapter XII, Section 4, we considered the measurement function Y_{0r}, which was defined to be the number of red balls in the sample, and we could also consider the function Y_{0g}, which can be taken to be the number of green balls in the sample. The simple sentence of this language $[Y_{0r} = r]$ says that the number of red balls in the sample is equal to r. As we saw above, we can express the sentence $[Y_{0r} \leq r]$, which says that the number of red balls is less that or equal to r, by a disjunction. The negation of this sentence says that the number of red balls is larger than r. The conjunction $[Y_{0r} \leq r] \wedge \neg[Y_{0r} \leq s]$ says that the number of red balls is in the half open interval $(s, r]$. The sentence $[Y_{0r} + Y_{0g} = r]$, says that the number of red balls plus the number of green balls is equal to r. This last sentence is allowed, since addition is an internal continuous functions from $^{2*}\mathbb{R}$ to $^*\mathbb{R}$. The formal definition of truth given below formalizes this interpretation of sentences.

Formally, we first define the *similarity type*, $\eta_\mathcal{K}$, of the simple probability structure $\mathcal{K} = \langle \mathbf{K}, \langle Y_k \mid k \in K \rangle \rangle$.[2] The similarity type of a structure only serves to code in a compact fashion which are the types of nonlogical symbols that we need in a language for the structure, i.e., which is to be interpreted in the structure. If a language is interpreted in a structure, then the structure should have objects corresponding to the nonlogical symbols, thus, the language and the structure must be of the same similarity type. In the case of the simple structures that we are discussing here, we only need to indicate the index of the measurement functions in the structure, and their possible values. Thus, in the language, we should have a symbol for each measurement function, and a symbol for each hyperreal number which is a possible value of a measurement function. For compound structures, the similarlty type is more complicated. A language for each type $\eta_\mathcal{K}$, $\mathcal{L}_{\eta_\mathcal{K}}$, can then be defined by giving a recursive definition of term and formula. We shall use the Greek letters τ for terms, and φ, ψ, and θ for formulas:

DEFINITION XVII.1 (TERMS AND FORMULAS).

(1) A *similarity type*, σ, is an internal sequence, $\langle \Lambda_k \mid k \in K \rangle$, of hyperfinite sets of hyperreal numbers. If $\mathcal{K} = \langle \mathbf{K}, \langle Y_k \mid k \in K \rangle \rangle$ is a probability

---

[2]We have already defined similarlity types for relational systems and chance structures. These old similarity types should not be confused with similarity types of probability structures introduced here.

structure, then the similarity type of $\mathcal{K}$ is

$$\sigma_{\mathcal{K}} = \langle \Lambda_{Y_k} \mid k \in K \rangle,$$

where Λ_{Y_k} is the set of values of Y_k.
(2) Definition of terms:
The class of terms for the language $\mathcal{L}_\sigma$, where $\sigma = \langle \Lambda_k \mid k \in K \rangle$ is a similarity type, is the least class containing $\bar{Y}_k$, for $k \in K$, and closed under the following operation: if g is an internal continuous function from $^{n*}\mathbb{R}$ to $^*\mathbb{R}$, and $\tau_1, \ldots, \tau_n$ are terms, then $g(\tau_1, \tau_2, \ldots, \tau_n)$ is a term.
(3) For τ a term, we define the possible set of values of τ, Λ_τ, by recursion on terms:
   (a) If τ is $\bar{Y}_k$, for a certain $k \in K$, then

$$\Lambda_\tau = \Lambda_k.$$

   (b) If τ is $g(\tau_1, \tau_2, \ldots, \tau_n)$, where g is an internal continuous function and $\tau_1, \ldots, \tau_n$ are terms, then

$$\Lambda_\tau = \{g(x_1, \ldots, x_n) \mid x_1 \in \Lambda_{\tau_1}, \ldots, x_n \in \Lambda_{\tau_n}\}.$$

(4) Definition of sentences:
   (a) If τ is a term and $r \in \Lambda_\tau$, then $[\tau = r]$ is an atomic sentence.
   (b) If φ is a sentence, then $\neg\varphi$ is also a sentence.
   (c) If Θ is a hyperfinite (and, hence, internal) set of sentences, then $\bigwedge \Theta$ and $\bigvee \Theta$ are sentences.

The formula $\neg\varphi$ is the *negation of* φ, $\bigwedge \Theta$, the *conjunction of* Θ, and $\bigvee \Theta$, the *disjunction of* Θ. Notice that we do not introduce quantifiers.

The definition of truth is similar to Tarski's definition. But in our case, truth is relative to a relational system ω in $\mathbf{K}$ and a measurement system Y.[3]

The expression "φ is true in $\langle \omega, Y \rangle$" will be symbolized "$\langle \omega, Y \rangle \models \varphi$", where $Y = \langle Y_k \mid k \in K \rangle$ is a measurement system for $\mathbf{K}$ and $\omega \in \mathbf{K}$. We first define the value τ^Y of a term τ according to the measurement system Y, where τ^Y is a measurement function. Since the expectation is computed according to a particular $\mathbf{K}$, if we had expectation in the terms, the value of τ would be relative to $\langle \mathbf{K}, Y \rangle$, and not just to Y.

DEFINITION XVII.2 (TRUTH). Let $\mathcal{K} = \langle \mathbf{K}, \langle Y_k \mid k \in K \rangle \rangle$ be a probability structure of similarity type σ, τ a term and φ a sentence of $\mathcal{L}_\sigma$, then:
(1) Definition of values of terms:
   (a) If $\tau = \bar{Y}_k$, then $\tau^Y = Y_k$, for $k \in K$.
   (b) If $\tau_1, \ldots, \tau_n$ are terms and g is an internal continuous function from $^{n*}\mathbb{R}$ to $^*\mathbb{R}$ with $\tau = g(\tau_1, \ldots, \tau_n)$, then

$$\tau^Y = g(\tau_1^Y, \ldots, \tau_n^Y).$$

(2) Definition of truth:

---

[3] If we had expectation in the terms, then truth would be also relative to the whole of $\mathbf{K}$.

(a) If φ is $[\tau = r]$, then $\langle \omega, Y \rangle \models \varphi$, if $\tau^Y(\omega) = r$.
(b) If φ is $\neg \psi$, then $\langle \omega, Y \rangle \models \varphi$, if not $\langle \omega, Y \rangle \models \psi$.
(c) If φ is $\bigwedge \Theta$, then $\langle \omega, Y \rangle \models \varphi$, if $\langle \omega, Y \rangle \models \psi$, for every $\psi \in \Theta$.
(d) If φ is $\bigvee \Theta$, then $\langle \omega, Y \rangle \models \varphi$, if $\langle \omega, Y \rangle \models \psi$, for at least one $\psi \in \Theta$.

We say that two sentences φ and ψ are *equivalent* if $\langle \omega, Y \rangle \models \varphi$ if and only if $\langle \omega, Y \rangle \models \psi$, for any structure $\langle \omega, Y \rangle$. Two equivalent sentences express, in fact, the same thing.

If S is an internal subset of $^*\mathbb{R}$, then we can express the fact that τ is in S by the hyperfinite disjunction

$$\bigvee \{[\tau = r] \mid r \in S \cap \Lambda_\tau\}.$$

(This expression is a sentence, because the set $\{[\tau = r] \mid r \in S \cap \Lambda_\tau\}$ is a hyperfinite set of sentences.) In fact, we have

$$\langle \omega, Y \rangle \models \bigvee \{[\tau = r] \mid r \in S \cap \Lambda_\tau\} \iff \tau^Y \in S.$$

Thus, we are justified in using the abbreviation $[\tau \in S]$ for this disjunction.

For instance, we could consider for Example XII.3, if we take it as modeling a roulette with a fixed force and a variable initial position, the measurement Y_2 introduced in Chapter XII, Section 4, which gives the final position of the roulette, given an initial position. Since we are considering only values in the hyperreal numbers, we need to modify a little Y_2. We define $Y_2(\omega_c) = r$ if the final position of the roulette, when the initial position is c, is e^{ir}.

The similarity type of the language for this measurement function, consists of the set Λ_{Y_2} of possible values of Y_2, that is, Λ_{Y_2} is a hyperfinite approximation to the interval $[0, 2\pi]$. Now, in the language for Y_2, as it was explained above, we could express the fact that $g(Y_2) \in B$, for any internal subset B, and any internal continuous function g. Since Y_2 is internal and continuous, the set $[g(Y_2) \in B]$, i.e., the inverse image of B by $g(Y_2)$, is also internal, and, hence, hyperfinite and measurable under $\Pr_{K_2}$. Likewise, the expectation is defined and finite for all random variables thus obtained. Thus, all sentences of the language for Y_2 correspond to events, i.e., to elements of $\mathcal{F}_{K_2}$.

The following definition introduces formally the notion that connects the language with subsets of $\mathbf{K}$, and hence with events, which we have been using informally.

DEFINITION XVII.3. *Let $\langle \mathbf{K}, Y \rangle$ be a simple probability structure and φ be a sentence of the language of the similarity type of $\langle \mathbf{K}, Y \rangle$. The set of models of φ in $\mathcal{K}$, denoted by $\mathrm{Mod}_\mathcal{K}(\varphi)$, is defined by*

$$\mathrm{Mod}_\mathcal{K}(\varphi) = \{\omega \in \mathbf{K} \mid \langle \omega, Y \rangle \models \varphi\}.$$

The class $\mathrm{Mod}_\mathcal{K}(\varphi)$ is the set of models of φ (i.e., of systems where φ is true) that are in $\mathbf{K}$. If $\mathrm{Mod}_\mathcal{K}(\varphi) = \mathbf{K}$, then we say that φ *is true in* $\langle \mathbf{K}, Y \rangle$. If $\mathrm{Mod}_\mathcal{K}(\varphi)$ is empty, then φ *is false in* $\langle \mathbf{K}, Y \rangle$. In all other cases, φ *is partially true in* $\langle \mathbf{K}, Y \rangle$. Thus, a measure of $\mathrm{Mod}_\mathcal{K}(\varphi)$ gives a measure of the degree of partial truth of φ in $\langle \mathbf{K}, Y \rangle$. These considerations justify the following definition of a probability interpretation of a language in a simple probability structure.

DEFINITION XVII.4.

(1) A simple probability structure $\mathcal{K} = \langle \mathbf{K}, Y \rangle$ is *adequate for a measurement language* $\mathcal{L}_\sigma$, if $\mathcal{K}$ is of similarity type σ, and $\mathcal{F}_\mathbf{K}$ includes the sets $\mathrm{Mod}_\mathcal{K}(\varphi)$ for any sentence φ of $\mathcal{L}_\sigma$.
(2) If $\mathcal{K}$ is adequate for $\mathcal{L}$, we define the degree of support of a sentence φ of $\mathcal{L}_\mathcal{K}$ in $\langle \mathbf{K}, Y \rangle$, $\mathrm{Pr}_\mathcal{K}(\varphi)$, by

$$\mathrm{Pr}_\mathcal{K}(\varphi) = \mathrm{Pr}_\mathbf{K}(\mathrm{Mod}_\mathcal{K}(\varphi)).$$

This definition agrees with the intuitive ideas given above, since a measure of $\mathrm{Mod}_\mathcal{K}(\varphi)$ gives a measure of the degree of possibility of truth, or of partial truth, of φ in $\langle \mathbf{K}, Y \rangle$. The definition satisfies the requirement for a logical interpretation of probability that $\mathrm{Pr}_\mathcal{K}(\varphi)$ be uniquely determined by the simple chance structure $\mathbf{K}$, the measurement system Y, and φ.

For Example XII.3, as completed with Y_2, the sets $\mathrm{Mod}_{\langle \mathbf{K}_2, Y_2 \rangle}(\varphi)$ correspond to the internal hyperfinite subsets of the approximation to the circle, C, and as was explained above, they are in the algebra of events. The measure of $\mathrm{Mod}_{\langle \mathbf{K}_2, Y_2 \rangle}[Y_2 \in A]$, where A is an interval, however, is not necessarily the length of the interval, because the roulette may be biased and an arc may be sent into a longer or shorter arc.

Notice that $\langle \mathbf{K}, Y \rangle$, just as the usual relational systems, plays a dual role. On the one hand, it is a possible model of reality, i.e., of a certain chance setup. On the other hand, it serves as the interpretation of a language. For relational systems, reality is completely specified, i.e., relational systems model completed possible worlds; hence, every sentence is either true or false in these systems. Probability structures, on the contrary, only assign probabilities to sentences.

The language also plays a dual role. On one hand, it is interpreted in a possible probability model. On the other hand, it is interpreted in reality. This dual role is important when discussing the techniques of statistical inference and decision theory. More will be said about these roles in the next section, after the languages for compound probability structures are introduced.

2. Languages for compound structures

Similarly as in the previous section we introduced for simple probability structures a language and the notion of degree of support, we do it here for general probability structures. Repeating what was said in the previous section, the notion we want to formalize is the "degree of support of φ given $\langle \mathbf{K}, \xi \rangle$" where φ is a sentence of a language appropriate for the probability structure $\langle \mathbf{K}, \xi \rangle$, where $\mathbf{K}$ is a chance structure, and ξ is a measurement system over $\mathbf{K}$. As it has been said in the previous section and earlier, this notion can be paraphrased as the "degree of possibility of (or of partial) truth of φ under the interpretation $\langle \mathbf{K}, \xi \rangle$". The probability structures are those introduced in Definition XV.2. The language of

φ is a language adequate for the measurement system ξ, which here is a system of measurement processes.

Just as for the simple case, we need to define a similarity type of probability structures. For the compound case we need to include the causal universe, T, of **K**, the arities of the different measurement processes in the system, ξ, and the sets of possible values of the processes at each tuple of elements of T. Hence, we need as symbols in our language, one for each measurement function of the corresponding arity, the elements of T, and the hyperreal real numbers which are possible values of the functions. We shall codify all the information about the symbols needed in the language by a triple $\mathcal{T} = \langle T, \delta, \Lambda \rangle$, where T is contained in the causal universe of **K**, δ is a function with domain K such that $\delta(k)$ is the arity of ξ_k, for $k \in K$, and Λ is a function such that $\Lambda(k, t_1, \ldots, t_{\delta(k)})$ is the set of values of $\xi_k(t_1, \ldots, t_{\delta(k)})$, for $k \in K$ and $t_1, \ldots, t_{\delta(k)} \in T$.

The language appropriate for the probability structure $\langle \mathbf{K}, \langle \xi_k : k \in K \rangle \rangle$ of type $\mathcal{T} = \langle T, \delta, \Lambda \rangle$ consists of sentences of the form $[\bar{\xi}_k(t_1, \ldots, t_m) = r]$ where $k \in K$, $\bar{\xi}_k$ is m–ary (i.e., $\delta(k) = m$), $t_1, \ldots, t_m \in T$, and r is a hyperreal number in $\Lambda_{\xi_k(t_1, \ldots, t_m)}$. As in the simple case, we also allow compositions with internal continuous functions, negations, and hyperfinite conjunctions and disjunctions.

For instance, for **IC** of Example XIII.1, we could consider the measurement process $\xi_0(i, j)$ that gives the length of the chord between the points in the circle chosen at i and j. A sentence of the language, $[\bar{\xi}_0(i, j) \leq r]$, says that this length is less than or equal to r, and, as in the simple case, it can be expressed by a hyperfinite disjunction.

As an example, we shall see the case, which will be the main case used later, with just one one-ary measurement function. We have, for this case, that $\xi = \langle \xi_0 \mid 0 \in \{0\} \rangle$ and $\delta(0) = 1$. We identify ξ_0 with ξ, and if $t \in T$, we sometimes write $\xi_t(\lambda)$ for $\xi(t, \lambda)$.

We now give the definition for any type $\mathcal{T} = \langle T, \delta, \Lambda \rangle$. Formally, the language, call it $\mathcal{L}_\mathcal{T}$, can be defined as follows.

DEFINITION XVII.5 (LOGICAL LANGUAGE).

(1) A *similarity type* is an internal triple $\mathcal{T} = \langle T, \delta, \Lambda \rangle$, where T is a hyperfinite set, δ is a function with domain, say, K, such that $\delta(k)$ is a natural number, for each $k \in K$, and Λ is a function such that, for $k \in K$ and $t_1, \ldots, t_{\delta(k)} \in T$, $\Lambda(k, t_1, \ldots, t_{\delta(k)})$ is a hyperfinite set of hyperreal numbers. The type $\mathcal{T}$ is a *similarity type appropriate for the probability structure*

$$\langle \mathbf{K}, \langle \xi_k \mid k \in K \rangle \rangle$$

if T is contained in the causal universe of **K**, $\delta(k)$ is the arity of ξ_k, for $k \in K$, and for $k \in K$ and $t_1, \ldots, t_{\delta(k)} \in T$, $\Lambda(k, t_1, \ldots, t_{\delta(k)})$ is the set of values, $\Lambda_{\xi_k(t_1, \ldots, t_{\delta(k)})}$, of the function $\xi_k(t_1, \ldots, t_{\delta(k)})$.

(2) The nonlogical symbols of the language, $\mathcal{L}_\mathcal{T}$, of similarity type $\mathcal{T}$, are the elements of T, and a $\delta(k)$–ary functions symbol $\bar{\xi}_k$, for each $k \in K$. The logical symbols include the brackets, $=$, a symbol r for each hyperreal number r, and a symbol g for each internal continuous function g from

$^{m*}\mathbb{R}$ to $^*\mathbb{R}$, for m a finite natural number, plus the symbols for negation, conjunction and disjunction.
  (a) Definition of terms:
    (i) For every $k \in K$, if $\delta(k) = n$, then $\bar{\xi}_k(t_1, \ldots, t_n)$ is a term, for each $t_1, \ldots, t_n \in T$.
    (ii) If $\tau_1, \tau_2, \ldots, \tau_n$ are terms and g is an internal continuous function from $^{n*}\mathbb{R}$ to $^*\mathbb{R}$, then $g(\tau_1, \ldots, \tau_n)$ is a term.
  (b) We define the possible set of values of the term τ, Λ_τ, by recursion on terms:
    (i) If τ is $\bar{\xi}_k(t_1, \ldots, t_m)$, for a certain $k \in K$, and $t_1, \ldots, t_m \in T$, then
    $$\Lambda_\tau = \Lambda(k, t_1, \ldots, t_m).$$
    (ii) If τ is $g(\tau_1, \tau_2, \ldots, \tau_n)$, where g is an internal continuous function and $\tau_1, \ldots, \tau_n$ are terms, then
    $$\Lambda_\tau = \{g(x_1, \ldots, x_n) \mid x_1 \in \Lambda_{\tau_1}, \ldots, x_n \in \Lambda_{\tau_n}\}.$$
  (c) Definition of sentences:
    (i) It τ is a term and $r \in \Lambda_\tau$, then $[\tau = r]$ is an atomic sentences.
    (ii) If φ is a sentence, then $\neg \varphi$ is a sentence.
    (iii) If Θ is a hyperfinite set of sentences, then $\bigwedge \Theta$ and $\bigvee \Theta$ are sentences.
  $\neg \varphi$ is the *negation of* φ, $\bigwedge \Theta$ is the *conjunction of* Θ, and $\bigvee \Theta$ is the *disjunction of* Θ.

The recursive definition of truth is similar to that for simple structures. Thus, we define the notion $\langle \lambda, \xi \rangle \models \varphi$, where ξ is a measurement system with domain contained in the causal universe of λ. The expression $\langle \lambda, \xi \rangle \models \varphi$ is read "φ is true in $\langle \lambda, \xi \rangle$" or "$\langle \lambda, \xi \rangle$ is a model of φ".

We first have to define the value τ^ξ of a term τ according to the measurement function ξ.

DEFINITION XVII.6 (TRUTH). Let τ be a term and φ a sentence of the language $\mathcal{L}_\mathcal{T}$, where $\mathcal{T} = \langle T, \delta, \Lambda \rangle$ is a similarity type with domain of δ, K.

(1) Definition of the value of terms:
  (a) If τ is $(\bar{\xi}_k(t_1, \ldots, t_n))$, for $k \in K$ and $t_1, \ldots, t_n \in T$, then τ^ξ is the measurement process defined by
  $$(\bar{\xi}_k(t_1, \ldots, t_n))^\xi(\lambda) = \xi_k(t_1, \ldots, t_n, \lambda)$$
  for any outcome λ with causal universe including T.
  (b) If τ is $g(\tau_1, \ldots, \tau_n)$, where g is an internal continuous function from $^{n*}\mathbb{R}$ to $^*\mathbb{R}$ and $\tau_1, \ldots, \tau_m$ are terms, then
  $$\tau^\xi = g(\tau_1^\xi, \ldots, \tau_n^\xi).$$

(2) Definition of truth for sentences:
  (a) If φ is $[\tau = r]$, where τ is a term and $r \in \Lambda_\tau$ a hyperreal number, then $\langle \lambda, \xi \rangle \models \varphi$, if $\tau^\xi(\lambda) = r$.

(b) If φ is $\neg\psi$, then $\langle\lambda,\xi\rangle \models \varphi$, if not $\langle\lambda,\xi\rangle \models \psi$.
(c) If φ is $\bigwedge\Theta$, then $\langle\lambda,\xi\rangle \models \varphi$, if $\langle\lambda,\xi\rangle \models \psi$, for every $\psi \in \Theta$.
(d) If φ is $\bigvee\Theta$, then $\langle\lambda,\xi\rangle \models \varphi$, if $\langle\lambda,\xi\rangle \models \psi$, for at least one $\psi \in \Theta$.

Just as for the case of simple structures, we have that if S is an internal subset of $^*\mathbb{R}$, then we can express the fact that τ is in S by an hyperfinite disjunction:

$$\bigvee_{r \in \Lambda_r \cap S} [\tau = r].$$

Thus, we are justified in using the abbreviation $[\tau \in S]$ for this disjunction.

The same chance structure admits different measurement systems. For instance, for **IC** we could take the measurement process η_{ic}, such that $\eta_{ic}(t,\lambda)$ is the hyperreal number associated with the point chosen at t. The language for $\langle\mathbf{IC},\eta_{ic}\rangle$ is different from that of $\langle\mathbf{IC},\xi_0\rangle$, mentioned before. For **DC**, we can have similar systems as for **IC**.

The chance structures **UB** and **D** admit measurement functions of the form $\xi(\lambda) =$ the number of red balls chosen. Here ξ is 0-ary, i.e., there are no elements of the causal relation as arguments of ξ.

A language $\mathcal{L}_\mathcal{T}$ admits many probability structures as its possible models. The only condition for $\langle\mathbf{K},\xi\rangle$ to be a possible model of $\mathcal{L}_\mathcal{T}$ is that the similarity type $\mathcal{T} = \langle T, \delta, \Lambda\rangle$ be appropriate for $\langle\mathbf{K},\xi\rangle$. That is, T must be contained in the causal universe of **K**, the function ξ_k must be $\delta(k)$-ary, for $k \in K$, and $\Lambda_{\xi_k(t_1,\ldots,t_m)} = \Lambda(k, t_1, \ldots, t_m)$, for $k \in K$ and $t_1, \ldots, t_m \in T$.

We now introduce the definitions of the models of a sentence in a structure and other definitions for compound structures, which are similar to the definitions of the previous section for simple structures. Part of the rest of this section is a repetition with obvious changes of the end of the previous section.

DEFINITION XVII.7. Let $\mathcal{L}_\mathcal{T}$ be a language. Then:

(1) The *possible probability models of* $\mathcal{L}_\mathcal{T}$ are the probability structures $\langle\mathbf{K},\xi\rangle$ of type $\mathcal{T}$.
(2) Let φ be a sentence of $\mathcal{L}_\mathcal{T}$. The set of models of φ in $\mathcal{K} = \langle\mathbf{K},\xi\rangle$, denoted by $\mathrm{Mod}_\mathcal{K}(\varphi)$, is defined by

$$\mathrm{Mod}_\mathcal{K}(\varphi) = \{\lambda \in \mathbf{K} \mid \langle\lambda,\xi\rangle \models \varphi\}.$$

The set $\mathrm{Mod}_{\langle\mathbf{K},\xi\rangle}(\varphi)$ is the set of models of φ, (i.e., of systems where φ is true) according to the interpretation $\langle\mathbf{K},\xi\rangle$, which are in **K**. If $\mathrm{Mod}_{\langle\mathbf{K},\xi\rangle}(\varphi) = \mathbf{K}$, then we say that φ is true in $\langle\mathbf{K},\xi\rangle$. If $\mathrm{Mod}_{\langle\mathbf{K},\xi\rangle}(\varphi)$ is empty, then φ is false in $\langle\mathbf{K},\xi\rangle$. In all other cases, φ is partially true in $\langle\mathbf{K},\xi\rangle$. Thus, a measure of $\mathrm{Mod}_{\langle\mathbf{K},\xi\rangle}(\varphi)$ gives a measure of the degree of partial truth (or of the possibility of truth) of φ in $\langle\mathbf{K},\xi\rangle$. These considerations justify the following definition of a probability interpretation of a language in a possible probability model.

DEFINITION XVII.8.

A probability structure $\langle\mathbf{K},\xi\rangle$ is *adequate for a measurement language* $\mathcal{L}$, if $\mathcal{F}_\mathbf{K}$ includes the sets $\mathrm{Mod}_{\langle\mathbf{K},\xi\rangle}(\varphi)$ for any sentence φ of $\mathcal{L}$.

2. LANGUAGES FOR COMPOUND STRUCTURES

(2) If $\langle \mathbf{K}, \xi \rangle$ is adequate for $\mathcal{L}$, we define the degree of support of a sentence φ of $\mathcal{L}$ in $\langle \mathbf{K}, \xi \rangle$, $\mathrm{Pr}_{\langle \mathbf{K},\xi \rangle}(\varphi)$, by

$$\mathrm{Pr}_{\langle \mathbf{K},\xi \rangle}(\varphi) = \mathrm{Pr}_{\mathbf{K}}(\mathrm{Mod}_{\langle \mathbf{K},\xi \rangle}(\varphi)).$$

This definition agrees with the intuitive ideas given above, since a measure of $\mathrm{Mod}_{\langle \mathbf{K},\xi \rangle}(\varphi)$ gives a measure of the degree of possibility of truth of φ in $\langle \mathbf{K}, \xi \rangle$. The definition satisfies the requirement for a logical interpretation of probability that $\mathrm{Pr}_{\langle \mathbf{K},\xi \rangle}(\varphi)$ be uniquely determined by $\mathbf{K}$, the measurement system ξ, and φ.

For a sentence φ we may give two probability interpretations, i.e., two probability structures for the language of φ, say $\langle \mathbf{K}, \xi \rangle$ and $\langle \mathbf{H}, \eta \rangle$, such that

$$\mathrm{Pr}_{\langle \mathbf{K},\xi \rangle}(\varphi) \neq \mathrm{Pr}_{\langle \mathbf{H},\eta \rangle}(\varphi).$$

This is similar to the situation with the usual relational systems. There may be two such systems which are adequate for the language of φ such that φ is true in one and false in the other. If φ obtains in reality, then the one in which it is false is rejected as a model of reality. In Chapter IV, we already discussed the related rules based on $\mathrm{Pr}_{\langle \mathbf{K},\xi \rangle}$ for acceptance and rejection of possible probability models. We shall discuss them further in Part 4.

I believe that the main importance of the logical interpretation is precisely this possibility of several probability models for the same sentence. This allows a clear formalization of statistical inference, as will be discussed in Part 4. So we have a parallel here. Usual logical languages are used to formalize deductive inferences. Probability logical languages, for formalizing statistical inference.

Notice that $\langle \mathbf{K}, \xi \rangle$, just as the usual relational systems, plays a dual role. On the one hand, it is a possible model of reality, i.e., of a certain chance setup to which certain measurements are applied. On the other hand, it serves as the interpretation of a language. For relational systems, reality is completely specified, i.e., relational systems model completed possible worlds; hence, every sentence is either true or false in these systems. Possible probability models, on the contrary, only assign probabilities to sentences.

The interpretations of the language can also be viewed in two ways. On one hand, a language is interpreted in a probability structure and a probability is assigned to its sentences. On the other hand, the language can be interpreted in reality, where its sentences may be either true or false. Let us take as an example the simple sentence $\varphi = [\xi_t \leq r]$. This is assigned a probability according to a possible probability model $\langle \mathbf{K}, \xi \rangle$. The function ξ_t can also be evaluated in reality, for instance by a certain measurement procedure, and, say, we obtain a value which is a real number y. Then $y \leq r$ or $y > r$, and hence φ is either true or false in reality. This interplay between the model and reality was discussed and used in Chapter IV, and will be further discussed in Part 4, in relation to statistical inference and decision theory.

It is important to notice an essential difference between the ordinary sample spaces $\langle \Omega, \mathcal{A}, \mathrm{Pr} \rangle$ of probability theory and the probability spaces $\langle \mathbf{K}, \mathcal{F}_{\mathbf{K}}, \mathrm{Pr}_{\mathbf{K}} \rangle$ we have discussed in this and the previous chapters. In our probability spaces,

K determines the family of events $\mathcal{F}_\mathbf{K}$ and the measure $\Pr_\mathbf{K}$, while in the usual sample spaces the set Ω may support different measures and algebras. Very often Ω and $\mathcal{A}$ are fixed and support different measures. For instance, when one is considering alternative probability distributions for a certain setup, the basic set of the sample space, Ω, and often the algebra $\mathcal{A}$, remain fixed, and what changes is the measure Pr.

A similar remark is in order with respect to random variables. If we are considering a random variable or random process Y with several possible probability distributions for Y, we customarily keep the domain of Y fixed as the base set of the sample space, Ω. The sets $[Y \in B]$, for B an internal set, are also fixed. What varies are the possible measures imposed on these sets. In our case, on the other hand the sets $\mathrm{Mod}_{\langle \mathbf{K}, Y \rangle}([Y \in B])$ change with a change in interpretation $\langle \mathbf{K}, Y \rangle$.

Examples. We now discuss a few examples. First, consider the selection of a random chord at random (Chapter XVI, Section 2). Here, we have three possible probability structures $\langle \mathbf{IC}, Y \rangle$, $\langle \mathbf{H}_2, Y \rangle$, and $\langle \mathbf{H}_3, Y \rangle$, where Y is 0-ary. We use the same letter Y for the three processes, since $Y(\lambda)$ is always the length of the chord and the set of possible values is the same. Thus, the same language is appropriate for these three structures. Hence, for any sentence φ of this language we can define $\Pr_{\langle \mathbf{IC}, Y \rangle}(\varphi)$, $\Pr_{\langle \mathbf{H}_2, Y \rangle}(\varphi)$, and $\Pr_{\langle \mathbf{H}_3, Y \rangle}(\varphi)$. These probabilities are in general different. The techniques of Part 4 help to decide which of the three is the right model for φ.

As a second example, let us examine the case of the theory of errors (Chapter XVI, Section 3). Suppose that the measurement we are interested in has a theoretical value in a interval, and also several possible standard errors. Thus, the language should contain a 0-ary measurement function ξ. The possible probability structures are of the form $\langle \mathbf{Er}, \xi \rangle$, which is defined in Chapter XVI, Section 3, but with possibly different mean and standard error. The different structures assign different probabilities to the sentences of the language. The techniques of Part 4 help to determine which is the right model.

For Brownian motion (Chapter XVI, Section 4), we can define a language which has $\langle \mathbf{BM}, B \rangle$ as a possible probability model. Since B is internal, the subsets

$$\mathrm{Mod}_{\langle \mathbf{BM}, B \rangle}([g(\bar{B}) \leq r])$$

of **BM** are internal for every $r \in {}^*\mathbb{R}$ and every internal continuous function g. We have the language closed under hyperfinite conjunctions and disjunctions. Since hyperfinite natural numbers are externally infinite, the effect of these hyperfinite conjunctions and disjunctions is similar to standard countable conjunctions and disjunctions. However, if we close the language under external countable conjunction and disjunctions, then there are sentences φ such that $\mathrm{Mod}_{\langle \mathbf{BM}, B \rangle}(\varphi)$ is not internal, because the internal power set of **BM** is not a σ-algebra, but only an algebra.

Hence, $\mathrm{Mod}_{\langle \mathbf{BM}, B \rangle}(\varphi) \notin \mathcal{F}_{\mathbf{BM}}$, for such sentences. As it was mentioned above, for the random process B, it is enough with the algebra $\mathcal{F}_{\mathbf{BM}}$. Thus, it is not

2. LANGUAGES FOR COMPOUND STRUCTURES

necessary to close the language to include infinite disjunctions and disjunctions, because hyperfinite formulas serve the same purpose as countable formulas.

For the possible probability model $\langle \mathbf{BM}, b \rangle$, on the other hand, we can have a language of type $\langle [0,1], \delta, \Lambda \rangle$, with the same δ as before, but with sets of possible values that include all real numbers. The language, in this case, should be closed under negations, and countable conjunctions and disjunctions, and contain the atomic sentences $[\bar{b}_t \leq r]$. These atomic sentences cannot be obtained from sentences of the form $[\bar{b}_t = s]$, since the set of possible values of b is not hyperfinite. We could construct an infinitary language, however, such that $\mathrm{Mod}_{\mathbf{BM},b}(\varphi) \in \mathcal{L}(\mathcal{F}_{\mathbf{BM}})$ for every sentence φ of the language.

We can also consider a language for the standard Brownian motion Z, defined above. For this purpose, we should extend a little the notion of possible probability model, which can be done with no difficulty. Since we shall not use this type of process in this book, I shall not pursue the matter further.

A natural way of formulating the laws of large numbers is with the probability languages we are discussing. For instance, the strong law of large numbers can be stated as follows. Let the probability structure be that defined in Chapter XVI, Section 5, i.e., $^\nu \mathbf{K}$, for $\mathbf{K}$ a probability structure and ν an infinite natural number. Consider the same 0-ary measurement process Y as in Chapter XVI, Section 5, and introduce the same definition of $\mathrm{Fr}_{Y,n}$ as in that section. Then, by Theorem XI.1, we obtain that the strong law of large numbers is equivalent to:

Suppose that $\mathrm{Pr}_{\langle \mathbf{K}, Y \rangle}([Y = 1]) = p$. *Then, for all* $\lambda \gg 0$ *(or, there is a* $\lambda \approx 0$ *such that)*

$$\mathrm{Pr}_{\langle \mathbf{K}, Y \rangle}[|\mathrm{Fr}_{Y,\mu} - p| \geq \lambda] \approx 0,$$

for every infinite $\mu \leq \nu$.

Here, the second occurrence of $\mathrm{Pr}_{\langle \mathbf{K}, Y \rangle}$ is really $\mathrm{Pr}_{\langle ^\nu \mathbf{K}, \mathrm{Fr}_Y \rangle}$, but the identification is usual.

We now analyze the statement of Bayes' formula. Suppose that we introduce a one-ary measurement process Y with field L_2 over $\mathbf{B}$, as defined in Chapter XVI, Section 5, and the corresponding language. In order to avoid subindexes I shall denote $\mathrm{Pr}_{\langle \mathbf{B}, Y \rangle}$ simply by Pr, $\mathrm{Pr}_\mathcal{L}$, by Pr_t, where $\mathcal{L} = \langle \mathbf{B}(t), Y_t \rangle$, and $\mathrm{Pr}_\mathcal{J}$ by Pr_{λ_t}, where $\mathcal{J} = \langle \mathbf{B}(\lambda(t)), Y_s \rangle$.[4] In Bayesian terminology, Pr_t is the a priori probability.

Recall that $\mathrm{Pr}\,\varphi = \mathrm{Pr}_\mathbf{B}(\mathrm{Mod}_{\langle \mathbf{B}, Y \rangle}(\varphi))$ where $\mathrm{Mod}_{\langle \mathbf{B}, Y \rangle}(\varphi)$ is the set of $\lambda \in \mathbf{B}$ where φ is true. The probability Pr_t is obtained in a similar way from $\mathrm{Pr}_{\mathbf{B}(t)}$, and Pr_{λ_t}, from $\mathrm{Pr}_{\mathbf{B}(\lambda(t))}$.

We can define conditional probability in the usual way, namely: if φ and ψ are sentence of the language with $\mathrm{Pr}\,\psi \neq 0$ then

$$\mathrm{Pr}(\varphi|\psi) = \frac{\mathrm{Pr}(\varphi \wedge \psi)}{\mathrm{Pr}\,\psi}.$$

---

[4]Recall that $Y_t(\lambda) = Y(t, \lambda)$ and $Y_s(\lambda) = Y(s, \lambda)$.

Bayes formula for a finite number of hypotheses can now be stated as follows:

Let $B_1, \ldots, B_n$ be hyperfinite sets which form a partition of the possible vaules of Y_t, and let A be a hyperfinite set. Then

$$\Pr([Y_t \in B_i]|[Y_s \in A]) = \frac{\Pr[Y_t \in B_i] \cdot \Pr([Y_s \in A]|[Y_t \in B_i])}{\sum_{j=1}^n \Pr[Y_t \in B_j] \cdot \Pr([Y_s \in A]|[Y_t \in B_j])}.$$

Bayes' formula, written above, gives the probability of the "hypothesis", $[Y_t \in B_i]$, given the "evidence", $[Y_s \in A]$, in terms of the "a priori probability"[5] of the hypothesis and the "a posteriori probability" of the evidence, given the hypothesis. The continuous version is similar. The proof of this theorem is obtained by standard methods of probability theory, given the definition of the probability Pr for $\langle \mathbf{B}, Y \rangle$ indicated above.

Although, we shall not discuss deeply in this book stochastic processes, we have an indication of how they can be modeled, through the examples in the previous chapter. A natural strengthening of our languages, then, is that defined in [54] by introducing in the language a conditional expectation operator. Then some of the theorems proved in that paper and in [42] can be applied to my models.

Another possible strengthening is to add regular existential and universal quantifiers over the causal variables t. An idea is to use for this purpose the definition of [82], as it was done in [23], for a particular case.

---

[5]Notice that $\Pr([Y_t \in B_i]) = \Pr_t([Y_t \in B_i])$.

Part 4

Statistical inference and decision theory

CHAPTER XVIII

Classical statistical inference

This chapter deals with some of the usual techniques of classical statistical inference[1] in the light of the principles laid out in Chapter IV. The discussion shall be brief and only touch on a few of the techniques, since it is only meant to give an idea of the use of the principles. A more detailed and precise discussion would take us too far here, and it is more appropriate for a separate publication. The next chapter contains a discussion of some other techniques that have been proposed, and justifications by other authors of the techniques discussed here.

As has been asserted in Chapter IV, Section 2, and elsewhere in the previous chapters, I consider statistical inference proper as a form of inference. That is, the theory of statistical inference gives rules to allow to proceed from premises accepted as true to a conclusion that is also to be accepted as true. Of course, I agree with the traditional view that these rules do not allow conclusive inference. That is, the conclusion is only accepted provisionally. If more evidence warrants a rejection of a previously accepted conclusion, then it is rejected. This is not to say that everything statisticians do under the name of statistical inference is inference. Some of their techniques belong properly to decision theory.

Classical statistical inference in the style of Fisher or Neyman and Pearson is used in two contexts. The first is the test of scientific hypotheses and the estimation of parameters for scientific purposes. I include in the term "scientific," all purposes whose primary aim is truth. The second context, more important in Neyman and Pearson than in Fisher, is decision theoretic, that is, the use of classical statistical techniques as a help for deciding on actions. In this chapter, I shall concentrate on the first context.

Another caveat is that I shall discuss, in this chapter, mainly the statistical techniques and not the philosophical ideas of the creators of these techniques, in particular of Neyman and Pearson. Neyman, especially, talks about statistical inference containing rules for inductive behavior, and not specifically for inference. I do not think that he would have agreed with my description of what statistical inference is. Although I believe that many of the Neyman-Pearson techniques are

---

[1] I am mainly using the terminology and descriptions of [4].

justified, I do not agree with the justification provided by these authors and some other classical statisticians. In the next chapter, I shall discuss Neyman's justification of his techniques of statistical inference, which is different from mine. In fact, I believe that my justification is related, although not identical, to Fisher's justification of significance test (but not to his fiducial probability), which we shall see in Section 3.

In this chapter, I will discuss, using the principles laid out in Chapter IV, especially in its Section 2, some of the methods of statistical inference and some of the different principles of statistical inference that have been proposed. The discussion of this chapter will be centered on inference proper, leaving decision theory for Chapter XX.

1. General framework

The basic principles that I propose for statistical inference are the rules for rejection and acceptance of hypotheses which were introduced in Chapter IV, Section 2, and discussed further in Chapter IV, Sections 4 and Chapter V, Section 3, with the name of the Inverse Inference Principle. The scenario for the application of this principle is the following. We are facing a chance setup, for instance, an experiment with several possible results. We consider alternative models $\mathbf{K}_i$, for $i \in I$, for this setup. In the general case, as was mentioned in Chapter IV, Section 2, these models are sequences $\langle \mathbf{K}_i^n \mid 1 \leq n \leq \nu \rangle$, with ν infinite. The models assign probabilities to the different events (i.e., sets of possible results) which can occur in the setup.

As was explained in Chapter IV, Section 2, the processes of rejection and acceptance are dialectical in nature. That is, part of the scenario is that it is supposed that we must convince somebody, may be ourselves, of the adequacy or nonadequacy of a certain model. In order to do this, the person that must be convinced must set a level of probability with which she would be convinced. And we must try to verify with the appropriate partial experiment, in the sequence that constitutes a discriminating experiment, the proposition with the right probability. This is why, we need to have the sequence of experiments at hand.

The acceptance and rejection rules, as have been stated in Chapter IV, Section 2, apply mainly to the rejection and acceptance of what are called simple hypotheses. Simple hypotheses, in our context, are the assertion that one particular probability model, say $\mathbf{K}_i$, is the model for the chance setup. That is, the hypothesis specifies completely the probability model for the setup. In statistics, composite hypotheses are also entertained. Let $\mathbf{K}_i$, for $i \in I$, be the set of possible probability models. A composite hypothesis asserts that the true model for the setup is among the $\mathbf{K}_j$, for $j \in J$, where J is a subset of I. Thus, a composite hypothesis does not specify completely the model for the setup.

The Inverse Inference Principle is applied in classical statistical inference in the following setting. We are analyzing a chance setup $\mathfrak{S}$, and we do not know which of several possible probability models describes the chance setup. Each

1. GENERAL FRAMEWORK

possible probability model is of the form $\langle \mathbf{K}, Y \rangle$ where $\mathbf{K}$ is a chance structure and Y a system of random variables.

We have a language for Y, as was introduced in Chapter III, Section 5, and Chapter XVII, with basic formulas of the form $[\tau = r]$ for r a real (or hyperreal) number and τ a term constructed from Y by the use of continuous functions. All the formulas in the language can be represented, in this case, by $[\tau \in B]$, where B is a hyperfinite subset of $^{n*}\mathbb{R}$. Each $\langle \mathbf{K}, Y \rangle$ determines a probability distribution $\Pr_{\langle \mathbf{K}, Y \rangle}$ for these formulas. We assume that this distribution can be obtained from a parameter θ, which may be of dimension one or higher (i.e., it may be a real number or an n-tuple of real numbers). So we have a space of parameters Ω, one parameter for each possible probability model of the setup.

As it was mentioned in Chapter III, Section 5, the idea of considering expressions of the form $[Y \in B]$ as sentences of a language is implicit in normal statistical practice. In logic, the essential feature of a language is the fact that a sentence is subject to several interpretations, such that the sentence is true under some interpretations and false under others. In our case, the interpretations assign probability (which is thought of as the degree of possibility of truth) and not, in general, truth or falsity. In normal statistical practice, the expression $[Y \in B]$ is assigned different probabilities by the different models being considered. Thus, each possible model can be taken as an interpretation of the sentence $[Y \in B]$ which assigns a probability to it.

In classical statistics, we assume, in most cases, that a setup can be repeated an indefinite number of times. The possible probability models for the sequence of these repetitions were discussed in Chapter III, Sections 4 and 5, and, with more details, in Chapter XVI, Section 5. I briefly review them here. If $\langle \mathbf{K}, Y \rangle$ is a possible probability model for the given setup, where Y is a random variable, then $\langle ^n\mathbf{K}, Y_1, \ldots, Y_n \rangle$ is the model for n independent repetitions of the setup, which is called the n-sampling probability model corresponding to $\langle \mathbf{K}, Y \rangle$. Here, we write $Y_i(\omega) = Y(i, \omega) = Y(\omega(i))$, where $\omega \in {}^n\mathbf{K}$ is a sequence of elements of $\mathbf{K}$. As it is usual in statistics, I am using the same letter Y for the function on $\mathbf{K}$ and on $^n\mathbf{K}$. The distribution of the n-sampling model is called the *n-sampling distribution*.

In general, the sampling distributions are functions of the parameter θ of the original distribution. When we want to consider the sampling distributions for an infinite number of n, we extend the probability structure to $^\nu\mathbf{K}$, with ν, an infinite natural number, and extend similarly the random variable Y.

Sometimes a more general situation is considered in which the sequence of probability models is not the product space $^\nu\mathbf{K}$ but just a sequence $\mathbf{K}^n$ such that the different elements of the sequence form a stochastic process. I shall mainly concentrate, however, on the situation described previously with independent repetitions.

There are two specific marks of classical statistical inference. First, different probability models $\langle \mathbf{K}, Y \rangle$, which determine the same probability distribution $\Pr_{\langle \mathbf{K}, Y \rangle}$, are not distinguished. That is, they cannot be differentiated by using these techniques. Second, the only evidence that is used is sampling evidence.

That is, the sequence of experiment of our principle consist of just of observations of the values of the random quantity τ where τ is a continuous functions of Y_1, $Y_2, \ldots, Y_\nu$. In our language, the only propositions whose probability determines acceptance or rejection are those expressed by sentences in the language for Y_1, $Y_2, Y_3, \ldots, Y_\nu$, which are supposed to be interpreted in repetitions of the same experiment $\mathfrak{S}$, which in general, but not always, are independent repetitions. The evidence is implied by sentences of the form

$$[Y_1 = r_1] \wedge [Y_2 = r_2] \wedge \cdots \wedge [Y_n = r_n],$$

for a certain n, which I shall call *complete sample descriptions*. That is, the sentences used for acceptance or rejection should be such that they or their negations are implied by complete sample descriptions.

In the medical example of Chapter V, I used the Inverse Inference Principle, without some of the limitations that were stated in the last paragraph. Thus, I believe that classical statistical inference is not the only form of inference that is based on the Inverse Inference Principle.

The two main problems considered in classical statistical inference are hypotheses testing and estimation of parameters. In hypotheses testing, we make the hypothesis that the random variable Y has a certain distribution in a given setup (the null or working hypothesis) against the hypotheses of alternative distributions. The null hypothesis either is rejected or passes the test. Notice that, as was said in the last paragraph, the hypotheses to be tested do not specify the exact probability model, but any probability model with the same distribution would pass the test or be rejected when a hypothesis passes the test or is rejected.

The same is true with estimation of parameters. In this case, we do not test a hypothesis but we decide which hypothesis to accept. That is, the parameters we estimate represent the distribution that we accept. Just as in the case of hypotheses testing, we do not accept a particular possible probability model, but just a probability distribution, i.e., a parameter.

So what we are testing or estimating is not the possible probability model $\langle \mathbf{K}, Y \rangle$ but the distribution $\Pr_{\langle \mathbf{K}, Y \rangle}$. We already discussed informally this situation when we studied the use of frequencies for estimating probabilities in Chapter IV, Section 4. Within classical statistical inference we cannot distinguish between two possible probability models $\langle \mathbf{K}, Y \rangle$ and $\langle \mathbf{J}, Z \rangle$ such that $\Pr_{\langle \mathbf{K}, Y \rangle} = \Pr_{\langle \mathbf{J}, Z \rangle}$. In order to distinguish between these two models we must use other evidence. I shall not discuss, however, this problem in the book.

The general framework for testing and estimation can then be described as follows. The possible probability models are represented by a family of probability distributions $\mathcal{P} = \{\Pr_\theta : \theta \in \Omega\}$, for a certain set Ω of parameters. These parameters may be one dimensional (i.e., each θ, a real number), or multidimensional (i.e., each θ, an n-tuple of real numbers). It is generally supposed, but not always, that from the parameter θ you can recover the distribution $\Pr_\theta$. The probability distributions $\Pr_\theta$ are applied to a language for a certain random variable X. That is, we can determine $\Pr_\theta[X \leq r]$, and the probabilities of all other sentences of the language.

1. GENERAL FRAMEWORK

We always assume that all elements of Ω are real numbers or sequences of real numbers. It is impossible to have discriminating experiments that distinguish between infinitely close hyperreal numbers or sequences of hyperreals, so that we require that if θ, $\theta' \in \Omega$, with $\theta \neq \theta'$, then $\theta \not\approx \theta'$. This is true if all elements in Ω are real.

We also assume that we may have ν repetitions of the experiment for an infinite natural number ν. The sampling distribution for $\Pr_\theta$ is also called $\Pr_\theta$, and the random variable arising from X, also written X, or $X_1, X_2, X_3, \ldots, X_\nu$ for the first, second, third, ..., νth repetition. The possible results of the experiment are just the possible values of $X_1, X_2, X_3, \ldots, X_\nu$, which are usually written $x_1, x_2, \ldots, x_\nu$.[2] All the inferences are only based on the family $\mathcal{P}$ and the values obtained for the random variables $X_1, X_2, \ldots, X_\nu$, namely, $x_1, x_2, \ldots, x_\nu$.

I shall not discuss extensively nonparametric statistics. For nonparametric statistics, we would have a family of probability distributions $\{\Pr_j : j \in J\}$, which are not identified by parameters. The study of nonparamentric statistics presents no difficulty from any standpoint. The considerations that will be made for parametric statistics can be modified to include the nonparametric case.

In my framework, $\Pr_\theta$ would be $\Pr_{\langle K, X \rangle}$ for a certain possible probability model $\langle K, X \rangle$, and the sampling distribution, $\Pr_{\langle {}^n K, X_1, \ldots, X_n \rangle}$, corresponds to the possible probability model $\langle {}^n K, X_1, X_1, \ldots, X_n \rangle$, with $X_i(\omega) = X(\omega(i))$. The set Ω represents a family of possible probability models. Since classical statistical inference does not differentiate between models with the same distribution, I shall not specify the probability models but just indicate the parameters which determine the distributions, as it is usual in statistics.

In hypotheses testing, the hypothesis H to be tested is of the form $\theta \in \Omega'$, where $\Omega' \subseteq \Omega$. That is, H asserts that the distribution for the setup in question is $\Pr_\theta$ with $\theta \in \Omega'$. A hypothesis H is called *simple*, if Ω' contains just one element, i.e., H asserts that $\theta = \theta_0$ for a certain $\theta_0 \in \Omega$. The hypothesis H is called *composite*, if Ω' contains more than one element.

For estimation, we estimate the value of the θ that is right for the setup, either by giving a particular value (point estimation) or indicating a subset of Ω where the parameter lies (region estimation).

Since Ω represents a family of possible probability models for the setup, a hypothesis H of the form $\theta \in \Omega'$ corresponds to a hypothesis φ about the right probability model for the setup. That is, φ asserts that the right probability model for the setup is a certain $\langle K, X \rangle$ with $\Pr_{\langle K, X \rangle} = \Pr_\theta$ for a certain $\theta \in \Omega'$. If H is rejected, then φ, i.e., that $\langle K, X \rangle$ is the right probability model for the setup, is also rejected.

On the other hand, the link between estimation and acceptance of possible probability models is more tenuous. If we estimate that the parameter belongs to a certain Ω', and we assume that the family of possible probability models for the setup is **H**, then we also accept that the probability model for the setup is among those members of **H** that determine a probability distribution with parameters

---

[2]See [4, Chapter 5], for instance, for the terminology. In [4], as in most books, Barnett does not use a nonstandard approach, as we do in this book.

in Ω'. But there may be another family of possible probability models with the same parameter family Ω adequate for the setup, which is different from **H**, and then the accepted probability models would be different.

2. Discriminating experiments

The notion of discriminating experiment introduced in Chapter IV, Section 2, Definition IV.3, owes its inspiration to Birnbaum, [6], although I had to modify his formalism considerably in order to accommodate my principles. I shall begin with Birnbaum's version of models of experiments. In this section, I will discuss what Birnbaum calls statistical models of experiments, and especify, for the case of classical statistical inference, the notion of discriminating experiment introduced in Chapter IV, Section 2.

Birnbaum denotes by E any specified model of an experiment. As examples of such models he mentions, for instance, tossing a newly bent coin 50 times, and observing the number of heads appearing; or observing the number of apparent cures in 50 clinical trials of a new drug. If x denotes any specified observed outcome of an experiment E, then $\langle E, x \rangle$ is a model of an experiment together with its observed result, which Birnbaum calls a *model of statistical evidence*. Referring to the examples mentioned, $\langle E, 40 \rangle$ represents that 40 heads were observed in 50 tosses of the coin, or that 40 apparent cures were observed in 50 clinical trials. In the notation of random variables, $\langle E, x \rangle$ is written $[E = x]$.

Even such simple cases of statistical evidence have to be modified, in order to satisfy my principles. The examples of models of experiments mentioned above, have to be modified to include the possibility of testing a sequence of propositions. Thus, the first example would be tossing a newly bent coin n times, for any n, and observing the number of heads appearing, for a particular n; the second example becomes the possible observations of the number of apparent cures in n clinical trials of a new drug, where n increases without bound, and actually observing them for a particular n. Thus, as introduced in Chapter IV, Section 2, an experiment E is really a sequence of experiments E_n, for n a natural number. The quantity E_n will be called a *partial experiment of E*. As before, I shall denote such sequences by

$$E = \langle E_n \mid 1 \leq n \leq \nu \rangle,$$

where ν is a infinite natural number.

In order to specify a result, we must also indicate the number n; thus, results will be indicated by pairs $\langle n, x \rangle$. Following Birnbaum with suitable modifications, the description of a model of statistical inference is, hence

$$\langle \langle E_n \mid 1 \leq n \leq \nu \rangle, \langle n, x \rangle \rangle,$$

or simply, $\langle E, \langle n, x \rangle \rangle$. In the notation of random variables, which we shall mainly use, $\langle E, \langle n, x \rangle \rangle$ is written $[E_n = x]$. For instance, $(\langle E_n : 1 \leq n \leq \nu \rangle, \langle 50, 40 \rangle)$ represents that 40 heads were observed in 50 tosses of the coin, or that 40 apparent cures were observed in 50 clinical trials. We write the whole sequence E_n to

2. DISCRIMINATING EXPERIMENTS

emphasize the fact that the experiment is a really a sequence of experiments. In our measurement function notation this result can be written $[E_{50} = 40]$.

According to Birnbaum, a *statistical model of an experiment* E is constituted by a specified set of alternative probability models of an experiment. Just as in [6], I shall consider as examples mainly discrete probability models. By S_n we denote the set of possible results of the partial experiment E_n. That is, if $x \in S_n$, then the sentence $[E_n = x]$ has a definite probability assigned to it in each of the possible probability models. In our old notation $S_n = \Lambda_{E_n}$. We assume that each one of the possible alternative probability models, $\mathbf{K}$, determines a probability density or frequency function f_n, for each $n \leq \nu$. We write, for $x \in S_n$, $f_n(x) = \Pr_{\mathbf{K}}[E_n = x]$. Since statistical inference only discriminates between probability distributions, for our purposes, the probability density f_n represents completely the probability model.

In our case, f_n is a discrete probability frequency function, which, abusing a little the language, we shall sometimes call a probability density function, or, briefly, a pdf. We have that each subset B of S_n has, in the model $\mathbf{K}$ that determines f_n a probability given by

$$\Pr_{\mathbf{K}} B = \Pr_{\mathbf{K}}[E_n \in B] = \sum_{x \in B} f_n(x).$$

Thus, we also write $\Pr_{\mathbf{K}}$ for the probability distribution determined by E_n.

Each of the alternative discrete probability models of a sequence of experiments can be represented by $\langle\langle S_n, f_n\rangle \mid 1 \leq n \leq \nu\rangle$, where S_n and f_n are specified for each $n \leq \nu$. For our example, $S_n = \{1, 2, \ldots, n\}$; and if the coin is unbiased (or if the probability of apparent cures is $1/2$), then the pdf (probability density function) has the binomial form

$$f_n(x) = \binom{n}{x} \frac{1}{2^n}, \quad \text{for } x = 0, 1, 2, \ldots, n.$$

In general, a set of pdfs, which stands for a set of alternative probability models, can be represented by a function $f = f(n, x, \theta)$, which, for each fixed value of n and θ, is a pdf. By Ω we denote the set of all θ. As we saw earlier, Ω is called the parameter space, and each element θ of Ω is called a parameter point. Thus, a statistical model of a sequence of experiments is represented by $\langle \Omega, S, f \rangle$, where $S = \langle S_n \mid 1 \leq n \leq \nu\rangle$ and $f = f(n, x, \theta)$ are specified. In our examples, a possible statistical model is the set of probability models represented by the binomial pdfs

$$f(n, x, \theta) = \binom{n}{x} \theta^x (1-\theta)^{n-x}, \quad x = 0, 1, \ldots, n, \text{ for } 0 \leq \theta \leq 1.$$

I shall base most of the discussion of this section on finite models of statistical experiments, that is, models in which both the spaces S_n and the parameter space Ω contain finite number of points. In this case, we may represent each partial experiment E_n by a matrix of probabilities. For example we may write

$$E_n = (p_{nij}) = \begin{bmatrix} p_{n11} & p_{n12} & \cdots & p_{n1J_n} \\ p_{n21} & p_{n22} & \cdots & p_{n2J_n} \\ \vdots & \vdots & \ddots & \vdots \\ p_{nI1} & p_{nI2} & \cdots & p_{nIJ_n} \end{bmatrix}$$

Here J_n is the number of points in S_n, I is the number of parameter points, and the generic probability p_{nij} denotes the probability of the point j in S_n under the assumption that the parameter point is i, i.e., $p_{nij} = \Pr_i[E_n = j]$. We are ambiguous in our notation. The expression E_n stands both for the random variable and for the matrix of probabilities. We could introduce a new symbol for the matrix of probabilities, but, since there is no danger of confusion, we shall not do it.

As an example, we can take a finite model mentioned by Birnbaum, which has found important use in linkage experiments in classical genetics. These are experiments to determine whether two genes are located in the same chromosome. When certain inbred strains of mice are crossbred, each of their progeny will have a certain trait A, with probability 1/4, if certain two genes lie on different chromosomes, but with probability 1/2, if the genes lie on the same chromosome. As it is usual in scientific practice, scientific theoretical and experimental judgement, in this case genetical, is involved in adopting such a simple model as an adequate basis for some interpretations of results.

The experiment E_n represents the observation of n progeny. Thus

$$E_1 = \begin{bmatrix} \frac{3}{4} & \frac{1}{4} \\ \frac{1}{2} & \frac{1}{2} \end{bmatrix}$$

Here row $i = 1$ denotes the hypothesis that the genes lie on different chromosomes, and row $i = 2$ denotes the alternative hypothesis; and $j = 0$ or 1 denotes respectively absence or presence of trait A. The experiment in which two progeny are observed has the model

$$E_2 = \begin{bmatrix} \frac{9}{16} & \frac{6}{16} & \frac{1}{16} \\ \frac{1}{4} & \frac{1}{2} & \frac{1}{4} \end{bmatrix} = \frac{1}{16} \cdot \begin{bmatrix} 9 & 6 & 1 \\ 4 & 8 & 4 \end{bmatrix};$$

and an experiment in which n progeny are observed has the model

$$E_n = \begin{bmatrix} p_{10} & \cdots & p_{1n} \\ p_{20} & \cdots & p_{2n} \end{bmatrix}$$

where

$$p_{1j} = \binom{n}{j}(0.25)^j(0.75)^{n-j}$$

and

$$p_{2j} = \binom{n}{j}(0.5)^n$$

2. DISCRIMINATING EXPERIMENTS

for $j = 0, 1, 2, \ldots, n$. Once more we have binomial probabilities, with j as the number of A's observed.

The experiment E is really the sequence of experiments $\langle E_n \mid 1 \leq n \leq \nu \rangle$, where ν is an infinite natural numbers. If, for instance, A is observed in one progeny, then the model of statistical evidence in Birnbaum's notation is $\langle E, \langle 1, 1 \rangle \rangle$, i.e., in measurement function notation, $[E_1 = 1]$. Recall that the second 1 in Birnbaum's notation denotes that A was observed and the first 1, that we are dealing with E_1. If A is observed 45 times in 60 progeny, then the model of statistical evidence is $\langle E, \langle 60, 45 \rangle \rangle$, i.e., $[E_{60} = 45]$.

Let us now explain in the new context what it means for an experiment to be discriminating for a set of possible models I. We use the same notion of discriminating experiment given in Definition IV.3, but here we do not distinguish between models assigning the same probability distribution. The following definitions are, in fact, almost a restatement of Definitions IV.1, IV.2, and IV.3.

Let E_n be as above, i.e.

$$E_n = (p_{nij}) = \begin{bmatrix} p_{n11} & p_{n12} & \cdots & p_{n1J_n} \\ p_{n21} & p_{n22} & \cdots & p_{n2J_n} \\ \vdots & \vdots & \ddots & \vdots \\ p_{nI1} & p_{nI2} & \cdots & p_{nIJ_n} \end{bmatrix}.$$

We recall the definition of a rejection set:

DEFINITION XVIII.1 (REJECTION SET). Let $j, k \in S_n$. Then $j \sim_n k$, if there is a $c > 0$ such that for all $i \in I$, $p_{nij} = cp_{nik}$. We write $j \preceq_{ni} k$, if

$$\sum_{a \sim j} p_{nia} \leq \sum_{b \sim k} p_{nib}.$$

. The rejection set R_{nij} is the set of all $k \in S_n$, wuch that

$$k \preceq_{ni} j.$$

The probability of this rejection set, P_{nij}, is

$$P_{nij} = \sum_{k \preceq j} p_{nik}.$$

For a result j, P_{nij} is what is called in usual statistical practice, the *p-value of the test with result j*.

We now explain more precisely what it means for a sequence of random variables to be eventually a.s. in a set. Let $r_1, r_2, \ldots, r_\nu$ be an internal ν-sequence of hyperreal numbers, and let $Y_1, Y_2, \ldots, Y_\nu$ be an internal ν-sequence of random variables defined over the same probability space. We say that the sequence $\{Y_i\}$ *eventually agrees with the sequence* $\{r_i\}$ if $Y_\eta \approx r_\eta$, for every infinite $\eta \leq \nu$.

Let A be a set (internal or external) of internal ν-sequences of hyperreal numbers. We say that the sequence of random variables $\{Y_i\}$ *is eventually a.s. in A* if a.s. there is a sequence $\{r_i\} \in A$ that eventually agrees with $\{Y_i\}$. That is, if the set N of ω such that, for every sequence $\{r_i\} \in A$, there is an infinite $\eta \leq \nu$ such that $r_\eta \not\approx Y_\eta$ is a null set.

On the other hand, the sequence $\{Y_i\}$ *is a.s. eventually never in* A if the set N' of ω such that there is a sequence $\{r_i\} \in A$ with $r_\eta \approx Y_\eta$, for every infinite $\eta \leq \nu$, is a null set.

If the set A is the set of internal ν-sequences that nearly converge to a certain number x, then the sequence $\{Y_i\}$ is a.s. eventually in A if and only if $\{Y_i\}$ nearly converges a.s. to x. On the other hand, if $\{Y_i\}$ nearly converges a.s. to a number $y \not\approx x$, then $\{Y_i\}$ is a.s. eventually never in A.

DEFINITION XVIII.2 (D. E. FOR PROBABILITY MATRICES). The system $E = \langle E_n : 1 \leq n \leq \nu \rangle$ is discriminating for I if

(1) The sequence $\{E_n\}$ is an internal sequence of random variables over $\Pr_n$.
(2) For every $k \in I$ there is an internal set A of ν-sequences of results such that $\{E_n\}_{n=1}^{\nu}$ is a.s. eventually in A, according to $\Pr_k$ and $\{E_n\}_{n=1}^{\nu}$ is a.s. not in A, according to $\Pr_j$, for $j \neq k$, and for every $r \in A$, $i \in I$ with $i \neq k$, and $\alpha \gg 0$, there is a finite n such that, for every $m \geq n$, we have

$$P_{m i r_m} \leq \alpha.$$

That is, the *p*–value of the test can be made as small as one wishes.

(3) It is possible to determine in principle for each $j \in J_n$ whether j is the actual result of the partial experiment E_n.

It is easy to show, by the use of the strong law of large numbers and the central limit theorem, as was done in Chapter IV, Section 2, that the examples of experiments given above are discriminating for the alternative models indicated. We shall prove a theorem, however, which includes these cases and others.

All experiments that are used in classical statistical inference are of the form $E_n = g(Y_1, \ldots, Y_n)$, where Y_1, ..., Y_n represent the n repetitions of the setup in question, and g is a continuous function. These are what are usually called *statistics*. The most usual statistics is the sample mean, for which we introduce a special symbol

$$\frac{1}{n}\sum_{i=1}^{n} Y_i = \overline{Y}_n.$$

For this statistics, we can prove a general condition under which it constitutes a discriminating experiment. We first introduce some notation, which I think is the usual. When having a set of possible model (i.e., parameters) Ω and a random variable X that makes sense for all models in Ω, we write $\mathbf{E}_\theta X$ and $\text{Var}_\theta(X)$ for the expectation and variance of X under the model θ. We shall only worry about Conditions (1) and (2) in the definition of discriminating experiment, when proving that an experiment is discriminating. The other condition is not mathematical in character.

THEOREM XVIII.1. *Let Ω represent the set of possible models for a certain setup $\mathfrak{S}$, and X a random variable for the setup with finite variance under every θ and such that for each $\theta \in \Omega$, $\mathbf{E}_\theta X = \theta$. Assume that all elements of Ω*

2. DISCRIMINATING EXPERIMENTS

are real. Suppose that $X_1, X_2, \ldots, X_i, \ldots$ are random variables representing independent repetitions of the setup $\mathfrak{S}$, i.e., these variable are independent and identically distributed with X. Then, for any infinite ν, $\overline{X} = \langle \overline{X}_n \mid 1 \leq n \leq \nu \rangle$ is a discriminating experiment for Ω.

In this theorem, as it will be the case for most applications of the notion of experiment to statistical inference, the set of alternative hypotheses represented by the set of parameters Ω is not a set of probability models, but a family of sets of probability models. Thus, a $\theta \in \Omega$ may represent a family of distributions with finite mean θ and finite variance. The notion of a discriminating experiment has to be modified for this case in the obvious way. We shall discuss situations where the composite character of the alternatives is more prominent in Section 4.

PROOF. Let A_θ, for each θ, be the set of sequences $\{r_n\}_{n=1}^{\nu}$ that nearly converge to θ. By the strong law of large numbers, Theorem XI.7, $\overline{X}$ is a.s. eventually in A_θ, according to $\Pr_\theta$, and a.s. eventually not in A_θ, according to $\Pr_{\theta'}$, for $\theta' \neq \theta$.

Let $\{r_n\} \in A_\theta$, $\theta' \neq \theta$, $\alpha \gg 0$, and $\mathrm{Var}_{\theta'}(X) = \sigma'$. Then, by the central limit theorem, when n is large and θ' is true then

$$\sqrt{n}\frac{\overline{X}_n - \theta'}{\sigma'}$$

is approximately normal. So that the rejection set, $R_{n\theta's}$, for θ' given s is approximately

$$[|\overline{X}_n - \theta'| \geq |s - \theta'|]$$

and, hence, its probability, $P_{n\theta's}$, is approximately

$$\Pr_{\theta'}[|\overline{X}_n - \theta'| \geq |s - \theta'|].$$

By the strong law of large numbers again, for every $\varepsilon \gg 0$, there is a finite n such that for all $m \geq n$

$$\Pr[|\overline{X}_m - \theta'| \geq \varepsilon] \leq \alpha. \tag{1}$$

Let $\varepsilon = |\theta - \theta'|/2$. We have that $\varepsilon \gg 0$, since θ and θ' are real. Let n be the smallest number that satisfies (1) and such that for every $m \geq n$, $|r_m - \theta'| > \varepsilon$. Then we have that for every $m \geq n$

$$\Pr[|\overline{X}_m - \theta'| \geq |r_m - \theta'|] \leq \alpha.$$

□

Birnbaum uses the notation $\mathrm{Ev}(E, \langle n, j \rangle)$, which I shall also use mainly written $\mathrm{Ev}[E_n = j]$, that may be read as "the content, the import, or the evidential meaning of the statistical evidence $\langle E, \langle n, j \rangle \rangle$ for the set of alternative models Γ". Notice that E is, in our case, a sequence $\langle E_n \mid 1 \leq n \leq \nu \rangle$, with $\nu \approx \infty$, and not just an experiment for a single n, as in Birnbaum. This introduces a very important difference, as we shall see later. I shall discuss some principles, in Section 5 and Chapter XIX, Section 4, also dealt by Birnbaum, that assert that certain instance of statistical evidence are equivalent (or have the same evidential import). I will discuss these principles on the light of the principles of statistical inference introduced in the previous chapters. Just as Birnbaum, we shall indicate an assertion of equivalence between given instances of statistical evidence $\langle E, \langle n, j \rangle \rangle$ and $\langle E', \langle n, j' \rangle \rangle$ by writing $\text{Ev} \langle E, \langle n, j \rangle \rangle = \text{Ev} \langle E', \langle n, j' \rangle \rangle$, or $\text{Ev}[E_n = j] = \text{Ev}[E'_n = j']$.

3. Significance tests

I shall discuss two main techniques of hypotheses testing: significance tests and hypotheses tests.[3] As it was explained above, the hypothesis to be tested, the null or working hypothesis H, is of the form $\theta \in \Omega'$, for some subset Ω' of the parameter space Ω.

In significance tests, we use the data x to measure inferential evidence against H, and if such contraindication of H is sufficiently strong we will judge H to be inappropriate, i.e., we reject H. Apparently, in this case, there is no consideration of hypotheses alternative to H. I will elucidate this matter later, which seems to go against my Inverse Inference Principle.

Hypotheses tests, in the form of Neyman and Pearson, explicitly consider an alternative hypothesis H_1. The tests are constructed and their properties assessed in regard to H_1. Although, the primary purpose of the test is to consider the rejection of H, the rejection of H implies, in these tests, that H_1 is consistent with the data, and thus, in a certain sense, the acceptance of H_1. Section 4 will be devoted to hypotheses tests, but many of the concepts developed in this section for significance tests are also useful for hypotheses tests.

From an inferential standpoint, it is clearly more convenient not to postulate alternative hypotheses, since we then could reject H outright, and not relative to alternatives. But, is this possible? In Chapter IV, Section 4, I discussed frequencies, and we saw that is was possible, in this case, to consider as alternative hypotheses all the possible probabilities assigned to a particular event. But even in this case, as it will be stressed a little later, not all logically possible hypotheses are entertained. In the case of significance tests, however, the definition of worse result can be simplified, so as to depend only on the hypothesis H. This is so, because for these cases, the discriminating experiments, E, are chosen in such a way that no two results are evidentially equivalent, and for any pair of results, a, b, if $\Pr_H[E_n = a] < \Pr_H[E_n = b]$, then there is always a alternative where the inequality is reversed. Thus, we can define a is at least as bad as b for H, in symbols, $a \preceq_{nH} b$, if

$$\Pr_H[E_n = a] \leq \Pr_H[E_n = b].$$

Thus, this definition seems not to depend on alternatives. The alternatives are implicit, however, because it is assume that there is enough of them, so as the experiment to work, and not all logically possible alternatives are entertained. I believe that this simple definition of rejection sets, which seems not to need the consideration of alternatives, is behind Fisher's justification of tail probabilities for significance tests.

---

[3] See, also, [4, pp. 129-132].

3. SIGNIFICANCE TESTS

As an example, I will now discuss again the case of frequencies stressing the rejection of hypotheses instead of their acceptance, as was done in Chapter IV, Section 4. We are, here, in the case of a random variable X with values 0 or 1. The parameter space Ω can be taken to be the real elements in the interval $[0,1]$. For $p \in [0,1]$, we assume that $\Pr_p[X = 1] = p$. Here, we thus consider what seems to be all the logically possible alternatives, namely all real p's in $[0,1]$. The n-sampling distribution corresponding to $\Pr_p$ is binomial $\langle n, p \rangle$. We also call it $\Pr_p$. The random variable for the n-sampling distribution is a random variable that can be considered as a sequence of random variables identically distributed to X, X_1, ..., X_n. The random variable giving the relative frequency of the event $[X=1]$ is $(1/n)\sum_{i=1}^{n} X_i$, which is expressible in the language for X_1, ..., X_n. In Chapter IV, Section 4, we defined it as

$$\mathrm{Fr}_n = \frac{1}{n}\sum_{i=1}^{n} X_i$$

This sequence of random variables, Fr_n, constitutes our experiment. It is clear that $\mathrm{Fr}_n = \overline{X}_n$ in the terminology of the previous section. Hence, by Theorem XVIII.1, Fr_n is discriminating.

If the result is $\mathrm{Fr}_n = x$, then we calculate the probability under p of a result x or worse, i.e.

$$\Pr_p[|\mathrm{Fr}_n - p| \geq |x - p|] = r.$$

If r, called the *probability* or *p-value* of the test, is small, say less than α, then we reject the hypothesis that the probability is p.

The dialectical process goes as follows: somebody sets the level α, and we try to get an $r \leq \alpha$. If p is not the true value, then in the long run we are almost sure to get an $r \leq \alpha$.

But for this case of frequencies, the dialectical process can be improved somewhat. As we saw in Chapter IV, Section 4, we can find a computable function, f, such that for each m and n we have

$$\Pr[|\mathrm{Fr}_n - p| \geq f(m,n)] < \frac{1}{m}.$$

Then, if $\alpha = 1/m$ is set, we can compute for each n, $f(m,n)$, and we can tell in advance how large the value of Fr_n must be.

Notice that $[|\mathrm{Fr}_n - p| \geq f(m,n)]$ or its negation is implied by each complete sample description of the form $[X_1 = r_1] \wedge \ldots \wedge [X_n = r_n]$, where r_i is 0 or 1, for every $i = 1, \ldots, n$.

Although we accepted as possible all real probabilities for $[X = 1]$ as our class of alternative models, the possible models do not include all logically possible models. First, we assume that we have a number of independent repetitions of the experiment. That is, the alternative models used for the Inverse Inference Principle are all of the form ^nK, for some K. We do not allow models that are not of the form ^nK. It is possible, with certain statistical techniques to test this independence, but this has to be done in a separate procedure where other alternative hypotheses are used.

Second, for the original set of alternative models for the experiment that is being repeated, we accept all possible probability distributions, but not all possible probability models, since we cannot distinguish by statistical methods alone between models with the same distribution.

Third, there are logically possible models, such as that of the coin tossing machine discussed in pages 94 and 104, which are never considered.

As was indicated above, however, the only assumption we need about alternatives is the existence of enough of them so that the definition of worse result depends only on the null hypothesis, H.

The general case for significance tests for simple hypotheses can then be described as follows. As was said above, a hypothesis H is simple if it fully specifies the distribution of X. For instance, the hypothesis that the probability of $[X = 1]$ is a given number p is simple. That is, Ω' is a set of one element. Suppose that H is simple, determining that the parameter is θ. Typically, $\mathbf{E}_\theta X = \theta$. We fist choose a *statistics* $E_n = g(X_1, \ldots, X_n)$, where g is a continuous function. Typically $E_n = \overline{X}_n$.

We then choose a *test statistics* (or *discrepancy measure*) $t(E_n)$. The function t is typically a continuous function of E_n and θ, and so should be written $t(E_n, \theta)$. If a result x is observed for E_n, then the sentence $[t(E_n) \geq t(x)]$ represents the set of results that are as least as bad as x for θ. We evaluate

$$r = \Pr_\theta[t(E_n) \geq t(x)]$$

where x is the value we have observed for E_n, and hence $t(x)$ is the value we have observed for $t(E_n)$. This is the *p*-value of the test. If r is small enough, we reject H. We choose a *significance level* α, observe x, and *reject* H *at level* α if $r \leq \alpha$, i.e., if the *p*-value is not greater than α.

Notice that we need to know enough of the distribution with parameter θ to be able to determine that the rejection set for θ is $[t(E_n) \geq t(x)]$, and to be able to calculate $\Pr_\theta[t(E_n) \geq t(x)]$.

Equivalently, we are choosing a critical value $t_0(\alpha)$ for $t(E_n)$ and rejecting H if $t_0(\alpha) < t(x)$. The number $t_0(\alpha)$ is defined as the least number s such that

$$\Pr_\theta[s < t(E_n)] \leq \alpha.$$

In the case we were considering above of a random variable X with the hypothesis H that asserts that $[X = 1]$ has probability p, the test statistics t can be taken to be $t(\text{Fr}_n) = |\text{Fr}_n - p|$. For any significance level of $\alpha = 1/m$ and fixed n, in this case, we can calculate $t_0(\alpha) = f(m, n)$. So, if x is the result in trials, with $f(m, n) < t(n, x) = |1/n \sum_{i=1}^n x_i - p|$, then we reject H.

We return to the general case. A sentence that rejects the hypothesis with parameter θ can be constructed as follows. We first fix an n and α and find t_0 as above. A rejecting sentence is then $[t(E_n) > t_0]$.

In order for the test statistics to make sense the following condition, which is usually left tacit, must be satisfied: the experiment E_n, on which t is based, must be discriminating.

3. SIGNIFICANCE TESTS

Another way of presenting these tests is by considering the space S of all possible values of X, for a certain fixed n. That is S consists of all n-tuples $x_1, \ldots, x_n$ of possible values of $X_1, \ldots, X_n$. The test of level α partitions S into two sets S_0 and S_1, a critical region S_0, where if $x \in S_0$ we reject H, and a non-critical region S_1, where if $x \in S_1$ we have no reason to reject H on the basis of the level α test. This way of presenting the test, hides even more the need for a discriminating infinite sequence of experiments. Since S depends on n, we should really write $S(n)$, and then the sequential character of the test becomes apparent.

If H is composite, asserting that $\theta \in \Omega' \subseteq \Omega$, the test of level α has a similar form. As was mentioned above, the test statistics t may be a function not only of E_n, which is a function of $X_1, X_2, \ldots, X_n$, but also of the parameter θ. Again we want a t_0 such that

$$\Pr_\theta[t_0 < t(E_n)] \leq \alpha,$$

for all $\theta \in \Omega'$, so the significance level α provides an upper bound to the maximum probability of incorrectly rejecting H.

Just as in the case of simple hypotheses, each of the sentences of the test sequence, $[t_0 < t(E_k, \theta)]$ or their negations are implied by complete sample descriptions $[X_0 = x_0] \wedge \cdots \wedge [X_{k-1} = x_{k-1}]$, i.e., their truth or falsity is determined by the observation of the values of X in the sample.

Returning to the example with a 0-1 random variable X, we can take as H the composite hypothesis that asserts that the probability of $[X = 1]$ is in an interval $[p, q]$. Then $t(\mathrm{Fr}_n, s) = |\mathrm{Fr}_n - s|$, where $s \in [p, q]$. All the other considerations made for a simple hypothesis discussed above can be then formulated without problems for this composite H.

There other random variables, besides the 0-1 ones, for which a significance test makes sense. For instance, suppose that we have a random variable X with finite mean and variance, and we want to test the hypothesis that it has a distribution with a certain finite mean μ. Then, by Theorem XVIII.1, we can take as alternative hypotheses all distributions with finite mean and variance. Here we cannot take all possible distributions, as in the 0-1 case, but we must require that they have finite mean and variance. This is a weak restriction, however.

We assume, again, that the experiment can be repeated an indefinite number of times and obtain the independent and identically distributed variables $X_1, X_2, \ldots, X_\nu$, with the same mean as X. By Theorem XVIII.1, we know that $\overline{X}_n$ is a discriminating experiment. Then we can construct a test statistics, similar to the one above for this case.

For significance tests, we consider the widest possible range of alternative hypotheses. We only limit them by supposing that the repetitions of the experiment are independent, and, frequently, by finiteness assumptions about the mean or the variance.

The requirement that the test statistics, $t(E_n)$ should be such that $[t(E_n) \geq t(x)]$ contain all results that are at least as bad as x for θ, serves to eliminate many possible functions of the discriminating experiment as test statistics. Let us return to the example of the frequency case. Suppose that we consider the same test statistics, $t(\text{Fr}_n) = |\text{Fr}_n - p|$, but with the formula $[t(\text{Fr}_n) < t_0]$. For a particular n, and for any α, we can certainly find a t_0 such that $\Pr_p[t(\text{Fr}_n) < t_0] < \alpha$. But the problem, here, is that $[t(\text{Fr}_n) < t_0]$ does not contain the bad results por p.

For another example, instead of considering the test statistics $t(\text{Fr}_n)$, we could entertain the test statistics $s(\text{Fr}_n) = \text{Fr}_n - p$. We then consider the test sentences $[t_0 < s(\text{Fr}_n)]$. It obviously has the following disadvantage. With this test sequence we would accept $\Pr_p$, if $[t_0 < s(\text{Fr}_n)]$ did not obtain in reality. Hence, we could have $p - (1/n)\sum_{i=1}^{n} x_i$ very large and still accept $\Pr_p$, although there would be, in this case better evidence for a $\Pr_q$ with $q < p$. It is clear, however, that $[t_0 < s(\text{Fr}_n)]$ does not contain all results that are worse than the given one.

4. Hypotheses tests

In hypotheses tests, we test a hypothesis H_0, called the *null* or *working hypothesis*, considering at the same time an alternative hypothesis H_1. The working hypothesis H_0 and the alternative hypothesis H_1 serve to partition the parameter space Ω. Under H_0, θ lies in a subspace Ω'; under H_1, θ lies in the complementary subspace $\Omega - \Omega'$.

$$H_0 : \theta \in \Omega' \quad \text{and} \quad H_1 : \theta \in \Omega - \Omega'.$$

The purpose of a hypothesis test is to determine whether H_0 or H_1 are consistent with the data. Thus, accepting H_0 means simply that we are not in a position of rejecting H_0, i.e., that H_0 is consistent with the data. Similarly, rejecting H_0, and hence, accepting H_1, means that H_0 is inconsistent with the data, and hence, with respect to H_1, it only means that H_1 is consistent with the data. Thus, hypotheses tests involve rejection and acceptance at the same time, but not on the same level of importance.

I now begin with the justification of hypotheses tests, according to the view spoused in this book. Although I consider hypotheses tests to be valid, my justification is different from that of the developers of this tests, Neyman and Pearson. In the next chapter, Section 1, we shall discuss the Neyman-Pearson justification for these tests.

Many of the notions introduced for significance tests are also used in hypothesis tests and the same conditions that apply to significance tests apply here. The most important, according to my point of view, is that the test statistics must be a function of a discriminating experiment. The notion of a discriminating experiment must be modified to take into account that we are only considering two alternative hypotheses, $H_0 : \theta \in \Omega'$ and $H_1 : \theta \in \Omega - \Omega'$. For simplicity, we write $\Omega_0 = \Omega'$ and $\Omega_1 = \Omega - \Omega'$. We assume that $E = \langle E_n \mid 1 \leq n \leq \nu \rangle$, for an infinte ν, is a sequence of random variables on each of the models in H_i, $i = 0, 1$. We begin by the notion of evidentially equivalent results, which is the same as the old notion in Chapter IV Section 2 and in Section 2 of this chapter, adapted to the new situation. Let r and s be possible results of E_n. Then $r \sim_n s$ if there is a $c > 0$ such that for every $\theta \in \Omega$

$$\Pr_\theta[E_n = r] = c \cdot \Pr_\theta[E_n = s].$$

4. HYPOTHESES TESTS

Similarly as before, $[r]$ is the equivalence class of r.

We say that $r \preceq_{\theta n} s$, for $\theta \in \Omega_i$, if

$$\Pr_\theta[E_n \in [r]] \leq \Pr_\theta[E_n \in [s]].$$

and there is $\theta \in \Omega_j$, with $j \neq i$, such that the inequality is reversed.

The rejection set for θ, with $\theta \in \Omega$, determined by the result r of E_n is then, as before

$$R_{n\theta r} = \{s \mid s \preceq_{\theta n} r\}.$$

The probability $\Pr_\theta R_{n\theta r}$ is the p-value of the test for θ with result r.

DEFINITION XVIII.3 (D. E. FOR HYPOTHESES TESTS). We say that the system $E = \langle E_n \mid 1 \leq n \leq \nu \rangle$ is a *discriminating experiment (d.e.) for* H_0 *against* H_1 if

(1) The sequence E_n is an internal sequence of random variable over the n–product space of the spaces in H_0 and H_1.

(2) For each $\theta \in \Omega_1$ there is an internal set A of sequences of results such that E is almost surely eventually in A according to θ, and E is almost surely eventually never in A according to θ', for $\theta' \in \Omega_0$, and such that for any $\langle r_n \mid 1 \leq n \leq \nu \rangle$ in A, and any $\alpha \gg 0$ there is a finite n such that for every $\theta' \in \Omega_0$ and every $m \geq n$

$$\Pr_{\theta'}[E_m \in R_{m\theta' r_m}] \leq \alpha.$$

That is, the p-value of the test can be made as small as one wishes.

(3) For each $\theta \in \Omega_0$ there is an internal set B of sequences of results such that E is a.s. eventually in B, according to θ, and a.s. eventually not in B, according to θ', for $\theta' \in \Omega_1$, and such that for any $\{r_n\}_{n=1}^{\nu}$ in B, any $\theta' \in \Omega_1$ and any $\alpha \gg 0$ there is a finite n such that for every $m \geq n$

$$\Pr_{\theta'}[E_m \in R_{m\theta' r_m}] \leq \alpha.$$

The main differences of this definition with Definitions IV.3 and XVIII.2 are:

(1) In the definition $r \preceq_{\theta n} s$, i.e., of r being at least as bad as s for θ, we require that the inequality be reversed for a θ' in the hypothesis which is alternative to that in which θ is.

(2) H_0 and H_1 are asymmetric in the sense that we require that sequences in the set A work for each α with the same n for all $\theta \in \Omega_0$, while for sequences of B, the n may be different for each $\theta' \in \Omega_1$.

The dialectical rule of rejection for this case is now:

RULE XVIII.1 (DIALECTICAL RULE OF REJECTION OF H_0 AGAINST H_1). *Let Ω be a set of parameters of possible probability models for a setup, and suppose that we are considering the null hypothesis* $H_0 : \theta \in \Omega'$ *against* $H_1 : \theta \in \Omega - \Omega'$. *We say that* H_0 *should be provisionally rejected at level of significance α, if there is a discriminating experiment* $\langle E_n \mid 1 \leq n \leq \nu \rangle$ *for* H_0 *against* H_1, *such that*

(1) $[E_n = a]$ *obtains, for some a and $n \in \mathbb{N}$.*

(2) $\Pr_\theta[E_n \in R_{n\theta a}] \leq \alpha$, for every $\theta \in \Omega'$, i.e., the p-value msut be small for every $\theta \in \Omega'$.

Since in the definition of discriminating experiment for H_0 against H_1 we require that the n be uniform for all elements of Ω_0, the old proof of Theorem XVIII.1 does not work because we do not know whether convergence to the normal distribution is uniform. If all sampling distributions for $\theta \in \Omega_0$ have the property that $R_{n\theta a}$ is an interval or union of intervals, then the theorem remains true:

THEOREM XVIII.2. *Let $\Omega \subseteq \mathbb{R}$ represent the set of possible models for a certain setup $\mathfrak{S}$, X a random variable for $\mathfrak{S}$, with finite variance under every θ, and such that for each $\theta \in \Omega$, $\mathbf{E}_\theta X = \theta$. Let $H_0 : \theta \in \Omega_0$ and $H_1 : \theta \in \Omega - \Omega_0 = \Omega_1$, with Ω_0 an interval of real numbers.*

Suppose that $X_1, X_2, \ldots, X_i, \ldots, X_\nu$, ν infinite, are random variables representing independent repetitions of $\mathfrak{S}$ (i.e., these variables are independent and identically distributed with X), such that $R_{n\theta a}$ is a union of intervals of real numbers when $\overline{X} = \langle \overline{X}_n \mid 1 \leq n \leq \nu \rangle$ is the experiment considered. Then $\overline{X}$ is a discriminating experiment for H_0 against H_1.

The proof, which I omit, is similar to that of Theorem XVIII.1.

We shall now discuss some examples:

EXAMPLE XVIII.1. Suppose that we are testing nearly normal distributions with known variance σ^2 and unknown mean μ. We are testing $H_0 : \mu = \mu_0$ against $H_1 : \mu \neq \mu_0$. Then Theorem XVIII.1 (and also Theorem XVIII.2) applies and hence, the sample mean $\overline{X}_n$ is a discriminating experiment. The usual test works. For one-sided tests, say $H_0 : \mu \leq \mu_0$ aginst $H_1 : \mu > \mu_0$, however, we must apply Theorem XVIII.2. This last theorem applies, since the rejection set determined by a result r, for this case, is the interval $[\overline{X}_n > r]$.

EXAMPLE XVIII.2. As a second example, assume that the alternative hypotheses are nearly normal distributions with mean $\mu \in \Omega$ and unknown variance. The experiment now should the be discriminating for $H_0 : \mu = \mu_0$ against $H_1 : \mu \neq \mu_0$, where μ_0 is a fixed number. We need, for this case, an experiment M with two results: the sample mean

$$\overline{X}_n = \frac{1}{n} \sum_{i=1}^n X_i,$$

and the sample variance

$$S_n^2 = \frac{1}{n-1} \sum_{i=1}^n (X_i - \overline{X}_n)^2.$$

Thus, M_n is the pair $\langle \overline{X}_n, S_n^2 \rangle$. The null hypothesis H_0 is then a composite hypothesis including all distributions with mean μ_0 and different variances. Here we cannot apply directly our theorems. We need some preliminary work to obtain the sets A and B of the definition of discriminating experiment. Besides the sample mean and variance we need the following function

$$\widehat{X}_n = \frac{1}{n} \sum_{i=1}^{n} (X_i - \mu)^2$$

where μ is the real mean.

Extend the sequence X_n to $^*\mathbb{N}$. Let $\nu \approx \infty$. Assume that μ is the true mean, and σ^2, the true variance. We have that $\overline{X}_\nu \approx \mu$, a.s. Also, a.s., $\widehat{X}_\nu \approx \sigma^2$, by the law of large numbers, Theorem XI.7, because $(X_1 - \mu)^2$, $(X_2 - \mu)^2$, ... are independent and identically distributed random variables with $\mathbf{E}_{\mu,\sigma^2}(X_i - \mu)^2 = \sigma^2$. Thus

$$S_\nu^2 = \frac{1}{\nu - 1} \sum_{i=1}^{\nu} (X_i - \overline{X}_\nu)^2$$

$$= \frac{1}{\nu - 1} \sum_{i=1}^{\nu} X_i^2 - \frac{\nu}{\nu - 1} \overline{X}_\nu^2$$

$$\approx \frac{1}{\nu} \sum_{i=1}^{\nu} X_i^2 - \mu^2$$

$$= \frac{1}{\nu} \sum_{i=1}^{\nu} X_i^2 - 2\mu^2 + \mu^2$$

$$\approx \frac{1}{\nu} \sum_{i=1}^{\nu} X_i^2 - \frac{2}{\nu} \sum_{i=1}^{\nu} X_i \mu + \mu^2$$

$$= \frac{1}{\nu} \sum_{i=1}^{\nu} (X_i - \mu)^2$$

$$\approx \sigma^2$$

Therefore, if μ_0 and σ^2 are the true mean and variance, then the set, A_σ, of pairs of infinite sequences, $\{r_n\}$ and $\{s_n\}$, such that $\langle \{r_n\}, \{s_n\} \rangle \in A$ if and only if r_n nearly converges to μ and s_n^2 nearly converges to σ^2, has the property that M is a.s. in A_σ. On the other hand, if the true mean or variance are different, then M is a.s. eventually not in A. We shall prove that this A_σ has the required properties of the definition of discriminating experiment.

By Theorem X.13

$$\sqrt{n} \frac{\overline{X}_n - \mu}{S_n}$$

has nearly t distribution with $n - 1$ degrees of freedom, i.e., it is $\mathcal{T}_{n-1}$, if μ is the true mean. Let $\langle m, s \rangle$ and $\langle m', s' \rangle$ be possible results of M_n. Then, because of the t distribution, $\langle m, s \rangle \preceq_{n,\mu} \langle m', s' \rangle$, if

$$\frac{|m - \mu|}{s} \geq \frac{|m' - \mu|}{s'}.$$

Let $|\mu-\mu_0| = \varepsilon$. Then, since μ and μ_0 are real, $\varepsilon \gg 0$. Assume that $\langle m, s\rangle \in A$. Then m_n S-converges to μ and s_n S-converges to σ^2. Let n_1 be such that for all $k \geq n_1$

$$|m_k - \mu_0| > \frac{\varepsilon}{2} \quad \text{and} \quad |s_k - \sigma| < \frac{\varepsilon}{2}.$$

Then, for $k \geq n_1$

$$\frac{|m_k - \mu_0|}{s_k} > K.$$

where

$$K = \frac{\varepsilon}{2\sigma + \varepsilon} \gg 0.$$

As is usual we define

$$\Pr([Z > z_\gamma]) \approx \gamma$$

where Z is a unit nearly normal random variable, and

$$\Pr([T_n > t_{\gamma,n}]) \approx \gamma,$$

where T_n is nearly a student t random variable with n degrees of freedom.

Let, now, $n_2 \geq n_1$ be such that

$$\sqrt{n_2} K > z_{\alpha/2},$$

and let $n_3 \geq n_1 + 4$ be a finite number such that

$$|t_{\alpha/2, n_2-1} - z_{\alpha/2}| < \frac{K}{2}.$$

Then

$$\sqrt{n_2} K > \sqrt{n_1} K + K > z_{\alpha/2} + K \geq t_{\alpha/2, n_2-1}.$$

Therefore, if $k \geq n_2$

$$\sqrt{n_2} \frac{|m_k - \mu_0|}{s_k} > t_{\alpha/2, n_2-1}.$$

Thus

$$\Pr_{\mu_0}([\sqrt{n_2} \frac{|\overline{X}_k - \mu_0|}{s_k} > \sqrt{n_2} \frac{|m_k - \mu_0|}{s_k}]) \leq \alpha.$$

The choice, for each $\mu \neq \mu_0$, of of the set B of the definition of discriminating experiment is similar. This shows that the experiment is discriminating.

EXAMPLE XVIII.3. Suppose, now, as another example, that we have as alternative hypotheses nearly normal distributions with a known and fixed mean μ and different variances. We would like to test the hypothesis $H_0 : \sigma^2 \geq \sigma_0^2$ against $H_1 : \sigma^2 < \sigma_0^2$, for a fixed σ_0, where σ^2 is the variance. We use as our statistics

$$\widehat{X}_n = \frac{1}{n} \sum_{i=1}^{n} (X_i - \mu)^2.$$

We have that $(X_1 - \mu)^2, (X_2 - \mu)^2, \ldots, (X_\nu - \mu)^2$, represent independent repetitions of a random variable $(X - \mu)^2$. Also

$$\mathbf{E}_{\sigma^2}(X - \mu)^2 = \sigma^2.$$

By Theorem X.12, the distribution of

$$\frac{n\widehat{X}_n}{\sigma^2}$$

is χ_n^2, if the true variance is σ^2. Then the rejection sets are intervals. Using Theorem XVIII.2, we get that $\widehat{X}_n$ is a discriminating experiment for H_0 against H_1. The rejection rule, then, is the same as the usual one.

EXAMPLE XVIII.4. As an example of a possible statistics that does not work, we have the following. I do not think that anybody has suggested this test, but in order to make the point, let us assume that we are in the same situation as before, i.e., testing $H_0 : \sigma^2 \geq \sigma_0^2$ against $H_1 : \sigma^2 < \sigma_0^2$, but that we choose as our test statistics

$$\overline{X}_n - \mu.$$

Since, if σ^2 is the true variance

$$\sqrt{n}\frac{\overline{X}_n - \mu}{\sigma}$$

has a normal distribution, the rejection sets have the appropriate form. However, $\overline{X}_n - \mu$ tends to 0, for every σ^2, and hence there is no discriminating set of sequences of probability approximating one. Thus, this possible statistics does not satisfy the requirements for a discriminating experiment.

EXAMPLE XVIII.5. Suppose, as a final example, that we are testing the same hypotheses as above, $H_0 : \sigma^2 \geq \sigma_0^2$ against $H_1 : \sigma^2 < \sigma_0^2$, but that now we do not know the mean μ. We can take as our statistics, as it is usual, the sample variance

$$S_n^2 = \frac{1}{n-1}\sum_{i=1}^n (X_i - \overline{X}_n)^2.$$

Here, the variable

$$(n-1)\frac{S_n^2}{\sigma^2}$$

is χ_{n-1}^2, when σ^2 is the true variance. Hence, the rejection sets have the right form, for the application of Theorem XVIII.2. We cannot obtain directly by the strong law of large numbers, however, the sets A or B of the definition of discriminating experiment. We have, as above

$$\Pr_{\mu,\sigma^2}([S_n^2 \to \sigma^2]) = 1. \tag{1}$$

and

$$\Pr_{\mu,\sigma_1^2}([S_n^2 \to \sigma^2]) = 0. \tag{2}$$

for $\sigma_1 \neq \sigma$.

Thus, by a similar argument as in the proof of Theorem XVIII.1, using (1) and (2), we can show that S_n^2 is a discriminating experiment for this case.

A few remarks about the differences between significance tests and hypothesis tests are in order. As we have seen, both types of tests need the consideration of a set of possible alternative hypotheses, which does not consist of all logically possible hypotheses. In significance tests, the widest reasonable set is entertained, but the alternative hypotheses are only considered in order to eliminate some tests that don't make sense. The only role of alternatives, for significance tests, is to insure that the definition of worse results is the correct one.

In hypothesis tests, the alternative hypothesis may be accepted. I think it is clearly better to have an alternative hypothesis that is to be accepted in case the main hypothesis is rejected. The other advantage of hypotheses tests is that, by restricting the class of possible hypotheses, the properties of the test are easier to study and more powerful tests are possible. These two extra advantages require the restriction of the class Ω of possible distributions. For decision theoretic purposes, this may pose no problems. But for strictly inferential purposes it may be better to have the widest possible class of alternative distributions.

If there are sufficient reasons to limit the possible alternatives, however, hypothesis tests are more flexible, since they allow, among other things, for minimization of, what are called, type II errors,[4] for all or some of the alternatives, as follows:

Let S_0 and S_1 be the critical and noncritical regions for the test and let

$$\beta(\theta) = \Pr_\theta[X \in S_1], \text{ for } \theta \in \Omega - \Omega'.$$

A test is better, for $\theta \in \Omega - \Omega'$, if $\beta(\theta)$ is small. Although there may be no test that makes $\beta(\theta)$ small for all $\theta \in \Omega - \Omega'$, in hypothesis tests we may privilege some of the $\theta \in \Omega - \Omega'$, and require $\beta(\theta)$ at some specific level β, for a certain $\theta \in \Omega - \Omega'$. This may involve us in another dialectical process: for each assigned β, a certain sample size n may be required. So if we are asked for a $\beta(\theta) = \beta$, for a certain θ, we must perform a test with at least this sample size.

5. The Sufficiency Principle

For the acceptance of hypotheses, we need to add another principle to the requirement of the discriminating character of experiments. In order to accept that a parameter is the true parameter, for instance, we have to be sure that we are using the full force of the evidence. Thus, we must determine conditions under which an experiment has the full evidential import of the data constituted by a complete sample description.

For this reason, in this section we discuss, again, in this new setting, conditions under which two instances of statistical evidence, $[E_n = j]$ and $[E'_n = j']$, are evidentially equivalent for a certain class of alternative models. Loosely speaking, both experiments E and E' must be discriminating in the same way and the results must correspond. We shall begin with an example.

Let us assume in the genetical example given in Section 2, that instead of considering as results the number of the progeny with trait A, we take the progeny as ordered and take the results as the sequence of the results for each progeny.

---

[4] See next Chapter, Section 1, for a discussion of types of error.

5. THE SUFFICIENCY PRINCIPLE

For instance, in the case of E_2 we have as possible results $\langle 0,0\rangle$, $\langle 1,0\rangle$, $\langle 0,1\rangle$, and $\langle 1,1\rangle$, where 0 denotes the absence of A and 1 the presence of A. The pair $\langle 1,0\rangle$, for example denotes the result that A was observed in the first progeny, but it was not observed in the second progeny. Let E' denote the new experiment. Then E'_2, as a matrix, is

$$E'_2 = (p'_{2ij}) = \frac{1}{16}\begin{bmatrix} 9 & 3 & 3 & 1 \\ 4 & 4 & 4 & 4 \end{bmatrix}.$$

As Birnbaum points out, this experiment could also be obtained by tossing a coin, in case the result of the original E_2 was one, and observing whether it falls heads or tails.

In general, in matrix form, E'_n is

$$E'_n = (p'_{nij}) = \begin{bmatrix} p'_{n11} & p'_{n12} & \cdots & p'_{n1J} \\ p'_{n21} & p'_{n22} & \cdots & p'_{n2J} \end{bmatrix}$$

where $J = 2^n$, and

$$p'_{n1j} = (0.25)^x (0.75)^{n-x}$$

$$p'_{n2j} = (0.5)^n$$

and x is the number of occurrences of A in the sequence j.

We see that for each n we can obtain E_n from E'_n by a function g_n that sends a sequence j' into the result of E_n, say $j = g_n(j')$, which represents the number of occurrences of A in the sequence. This means, in the random variable notation, that $E_n = g_n(E'_n)$. The likelihood ratio for results j' of the two parameters is

$$\frac{p'_{n1j'}}{p'_{n2j'}} = \frac{\Pr_1[E'_n = j']}{\Pr_2[E'_n = j']}.$$

This ratio is the same as the likelihood ratio for the corresponding result j of E_n, namely

$$\frac{p_{n1j}}{p_{n2j}} = \frac{\Pr_1[E_n = j]}{\Pr_2[E_n = j]} = \frac{\Pr_1[g_n(E'_n) = g_n(j')]}{\Pr_2[g_n(E'_n) = g_n(j')]},$$

for every natural number n.

The definition of discriminating experiments that have been given so far depend on the notion of a rejection set. If E is related to E' in the manner described above, the probabilities, under any i, for the rejection sets for both experiments are the same, as we proceed to explain:

For simplicity, we assume that $i = 1, 2$. We have:

$$\Pr_i[E_n \in [g(j')]] = \Pr_i[g(E'_n) \in [g(j')]]$$
$$= \Pr_i[E'_n \in g^{-1}[g(j')]]$$

So, we must prove

$$g^{-1}[g(j')] = j'.$$

We have

$$k' \in [j'] \iff \frac{p_{n1j'}}{p_{n2j'}} = \frac{p_{n1k'}}{p_{n2k'}}$$

$$\iff \frac{p_{n1g(j')}}{p_{n2g(j')}} = \frac{p_{n1g(k')}}{p_{n2g(k')}}$$

$$\iff g(k') \in [g(j')]$$

$$\iff k' \in g^{-1}[g(j')]$$

This means that

$$k' \preceq_{ni} j' \iff g(k') \preceq g(j'),$$

and, hence

$$[E_n \in R_{nig(j')}] = [E'_n \in R'_{nij'}],$$

where $R_{nig(j')}$ and $R'_{nij'}$ are the respective rejection sets.

So it is clear, that the two results and experiments discriminate in the same way. In other words, the evidential content of both experiments is the same, in symbols

$$\mathrm{Ev}[E_n = g_n(j')] = \mathrm{Ev}[E'_n = j'].$$

This is an example of what has been called the *Sufficiency Principle*. A general formulation of the Sufficiency Principle is the following:

The Sufficiency Principle. Let $E = (p_{nij})$ for $i \in I$ and $j \in J_n$ and $E' = (p'_{nij'})$, for $i \in I$ and $j \in J'_n$, be discriminating experiments for the same set of alternative probability models I. Suppose that

(S): *there is a function g such that $E_n = g_n(E'_n)$, (that is, for each $n \in \mathbb{N}$, g_n is a function from J'_n onto J_n) such that there is a $c > 0$ satisfying*

$$p'_{nij'} = c\, p_{nig_n(j')}$$

for every $i \in I$,

then $\mathrm{Ev}[E'_n = j'] = \mathrm{Ev}[E_n = g_n(j')]$, for every $j' \in J'_n$.

This general principle can be justified as above. There are several different formulations of the Sufficiency Priciple. As it was mentioned before, any function g which depends only upon a sample point j and n is called a *statistics*. Examples are the functions $j = g_n(j')$ considered above. Any statistics meeting condition (S) is called a *sufficient statitistics* in E'. Thus, the Sufficiency Principle may be stated as follows:

If $j = g_n(j')$ is a sufficient statistics in E', then

$$\mathrm{Ev}[E'_n = j'] = \mathrm{Ev}[E_n = g_n(j')].$$

We shall discuss, now, a more customary formulation that is well known to be equivalent for a statistics to be sufficient.[5] Suppose that E', E, and g are as above. Let i, and $k \in I$, and assume that Pr_i and Pr_k are the probability

---

[5] See [6, p. 121], for references.

distributions determined by the alternatives i and k, respectively. We have $[E'_n = j']$ implies that $[g_n(E'_n) = g_n(j')]$, i.e, $[E_n = g_n(j')]$. Thus, $[E'_n = j'] \wedge [E_n = g_n(j')]$ is equivalent to $[E'_n = j']$. Then

$$\Pr_i([E'_n = j']|[E_n = g_n(j')]) = \frac{p'_{nij'}}{p_{nig_n(j')}}$$

and

$$\Pr_k([E'_n = j']|[E_n = g_n(j')]) = \frac{p'_{nkj'}}{p_{nkg_n(j')}}.$$

Thus, it is clear that *g is sufficient in E' if and only if for every* $i, k \in I$

$$\Pr_i([E'_n = j']|[E_n = g_n(j')]) = \Pr_k([E'_n = j']|[E_n = g_n(j')]).$$

That is, the probability of $[E'_n = j']$ *given* $[E_n = g_n(j')]$ *is independent of the parameter* $i \in I$.

In the case of acceptance of hypotheses, what is called in statitistics estimation of parameters, it is important that all the available evidence be used. For statistical inference, the total evidence is contained in the complete sample description $X_1 = x_1, X_2 = x_2, \ldots, X_n = x_n$. Thus, in the binomial example mentioned above, the experiment E' whose results are the complete sample descriptions includes all the evidence. So we must require that any discriminating experiment that we may use for acceptance be sufficient in E'. We have included the requirement of sufficiency in the discriminating experiments defined up to now by lumping together in our probability calculations all evidentially equivalent results. We must be careful that the new experiments that we shall introduce be also sufficient.

6. Point estimation

This section is devoted to pure acceptance of hypotheses in the form of point estimation. I shall concentrate on the estimation of one dimensional parameters.

The general enquiry for estimation is the following. I have observed $\vec{x}$; what should I infer about θ? The framework for estimation is the same that we have been considering. We have a family $\mathcal{P} = \{\Pr_\theta \mid \theta \in \Omega\}$ of possible probability distributions for a random variable X in a certain setup, $\mathfrak{S}$. We repeat independently $\mathfrak{S}$ n times and obtain possible values for X, $x_1, \ldots, x_n$. Which θ in Ω should we accept as value for the parameter? This is point estimation. If we ask for a subset Ω' of Ω in which θ should be, we are in the case of region estimation. In this section, I shall discuss point estimation and in Section 7, region estimation.

In point estimation, we have a continuous function $\hat{\theta}$ such that if the data are the sequence $\vec{x}$, $\hat{\theta}(\vec{x})$ ($= \hat{\theta}(x_1, \ldots, x_n)$) is accepted as the value of θ. In multidimensional cases, we need several functions, one for each component of the parameter. The estimate $\hat{\theta}$ is clearly a function of $\vec{x}$ and n. When I want to stress its dependence on n, I shall write, as is usual $\hat{\theta}_n$. The random variable $\hat{\theta}(\vec{X})$ ($= \hat{\theta}(X_1, \ldots, X_n)$) is called the *estimator*. I shall analyze, in the light of the Inverse Inference Principle, the main conditions that an estimator should satisfy.

In general, $\langle \widehat{\theta}_n(\vec{X}) \mid 1 \leq n \leq \nu \rangle$ is the discriminating experiment. We now discuss the conditions usually imposed on an estimator from my perspective.

The main condition is consistency. We say that $\widehat{\theta}_n$ is *weakly consistent for* θ, if $\widehat{\theta}_n$ nearly converges in probability to θ (according to the measure $\Pr_\theta$). The estimator $\widehat{\theta}_n$ is *strongly consistent* if $\widehat{\theta}_n$ nearly converges almost surely to θ, according to $\Pr_\theta$. Weak consistency would be enough for our purposes, but strong consistency is easier to state, so we shall assume it in the definition of discriminating experiment.

The second important condition is sufficiency. That is, the experiment $\widehat{\theta}_n$ should be sufficient in the experiment E' of complete sample descriptions.

With these two conditions we obtain the following definition of a discriminating experiment for point estimation.

DEFINITION XVIII.4 (D. E. FOR POINT ESTIMATION). The sequence of random variables $\widehat{\theta}_n$ is a discriminating experiment for point estimation with respect to the set of possible parameters Ω, if

(1) The sequence $\widehat{\theta}_n$ is an internal sequence of random variables over the n-product spaces.
(2) For every $\theta \in \Omega$, there is an internal set A of infinite sequences, such that $\widehat{\theta}$ is a.s. eventually in A, according to $\Pr_\theta$, and $\widehat{\theta}$ is a.s. eventually not in A, according to $\Pr_{\theta'}$ for $\theta' \neq \theta$, satisfying:
For every sequence $\{r_n\} \in A$, r_n S-converges to θ.
(3) Let E' be the experiment with results the complete sample descriptions. Then $\widehat{\theta}_n$ is sufficient in E'.

We have a theorem similar to Theorem XVIII.1, with an easier proof, which we omit, for these discriminating experiments:

THEOREM XVIII.3. *Let $\Omega \subseteq \mathbb{R}$ represent the set of possible models for a certain setup $\mathfrak{S}$, and X a random variable for $\mathfrak{S}$, such that for each $\theta \in \Omega$, $\mathbf{E}_\theta X = \theta$. Suppose that $X_1, X_2, \ldots, X_i, \ldots, X_\nu$, ν infinite, are random variables representing independent repetitions of $\mathfrak{S}$, i.e., these variable are independent and identically distributed with X. Then $\overline{X} = \langle \overline{X}_n \mid 1 \leq n \leq \nu \rangle$ is a discriminating experiment for point estimation with respect to Ω.*

Ordinarily there is more than one consistent estimator. For instance, under very general assumptions, both the moment and the maximum likelihood estimators are consistent. The moment estimator is obtained as follows. Suppose that the parameter θ depends on the first r moments and so we can express it as

$$\theta = g(\mathbf{E}[X], \mathbf{E}[X^2], \ldots, \mathbf{E}[X^r]).$$

We let

$$M_k = \frac{\sum_{i=1}^n X_i^k}{n}, \qquad k = 1, \ldots, r,$$

where $X_1, \ldots, X_n$ is a random sample from the distribution of X. We call M_k the kth sample moment. The estimate of θ, $\widehat{\theta}$, is then given by

$$\widehat{\theta} = g(M_1, \ldots, M_n).$$

Under very general conditions, M_k converges in probability to $E[X^k]$, and thus $\widehat{\theta}$ to θ. A particular example of this general result is Theorem XVIII.3.

The maximum likelihood estimator is obtained from the density or frequency function of X. Suppose that the density function of the distribution with parmeter θ is f_θ and that $X_1, \ldots, X_n$ is a random sample. The joint density with parameter θ is given by

$$f(x_1, x_2, \ldots, x_n | \theta) = f_\theta(x_1) \cdot f_\theta(x_2) \cdots f_\theta(x_n).$$

We consider $f(x_1, x_2, \ldots, x_n | \theta)$ as a function of θ for fixed $x_1, \ldots, x_n$. The maximum likelihood estimate $\widehat{\theta}$ is the value of θ maximizing $f(x_1, x_2, \ldots, x_n | \theta)$, where $x_1, x_2, \ldots, x_n$ are the observed values. Again, under very general conditions the maximum likelihood estimator is consistent and sufficient.

The maximum likelihood estimator has an added intuitive appeal over the other estimators. We have that $\widehat{\theta}$ is the θ such that there is an infinitesimal cube, C, containing the result $\vec{x}$, such that

$$\frac{\Pr_\theta[\vec{X} \in C]}{V(C)}$$

is maximum. Thus, the ratio of the probabilities that the observed values are true between $\widehat{\theta}$ and any other θ is highest. This means that the observed values are closer to truth if $\widehat{\theta}$ is the real parameter, than with any other θ. Consistency also implies that we can get closer and closer to truth. Thus, it seems to me that maximum likelihood estimators, if consistent, are very well founded.

I shall interlope a short discussion of sufficiency, extending that in Section 5. As we saw there, it seems that sufficiency is an esential condition that an estimator should satisfy. Let $\widehat{\theta}_n$ be an experiment defined from $E'_n = \langle X_1, X_2, \ldots, X_n \rangle$. Recall that we proved that $\widehat{\theta}$ is sufficient in E' for θ within the family Ω, if the conditional distribution of X, given $\widehat{\theta}$, does not depend on θ. In other words, $\widehat{\theta}$ is sufficient within the family $\mathcal{P} = \{\Pr_\theta : \theta \in \Omega\}$, if

$$\Pr_\theta(\bigwedge_{i=1}^n [X_i = x_i] | [\widehat{\theta}_n(X) = s]) = \Pr_{\theta'}(\bigwedge_{i=1}^n [X_i = x_i] | [\widehat{\theta}_n(X) = s]),$$

for any θ, $\theta' \in \Omega$, and any numbers $x_1, x_2, \ldots, x_n$ and s.

Thus, all the information provided by the sample which permits us to discriminate between θ and θ' within the family Ω is contained in the value $\widehat{\theta}(x)$.[6] In most cases, both the moment and the maximum likelihood estimator are sufficient.

---

[6]The Fisher-Neyman Factorization Criterion also gives us an insight into the nature of this condition. See [4, p. 137].

This sufficiency condition is likewise important, but not esential, for significance or hypotheses tests. It is not essential, because we have included the sufficiency condition in the definition of evidentially equivalent results, and, hence, of the rejection set. This means that all the evidence provided by the experiment is included, even though the statistics may not be sufficient. The definition of a sufficient statistics is similar to that of a sufficient estimator. It is convenient that a test statistics t be sufficient, because the calculations may be easier: we can always obtain a sufficient statistics with no two different results being evidentially equivalent. For point estimating a parameter, on the other hand, we do not use the rejection set. Thus, if the estimator is not sufficient, we may be not using part of the evidence provided by the sample.

Since consistency and sufficiency do not always determine uniquely an estimator, other properties are required in order to insure its uniqueness. I shall now discuss these other condition, some of which do not seem so crucial.

The first is unbiasedness. An estimator $\widehat{\theta}$ is called unbiased if the expectation $E_\theta(\widehat{\theta}(X))$ of $\widehat{\theta}(X)$ according to $\Pr_\theta$ is θ. Thus, $\widehat{\theta}$ is unbiased if, on average, $\widehat{\theta}$ is θ.

In order to explain why unbiasedness is a desirable property of estimators, we have to analyze what is the meaning of the expectation of a random variable. I shall take as an example a discrete random variable Y. The expectation of Y is then defined $\sum_{\text{all } y} y \Pr[Y = y]$ Since $\Pr[Y = y]$ is the degree of possibility of truth that $Y = y$, then we might say that EY is the truth-average of the value of Y. Thus, unbiasedness requires that the truth-average of $\widehat{\theta}(X)$ be θ. It seems to me that unbiasedness is not an important property of estimators, because the truth-average is a very weak requirement. It is clear, however, that consistent estimators are asymptotically unbiased. That is, the expectaction of the estimator tends to θ, when n tends to infinity. This condition seems to me much more important, since the truth-average should really tend to θ.

The second condition is efficiency. Roughly speaking, an estimator $\widehat{\theta}_1$ is more efficient than $\widehat{\theta}_2$ if the dispersion of $\widehat{\theta}_1(X)$ is smaller that the dispersion of $\widehat{\theta}_2(X)$, according to the distribution $\Pr_\theta$. The dispersion can be measured by the variance, if the estimator is unbiased. The variance, in this case is the truth-average of the dispersion from the true parameter. Thus, a more efficient estimator permits on average a faster approximation to the true parameter. A similar condition can be imposed on test statistics.

A large part of mathematical statistics is concerned with finding conditions for the existence of optimum estimators and tests. I shall not discuss further these technical matters further here.

7. Confidence region estimation

Region estimation uses the same principles as hypotheses tests, but here the main emphasis is in acceptance and not in rejection. In region estimation, instead of estimating the parameter by one value, a region R is given such that if the parameter is outside of R, then the observation leads to rejection of the parameter. This rejection is obtained according to the rule of rejection of hypotheses testing. Thus, hypotheses testing and confidence regions are intimately connected. In case the parameter is one dimensional, the region is usually an interval, and we are in the domain of interval estimation or confidence intervals.

I shall discuss in this section what Birnbaum calls in [6] the "confidence concept of statistical evidence". This is, according to him, the concept, or as I prefer to call it, the principle of statistical evidence by which estimates having the confidence region form are usually interpreted. The main problem to which I will refer in the next chapter is the contradiction between the Confidence Principle and the Likelihood Principle, and the fact that Birnbaum could deduce the Likelihood Principle from the Sufficiency Principle and another principle that he considered valid, the Conditionality or Ancillarity Principle. For Birnbaum this was a major problem, since he accepted as valid the Sufficiency, Conditionality, and Confidence Principles.

As I have shown in Section 5, the Sufficiency Principle is valid in my approach, and as I will show here, the Confidence Principle seems also to be valid. If it were possible to derive the Likelihood Principles from accepted principles, then there would be a contradiction in my theory. What I will show, in the next chapter, is that the Conditionality Principle in its general form is not acceptable, and hence, that there is no way of deducing the Likelihood Principle, which is also not acceptable.

As pointed out by Birnbaum, the formal theory of confidence region estimation as developed by Neyman does not include reference to any concept of statistical evidence. The main thrust of Neyman's confidence estimation is decision theoretical. Confidence region estimation, however, *is* used as representing statistical evidence about a parameter point. When this is done, the investigator or expositor has added to the formal theory a principle of statistical evidence, which Birnbaum calls, the Confidence Concept, and which I shall call, the Confidence Principle. As we shall see next, the Confidence Principle can be deduced from my Inverse Inference Principle.

We shall begin with an example. I shall modify the interpretation of the binomial example presented by Birnbaum in [6], accommodating it to my principles. A lower 99 per cent confidence limit estimator of the binomial parameter θ is by definition any function of the observed result $\langle n, x \rangle$,[7] denoted by $\widehat{\theta}(0.99, n, x)$, satisfying

$$0.99 \leq \Pr_\theta[\widehat{\theta}(0.99, n, X) \leq \theta].$$

We are here in the case of Bernoulli trials, X_1, X_2, ..., X_ν, with real parameters $\theta \in [0, 1]$ such that $\Pr_\theta[X_i = 1] = \theta$. We can take as our experiment $\overline{X}_n = 1/n \sum_{i=1}^n X_i$, i.e., the average number of successes in n trials. As we know already, this is a discriminating experiment for this case. Although $\widehat{\theta}$ should be a function of the discriminating experiment $\overline{X}_n$, for simplicity we shall write it as a function of the number of successes, $X = n\overline{X}_n$, and write it E_n, to emphasize that it is an experiment.

---

[7]Notice that I include the number of trials n in the description of the result. Birnbaum does not include it. To be consistent with my previous approach, I write the estimator as $\widehat{\theta}(0.99, n, X)$, where X is a random variable. I am using here normal statistical practice.

We also use the rejection Rule XVIII.1, considering E as a discriminating experiment for rejection $H_0 : \theta = \theta_0$ against $H_1 : \theta > \theta_0$. Here, if we have a result x of E_n, the rejection set for any θ is the set of y such that $n \geq y \geq x$. Recall that the probability of this rejection set is the p-value of the test. We have that $\widehat{\theta}(0.99, n, x)$ is that value of θ such that any $\theta_0 \leq \widehat{\theta}(0.99, n, x)$ would be rejected at a $\alpha = 0.01$ level of significance, because the p-value of the test would be not greater than 0.01. We shall see later that this is the case. Thus, we say that we accept that $\theta > \widehat{\theta}(0.99, n, x)$ with 0.99 level of confidence.

Application of such an estimator to the statistical evidence $[E_{50} = 40]$, for example, might give the result $\widehat{\theta}(0.99, 50, 40) = 0.63$. Such a lower confidence limit estimate is usually interpreted as a lower bound on the unknown true value θ, whose correctness is supported by fairly strong statistical evidence, indexed by 0.99.

According to my interpretation, which I think it is consistent with the usual one, n, in this case 50, and the level, in this case 0.99, are decided dialectically. Then, we calculate lower bound estimation in accordance with result of the experiment E_n. Since the result of E_{50} is 40, we obtain that $\widehat{\theta}(0.99, 50, 40) = 0.63$. It is important for this dialectical decision that we have available a discriminating experiment for the set of alternative possible probability models.

Another important property of confidence region estimates is that, as n tends to infinity, the lower, and also the corresponding upper, confidence estimator tends to the true value of θ.

An approximate formula for this estimator is given by using the normal approximation to the binomial distribution, namely for any number of successes x

$$\widehat{\theta}(0.99, n, x) \approx \frac{x}{n} - z_{0.01}\sqrt{\frac{x}{n}(1 - \frac{x}{n})/n}.$$

Thus, for instance

$$\widehat{\theta}(0.99, 50, 40) \approx \frac{40}{50} - z_{0.01}\sqrt{\frac{40}{50}(1 - \frac{40}{50})/50} \approx 0.67.$$

Here x/n is of course the usual point estimator of a binomial parameter, i.e., the sample mean, and the second term is a multiple of the usual estimator of its standard deviation.

A precise formula for such an estimator is $\widehat{\theta}(0.99, n, x) =$ the smallest value θ for which

$$\sum_{y \leq x} f(n, y, \theta)) \leq 0.99, \tag{1}$$

where $f(n, y, \theta) = \binom{n}{y}\theta^y(1-\theta)^{n-y}$, the binomial pdf of our examples. Using the appropriate tables or computer programs, we obtain that $\widehat{\theta}(0.99, 50, 40) = 0.63$.

It is clear that, given a result with x successes, we have that if $n \geq y \geq x$, then

$$\widehat{\theta}(0.99, n, x) \leq \widehat{\theta}(0.99, n, y).$$

Thus, the corresponding rejection set $R_{n\theta x}$, for any θ, satisfies

$$R_{n\theta x} \subseteq [\widehat{\theta}(0.99, n, x) \leq \widehat{\theta}(0.99, n, X)]$$

and thus

$$\Pr_\theta[E_n \in R_{n\theta x}] \leq \Pr_\theta[\widehat{\theta}(0.99, n, x) \leq \widehat{\theta}(0.99, n, X)].$$

Suppose, now, that $\theta < \widehat{\theta}(0.99, n, x)$. We have

$$\Pr_\theta[\widehat{\theta}(0.99, n, X) \leq \theta] \geq 0.99.$$

Thus

$$\Pr_\theta([\theta < \widehat{\theta}(0.99, n, X)]) < 0.01$$

and hence

$$\Pr_\theta[\widehat{\theta}(0.99, n, x) \leq \widehat{\theta}(0.99, n, X)] < 0.01.$$

Therefore

$$\Pr_\theta[E_n \in R_{n\theta x}] < 0.01$$

i.e., the p-value for θ of the test with result x is not greater than 0.01, and θ is rejected at level $\alpha = 0.01$. Thus we have shown that any θ not in the confidence interval is rejected.

We now introduce the formal dialectical rule for acceptance of confidence intervals:

RULE XVIII.2 (D. R. FOR ONE-SIDED CONFIDENCE INTERVALS). *Let Ω be a set of one dimensional parameters of probability distributions and let E be a discriminating experiment for $H_0 : \theta = \theta_0$ against $H_1 : \theta > \theta_0$ (according to Definition XVIII.3). A lower bound confidence estimator is a function $\widehat{\theta}(1 - \alpha, n, E_n)$ such that*

(1) *For any $\theta \in \Omega$, $\alpha \gg 0$*

$$\Pr_\theta([\widehat{\theta}(1 - \alpha, n, E_n) \leq \theta]) \gtrsim 1 - \alpha.$$

(2) *For any $\theta \in \Omega$, any $\alpha \gg 0$, and any possible result x of E_n*

$$R_{n\theta x} \subseteq [\widehat{\theta}(1 - \alpha, n, x) \leq \widehat{\theta}(1 - \alpha, n, E_n)].$$

(3) *For every $\theta \in \Omega$, there is a set of infinite sequences of results, C, such that E_n is a.s. eventually in C, according to $\Pr_\theta$, and for any $\{r_n\} \in C$ and any $\alpha \gg 0$ $\widehat{\theta}(1 - \alpha, n, r_n)$, S-converges to θ.*

Then, the rule based on this estimator is:
Accept that $\theta \geq \widehat{\theta}(1 - \alpha, n, x)$ with confidence $1 - \alpha$, if $\widehat{\theta}$ is a lower confidence estimator and x is the result of E_n.

For two-sided confidence interval we need two estimators, say $t(1-\alpha, n, E_n)$ and $s(1-\alpha, n, E_n)$. For a two-sided confidence interval with confidence coefficient $1-\alpha$, we need that t and s satisfy

$$1 - \alpha \leq \Pr_\theta([t(1-\alpha, n, E_n) < \theta < s(1-\alpha, n, E_n)]), \text{ for all } \theta \in \Omega.$$

The functions t and s must be defined from an experiment E that is discriminating for the alternatives $\theta = \theta_0$ against $\theta \neq \theta_0$.

If we get the observed value x of the experiment E_n, then the event $[s(1-\alpha, n, x) \leq s(1-\alpha, n, E_n)] \cup [t(1-\alpha, n, x) \geq t(1-\alpha, n, E_n)]$ must contain the corresponding rejection set. The rest of the conditions are similar to those for one-sided intervals. In particular, both t and s must tend to θ almost surely under θ.

The provisional acceptance is again dialectical. In this case, our opponent fixes the n and the α, and we perform the experment E_n. Then the confidence region is constituted by all those $\theta \in \Omega$ which are not rejected at level α.

Other examples are simple to construct. For instance, suppose that we have as alternative hypotheses nearly normal distributions with mean $\mu \in \Omega$ and unknown variance. We would like to derive a two-sided confidence interval for this case. The experiment should be discriminating for $H_0 : \mu = \mu_0$ against $H_1 : \mu \neq \mu_0$, where μ_0 is a fixed number. We already discussed the rejection discriminating experiment M for this case in Section 4. This experiment has results which are pairs composed of the sample mean, $\overline{X}_n = 1/n \sum_{i=1}^n X_i$, and the sample variance, $S_n^2 = 1/(n-1) \sum_{i=1}^n (X_i - \overline{X}_n)^2$. The null hypothesis H_0 is then a composite hypothesis including all distributions with mean μ_0 and different variances.

As it was established in Section 4, given a result $\langle \bar{x}_n, s_n \rangle$, the rejection set is constituted by all $(\bar{y}, t_n)$ with

$$\frac{|\bar{y} - \mu_0|}{t_n} \geq \frac{|\bar{x} - \mu_0|}{s_n}.$$

We have for any μ

$$\Pr_\mu([t_{-\alpha/2, n-1} \leq \sqrt{n}\frac{|\overline{X}_n - \mu|}{S_n} \leq t_{\alpha/2, n-1}]) \approx 1 - \alpha.$$

Thus, if $(\bar{x}_n, s_n)$ is the result of M_n, any

$$\mu \notin (\bar{x}_n - \frac{s_n}{\sqrt{n}} t_{\alpha/2, n-1}, \bar{x}_n + \frac{s_n}{\sqrt{n}} t_{\alpha/2, n-1})$$

is rejected at level α, because

$$\Pr_\mu([\frac{|\overline{X}_n - \mu|}{S_n} \geq \frac{|\bar{x} - \mu_0|}{s_n}]) \leq \alpha.$$

Thus, for this case,

$$t(\overline{X}_n, S_n) = \overline{X}_n - \frac{S_n}{\sqrt{n}} t_{\alpha/2, n-1} \text{ and } s(\overline{X}_n, S_n) = \overline{X}_n + \frac{S_n}{\sqrt{n}} t_{\alpha/2, n-1}.$$

We shall see other examples in the next chapter, when discussing other approaches to statistical inference.

CHAPTER XIX

Problems in statistical inference

We shall discuss, in this chapter, several issues related to statistical inference, in particular, the usual justification of Neyman-Pearson theory given by these authors and its difference with my justification. This will lead to the discussion of techniques spoused by classical statistians which I think are not justified.

We shall also discuss some other proposals for inference, such as Gillies' falsifying rule, [37], and Bayesian estimation.

1. The Neyman-Pearson theory

Outline of the Neyman-Pearson justification. According to the Neyman-Pearson theory, we devise hypotheses test so as to minimize two errors. The main error we want to minimize, called error of type I, is to reject H_0, when in fact H_0 is true. If we have achieved a reasonable minimization of this type of error, we also try to minimize the error, called error of type II, of accepting H_0, when it is false, i.e., when H_1 is true. I do not consider, however, the minimization of type II error as a justification for the tests, but as we saw at the end of Section 4 in the previous chapter, only as a convenient feature that may be available.

The Neyman-Pearson tests are usually presented in the following way: The working hypothesis H_0 and the alternative hypothesis H_1 serve to partition the parameter space Ω. Under H_0, θ lies in a subspace Ω'; under H_1, θ lies in the complementary subspace $\Omega - \Omega'$.

$$H_0 : \theta \in \Omega' \quad \text{and} \quad H_1 : \theta \in \Omega - \Omega'.$$

The sample space S of possible values of X_1, ..., X_n, is partitioned by the test procedure into two complementary subspaces S_0 and S_1: the critical region and the noncritical region, respectively. The test is expressed by a sentence $\varphi = [X \in S_0]$, of the language with X, such that if $x \in S_0$ occurs in reality, then φ is true, and if $x \in S_1$ obtains, then φ is false. The test proceeds on the policy:

reject H_0, if $x \in S_0$; accept H_0, if $x \in S_1$.

The usual justification of hypotheses tests, by Neyman and his school can be summarized as follows. The choice of test $\{S_0, S_1\}$ is mainly guided by the desire to safeguard the incorrect rejection of H_0, by assigning a significance level α, to be an upper limit to the probability of incorrectly rejecting H_0. That is, $\Pr_\theta[X \in S_0] \leq \alpha$, for all $\theta \in \Omega'$. Thus, the probability assigned by $\Pr_\theta$, for $\theta \in \Omega'$, to the sentence $[X \in S_0]$ is low. A new measure of the test is introduced, its power. The power of the test for θ is defined as $1 - \beta(\theta)$, where

$$\beta(\theta) = \Pr_\theta([X \in S_1]), \text{ for } \theta \in \Omega - \Omega'.$$

In the traditional Neyman-Pearson theory, among all the tests with significance level α, the one with greatest power is chosen. In fact, since there always many tests of level α, in order to choose among them, one of the criteria is the power of the test. One of the problems is that there may not be one test which has the greatest power for all $\theta \in \Omega - \Omega'$. So one of the important problems in mathematical statistics for the Neyman-Pearson school, is to find conditions under which these uniformly most powerful tests exist.

Let us continue with an example. Suppose that we want to test the hypothesis H_0 that a certain random variable X is nearly normally distributed with mean μ_0 and variance σ_0^2. We suppose that the evidence consists of n independent observed values of X, $\vec{x} = \langle x_1, x_2, \ldots, x_n \rangle$. The vector $\vec{x}$ is considered as a point in the samples space of possible values, S. The testing problem, as conceived by Neyman and Pearson, is that of choosing a subset S_0 of S, such that, if the observed value $\vec{x} \in S_0$, then we should reject H_0. The set S_0 is the critical region. As we mentioned above, Neyman and Pearson require that S_0 should be chosen so as to minimize two types of errors (see, for instance, the first joint paper of Neyman and Pearson, [73, p. 3]). Type I error is the error of rejecting H_0 when it is in fact true. Type II error is the error of accepting H_0 when it is in fact false.

The treatment of type I error, which is the more important of the two, is the following. The hypothesis H_0 induces a probability distribution over S, $\Pr_{H_0}$. We control type I error by finding a low value for $\Pr_{H_0} S_0$. If

$$\Pr_{H_0} S_0 = \alpha,$$

we say that we are using a test of *size* α. The trouble, here, is that we can choose a large number of sets S_0 such that $\Pr_{H_0} S_0 = \alpha$. Of these, many will be intuitively unsatisfactory as critical regions. In order to eliminate these unsatisfactory regions, and obtain a good critical region, we must, according to Neyman and Pearson, consider type II error. Here lies the main difference with my theory. Instead of considering type II error, I propose the use of rejection sets.

It is in order to treat type II error, that Neyman and Pearson introduce a set of alternative hypotheses such that, if H_0 is false, one of these alternatives hold. We, thus, begin with a set Ω of hypotheses. Each hypothesis specifies completely the probability distribution of a certain random variable, X. We assume, for simplicity, that each hypothesis is specified by values of a set of parameters. In the example, we are considering, these are the mean, μ and variance, σ^2. Thus, in this case, we consider nearly normal distributions with mean μ and known variance σ^2, where $\sigma > 0$. As before, we assume a sample space, S, and take

1. THE NEYMAN-PEARSON THEORY

the problem to be that of choosing an $S_0 \subseteq S$ for testing our hypothesis, say that the mean is μ_0. It is required that S_0 should have some fixed size α, i.e., $\Pr_{H_0} S_0 = \alpha$. As we mentioned above, we reduce the type II error by making $\beta(\mu)$ as small as possible, for $\mu \neq \mu_0$, where $1 - \beta$ is the power or the test. Thus, we must make the power as big as possible.

Suppose that there are only two simple alternatives, μ_0 and μ_1. We need, then, a set S_0 such that

$$\Pr_{\mu_0} S_0 = \alpha,$$

and

$$\Pr_{\mu_1} S_0 \geq \Pr_{\mu_1} C,$$

for any other set C, such that $\Pr_{\mu_0} C = \alpha$. In the case of two simple alternatives with nearly normal distributions, Neyman and Pearson have shown that it is always possible to find such a set.

The problem arises when there are many alternatives, μ to μ_0. We need, then, a set S_0 such that $\Pr_{\mu_0} S_0 = \alpha$, and

$$\Pr_\mu S_0 \geq \Pr_\mu C,$$

for any other set C such that $\Pr_{\mu_0} C = \alpha$, and for every $\mu \neq \mu_0$. This is what is called a *uniformly most powerful test* (or UMP). This does not always exists, and, then, the requirement is usually weakened to the existence of a UMP test among unbiased test, or UMPU. We shall not discuss the matter further, but present two examples due to Gillies, [37, pp. 204, 205], which I think show that the procedure is not sound.

Gillies' examples. Suppose that we are testing the null hypothesis H_0 that the random variable X has the nearly uniform distribution with density

$$f_1(x) = \begin{cases} \frac{1}{2}, & \text{for } -1 \leq x \leq 1 \\ 0, & \text{otherwise,} \end{cases}$$

against the alternatives

$$f_\varepsilon(x) = \begin{cases} \frac{1}{2\varepsilon}, & \text{for } -1 \leq x \leq -1+\varepsilon \text{ and } 1-\varepsilon \leq x \leq 1, \\ 0 & \text{otherwise,} \end{cases}$$

where $0 < \varepsilon \leq 1$. Here, the problem is that of testing $\varepsilon = 1$ against the alternatives $0 < \varepsilon < 1$, ε real. Fig. XIX.11 shows the distributions.

According to the Neyman-Pearson theory, if we set

$$S_0 = [-1, -1+\alpha] \cup [1-\alpha, 1],$$

we have that S_0 is an UMP test of size α. This result is unnatural, because, since the null hypothesis, H_0, is of a uniform distribution, no result in the interval $[-1, 1]$ should count against it.

Let us analyze this example according to my principles. In the first place, we have that no two different results are evidentially equivalent. Also, the rejection set for f_1, determined by any result, is the whole interval $[-1, 1]$, because all

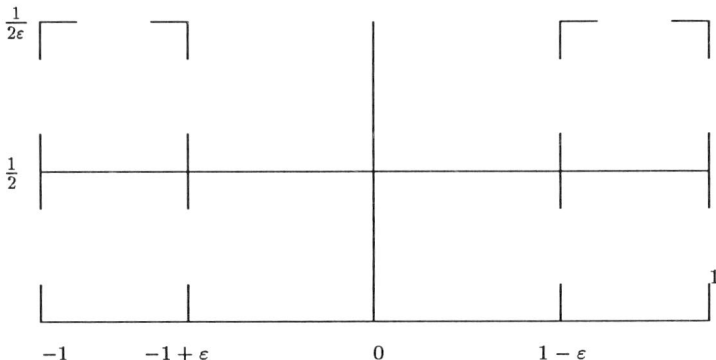

FIGURE XIX.11. Gillies' first example

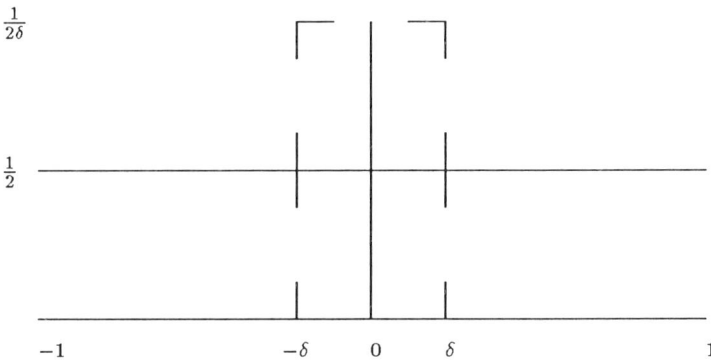

FIGURE XIX.12. Gillies' second example

results are equiprobable. Thus, there is no discriminating experiment. Besides, the set S_0 is not a rejection set, and, hence, the test is not justified.

A similar analisis can be given to Gillies' second example, where we consider as alternatives to f_1 the distributions with densities

$$f_\delta(x) = \begin{cases} \frac{1}{2}\delta, & \text{for } -\delta \leq x \leq \delta \\ 0. & \text{otherwise,} \end{cases}$$

where $0 < \delta \leq 1$, given by Fig. XIX.12.

In this case, the set

$$S_0 = [-\delta, \delta]$$

is an UMP test. As above, the rejection set for f_1 is constituted by the whole interval, and this is not a good test according to my principles.

2. Alternative hypotheses

One of the main criticism to the Neyman-Pearson theory has been to their requirement that in testing a hypothesis H_0 one must consider alternatives to it. That is, one cannot test H_0 in isolation, but only with respect to several alternatives. The position that one can test hypotheses by themselves has been spoused, among others, by Fisher, [34, p. 42], and Gillies, [37, pp. 206–216].

It is true that sometimes in the Neyman-Pearson theory it seems that one should consider only a very restricted class of alternative hypotheses, that is, alternatives that only differ in their parameter values. For instance, if the hypothesis to be tested is that the random variable, X, is nearly normally distributed with mean μ_0 and standard deviation σ_0, then the alternatives considered are also normal distributions, but with differente parameters, μ and σ. In order to make the selection of tests based on type II error work, one must have such restricted class of alternatives, because, if this is not the case, the mathematics becomes too complicated.

In my view, I consider any sort of alternatives, as the example on page 104 shows. (This is the example of the coin and the machine that produces a fixed sequence.) Instead of using a justification based on type II error, the method advocated in this book uses rejection sets. Thus, one can get the right tests.

As I have shown before, especially in Chapter XVIII, Section 3, in some cases the class of alternative hypotheses is so wide that it seems that we are not considering alternatives. But, as the example on page 104 shows, one cannot entertain all logically possible hypotheses, although one considers only all "reasonable" hypotheses.

In this section, I shall discuss an example given by Gillies in [37, pp. 209–214], where he attempts to show that no alternative hypotheses are entertained in a case proposed by Neyman himself in [72, pp. 33–37]. The problem arises in the field of biology. An experimental field was divided into small squares, and counts of larvae in each of these squares were made. The problem was to find the probability distribution of the number n of larvae in a square. The first hypothesis suggested was that this random variable had a Poisson distribution, $p_n = e^{-\lambda}\lambda^n/n!$, for some value of the parameter λ. This was tested by the following method: the possible results were divided into ten classes corresponding to 0, 1, ..., 8, and 9 or more, observed larvae. The number m_s, for $s = 0, 1, \ldots, 9$, observed in each class, was noted and the expected number, m'_s, was calculated, given the hypothesis: first, the parameter λ is estimated in the usual way by point estimation as $\hat{\lambda}$. Then the number m'_s is $n\hat{p}_s$, where n is the number of elements of the sample and

$$\hat{p}_s = e^{-\hat{\lambda}}\frac{\hat{\lambda}^s}{s!},$$

for $s = 0, 1, \ldots, 8$, and

$$\hat{p}_9 = 1 - \sum_{i=0}^{8} e^{-\hat{\lambda}}\frac{\hat{\lambda}^n}{n!}.$$

We consider n independent random variables, $Y_0, Y_1, \ldots, Y_n$, such that the possible values of Y_k are the numbers $0, 1, \ldots, 9$. These random variables are considered independent repetitions of a random variable Y which indicates the number of larvae obtained in each square. Hence, we are testing the hypothesis

$$H_0 : \Pr[Y = i] = \hat{p}_i \qquad i = 1, \ldots, 9,$$

against the hypothesis

$$H_1 : \Pr[Y = i] \neq \hat{p}_i \qquad \text{for some } i = 1, \ldots, 9.$$

Gillies considers that there are no alternative hypotheses, in this case, but, as we can see, the alternatives are clear. It is true, however, that the type II error analysis that would justify the choice of critical region is not clear. With my analysis of rejection sets, however, there is no problem:

Let X_i denote the number of Y_j's that equal i. Then, as each Y_j will, under H_0, independently equal i with probability $\Pr[Y = i]$, it follows that X_i is binomial $\langle n, \hat{p}_i \rangle$. Hence, if H_0 is true

$$\mathrm{E}\, X_i = n\hat{p}_i$$

and so $(X_i - n\hat{p}_i)^2$ will be an indication as to how likely it appears that $\hat{p}_i$ indeed equals the probability that $Y = i$. That is, a large result for $(X_i - n\hat{p}_i)^2$ is bad for H_0. Such reasoning leads us to consider the following test statistic

$$T_n = \sum_{i=0}^{9} \frac{(X_i - n\hat{p}_i)^2}{n\hat{p}_i}.$$

It is clear that if result a of T is larger than other result, b, then a is worse than b for H_0. The statistics T is nearly χ^2_{k-1-m}, where k is the number of classes, in our example, ten, and m is the number of parameters that have to be estimated, in our case only λ, that is, one.

Thus, we consider the experiment $\langle T_n \mid 1 \leq n \leq \nu \rangle$, which is not difficult to show discriminating. If we obtain result t for T_n, then the rejection set is

$$[T_n \geq t],$$

whose probability can be calculated using the χ^2 distribution. The value t obtained by Neyman was 46.8, and thus, we get a clear rejection of H_0.

We see that there is no problem in dealing with this example, according to my principles. This is a case where all "reasonable" hypotheses are considered as alternatives, and so, there are no different evidentially equivalent results for T_n, and the inequalities of probabilities are always reversed in some alternative. Therefore, it is one of the situations that we have already encountered, where the hypothesis H_0 seems to be tested in isolation, without the consideration of alternatives. As we have seen from previous examples, however, this is not, strictly speaking, the case.

It is to be remarked, that the distribtuion of T_n is sometimes obtained by simulation, instead of using the χ^2 distribution. Such an analysis can also be placed in my framework. I shall not do it here, however.

3. Estimation techniques that are unsound.

We include, in this section, a couple of techniques of inference, which have been proposed by members of the Neyman-Pearson school, but which can be proved unsound, according to my principles.

Randomized tests. We continue with the example of a lower 99% confidence limit estimator of the binomial parameter, which was discussed at the beginning of Section 7 in Chapter XVIII. In this example, the estimator $\widehat{\theta}(0.99, n, x)$ is the sharpest among all possible 99 per cent lower confidence limit estimators, with one significant qualification. This estimator is not the sharpest, if we allow consideration also of estimators of the more general randomized form $\widehat{\theta}(0.99, n, X, Y)$, depending not only on the binomial observation X but possibly also on an auxiliary randomization variable Y. The latter phrase denotes just the observation of a random variable Y with a uniform distribution on the unit interval $[0, 1]$. In order to maintain our discrete pdfs, just as Birnbaum, [6], I shall illustrate this procedure by using a discrete $y = 0, 0.1, 0.2, \ldots, 0.9$. A continuous distribution gives the fullest improvement.

We change the original pdf f to

$$g(n, x, y, \theta) = 0.1 \cdot f(n, x, \theta).$$

Then the formula for the optimal randomized confidence lower limit estimator becomes: $\widehat{\theta}(0.99, n, x, y) =$ is the smallest value of θ such that

$$\sum_{\langle t,w \rangle \preceq \langle x,y \rangle} g(n, t, w, \theta) \leq 0.99 \tag{1}$$

where $\langle t, w \rangle \preceq \langle x, y \rangle$ if and only if $t < x$ or $t = x$ and $w \leq y$. But

$$\sum_{\langle t,w \rangle \preceq \langle x,y \rangle} g(n, t, y, \theta) = \sum_{t<x} f(n, t, \theta)) + y f(n, x, \theta).$$

In our binomial example, if we adjoin an observed random number $y = 0.6$, we obtain $\widehat{\theta}(0.99, 50, 40, 0.6) = 0.65$, larger (and hence sharper) than our original estimate 0.63. With $y = 0.2$, we would obtain 0.64, a number still larger than 0.63, but different from 0.65. The smaller lower bound for θ in case $y = 0.2$ represents a strictly weaker conclusion than with $y = 0.6$.

It does not seem appropriate that such an irrelevant addition to our evidence as a random number would change our conclusion. Even the feature of nonuniqueness of estimates $\widehat{\theta}(0.99, 50, 40, y)$ based on a given evidence $[E_{50} = 40]$ similarly seems inappropriate. The inappropriateness can be verified by the fact that randomized estimation represents the violation of the Sufficiency Principle. Let E' be the randomized experiment. E' and E are both discriminating experiments for the binomial alternatives. It is easy to show, that according to the Sufficiency Principle, $\mathrm{Ev}[E_n = x] = \mathrm{Ev}[E'_n = \langle x, y \rangle]$, for the binomial alternatives. Hence, the estimates should be the same. We now show that because of the way we defined rejection sets, anything rejected by E' would also be rejected by E and so the confidence intervals should be the same:

It is clear that any two results of E'_n, $\langle x, y_1 \rangle$ and $\langle x, y_2 \rangle$, with the same result, x, but maybe with different random numbers, y_1 and y_2, are evidentially equivalent, according to Definition XVIII.1. Thus, if the rejection set, $R'_{ni\langle u,y \rangle}$ for E' includes $\langle x, y_1 \rangle$, for a certain random number, it also must include $\langle x, y_2 \rangle$, for any other random number, y_2. Thus

$$\Pr_i[E'_n \in R'_{ni\langle u,y \rangle}] = \Pr_i[E_n \in R_{niu}].$$

Thus, since we have incorporated the Sufficiency Principle in our definition of rejection set, a more accurate calculation of the estimator shows that in fact we do obtain the same result with E and E'. The formula for $\hat{\theta}$ is $\hat{\theta}(0.99, n, x, y) = $ the smallest value θ for which

$$\sum_{t \leq x} f(n, t, \theta)) \leq 0.99,$$

and not (1), even when applying E'. This is so, as we indicated above, because the same results of E coupled with different results for Y are evidentially equivalent, and hence, should be added together when calculating the rejection sets.

I conclude that, according to my principles, confidence region estimation is valid, but randomized confidence region estimation is not.

Prediction intervals. Assume that we are trying to get a "prediction interval" for X_{n+1}, where $X_1, X_2, \ldots, X_n, X_{n+1}$ is a random sample from a nearly normal population with unknown mean and variance, and we have observed the first n results. We have that $X_{n+1} - \overline{X}_n$ is normal with mean 0 and variance $\sigma^2 + (\sigma^2/n)$. Thus, it is not difficult to show that

$$\sqrt{\frac{n}{n+1}} \frac{X_{n+1} - \overline{X}_n}{S_n}$$

has a student's T_{n-1}. Therefore

$$\Pr[-t_{\alpha/2, n-1} \leq \sqrt{\frac{n}{n+1}} \frac{X_{n+1} - \overline{X}_n}{S_n} \leq t_{\alpha/2, n-1}] \approx 1 - \alpha.$$

Thus, if $\bar{x}_n$ and s_n are observed, we could think of a prediction interval of the form

$$(\bar{x}_n - \sqrt{\frac{n+1}{n}} s_n t_{\alpha/2, n-1}, \bar{x}_n + \sqrt{\frac{n+1}{n}} s_n t_{\alpha/2, n-1}).$$

We have, however, that X_{n+1} is independent of $\overline{X}_n$ and S_n, and, hence, we cannot conclude that

$$\Pr[\bar{x}_n - \sqrt{\frac{n+1}{n}} s_n t_{\alpha/2, n-1} \leq X_{n+1} \leq \bar{x}_n + \sqrt{\frac{n+1}{n}} s_n t_{\alpha/2, n-1}] \approx 1 - \alpha.$$

Only the joint distribution of $\sqrt{(n/n+1)}((X_{n+1} - \overline{X}_n)/S_n)$ is T_{n-1}. It is also clear, that there are no set of alternative hypotheses to be accepted or rejected.

In order to get something similar to a prediction interval, we can use the technique of tolerance limits, which can be easily accommodated in my framework,

although not using the estimators of Rule XVIII.2. Suppose, as before, that the possible distributions for the variable X are normal with unknown mean and variance, and that $X_1, X_2, \ldots, X_n, X_{n+1}$ are independent repetitions of X. With knowledge of the first n variables, we would like to have an interval such that X is in this interval with probability of $1 - \gamma$, this with confidence $1 - \alpha$. Let us call F_μ any cumulative distribution function when μ is the mean and some variance. That is
$$F_\mu(x) = \Pr_\mu[X_{n+1} \leq x].$$
There are tables that give the number k such that
$$\Pr_\mu[F_\mu(\overline{X}_n + kS_n) - F_\mu(\overline{X}_n - kS_n) \geq 1 - \gamma] = 1 - \alpha$$
for every μ. Our experiment is the same M_n as in Example XVIII.2. Suppose that we obtain the result $\bar{x}_n$ and s_n and assume that
$$F_\mu(\bar{x}_n + ks_n) - F_\mu(\bar{x}_n - ks_n) < 1 - \gamma$$
for a certain μ.

A result (m_n, r_n) is at least as bad for μ as $(\bar{x}_n, s_n)$, if
$$\frac{|m_n - \mu|}{r_n} \geq \frac{|\bar{x}_n - \mu|}{s_n}.$$
and we have for a worse result
$$F_\mu(m_n + kr_n) - F_\mu(m_n - kr_n) < 1 - \gamma.$$
But
$$\Pr_\mu([F_\mu(\overline{X}_n + kS_n) - F_\mu(\overline{X}_n - kS_n) < 1 - \gamma]) = \alpha$$
so that the probability of the rejection set is less than α. This means that we are able to reject any μ such that
$$F_\mu(\bar{x}_n + ks_n) - F_\mu(\bar{x}_n - ks_n) < 1 - \gamma$$
and, thus, we accept with confidence $1 - \alpha$ that the next repetition X_{n+1} has a cumulative distribution F with
$$F(\bar{x}_n + ks_n) - F(\bar{x}_n - ks_n) \geq 1 - \gamma$$
and, hence, that with probability $1 - \gamma$, the result of X_{n+1} will be in the interval
$$(\bar{x}_n - ks_n, \bar{x}_n + ks_n).$$

4. The Likelihood and Conditionality Principles

In this section, we shall discuss two principles that have been proposed for statistical evidence, and we shall show that they are not justified. For this discussion, we return to experiments with a finite number of results and a finite number of possible probability models defined on Chapter XVIII, Section 2.

The Likelihood Principle. The Likelihood Principle is based on the statement that the evidence given by the result $\langle n, j \rangle$ supporting one parameter point i against another i' is represented just by the numerical value of the likelihood ratio
$$L(n, i, i') = \frac{p_{nij}}{p_{ni'j}}.$$
This is apparently similar to my principles, but there is an important difference. According to what has been laid out in this book, the evidence is given not just by a result j, but by the pair $\langle n, j \rangle$, and what matters is not just $L(n, i, i')$, but also its limit when n tends to infinity. Thus, the Likelihood Principle, which I shall presently state, is not valid in my framework.

Before stating the principle, we need to introduce the likelihood function. For each model $[E_n = j]$ of statistical evidence $L_{nj}(i) = p_{nij}$ as a function of $i \in I$, is called the likelihood function. Here j (and also n) is fixed. More precisely, the function $L_{nj}(i)$ is specified as one among many alternative, equivalent representations of the same function, all having the form $c\, p_{nij}$, for every $i \in I$, where c denotes an arbitrary positive number. Now, the Likelihood Principle, as formulated by Birnbaum, is the following:

Likelihood Principle. If two models of statistical evidence $[E_n = j]$ and $[E'_m = j']$ determine the same likelihood function, then they represent the same evidential meaning. That is, if for some positive c we have $p_{nij} = c\, p'_{mij'}$, for each $i \in I$, and for fixed j, j', m, and n, then $\mathrm{Ev}[E_n = j] = \mathrm{Ev}[E'_m = j']$ (where $E = (p_{nij})$ and $E' = (p'_{mij'})$).

As we saw above, the Likelihood Principle is not justified, in accordance with my principles. It is accepted by Bayesians, but many other statisticians, such us Fisher and Barnard, also accept it.

The Likelihood Principle is incompatible with the Confidence Principle (and indeed with usual evidential interpretations of most standard classical statistical techniques). For example, the lower 99 per cent confidence limit estimate of θ in the binomial example $[E_{50} = 40]$ is determined by a formula which requires that the distribution determined by θ be given in considerable detail. In particular, the method cannot be applied if $[E_{50} = 40]$ is represented only by the likelihood function $\theta^{40}(1 - \theta)^{10}$, for $0 \le \theta \le 1$, as would be required by compatibility of confidence estimates with the likelihood principle.

It is also true that if we perform the experiment E' by waiting until there are 40 favorable occurrences of a trait and it happens to be in 50 trials, then this result has the same likelihood function. Since the distribution is the negative binomial, however, the lower 99 per cent confidence limit estimate is very different.[1]

The Conditionality Principle. Still another broad concept which is expressible as a principle has played a significant role. This is the *Conditionality Principle* (sometimes called the Ancillarity Principle). Birnbaum was able from this principle and his Sufficiency Principle (see [6]) to deduce the Likelihood

---

[1] For a nice example, where by applying the Likelihood Principle we arrive at unintuitive results, see [6, pp. 127-129].

4. LIKELIHOOD AND CONDITIONALITY

Principle. If the Conditionality Principle were acceptable, then even with my slight modification of the Sufficiency Principle it would be possible to deduce the Likelihood Principle. This would show that my principles are contradictory, since I accept confidence estimation, and the Likelihood Principle is incompatible with it. So it is important for me to formulate and analyze the Conditionality Principle, and show why it is not valid.

I indicate the principle, just as Birnbaum, by reference to our binomial example of Chapter XVIII, Section 2. Suppose that the number of tosses of the coin (or the number of progeny to be bred and observed for trait A, or the number of clinical patients to be observed) is not certain to be 50, but it will be either 50 or 200, depending upon certain uncontrollable unpredictable conditions. We assume that these conditions are unrelated to the parameter θ of interest, and that it is known to be equally probable that the number will be 50 or 200. I, now quote from [6]:

> If the number of observations turns out to be 50, then the experimental situation is represented by the binomial model E_{50} as before; and if it turns out to be 200, then another binomial model will apply. But an accurate and more complete model of the experimental situation is available, within which the models E_{50} and E_{200} are embraced as possible models, with probability $1/2$ that each will turn out to be the one applicable. The experimental situation represented by this broader model E will result first in either E_{50} or E_{200} being realized and then in the observation of an outcome. Thus each sample point of E has the form (E_{50}, x), with $x = 0, 1, 2, \ldots, 50$, or (E_{200}, x), with $x = 0, 1, 2, \ldots, 200$. ... Hence a model of statistical evidence represented by an outcome of E has one of the forms
>
> $$(E, (E_{50}, x)) \text{ or } (E, (E_{200}, x)).$$
>
> The pdfs of E are readily determined.
>
> Let us consider in this context E the outcome $(E_{50}, 40)$, which represents the statistical evidence $(E, (E_{50}, 40))$. The various concepts and techniques of evidential interpretation discussed above, and others, are applicable here in their usual general ways. For example an optimal 99 per cent lower confidence limit estimator of the parameter can be given as before ...
>
> However it has seemed to many theoretical and applied statisticians that a model of evidence such as $(E, (E_{50}, 40))$ here, comprehensive and accurate though it is, contains parts which are clearly irrelevant to evidential meaning: given the bad luck of getting just 50 rather than 200 observations, the evidential meaning of the outcome of the fifty observations is fully represented by (E_{50}, x); and the hypothetical possibility that an additional 150 observations might have become available but in fact did not, seems irrelevant to the evidential meaning of the result.

Formally, this is represented by adopting $(E_{50}, 40)$ in place of $(E, (E_{50}, 40))$ as the appropriate model of the statistical evidence obtained (or as a more parsimonius, though equivalent model). Within this "conditional" model any chosen model of evidential interpretation could be applied.

Birnbaum's reasoning does not take into account that an experiment E should be represented by a sequence $\langle E_n \mid 1 \leq n \leq \nu \rangle$. In fact, in all his paper, he never allows for this representation of experiments as sequences. The example he gives is, then, according to my principles, not fully determined. We do not know which is this sequence. In any case, whatever is this sequence, it is clear that $(E, (E_{50}, 40))$, which in my notation should be written $[E_n = [E_{50} = 40]]$, for a certain n, is not evidentially equivalent to $(E_{50}, 40)$, i.e., $[E_{50} = 40]$, because E is not just an experiment, but a sequence of experiments.

Without recourse to a sequence of experiments, but only considering rejection sets, it is possible to give couterexamples to the Conditionality Principle. Suppose that we are in the coin tossing case, testing the hypothesis H_0 that the coin is fair against the alternatives of all possible biases, and we set $\alpha = 0.01$. Assume, first, that we are testing with a sample of 25 tossings and obtain 19 heads. It is clear that the rejection set is constituted by results with at least 19 heads or at most 6 heads. The probability of this rejection set, i.e., the p-value of the test, is easily obtained using tables or computer programs to be 0.01444 and, hence, H_0 passes the test at $\alpha = 0.01$.

Assume, now, that instead of 25 tossings, we could obtain 25 or 30 tosses, each with probability $\frac{1}{2}$, and that we happen to obtain 25 tosses, again with 19 heads. Looking at tables or using computers, it is easy to check that the rejection set (i.e., the results as least as bad as 19 for H_0) is now constituted by the following possible results: 25 tosses with at least 19 or at most 6 heads, or 30 tosses with at least 23 or at most 7 heads. The p-value of the test is now 0.00983, and, hence, H_0 does not pass the test at a 0.01 level.

According to the Conditionality Principle, the two experiments are evidentially equivalent. We see, however, that they lead to different results when one uses rejections sets. Thus, we conclude that the Conditionality Principle is not justified.

5. Initial and final precision

In this section, I shall analyze the concepts of initial and final precision, which have been touched upon in the preceding sections. One of the criticism that has been leveled at the classical techniques of statistical inference is that they take into account what is called the initial precision of the technique, but not its final precision. These concepts are better explained with an example. Consider a simple problem in which we take a random sample $x_1, x_2, \ldots, x_n$ of observations from a distribution with mean μ. Our purpose is to estimate the unknown mean, μ. To this end we consider different ways of reducing the data to a single quantity $\hat{\theta}(x)$, which is to be our estimate of μ. We might for example use the sample mean $\bar{x} = (1/n) \sum_{i=1}^{n} x_i$, or the sample median, m, or even the mid-range.

5. INITIAL AND FINAL PRECISION

The question that is usually raised is how are we going to assess 'accuracy' or 'precision' in estimation. On the one hand, the procedure of using the sample mean, for instance, to estimate μ could be assessed in the light of different possible sets of data which might be encountered. It will have some characteristics which express the precision we initially expect, i.e., before we take our data. This is what is called initial precision. But it is sometimes raised as a criticism that this is the only aspect of assessment applied in classical statistical procedures, in that the sampling distribution of the estimator $\hat{\theta}(X)$ is the sole basis for measuring properties of the estimator.

An alternative concept of final precision has been proposed, in which the aim is to express the precision of an inference in the specific situation that is being studied. In the classical approach, if a sample is actually taken and it is found that $\bar{x} = 25.3$, then if the random variable $\overline{X} = (1/n)\sum_{i<n} X_i$ has high probability of being close to μ, we are encouraged to feel that 25.3 should be close to μ. But not everyone is satisfied by this attitude. The main reason for this dissatisfaction hinges on the frequency interpretation of probability spoused by most classical statisticians. Suppose that from data we obtain a 95 per cent confidence interval for μ as $25.1 < \mu < 26.0$. It is suggested that it is of little value to someone who is to accept of reject the hypothesis in this situation to know that the '95 per cent confidence' attaches not to the specific statement made, but to what proportion of such similarly based statements will be correct in the long run.

In the interpretation of probability that has been offered in this book, the '95 per cent confidence' does attach to the specific statement. Since a 95 per cent confidence interval is determined from the fact that $\Pr_\mu([X - s \leq \mu \leq X + s])$ is 0.95, then the degree of possibility of truth of the particular result obtained is high, if μ is in the interval. Also, which is a more important feature, any μ outside the interval is rejected at level $\alpha = 0.05$, i.e., the result obtained or worse, has a possibility of being true according to any such μ of less than $\alpha = 0.05$, and, hence, μ is rejected.

Another line of criticism to the foundations of classical statistics disputes its relevance in a more fundamental way. Some critics claim that final precision is all that matters and cannot be sought through principles concerned only with initial precision. For instance, tail area probabilities in a test of significance are claimed to be irrelevant to the rejection or acceptance of an hypothesis. The argument goes as follows. If we are testing $H_0 : \mu = 30$ and find $\bar{x} = 31.6$, what sense can there be in considering values of $\bar{x}$ in excess of 31.6 (or even less than 28.4) that may have occurred, but have not. Jeffreys, [48, p. 385], for instance, asserts:

> ... a hypothesis which may be true may be rejected because it has not predicted observable results which have not occurred. This seems a remarkable procedure.

It is clear that, according to my point of view, the possibilities under a hypothesis are of crucial importance for its rejection or acceptance. In order to reject a hypothesis, the results of the experiment that matter are those whose probability is low under the hypothesis and high under an alternative. In deterministic situations, what is possible is also important. In fact, we reject a hypothesis if something happens which is impossible if the hypothesis were true, but possible according to an alternative hypothesis.

The main reason why we must base our decisions on initial precision alone, however, is that we cannot reject hypotheses definitively, but only provisionally following a dialectical process. Thus, we must have a sequence of experiments deciding among the hypotheses. The result that actually obtains is not the only thing that matters. If we are challenged in our rejection (maybe by ourselves), we must be able to meet the challenge by performing an experiment which validates a proposition with the appropriate low probability. Hence, we are always in an initial position, i.e., we are always before a possible experiment. In order to describe a testing experiment, then, we must include a sequence of possible experiments, besides the particular result of one of the experiments in the sequence.

Some of the criticisms to basing rejection of hypotheses on initial precision do not apply to my justification of test by the use of rejection sets. Several examples of Hacking, [39, pp. 94–99], which has been one of the main critics of the Neyman-Pearson theory on this matter of initial precision do not yield with rejection sets the same results as the Neyman-Pearson theory, and, hence, Hacking's criticism do not apply. I shall discuss one of Hacking's examples, [39, pp. 97–98].

Let the setup $\mathfrak{S}$ have 101 possible results, labeled from 0 to 100. There are also, let us suppose, 101 possible hypotheses about the setup. H is the hypothesis under test; according to it the chance of result number 0 shall be 0.90, and of every other possible result, 0.001. The rival hypothesis J includes 100 simplie hypothses $J_1, J_2, \ldots, J_{100}$. According to J_n, the chance of getting reualt 0 is 0.91, and of getting reuslt numbered n is 0.09; the chance of getting a result not numbered 0 or n is, according to J_n, absolutely zero. Suppose that we set the level at $\alpha = 0.1$.

We can easily see that no two different results are evidentially equivalent. If the result of a single trial is any number $n > 0$, then the rejection set for H consists of all $m > 0$, and, thus, it has probability 0.1, and H would be rejected. If the result is 0, on the other hand, the rejection set has probability 0.9, and H would not be rejected. Hacking points out, that rejecting if the result is an $n > 0$ is not the most powerful unbiased invariant test.[2] In fact, the test is not even unbiased, since its size is 0.1 , while its power is 0.09.

There is a randomized test which is most powerful unbiased invariant, and which advises to reject H if the result is exactly 0. We take some auxilliary unrelated setup with two possible results, A and B, the probability of B being $\frac{1}{9}$. This is called the randomizer. Now consider the test which rejects H if and only if 0 does occur and the randomizer has result B. The size of this test, according to the Neyman-Pearson theory, is again 0.1 and its power a little more than 0.1. So the test is unbiased, and, in fact, it is uniformly morst powerful among invariant tests. As in the case with a randomizer that was discussed in Section 3, however, the result 0 and B is evidentially equivalent to the result 0 and A. So that the

---

[2]Invariance is another requirement of the Neyman-Pearson test, which we have not discussed.

6. GILLIES' FALSIFICATION RULE

rejection set determined by 0 has probability 0.9, and, thus, the size of the test, using rejections sets, is 0.9 and not 0.1 as the Neyman-Pearson theory demands. Thus, a result zero does not imply rejection.

Thus, it seems that at least some of the criticisms leveled at the Neyman-Pearson theory because it does not take into account final precision, do not apply to the rejection set theory. So that it seems that, at least in part, rejection sets do consider final precision.

6. Gillies' falsification rule

This chapter concludes with brief analyses of two rules that have been offered for inference. In this section, we discuss a falsification rule proposed by Gillies, [37, Chapter 9], and in the next, Bayesian estimates.

I will state Gillies' rule in my notation. We need, however, some definitions. We assume that from a statistical hypothesis H we deduce that a random variable, X has a distribution D. Gillies further requires that D be either discrete or continuous, and he assumes that the density or frequency function, f, be finite, that is, that its maximum value be finite.

Gillies, then, introduces the concept of the *relative likelihood* $l(x)$ of a possible result $x \in \Lambda_X$. This is defined as follows. Let $f_{\max}$ be the maximum value of $f(x)$. We then set

$$l(x) = \frac{f(x)}{f_{\max}}.$$

According to Gillies, $l(x)$ gives the measure of the probability of an observed event in relation to other possible events.

After analyzing some examples, Gillies is led to a concept of critical region. Thus, he looks at the problem as that of finding a subset C of Λ_X, such that, if the observed value is in C, we regard H as falsified. He calls the probability that we get a result $x \in C$, $k(C)$ (or simply k). He also defines the relative likelihood of any arbitrary subset B of Λ_X by the formula

$$l(B) = \max_{x \in B} l(x).$$

Putting all this together, we have the final version of an F.R.S.P. (falsifying rule for probability statements).

RULE XIX.1. *Suppose that the range Λ_X of X can be partitioned into two disjoint subsets A and C, with $C \cup A = \Lambda_X$, and such that*

(1) $\Pr[X \in C] = k < k_0$;
(2) $l(C) = l < l_0$, *where l_0 and k_0 are suitably small constants; and*
(3) $l(x) > l$, *for all $x \in A$.*

We shall regard H as falasified if the observed value of X, x say, lies in C.

Gillies then adds a requirement for a distribution to be falsifiable, that is, so that the rule may be applicable. He says that a random variable X has a *falsifiable* distribution if it is possible to partition Λ_X into disjoint sets A and C, with $A \cap C = \Lambda_X$, where

(1) $\Pr[X \in C] = k < k_0$
(2) $l(C) = l > l_0$, where l_0 and k_0 are suitably small constants, and
(3) the value $f_{\max}$ is in some sense representative of the probabilities of points $x \in A$.

The rule is, then, only applied to falsifiable distributions. It seems to me that the last clause of the definition of falsifiable distribution is too vague for precise applications.

One of the main differences with my rules is that Gillies does not consider an indefinite sequence of experiments, and, thus, does not provide for revision of falsification. Since rejection should always be provisional, one should have means in the rule itself for revising decisions. Besides this fact, the numbers l_0 and k_0 are arbitrary. In my case, the number α, which is also arbitrary, is decided dialectically, that is, by agreement. This is possible, because by going up in the sequence of experiments, one can always obtain a small enough α. In Gillies' case, since he does not have a sequence of experiments, he cannot decide on k_0 and l_0 dialectally. I believe, however, that one could complicate Gillies rule and demand a sequence of experiments.

The critical region, C, in Gillies rule, is similar to my rejection sets. In fact, if no two results are evidentially equivalent and there are enough alternatives so that the probability inequalities are always reversed, then C is just the rejection set determined by the result x, such that $l(x) = l$. If there are evidentially equivalent results, however, the situation is different.

As an example, consider the following case. Suppose that we have a coin and that we are testing the hypothesis H that the coin is symmetric, that is, that the probability of heads is $1/2$. Suppose that we toss the coin 20 times, and then select at random a ball from an urn with four balls, one black and three white. Let X be the random variable defined by

$$X = \begin{cases} \langle 0, i \rangle, & \text{if } i \text{ heads and black ball,} \\ \langle 1, i \rangle, & \text{if } i \text{ heads and white ball.} \end{cases}$$

for $i = 0, 1, \ldots, 20$. Let C be the set of the results $\langle 0, i \rangle$ such that $i \geq 14$ or $i \leq 6$. Then, according to H, $k = \Pr[X \in C] \approx 0.029$, and thus, $k < 0.05$ and $l(C) = 0,076$, sufficiently low values. Thus, a result in C, for instance $\langle 0, 6 \rangle$, would be enough to reject H. It seems, however, that the selection of a ball is irrelevant for the hypothesis H. Thus, results $\langle 0, i \rangle$ and $\langle 1, i \rangle$ should be evidentially equivalent. Hence, the rejection set, $R_{\langle 0,6 \rangle}$, determined by $\langle 0, 6 \rangle$ contains all tuples with $i \leq 6$ or $i \geq 14$, with first component either zero or one. Thus, $\Pr[X \in R_{\langle 0,6 \rangle}] \approx 0.116$, and hence, we should not reject H.

I do not believe that it is possible to define evidentially equivalent results without alternatives. Therefore, I do no think that Gillies' rule, without alternatives, can be ammended to take care of evidentially equivalent results.

7. Bayesian estimation

Before going into decision theory, I shall dispose of Bayes estimators as a form of inference. In this case, the unknown parameter θ is regarded as the

7. BAYESIAN ESTIMATION

value of a random variable with a given probability distribution, called the *prior distribution of θ*.

In order to be more definite, suppose that we believe that θ can be regarded as the value of a continuous distribution having probability density $p(\theta)$; and suppose that we are observing the value of a sample whose distribution depends on θ. Specifiacally, assume that $f(x|\theta)$ is the density of the value x when θ is the value of the parameter. If the observed data values are $X_i = x_i$, for $i = 1, \ldots, n$ then the updated, or conditional probability density function of θ is

$$f(\theta|x_1, \ldots, x_n) = \frac{f(x_1, \ldots, x_n|\theta) p(\theta)}{\int_\Omega f(x_1, \ldots, x_n|\theta) p(\theta) \, d\theta}$$

The *Bayes estimator* is the mean of the posterior distribution $f(\theta|x_1, \ldots, x_n)$, which is written as $\mathbf{E}(\theta|X_1, \ldots, X_n)$. That is, if $X_i = x_i$, for $i = 1, \ldots, n$, then the value of the Bayes estimator is

$$\mathbf{E}(\theta|X_1 = x_1, \ldots, X_n = x_n) = \int_\Omega \theta f(\theta|x_1, \ldots, x_n) \, d\theta.$$

The justification of this sort of estimator is that, on average, we shall be right most of the time by adopting it. That is, in the long run, we shall be more often successful than not. But when we have to accept or reject a scientific hypothesis, what happens on average or on the long run is not important. Thus, for acceptance of hypothesis outright, this technique does not seem adequate.

Bayesian inference uses Bayes theorem to obtain the posterior probability of a hypothesis, given the probability of the evidence and the prior probability distribution of the hypotheses. But do Bayesians really accept a hypothesis? In fact, can we really accept a hypothesis under such conditions? I believe that the answer to both questions is no. I do not believe that Bayesians accept hypotheses, but only assign probabilities to them and act using these probabilities. This is an assertion about the behavior of Bayesians that should be substantiated, but I shall not do it here, because it is not essential to my argument. If I am right on this point, then Bayesians reject the notion of statistical inference as I propose it.

I believe[3] that the deeper reason for the inadequacy of the Bayes estimator for the acceptance of hypotheses, is that it is not possible to accept hypotheses on the basis of their probabilities. When you assign a probability different from one to a proposition, it means that you believe that both the proposition and its negation (which has a probability different from zero) are really possible. That is, you believe that neither the proposition nor its negation are true in reality at the moment you are computing the probabilities. You believe that one of them will become true later, but at the moment, neither of them is. So, you should not accept it. Although you can assign probabilities to propositions that are not really possible, but only antecedently really possible, on assigning these probabilities you put yourself in the position of the propositions being really possible, and hence you can neither reject them nor accept them.

---

[3] A more thorough discussion of this matter appears in Chapter IV, Section 2.

Bayesian techniques are important and justified for decision theory, as we shall see in Chapter XX, but I do not think they are adequate for what I am calling inference. I believe that Bayesians take inference to be a form of decision theory.

CHAPTER XX

Decision theory

This chapter contains a somewhat brief discussion of decision theory. The approach that I shall adopt is essentially Bayesian decision theory.

When one is confronted with a decision for some action, the situation is very different than when deciding whether to accept or reject a hypothesis. When deciding on an action, one must act, and even the option of not acting is an action. In order to decide on the best action, the advantages or disadvantages of the consequences of each action must be evaluated. On the other hand if there is not enough evidence, one can decide neither to accept nor to reject a certain hypothesis. More importantly, when one is only interested in the truth or falsity of a hypothesis, one should, as a matter of policy, disregard the advantages and disadvantages of the consequences of accepting or rejecting the hypothesis. As a practical matter, because of this pressure for action, the Inverse Inference Principle, with its demand for an infinite sequence of propositions, is not adequate for decision theory. As we have seen and shall expand later, the simple use of the Direct Inference Principle is perfectly justified for this theory.

Before delving into the analysis and justification of decision theoretic techniques, I shall discuss what it means to justify these techniques. I am taking here decision theory to be a normative theory. That is, decision theory guides us in our quest for better actions. Decision theory does not analyze how we actually make decisions, but advises us how to arrive at better decisions. A decision is better, if the action taken leads to better consequences. Decision theory does not analyze what is a better consequence, but only takes it as given beforehand in each particular situation. If it were determined which consequence arises from each possible action, one would simply decide on the action that leads to the best consequence. The problem is complicated when the consequences are not determined by the actions, but also depend on the uncertain state of the world.

1. Decision Theory and Consequentialism

Although the proposal presented here has some similarities to the ethical doctrine of consequentialism, there are basic differences between the two systems, which make decision theory, in the version presented in this chapter, immune to the usual criticism leveled at this doctrine. I understand, here, by consequentialism the following ethical doctrine: the possible consequences of an act are ranked from best to worst in accordance with certain criterion which varies according to the different brands of consequentialism. This ranking should be objective, that is, independent of the agent performing the act. Then, the agent should decide in favor of the act with the best overall consequences. There has been, lately, a vivid discussion of consequentialism,[1] so that I shall include a brief analysis of the relation between this theory and the framework for decisions proposed here.

In the framework for decisions presented here, which I think is the usual Bayesian framework, as in consequentialism, it is assumed a ranking of the possible consequences of the available acts from best to worst, and the decisiin is on the action with the best overall consequences. This framework, however, is not offered as an ethical theory, that is, as a theory for deciding on any moral act, but only as a framework for decisions in a restricted situation where all the actions are morally acceptable. The theory that is proposed here, which I think is the usual Bayesian decision theory, may be called *local consequentialism*, because the consequences are clearly delimited and do not embrace the whole moral life of a person. This fact has many implications that differentiate the theory from consequentialism:

(1) The possible consequences of the acts are restricted in number, and are of a very specific form, depending on the situation at hand. For instance, in the case of a criminal trial, examples of consequences are:
- condemn a defendant who is guilty of murder in the first degree to 15 years imprisonment.
- condemn an innocent defendant to 10 years.
- condemn a defendant who is guilty of murder in the second degree to be executed.
- set free a defendant guilty of murder in the first degree.

In case of medical decisions, the consequences of a certain treatment are, for instance: to obtain a cure, or a partial cure, or death.

(2) In general, we only take into consideration the immediate consequences, and not consequences of these consequences, as in most forms of consequentialism.[2] For instance, in the case of judicial decisions, the theory does not, and I think it should not, consider consequences such as the suffering of the family of the defendant (which may be a consequence of hanging him). In consequentialism as an ethical theory, one should take into account all these remote consequences.

(3) Consequences, in these restricted situations, are relatively easy to order from best to worst. It is somewhat more difficult to assign numbers (utilities) to them, but I don't think that this is an unsurmountable difficulty. In consequentialism, on the other hand, the consequences that must be considered are so numerous and varied, that it is doubtful whether it is

---
[1] See, for instance, [86], [80], and [81].
[2] See, for instance, for the case of utilitarianism, the work of Smart in [86].

possible or not, in all cases, to rank them from best to worst.
(4) One of the main criticisms of consequentialism is that it forces, in certain cases, actions which go against the "integrity" of the agent.[3] The most serious cases of this type of conflict arises when an individual is required to perform an act that he would regard, or that is commonly regarded, as immoral, because the overall consequences of this immoral act are the best, among all possible acts. For instance, it may be the case that the consequences of killing somebody who is innocent are better than the consequences of not killing him.

In the context of decision theory that is adopted here, this possibility does not arise. The only actions that are considered are those which are morally permissible. Since we are only taking into account a restricted set of actions, and not, necessarily, all possible actions, it is natural to exclude all those that are not morally permissible. So, in the context presented here, possible actions, means physically possible to perform and morally permissible.

(5) It must be clear from the previous discussion, that I do not advocate the ranking of consequences according to the "pleasure" or "happiness" that they produce. Thus, I am far removed from utilitarianism, which is one of the most important versions of consequentialism.

Although the criticisms of consequentialism do not apply to my theory, the intuitive appeal of consequentialims does apply. I think, as does Scheffler, [80], that the main appeal of consequentialism is that it embodies the deeply plausible-sounding feature that one should always do what would lead to the best available outcome overall. This, I believe, is the main positive feature of the Bayesian framework for decisions.

2. Maximizing expected utility

If the state of the world is not determined, since we do not know which is the true state of the world when we perform a certain action, there are several possible consequences of the action. Hence, the decision cannot be based on a single consequence, but on the whole set of possible consequences. We must find a way of evaluating the goodness of the set of possible consequences of each action. We then perform the action whose set of possible consequences is best. So we assume that we have a set of possible consequences Γ and that each action determines a subset of Γ as the set of possible consequences of the action. If we are able to define a measure of "goodness" of each subset of Γ determined by an action, then we should certainly choose the action whose set of possible consequences has the maximum goodness.

There is one situation when I think we can clearly define a measure of goodness on the set of possible consequences. We assume that we have a set of possible actions and possible states of the world, and that a consequence is completely determined by the action and the state of the world. So that consequences can be

---

[3]There is a vivid discussion of this problem in the literature cited before. For instance, the work of Williams in [86] is mostly devoted to the problem.

identified with pairs formed by an action and a state of the world. Suppose we can assign a measure $u(c)$ of goodness, the utility of c, to each possible consequence c. We must also be able, or willing, to assign a probability $\Pr(s)$ to each possible state of the world s. An action can then be identified with the set of consequences, i.e., pairs, which have the action as first component. Then the natural measure of goodness of each action a is given by

$$\mathbf{U}(a) = \sum_{\text{all } s} u(a,s) \cdot \Pr(s)$$

We then perform the action whose expected utility, $\mathbf{U}(a)$, is maximum.

When one of these conditions fail, we cannot assign a measure to the actions, and, hence, we must proceed otherwise. Before going into the analysis of incomplete cases, I shall discuss the requirements for the perfect case.

The principle behind maximizing $\mathbf{U}(a)$, which is called maximizing the expected utility, was first formulated by J. Bernouilli in the 17th century. It can be justified by using the analysis of the expected value of a random variable as a truth-average.[4] Given the action a, the random variable $X_a(s) = u(a,s)$ measures the "goodness" of a, if s occurs. Then, since $\Pr[X_a = u(a,s)] = \Pr(s)$,

$$\mathbf{U}(a) = \sum_{\text{all } s} u(a,s) \cdot \Pr[X_a = u(a,s)] = \mathbf{E}\, X_a,$$

measures the truth-average of the goodness of a, i.e., the average goodness of a. We are then justified in performing the action that has a maximum average goodness.

I believe that the justification based on truth-averages, which is similar to the appeal of unbiasedness for estimators, is not very strong. In the case of estimators, we had the more important conditions of consistency. For decision theory we cannot appeal to consistency, since it involves and infinity of experiments and we cannot wait to act. There is, however, a justification for maximizing the expected utility which is similar to consistency. If we act maximizing the expected utility, in the long run we shall be most probably better off. This justification is proper only in case we are performing a series of similar actions.

This justification by success in the long run is not a perfect justification, since what is guaranteed is a high probability of success in the long run and not the certainty of success. But by increasing the length of the sequence we can increase the probability of success as much as we want. Thus, high probability of success in the long run has some similarities to consistency. The long run justification has also been used, by other authors, for the inferential procedures discussed in the previous two chapters. But when we are interested in truth and not in decision, as it is the case in inference, that in the long run we shall be right most of the time is a small justification for accepting as true a hypothesis in the particular case at hand.

The problem we are most concerned in this book is how to assign the probabilities $\Pr(s)$. I believe that in these cases we accept one probability model

---

[4] See also Chapter XVIII, Section 6, discussion of unbiasedness.

and we assign the probabilities accordingly. We are not involved with testing the model, but we just accept it in order to base our decisions. The model may be very vaguely formulated. For instance, one may assume that the probability $\Pr(s)$ is the same as that of tossing a fair coin n times and obtaining all heads. It is possible to assign probabilities without having a very clear mechanism for the setup by comparing one's degree of belief with the degree which one has in the occurrence of a sequence of m heads in n tosses of a fair coin.

There is an important argument that have been offered for the use of the maximizing expected utility in decision theory. This argument, due to Ramsey and completed by Savage, uses a theorem which states that given some conditions on a preference relation one can obtain a probability and a utility such that the preference relation coincides with the ordering of preferences given by maximizing expected utilities. The conditions are not weak, and I shall discuss a case in Section 5 where they are not satisfied. This argument also justifies, in a sense, the subjective assignment of probabilities for decision theoretic purposes. It must be noted that Savage's hypotheses for his theorem includes in fact what amounts to the possibility of comparing one's preferences with the preference of obtaining a certain outcome in the tosses of n fair coins.

Another approach for the justification of the assignment of subjective probabilities has already been discussed in Chapter IV, Section 3. I am referring to the approach due to de Finetti, which uses betting. I shall not discuss it further here.

Although I believe that we are justified in assigning subjective probabilities for decision purposes, I think that we should do it by comparing the situation at hand with a setup which has a known probability model. For instance, when assigning a probability to a state s, we should think how our degree of belief in s compares with the degree of belief we would have in the occurrence, say, of m heads in n tosses of a fair coin. If we believe that the degree of possibility of truth of s is the same as that of the occurrence of m heads, then our degree of belief should also be the same.

I believe that if we have some grounds for assigning probabilities and utilities, then it is better to proceed by maximizing expected utilities than just by hunches. How much grounds we should have is open to discussion and I do not think that a definite answer can be given to this question. Some of my ideas about it are given in the rest of this chapter.

3. Decisions based on evidence

I shall illustrate this problem with an example slightly modified from [4, pp. 28-63]. Suppose we have a company called TWEECO that makes a special type of electronic component that is used by another company, called THEECO, which utilizes this component for making a piece of hospital equipment.

The batches consist of the output from uninterrupted production runs between one monthly overhaul and the next. A large number of batches of the product are produced each year. A basic fault can sometimes arise during this complex process; this renders an individual component useless so it has to be discarded. The fault is readily detected by an appropriate test.

For a given batch of components it is important to measure the proportion of components suffering from the basic manufacturing fault. Suppose it is worthwhile for TWEECO to operate some inspection procedure to try to ensure that batches with a high proportion defective are not sent on to THEECO.

In order to build a model for the situation, we can assume that the quality of any particular component is independent of the quality of any other components either in the same or in some other batch; the proportion of defective components, θ, however, may vary from batch to batch. To know the value of θ for a particular batch currently under study is to have of the quality of that batch of components. This knowledge of θ is important in any decision to use or scrap a particular batch of components, or maintain or modify the style of manufacture of the components. But of course, θ will not be known for some particular batch currently being examined.

The question of the validation of this model is important, but in real life problems is often not feasible to carry out a thorough validation. A validation would have to be made with the techniques discussed in the previous two chapters. It is unlikely that adequate information would be available, and the model might be justified by a combination of statistical and non-statistical techniques. Independence, for instance, might be justified by arguments about the physical properties of the manufacturing process.

Suppose a current batch of the electronic components is available, we draw a random sample from the large number in the batch, and examine their individual properties. A standard test is available to check for the basic manufacturing fault and this is applied to all components in the sample. As a result we find that a number r, out of the n components in the sample are defective. By using the normal techniques of estimation discussed in Chapter XVIII, Section 6, we can estimate the actual value of θ for the batch, call it θ_0, by $\hat{\theta} = r/n$. We know that such an estimator is consistent, unbiased, and has the highest efficiency among unbiased consistent estimators.

In order to use the procedure of maximizing the expected utility, it is not enough to obtain this estimation of θ_0. We accuually need to assign probabilities and utilities in an appropriate decision theoretic model. Here we have two possible actions for TWEECO: supply the batch to THECO, or scrap the batch. Call these actions A and B, respectively. The states of the world are the possible values of θ for the batch. The consequences are the pairs $\langle A, \theta \rangle$ and $\langle B, \theta \rangle$ for these values of θ. The pair $\langle A, \theta \rangle$ is the consequence of supplying the batch when it has a proportion of θ defectives. Here we have another element in the model, namely the data r/n. So we need a rule that tells us what to do in case the data is r/n. This rule has the form of a function δ from the set of possible data S to $\{A, B\}$. That is, if r/n is observed, then our action according to rule δ is $\delta(r/n)$, which may be A or B. We write $u(A, \theta)$ and $u(B, \theta)$ for the utility of consequences $\langle A, \theta \rangle$ and $\langle B, \theta \rangle$. Here, as usual, a consequence is a pair of an action and a state of the world. For each θ, we can compute the expected utility of δ by

$$\mathbf{U}(\delta,\theta) = \sum_{\text{all } x} u(\delta(x),\theta) \cdot \Pr_\theta(x)$$

I think it is not too difficult in this case to assign utilities $u(A,\theta)$ and $u(B,\theta)$, since the possible losses can be evaluated in money, and the probabilities $\Pr_\theta$ are clearly given. So it is uncontroversial to assign utilities to pairs of action rules and states of the world. But for deciding which action rule to follow, we need the expected utility $\mathbf{U}(\delta)$ of each action rule δ independently of θ, because we do not know the real value of θ. The expected utility, $\mathbf{U}(\delta)$, is given by

$$\mathbf{U}(\delta) = \sum_{\text{all } \theta} \mathbf{U}(\delta,\theta) \cdot \Pr\theta.$$

Thus, we need the a priori probability of θ, $\Pr\theta$.

Combining the two last formulas and writing $\Pr_\theta(x) = \Pr(x|\theta)$, we get

$$\mathbf{U}(\delta) = \sum_{\text{all } \theta} \sum_{\text{all } x} u(\delta(x),\theta) \cdot \Pr(x|\theta) \cdot \Pr(\theta),$$

which looks similar to Bayes formula for probabilities. The decision rule that selects the action with the maximum expected utility is called *the Bayes decision rule*.

Here we assume just one model for assigning probabilities similar to the urn model discussed in Chapter III, Section 4 and Chapter XVI, Section 5. There are two causal moments t and s with t preceding s, at t a parameter θ is chosen and at s the sample is obtained. As we saw in Chapter XVI, Section 5, the probability of θ given x is obtained by using Bayes formula, and the weighted average of the utilities is the expected utility expressed by the formula above.

The problem with this formula is that we need the a priori probabilities of the different possible θ's. We may have data from previous batches that allow us to estimate these probabilities. If we have enough information for obtaining reliable a priori probabilities, that is, for supposing that a particular model holds, then we are much better off using the expected utility formula. If we decide not to use a priori probabilities, that is, if we assume several alternative models and not just one, then we cannot use utilities. The only possibility then is to use the estimate $\widehat{\theta}(x)$, given the data x. We must informally assess the situation without assigning numbers as utilities and decide just on the basis of this estimate. For instance, we might choose the rule that indicates that is the proportion of defectives in the sample is above 0.3, the we reject the batch. We base this decision on the fact that the proportion of defectives in the whole batch is also approximately 0.3 and this proportion leads to a loss that is too high. The justification for estimating this proportion in the whole batch lies in the fact that in the long run it is very probable that we shall be right most of the time. However, there is only a high probability that we shall be right most of the time and not certainty.

How reliable has to be the a priori probability of θ for using the expected utility decision procedure is a matter of subjective judgement. In general, we do not have a measure of the reliability of the prior probabilities. One could combine, in some cases, the inference techniques of the previous sections with decision theoretic techniques. For instance, it may be possible to apply the inferential techniques in order to determine which prior probabilities to accept.

A problem that has been widely discussed in the literature is what to do when there is no information about the prior probability, that is, in the face of complete ignorance. It has been suggested that in this case we must assume a uniform distribution. I believe that this is a remnant of the Principle of Indifference, which was discussed in Chapter III, Section 2. As it was asserted there, the assignment of equal probabilities, however, should not be on the basis of lack of knowledge, but on the knowledge or, at least acceptance, of some facts about the symmetries of the situation. Thus, if we assume that all prior probabilities are equally likely, this assumption must be based on our belief that the setup that produces the priors is similar to that of a roulette in the symmetries that it possesses. Obviously, it is not necessary to assume that there is a physical process similar to a roulette, but just that there is some process whose symmetries are similar.

In some cases the space of parameters, call it Ω, is the whole real line, or the positive real numbers. The supposition of a uniform distribution gives rise to an improper prior, that is, to a measure which is not finite. Since I consider probability as a measure of the degree of possibility, I do not consider the requirement of finiteness as essential. It is only important to have a measure of the relative sizes of the different sets of possibilities. So I do not have any objections to the use of improper priors in this regard. I have not been able to find a situation in nature, however, where improper probabilities might arise, although they are theoretically possible. Since I believe that subjective probabilities should be assigned by comparing the situation with known objective setups, the assignment of improper priors seems to me of doubtful value.

4. Decisions in the medical example

We continue, here, with the discussion of the medical example of Chapter V, in the Bayesian approach of its Section 2. We have, as in the last mentioned section, m possible deseases, $D_1, \ldots, D_m$, and a cluster of symptoms, S. We discussed in Chapter V, Section 2, how to obtain $\Pr(D_i|S)$, and we shall use these probabilities here.

We assume that we have n possible treatments, $T_1, \ldots, T_n$, and p possible consequences of these treatments, $C_1, \ldots, C_p$. These consequences may be cure, partial cure, death, etc. To each consequence, C_j we assign a number, $u(C_j)$, its *utility*, which measures the goodness of the consequence. I think that, in this case, it is easy to rank the consequences from best to worst. It may be somewhat more difficult to assign numbers to them, but I do think that it is perfectly possible.

Notice that is this case we do not consider consequences as pairs of an action, i.e., a treatment, and a state of the world, i.e., a disease. This is so because the connection between treatment, consequence, and disease is probabilistic. Each treatment, T_i determines distribution of probabilities, $\Pr_i$, such that $\Pr_i(C_j|D_k)$

is the probability of C_j, given that the patient has disease D_k, and is treated with T_i.

We assume that the probability of a consequence with a certain treatment, does not depend on the symptoms, S, but only on the disease. That is

$$\Pr{}_i(C_j|D_k S) = \Pr{}_i(C_j|D_k).$$

Then, we have

$$\Pr{}_i(C_j|S) = \sum_{k=1}^{m} \Pr{}_i(C_j|D_k S) \Pr{}_i(D_k|S)$$
$$= \sum_{k=1}^{m} \Pr(C_j|D_k) \cdot \Pr(D_k|S).$$

Perhaps this assumption is an oversimplification, but it is in most of the cases acceptable. The formalism could be changed to avoid this simplification.

Given S, we can consider T_i as a random variable such that

$$\Pr[T_i = u(C_j)] = \Pr{}_i(C_j|S).$$

Then, the *expected utility of* T_i, $\mathbf{U}(T_i)$, is the expectation of T_i as a random variable, that is

$$\mathbf{U}(T_i) = \mathbf{E}\, T_i$$
$$= \sum_{j=1}^{p} u(C_j) \Pr{}_i(C_j|S)$$
$$= \sum_{j=1}^{p}\sum_{k=1}^{m} u(C_j) \Pr(C_j|D_k) \Pr(D_k|S).$$

The Bayesian decision rule advises us to choose the treatment with the maximum expeted utility, that is, the treatment that, on average, will give us the best results. This decision rule has as a consequence that in certain cases we have to act using a treatment for a disease that is not the most probable. For instance, in case appendicitis is a possible disease, although with low probability, it may be advisable to operate, in spite of the fact that another disease that does not require an operation is more probable.

5. Decision Theory in the law

I shall end my short discussion of decision theory with an analysis of its possible applications to judicial decisions, especially in criminal cases. The discussion is based on joint work with J. Malitz, [24].

When trying a person for a criminal offense, there are usually two considerations to be made. The consideration of punishment for some wrongdoing and the protection of society. If punishment is considered of the highest importance, then one should be sure that the person really committed the crime. On the other hand, if the benefit of society is of primary concern, then one could allow judgement based on probabilities and utilities. I am not going to discuss the fundamental issues involved in this problem, but only consider how one can arrive to a decision under either premise, when the evidence is only probabilistic in nature.

Suppose first that we must arrive at a complete conviction that the person committed the crime. It is not very likely that the evidence is absolutely conclusive. In fact, the laws most countries convict a criminal when the jurors or the judge are convinced "beyond a reasonable doubt", not "beyond any doubt". What does this mean? Since we should not assign probabilities, but only accept or reject the guilt, we are in the evidential situation of statistical inference. But, when evidence is not absolute, as very rarely is, we have seen that we only are entitled to a provisional acceptance, and not to absolute acceptance of a hypothesis, in this case, the hypothesis of guilt. To provisionally accept a hypothesis means, as we have seen, to be ready to abandon this acceptance if new evidence arises which warrants the rejection of the hypothesis. Moreover, we must be able to know under which conditions the new evidence forces our rejection. That is, we must have a discriminating experiment which is an infinite sequence of possible experiments. We could, then, provisionally accept guilt at a certain stage, and then later, if new evidence arises, reject guilt. For this case, the "experiments" are just the elements of proof brought into the trial.

In a court of law, provisional acceptance is not allowed. Judgements are supposed to be definitive and not subject to revision after all appeals have been made. Although in some cases of guilty verdicts, revisions of judgements are allowed, this is an uncommon practice. And, revisions of judgement in case of an innocence verdict are almost never allowed. Hence, "guilty beyond a reasonable doubt" cannot be construed as provisional acceptance. But I think that absolute acceptance is only justified in very restricted situations, if at all. Thus, the literal acceptance of the principle "innocent unless proven guilty beyond a reasonable doubt" would imply that very few criminals would be convicted and this would be prejudicial to the need of protecting society. Hence, on the model presented in this book, if acceptance is what is required, court trials should be drastically modified, in order to allow for provisional acceptance.

The other possibility is to decide on the penalty on the basis of the probability of guilt and the utilities of the different consequences. The decision theoretic model that Malitz and I have envisioned is the following. Suppose that we are describing a trial where there is a defendant who may be guilty, say, of first degree murder, second degree murder, or not guilty. These possibilities may be called the possible states of the defendant, and correspond to what are called the states of the world of the usual utility model.

We assume that the purpose of the first stage of the trial, the fact finding stage, is to present the court with a probability for each of these states, i.e., for instance, a probability that the defendant is guilty of first degree murder, second degree murder, and not guilty. The alternative that we discussed before is that the court is presented, say by a jury or just by the judge, with an acceptance of one of these possibilities, but since we assume that we only can have provisional acceptance, this is not quite satisfactory. In the utility model, probabilities are assigned, instead of just accepting one of the states of the defendant.

I shall not discuss extensively how these probabilities may be assigned, and I realize that there are many difficult questions regarding this topic. One of the difficulties is the following. If a court has to decide on the guilt of a person, given some evidence of the crime, is it justified in using the frequency of guilt among the accused? Notice that a natural way of assigning prior probabilities is by using these frequencies, as in the case of medical diagnosis analyzed in Chapter V, Section 2. I believe the court would be considered immoral, if it did use this frequency. Why? Because an accused should be presumed innocent, unless proven guilty, and no accused should start the trial with a higher probability. We cannot use zero probability as a prior, because then Bayes formula would not allow to change it. A possible solution to this problem is to assign a uniform prior, or to assign positive, but infinitesimal probabilities to guilt. I shall not discuss the matter further and just assume that the probabilities may be given by a jury as a consensus estimate based on the evidence presented, or by a judge.

We assume that utilities are also available to the court. These utilities are numbers assigned to each consequence, which are pairs consisting of an action and a state of the defendant. Actions, here, are penalties imposed. In this case, we may assume that consequences as pairs of actions and states, since each consequence is completely determined by the corresponding pair. Thus, we may assign in the example we are considering, utilities to hanging a person guilty of first degree murder, to hanging a person guilty of second degree murder, to letting go free a person guilty of first degree murder, etc. The utilities may be fixed by law and should reflect the concerns of protecting the society and the individual rights.

As is usual with the utility model, these utilities are used in conjunction with probabilities to obtain the expected utility of each penalty and determine the penalty which has the highest expected utility.

The basic notions, introduced in [24], of a decision making model which can be used in legal contexts will be summarized next. We adopted, in the paper, and I follow here, a somewhat non-traditional notation in the hope that it will be more suggestive in the context of legal reasoning.

The model has several components:

(1) *The states of the defendant* **S** is for our purposes a finite set with, asy, m elements. For example **S** may consist of the members "not guilty", "guilty of murder in the second degree", "guilty of murder in the first degree". These are what in other decision making contexts are called the "states of the world".

(2) A *verdict* is a distribution of probabilities pr_α to the members of **S**, i.e., to the states of the defendant, which are our states of the world.
We assume that such a vector $\alpha \in \mathbf{V}$ is made available to the court as the result of the weighing of the evidence. It may represent a consensus of opinion of the jury or an evaluation of the judge. As it has

been mentioned before, a natural way of assigning these probabilities is by comparing the possibilities of truth of each of the alternatives with possibilities in known chance setups.

(3) The *set of penalties*, **P**, is a finite set with elements $p_1, p_2, \ldots, p_n$ ordered according to increasing severity. For example, **P** might be "free", "incarcerate for ten years", "hang". Here $n = 3$ and p_3 is hang. We write $p_i < p_j$ when p_i is less severe than p_j. The penalties are the actions, in this context.

The *consequences* of penalty p_i are then the pairs $\langle p_i, s \rangle$, where $s \in \mathbf{S}$.

(4) A *utility function* is a function β which assigns a number to each consequence, i.e., to each pair $\langle p, s \rangle$, where $p \in \mathbf{P}$ and $s \in \mathbf{S}$. For each $p_i \in \mathbf{P}$, the function $\beta(p_i, \ )$ is in fact a random variable over the probability space $\langle \mathbf{S}, \mathrm{pr}_\alpha \rangle$, for each verdict α.

The utility is a measure of the value of this outcome. I shall not discuss how it is obtained, but merely assume that it is part of the law available to the court before the trial.

(5) The *expected utility of penalty p_i with respect to the verdict α*, $\mathbf{U}(p_i, \alpha)$ is the expectation of the random variable $\beta(p_i, \ )$ with respect to the probability Pr_α, i.e.

$$\mathbf{U}(p_i, \alpha) = \sum_{s \in \mathbf{S}} \beta(p_i, s) \, \mathrm{pr}_\alpha(s).$$

We don't need, in this case, the expected utility of p_i independent of verdict α, because we always are supposed to have a verdict, in order to make a decision on the penalty.

According to the Bayesian decision model, the court having received the verdict α should assign the penalty whose expected utility is greatest, i.e., it assigns that p_i such that $\mathbf{U}(\alpha, \beta, i) > U(\alpha, \beta, j)$ for all $j \neq i$.

There are some differences between this model for trials and the usual decision theoretic models. In the first place, not every set of consequences is a possible action. Only those sets that impose the same penalty are possible actions, i.e., those set of pairs which have the same first member. For most of the theorems proved in decision theory concerning the existence of utilities and probabilities, it is an assumption that we have available all sets of consequences, or at least a very rich family of sets of consequences, as actions. In the second place, the penalties are ordered according to increasing severity and there is an order preserving function from the states of the defendant to the penalties. This function imposes some restrictions upon the possible utility functions. Malitz and I[5] have found that under these natural restrictions, deciding on a penalty according to the expected utility model is equivalent to dividing the space of probabilities into convex disjoint sets and a deciding just on the basis of the set in which the vector of probabilities in the trial happens to be.

---

[5] For further details see [24].

APPENDIX A

Foundations of nonstandard analysis

In all versions of nonstandard analysis that I know of, one of the principles or axioms used is the *Transfer Principle*. This is true both for the method of enlargements, for instance [78], [52], or [1], and for variants of set theory, such as [70] and [43].

In order to use this principle with the method of enlargements, it is necessary to determine the truth or falsity of certain sentences in two models: the standard and the nonstandard universes. For most mathematicians who are not logicians, checking the truth or falsity of a sentence in a model and shifting back and forth from one universe to another, is foreign to their training, and thus, it is difficult for them to use nonstandard analysis.

Although in the variants of set theory this shifting is not necessary, it is still necessary to check whether a sentence is true in a model (or more precisely in a family of sets) or not. Besides, nonstandard set theories have other inconvenients. Nelson's internal set theory [70] has the disadvantage that external sets, for instance the set of real numbers, are not sets in the system, which seems awkward. For instance, the construction sketched in the next appendix cannot be carried out in Nelson's internal set theory. On the other hand, in Hrbáček's theory, [43, 44], the saturation principle is inconsistent with the power set and replacement axioms for external sets. The attempts in [50] and [35] to avoid these problems seem to me too complicated. Anyway, the method of enlargements is sufficient for most purposes and also simpler, so as to be preferable.

The purpose of this appendix is to present axioms, within the method of enlargements, that substitute the Transfer Principle by other axioms which are more set theoretical in character. With these axioms, it is possible to do all the work in the nonstandard universe, without shifting back to the real numbers. The equivalence of the new axioms with the Transfer Principle will also be proved. Finally, we shall indicate how to construct a model for these axioms.

1. Axioms

The axiomatization[1] that I shall introduce here is enough for developing all nonstandard analysis required for this book. For our axiomatization, it is enough to add to the usual mathematical notions, the notions of standard number and set. Standard sets are certain subsets of $^*\mathbb{R}$, of the power set of $^*\mathbb{R}$, of the power set of the power set of $^*\mathbb{R}$, etc. In order to have a comprehensive class of sets to work with, we define, as in page 176, by induction, for any set X

$$V_0(X) = X,$$
$$V_{n+1}(X) = V_n(X) \cup \mathcal{P}(V_n(X)),$$

and

$$V(X) = \bigcup_{n=0}^{\infty} V_n(X).$$

In page 176, there are examples of sets in $V_m(^*\mathbb{R})$.

We call the usual mathematical notions *standard notions*. In fact, we shall just have the membership relation as the only primitive nonlogical standard notion (with the identity as a (standard) logical notion), and the notion of 'being standard', as the only primitive nonstandard notion. All other nonstandard notions can be defined with the latter notion. For instance, we can define 'x is infinitesimal' by '$x < r$ for every standard positive r'.

We now describe the formal theory. The only standard nonlogical symbol is the binary predicate $\in$ and we add a new, nonstandard individual constant, say $\mathbb{S}$, which denotes the class of standard objects. The elements of standard sets are called *internal*. That is, the class of internal objects is

$$\bigcup \mathbb{S} = \{x \mid \exists Y (x \in Y \text{ and } Y \in \mathbb{S}\}.$$

We use as logical notions the connectives: $\neg$ (negation), $\wedge$ (and), $\vee$ (or), $\Longrightarrow$ (implies), $\Longleftrightarrow$ (if and only if); quantifiers: $\forall$ (for all), $\exists$ (exists); and the identity symbol: $=$. All logical notions are standard.

We call a formula *internal*, if $\mathbb{S}$ does not occur in the formula and all its quantifiers are relativized to other sets. More precisely, an *internal formula* is a formula built from the atomic formulas $X \in Y$ and $X = Y$, using the logical connectives $\neg, \wedge, \vee, \Longrightarrow$, and $\Longleftrightarrow$, and the relativized quantifiers

$$\forall X \in Y \quad \text{and} \quad \exists X \in Y.$$

That is, an internal formula is just a formula of set theory with only bounded quantification.

The axioms for the class $\mathbb{S}$ of standard objects are the following:

---

[1] This axioms were introduced for the first time in [12].

1. AXIOMS

Statement of the axioms.

(1) We assume that $\mathbb{R}$ is a proper subset of $^*\mathbb{R}$ and that $^*\mathbb{R}$ does not contain any elements in $V(\mathbb{R})$ that are not in $\mathbb{R}$, in symbols

$$\mathbb{R} \subset {^*\mathbb{R}} \wedge {^*\mathbb{R}} \cap V(\mathbb{R}) = \mathbb{R}.$$

(2) The second axiom is

$$\mathbb{S} \subseteq V(^*\mathbb{R}).$$

(3) Real numbers (i.e., elements of $\mathbb{R}$, but not $\mathbb{R}$ itself) constitute the standard elements of $^*\mathbb{R}$. In symbols

$$^*\mathbb{R} \cap \mathbb{S} = \mathbb{R}.$$

(4) $^*\mathbb{R}$ is the smallest standard set containing $\mathbb{R}$, i.e.

$$^*\mathbb{R} = \bigcap \{X \mid X \in \mathbb{S} \wedge \mathbb{R} \subseteq X\} \in \mathbb{S}.$$

(5) If a set A can be defined as the subset of a standard set that satisfies a condition that involves only standard notions and other standard objects, then A is also standard. In symbols, this postulate is an axiom schema of the following form:
Let $\varphi(x, X_1, X_2, \ldots, X_n)$ be an *internal formula with free variables among* $x, X_1, X_2, \ldots, X_n$. Then

$$X_1, X_2, \ldots, X_n, Y \in \mathbb{S} \implies \{x \in Y \mid \varphi(x, X_1, \ldots, X_n)\} \in \mathbb{S}$$

is an axiom.

(6) This property says intuitively that a standard set is determined by its standard elements. That is, if A and B are standard sets which contain the same standard elements, then $A = B$. In particular, if A and B are standard subsets of $^*\mathbb{R}$ such that $A \cap \mathbb{R} = B \cap \mathbb{R}$, then $A = B$. In symbols

$$A, B \in \mathbb{S} \wedge A \cap \mathbb{S} = B \cap \mathbb{S} \implies A = B.$$

(7) Every element of $V(^*\mathbb{R})$ which only contains standard elements has a standard extension. It is clear, by Axiom (6), that the standard extension of a set of standard objects is unique. In symbols, this postulate reads

$$A \in V(^*\mathbb{R}) \wedge A \subseteq \mathbb{S} \implies \exists B (B \in \mathbb{S} \wedge B \cap \mathbb{S} = A).$$

(8) Finally, we need an axiom that implies denumerable comprehension, which was introduced in Chapter VIII, Section 3, in page 183. We formulate the axiom, which is called *countable saturation*, in this form, because it is easier to prove in the model we shall construct.

If A_1, A_2, ..., A_n, ..., for $n \in \mathbb{N}$, is a sequence of nonempty sets, elements of a standard set B, such that $A_1 \supseteq A_2 \supseteq \cdots \supseteq A_n \supseteq \ldots$, then

$$\bigcap_{n=1}^{\infty} A_n \neq \emptyset.$$

If A is any set in $V(^*\mathbb{R})$, then $A \in V_n(^*\mathbb{R})$ for a certain n. Thus, $A \cap \mathbb{S} \in V_{n+1}(^*\mathbb{R})$, and hence $A \cap \mathbb{S} \in V(^*\mathbb{R})$. Therefore, for any $A \in V(^*\mathbb{R})$, there is a unique standard set whose only standard elements are the standard elements of A. We introduce the notation $^*A$ for this set.

Notice that since

$$A \subseteq B \iff \forall x (x \in A \implies x \in B)$$
$$\iff (\forall x \in A)(x \in B),$$

we have that the formula $A \subseteq B$ is internal. We have the following elementary properties of standard sets.

PROPOSITION A.1. *Let A_1, A_2, ..., A_n be standard. Then*

(1) $A_1 \cap A_2 \cap \cdots \cap A_n$ *is standard.*
(2) *If $A_1 \cap \mathbb{S} \subseteq A_2$, then $A_1 \subseteq A_2$.*
(3) $\{A_1, A_2, \ldots, A_n\}$ *is standard. That is, a finite set of standard objects is standard. In fact, we have*
$\{A_1, A_2, \ldots, A_n\}$ *is standard if and only if $A_1, A_2, \ldots, A_n$ are standard. Hence, we also have:*
$\langle A_1, A_2, \ldots, A_n \rangle$ *is standard if and only if $A_1, A_2, \ldots, A_n$ are standard.*
(4) $A_1 \cup A_2 \cup \cdots \cup A_n$ *is standard.*
(5) *If A is standard, then $\bigcup A = \{x \mid \exists y (x \in y \in A)\}$ is also standard.*
(6) $A_1 \times A_2 \times \cdots \times A_n = \{\langle x_1, x_2, \ldots, x_n \rangle \mid x_1 \in A_1, x_2 \in A_2, \ldots, x_n \in A_n\}$ *is standard.*

PROOF OF (1). We prove the properties for two standard sets A and B and extend it to a finite n by induction. So assume in the proof of all properties that A and B are standard. We have, by Axiom (5), that

$$A \cap B = \{x \in A \mid x \in B\}$$

is standard. □

PROOF OF (2). Suppose that $A \cap \mathbb{S} \subseteq B$. Then $A \cap \mathbb{S} = A \cap B \cap \mathbb{S}$. Since, by (1), $A \cap B$ is standard, we have, by Axiom (6), that $A = A \cap B$, i.e., $A \subseteq B$. □

PROOF OF (3). We have

$$\{A, B\} = \{x \in {}^*\{A, B\} \mid x = A \vee x = B\},$$

1. AXIOMS

which is standard, by Axiom (5).

On the other hand, suppose that $C = \{A, B\}$ is standard. By Axioms (7) and (6), C is the standard set containing the standard elements of C, i.e., $C = {}^*C$. Suppose that A is not standard, but B is standard. We have that $\{B\}$ is standard, so that $\{B\} = {}^*C$. But $C \neq \{B\}$, which is a contradiction. The other cases are treated similarly. $\square$

PROOF OF (4). Let $C = {}^*((A \cap \mathbb{S}) \cup (B \cap \mathbb{S}))$. By 2, $C \supseteq A \cup B$. Thus, by Axiom (5), the set

$$A \cup B = \{x \in C \mid x \in A \vee x \in B\}$$

is standard. $\square$

PROOF OF (5). Let $C = {}^*(\mathbb{S} \cap \bigcup A)$. Then if $X \in A$ is standard, $X \subseteq C$. Thus the subset B of A

$$B = \{X \in A \mid X \subseteq C\}$$

is standard and contains all standard elements of A. Hence, by Axiom (6), $A = B$. Thus, for every $X \in A$, $X \subseteq C$. Hence, $\bigcup A \subseteq C$. Therefore, by Axiom (5)

$$\bigcup A = \{x \in C \mid \exists y (y \in A \wedge x \in y)\}$$

is standard. $\square$

PROOF OF (6). We shall prove this property for $n = 2$. The general case is proved by induction. Let $C = {}^*(A_1 \times A_2)$. Let $x \in A_1$. For x standard, by (3), the set

$$B_x = \{y \in A_2 \mid \langle x, y \rangle \in C\}$$

is standard and contains all standard objects in A_2. Thus, $B_x = A_2$. Now, the set

$$B = \{x \in A_1 \mid B_x = A_2\}$$

is standard and contains all standard elements of A_1. Thus, $B = A_1$. Therefore, $A_1 \times A_2 \subseteq C$. Hence

$$A_1 \times A_2 = \{\langle x, y \rangle \in C \mid x \in A_1 \wedge y \in A_2\}$$

is standard. $\square$

We also have:

PROPOSITION A.2. *Let* $A_1, A_2, \ldots, A_n \in V({}^*\mathbb{R})$. *Then*
(1) *If* $A_1 \subseteq A_2$, *then* ${}^*A_1 \subseteq {}^*A_2$.
(2) ${}^*\{A_1, A_2, \ldots, A_n\} = \{{}^*A_1, {}^*A_2, \ldots, {}^*A_n\}$.
(3) ${}^*\langle A_1, A_2, \ldots, A_n \rangle = \langle {}^*A_1, {}^*A_2, \ldots, {}^*A_n \rangle$.
(4) ${}^*(A_1 \cup A_2 \cup \cdots \cup A_n) = {}^*A_1 \cup {}^*A_2 \cup \cdots \cup {}^*A_n$.
(5) ${}^*(A_1 \cap A_2 \cap \cdots \cap A_n) = {}^*A_1 \cap {}^*A_2 \cap \cdots \cap {}^*A_n$.
(6) ${}^*(A_1 \times A_2 \times \cdots \times A_n) = {}^*A_1 \times {}^*A_2 \times \cdots \times {}^*A_n$.

PROOF. We prove (6), leaving the rest to the reader. By the previous theorem, $^*A_1 \times {^*A_2} \times \cdots \times {^*A_n}$ is standard. Since an n-tuple is standard if and only if its components are standard, this set contains all standard elements in $A_1 \times A_2 \times \cdots \times A_n$, and hence (6) is proved. □

We also have the following theorem.

THEOREM A.3. *Let $A \in V(^*\mathbb{R}) \cap \mathbb{S}$ and let B be the set of standard elements of A. Then B is finite if and only if $A = {^*A} = B$.*

PROOF. Suppose that B is finite. Then, by Proposition A.1, $B = {^*B}$ is standard. Since it contains all standard elements of A, we have, by Axiom (6), $A = B$. The proof in the other direction is obtained from Proposition A.1. □

We can easily prove for functions:

PROPOSITION A.4.

(1) *If f is standard, then its domain is standard and for every standard x in its domain, $f(x)$ is also standard.*
(2) *If f and g are standard functions with the same domain and such that for every x standard in its domain, $f(x) = g(x)$, then $f = g$.*
(3) *If f is a function in $V(^*\mathbb{R})$, then $^*f$ is also a function whose domain is the standard extension of the domain of f.*

PROOF OF (1). This is proved from the fact that the domain and the value of a function have a standard definition. □

PROOF OF (2). Let D be the common domain of f and g, which is standard, and let
$$A = \{x \in D \mid f(x) = g(x)\}.$$
Then A is standard subset of D and contains all standard elements of D. Hence $A = D$. □

PROOF OF (3). Let $^*f$ be the standard extension of f, A the domain of f. Then $f \subseteq A \times B$ for a certain B. Hence, by the previous propositions
$$^*f \subseteq {^*A} \times {^*B}.$$
We have that the domain of $^*f$, which is
$$\{x \in {^*A} \mid (\exists y \in {^*B}) \langle x, y \rangle \in {^*f}\},$$
is standard and contains all standard elements in $^*A$. Hence, it is equal to $^*A$.

On the other hand, suppose that $x \in {^*A}$ is standard. Then the set
$$\{y \in {^*B} \mid \langle x, y \rangle \in {^*f}\}$$
is standard and contains exactly one standard element. Hence, by the previous theorem, it contains exactly one element. Thus, the set
$$\{x \in {^*A} \mid (\exists! y \in {^*B}) \langle x, y \rangle \in {^*f}\}$$

is standard and contains all standard elements of $^*A$. Hence, it is equal to $^*A$, and therefore $^*f$ is a function. □

In $V(^*\mathbb{R})$ we have the field operations $+$ and $\times$, and the ordering $\leq$ on $\mathbb{R}$. By Proposition A.1, (3), an ordered pair is standard if and only if its components are standard. Then $+$, $\times$ and $\leq$ contain only standard elements. Hence, they have standard extensions $^*+$, $^*\times$ and $^*\leq$. We have

THEOREM A.5. $^*\mathbb{R}$ *with the operations* $^*+$ *and* $^*\times$ *and the relation* $^*\leq$ *is an ordered field.*

PROOF. By Proposition A.4, $^*+$ and $^*\times$ are binary operations on $^*\mathbb{R}$, and, by Proposition A.2, $^*\leq$ is a binary relation on $^*\mathbb{R}$. The rest of the proof consists on easy verifications of the axioms. We shall prove one of the field axioms, commutativity of addition, as example. The set

$$\{\langle x, y\rangle \in {}^*\mathbb{R} \times {}^*\mathbb{R} \mid x \,{}^*{+}\, y = y \,{}^*{+}\, x\}$$

is standard and contains all standard elements of $^*\mathbb{R} \times {}^*\mathbb{R}$. Hence, it is equal to $^*\mathbb{R} \times {}^*\mathbb{R}$. □

We define internal objects as elements of standard sets. Thus, for instance, the elements of $^*\mathbb{R}$ are internal. The class of all internal sets in $V(^*\mathbb{R})$ is symbolized $^*V(\mathbb{R})$, which, as was mentioned before, is formally defined by

$$^*V(\mathbb{R}) = \bigcup \mathbb{S} = \{X \mid \exists Y (Y \in \mathbb{S} \wedge X \in Y)\}.$$

An element of $V(^*\mathbb{R})$ that is not internal is called *external*. From this definition and the properties of standard objects, we shall deduce the main properties of internal objects.

The next theorem collects some properties of internal sets.

THEOREM A.6.
 (1) $\mathbb{S} \subseteq {}^*V(\mathbb{R})$.
 (2) *If* Y *is internal and* $X \in Y$*, then* X *is also internal, i.e.,* $^*V(\mathbb{R}) = \bigcup {}^*V(\mathbb{R})$*, that is,* $^*V(\mathbb{R})$ *is transitive (in the set-theoretic sense).*
 (3) *If* $\varphi(x, X_1, \ldots, X_n)$ *is an internal formula and* Y*,* X_1*,* X_2*,* $\ldots$*,* X_n *are internal, then*

$$\{x \in Y \mid \varphi(x, X_1, X_2, \ldots, X_n)\}$$

is internal.
 (4) *If* Y *is internal, then the set of internal subsets of* Y*, called the* internal power set of Y*, is also internal. In case* Y *is standard, then the set of internal subsets of* Y *is also standard.*

PROOF OF (1). Let A be standard. We have, by Proposition A.1, that $\{A\}$ is also standard, and thus, $A \in \{A\}$ is internal. □

PROOF OF (2). Let $X \in Y$ with Y internal. We have that $Y \in A$ with A standard. By Proposition A.1, $\bigcup A$ is standard and $X \in \bigcup A$. Thus, X is internal. □

PROOF OF (3). Let A be a standard set containing Y, X_1, X_2, ..., X_n (by the definition of internal sets and Proposition A.1, there is such a set A). Then, by Proposition A.1, $^n A$, the set of n-tuples of elements of A, is standard. Let C be the standard set containing all standard subsets of $\bigcup A$. Then, by Axiom (5) and Proposition A.1

$$D = \{\langle Y, X_1, X_2, \ldots, X_n \rangle \in {}^n A \mid \{x \in Y \mid \varphi(x, X_1, X_2, \ldots, X_n)\} \in C\}$$

is standard. Again, by Axiom (5), D contains all standard tuples in $^n A$, and, hence, by Axiom (6), $D = {}^n A$. Therefore, for the given internal sets Y, X_1, ..., $X_n \in A$

$$\{x \in Y \mid \varphi(x, X_1, \ldots, X_n)\} \in C$$

and thus, it is internal. □

PROOF OF (4). Let $Y \in A$, where A is standard. Then, by Proosition A.1, $\bigcup A$ is also standard. Let C be the standard set containing all standard $X \subseteq \bigcup A$. We claim that every internal $X \subseteq \bigcup A$ belongs to C. In order to prove the claim, let $X \subseteq \bigcup A$ be internal. Then $X \in B$ with B standard. The set

$$D = \{X \in B \mid X \subseteq \bigcup A\}$$

is standard. Suppose that $Z \in D$ is standard. Then $Z \subseteq \bigcup A$, and, hence, $Z \in C$. Therefore, $D \subseteq C$. Since $X \in B$ and $X \subseteq \bigcup A$, $X \in D$, and so $X \in C$. Thus, we have proved the claim.

The set

$$E = \{X \in C \mid X \subseteq Y\}$$

is internal (E is standard, if Y is standard), by (3) and the claim, it contains all internal subsets of Y. By (2), E is exactly the set of internal subsets of Y. □

We denote the internal power set of a set Y by $^*\mathcal{P}Y$.

The least upper bound principle is not true for external subsets of $^*\mathbb{R}$. For instance, $\mathbb{R}$ is a bounded (in $^*\mathbb{R}$) subset of $^*\mathbb{R}$ which does not have a least upper bound. On the other hand, the least upper bound principle is valid for internal subsets. Thus, $\mathbb{R}$ is external.

We also state some properties of $^*\mathbb{N}$, the standard set containing $\mathbb{N}$, i.e., containing all finite natural numbers. It is clear, by Proposition A.1, since $\mathbb{N} \subseteq \mathbb{R}$, that $^*\mathbb{N} \subseteq {}^*\mathbb{R}$, and, by the definition of $^*\mathbb{N}$, that $^*\mathbb{N} \cap \mathbb{S} = \mathbb{N}$ and $^*\mathbb{N} \in \mathbb{S}$.

THEOREM A.7.
(1) *The ordering of $^*\mathbb{N}$ is discrete, i.e., if $n \in {}^*\mathbb{N}$ then there is no element m of $^*\mathbb{N}$ such that $n < m < n + 1$.*
(2) *The finite elements of $^*\mathbb{N}$ are exactly the elements of $\mathbb{N}$. The set $^*\mathbb{N}$ is a proper extension of $\mathbb{N}$. That is, there are elements in $^*\mathbb{N}$ which are not in $\mathbb{N}$. These elements of $^*\mathbb{N} - \mathbb{N}$ are infinite.*
(3) *$^*\mathbb{N}$ satisfies the Internal Induction Principle, that is*
   Internal Induction Principle.: *If S is an internal subset of $^*\mathbb{N}$ that satisfies the following conditions:*
   (a) $1 \in S$;

(b) $n \in S \implies n+1 \in S$;
then $S = {}^*\mathbb{N}$.

We call the usual induction principle for $\mathbb{N}$ the *external induction principle*.

PROOF OF (1). Let $n \in {}^*\mathbb{N}$ be standard. The set
$$\{m \in {}^*\mathbb{N} \mid n < m < n+1\}$$
is standard and, since ${}^*\mathbb{N} \cap \mathbb{S} = \mathbb{N}$, it contains no standard element. Hence, by Theorem A.3, it is empty.

Thus, the set
$$\{n \in {}^*\mathbb{N} \mid \neg(\exists m \in {}^*\mathbb{N}) n < m < n+1\}$$
is standard and contains all standard elements of ${}^*\mathbb{N}$. Hence, it is equal to ${}^*\mathbb{N}$. □

PROOF OF (2). It is clear that all elements of $\mathbb{N}$ are finite. Suppose, on the other hand, that $n \in {}^*\mathbb{N}$ is finite. Then, since there is an $m \in \mathbb{N}$ with $m > n$, by the well ordering of $\mathbb{N}$, there is an $m \in \mathbb{N}$ such that $m \leq n < m+1$. By (1), $m = n$. □

PROOF OF (3). Let C be the standard set containing all standard S satisfying the Internal Induction Principle. By Theorem A.6, ${}^*\mathcal{P}{}^*\mathbb{N}$ is standard. Let $S \in {}^*\mathcal{P}{}^*\mathbb{N}$ be standard and suppose that $1 \in S$ and $n \in S \implies n+1 \in S$. Then, by the external induction principle, $S \supseteq \mathbb{N}$. Thus, $S \cap \mathbb{S} = {}^*\mathbb{N} \cap \mathbb{S}$. Therefore, $S = {}^*\mathbb{N}$, and, hence, $S \in C$. Thus
$$ {}^*\mathcal{P}{}^*\mathbb{N} = C$$
and we have proved the induction principle. □

Notice that for all the proofs so far we have not used saturation, Axiom 8. Finally, we shall prove, from Axioms 8, denumerable comprehension.

THEOREM A.8 (DENUMERABLE COMPREHENSION). *If $x_1, x_2, \ldots, x_n, \ldots,$ for $n \in \mathbb{N}$, is a sequence of elements of an internal set A, then there is an internal sequence of elements of A, $x_1, x_2, \ldots, x_\nu, \ldots,$ for $\nu \in {}^*\mathbb{N}$, which extends it.*

PROOF. We have that ${}^{*\mathbb{N}}A$, the set of all sequences of elements of A is internal. Let A_k be the set of all elements of ${}^{*\mathbb{N}}A$ that extend $x_1, x_2, \ldots, x_k$. Then the hypothesis of Axiom 8 is satisfied, since the A_k are internal nonempty sets, elements of the internal power set of ${}^{*\mathbb{N}}A$, and $A_1 \supseteq A_2 \supseteq \cdots \supseteq A_k \supseteq \ldots$. Thus, $\bigcap_{k=1}^\infty A_k \neq \emptyset$. Any element of this intersection is an internal sequence extending $\langle x_n \mid n \in \mathbb{N}\rangle$. □

2. Proof of the equivalence of Axioms (2)–(7) to Transfer Principle

We first state precisely the Transfer Principle.

Extension and Transfer Principles. We add to set theory a unary function symbol * satisfying

(1) $* : V(\mathbb{R}) \to V(*\mathbb{R})$.
(2) (**Extension Principle**) $*\mathbb{R}$ is a proper extension of $\mathbb{R}$, and for every $r \in \mathbb{R}$, $*r = r$.
(3) (**Transfer Principle**) For every internal formula (i.e., a formula of set theory with bounded quantification) $\varphi(X_1, \ldots, X_n)$, and $X_1, \ldots, X_n \in V(\mathbb{R})$ we have:

$\varphi(X_1, \ldots, X_n)$ is true in $V(\mathbb{R})$ if and only if $\varphi(*X_1, \ldots, *X_n)$ is true in $V(*\mathbb{R})$.

Proof of the Extension and Transfer Principles from Axioms (1)–(7). We, first, have to define, on the basis of Axioms 2–7, the function * of the Transfer Principle. We define by external induction $* : V(\mathbb{R}) \to \mathbb{S} \subseteq V(*\mathbb{R})$.

(1) If $r \in \mathbb{R}$, then $*r = r$.
(2) Suppose that * is defined for elements of $V_n(\mathbb{R})$ and let $X \subseteq V_n(\mathbb{R})$. Then

$$*X = \text{ the standard set that contains } \{*Y \mid Y \in X\}.$$

It is clear that the Extension Principle follows from Axiom (1). The following theorem follows from Axioms (2)–(7).

THEOREM A.9. *For each natural number n, * is a one-one function from $V_n(\mathbb{R})$ onto $\mathbb{S} \cap V_n(*\mathbb{R})$, and, thus, * is one-one from $V(\mathbb{R})$ onto $\mathbb{S}$.*

PROOF. The proof is by external induction on n. It is clear from the definition of * that the theorem is true for $n = 0$. Suppose that the theorem is true for n and we prove it for $n+1$. We prove, first, that * is one-one. Let $*A = *B$, for $A, B \in V_{n+1}(\mathbb{R})$. We then have

$$\{*X \mid X \in A\} = \{*X \mid X \in B\}.$$

Since, by the induction hypothesis, * is one-one on elements of A and B, we get that $A = B$.

We now prove that * is onto $\mathbb{S} \cap V_{n+1}(*\mathbb{R})$. Let A be a standard element of $V_{n+1}(*\mathbb{R})$. By the induction hypothesis, every standard element X of A is of the form $X = *Y$, for a certain $Y \in V_n(\mathbb{R})$. Let

$$B = \{X \mid *X \in A\}.$$

Then $*B = A$. □

We can now show the Transfer Principle, from Axioms (2)–(7).

THEOREM A.10 (TRANSFER PRINCIPLE). *Let $X_1, X_2, \ldots, X_n \in V(\mathbb{R})$ and let φ be an internal formula. Then*

$$\varphi(X_1, X_2, \ldots, X_n)$$

is true in $V(\mathbb{R})$ if and only if

$$\varphi(*X_1, *X_2, \ldots, *X_n)$$

2. EQUIVALENCE OF AXIOMS TO TRANSFER

is true in $V(^*\mathbb{R})$.

PROOF. The proof is by induction on the complexity of the formula φ. We begin the induction with φ is atomic of the form $X_1 \in X_2$. Then, by the definition of the function *

$$X_1 \in X_2 \iff {^*X_1} \in {^*X_2}.$$

The other interesting case is that of the quantifier. Let φ be $(\forall x \in X)\psi$ and assume the theorem proved for ψ. Suppose that

$$(\forall x \in X)\psi(x, X_1, \ldots, X_n)$$

is true in $V(\mathbb{R})$. Then, by the induction hypothesis, for every $x \in X$,

$$\psi(^*x, {^*X_1}, \ldots, {^*X_n})$$

is true in $V(^*\mathbb{R})$, and since $^*x, {^*X_1}, \ldots, {^*X_n}$ are standard, it is equivalent to an internal formula. Let

$$Y = \{y \in {^*X} \mid \psi(y, {^*X_1}, \ldots, {^*X_n})\}.$$

We have that Y is a standard subset of $^*X$ and it contains all the standard elements of $^*X$. Thus $Y = {^*X}$. Therefore

$$(\forall x \in {^*X})\psi(x, {^*X_1}, \ldots, {^*X_n})$$

is true in $V(^*\mathbb{R})$.

Assume now that in $V(^*\mathbb{R})$ we have

$$(\forall x \in {^*X})\psi(x, {^*X_1}, \ldots, {^*X_n}).$$

Then

$$\psi(^*x, {^*X_1}, \ldots, {^*X_n})$$

is true in $V(^*\mathbb{R})$ for every $x \in X$. By the induction hypothesis

$$\psi(x, X_1, \ldots, X_n)$$

is true in $V(\mathbb{R})$ for every $x \in X$. Thus

$$(\forall x \in X)\psi(x, X_1, \ldots, X_n)$$

is true in $V(\mathbb{R})$. □

Proof of Axioms (1)–(7) from the Extension and Transfer Principles.
We assume the Extension and Transfer Principles and prove our axioms (1)–(7).

We define the standard sets $\mathbb{S}$ as those that are in the image of *. Axiom (1) is clear from the Extension Principle. Axioms (2), (3), and (4) are clear from the Extension and Transfer Principles, and the definition of $\mathbb{S}$.

We now prove Axiom (5). Let $\varphi(x, X_1, \ldots, X_n)$ be an internal formula, Y, $X_1, \ldots, X_n$ standard and

$$A = \{x \in Y \mid \varphi(x, X_1, \ldots, X_n)\}.$$

Let $Y = {}^*U$ and $X_i = {}^*Z_i$ for $i = 1, \ldots, n$ and let

$$B = \{x \in U \mid \varphi(x, Z_1, \ldots, Z_n)\}.$$

We shall prove that $A = {}^*B$ and, hence, that A is standard. We have

$$\forall x (x \in B \iff x \in U \wedge \varphi(x, Z_1, \ldots, Z_n)).$$

This formula is equivalent to the conjunction of

$$\forall x (x \in B \implies x \in U \wedge \varphi(x, Z_1, \ldots, Z_n)),$$

and

$$\forall x (x \in U \implies (\varphi(x, Z_1, \ldots, Z_n) \implies x \in B)).$$

Thus, it is an internal formula and we can apply transfer to it. By transfer we get

$$\forall x (x \in {}^*B \iff x \in Y \wedge \varphi(x, X_1, \ldots, X_n)),$$

and this proves that ${}^*B = A$.

Next, we prove Axiom (6). Let ${}^*A$ and ${}^*B$ be standard sets that contain the same standard elements. Then we have

$${}^*x \in {}^*A \iff {}^*x \in {}^*B.$$

Thus, by transfer, we get

$$x \in A \iff x \in B$$

and hence

$$\forall x (x \in A \iff x \in B),$$

that is, $A = B$. Therefore, ${}^*A = {}^*B$.

Axiom (7) is proved as follows. If one has a set C of standard elements, then one can take the set

$$D = \{x \mid {}^*x \in C\}.$$

Thus ${}^*D$ is a standard set whose only standard elements are the elements of C.

Thus, the Extension, Transfer and Countable Saturation Principles are equivalent to our axioms. □

3. Construction of the nonstandard universe

In this section, I shall sketch the construction of the nonstandard universe and prove that the Extension and Transfer principles and the Saturation Axiom (8) are satisfied. Therefore, we can conclude that Axioms (1)–(8) are also satisfied in this universe.[2]

Ultrafilters.. We begin with a few definitions. A *filter* $\mathcal{F}$ is a family of subsets of $\mathbb{N}$ satisfying the following properties:

(1) $\mathbb{N} \in \mathcal{F}$, $\quad \emptyset \notin \mathcal{F}$.
(2) $A_1, \ldots, A_n \in \mathcal{F} \implies A_1 \cap \cdots \cap A_n \in \mathcal{F}$.
(3) If $A \in \mathcal{F}$ and $A \subseteq B$, then $B \in \mathcal{F}$.

The first property is a property of nontriviality. the second, is the *finite intersection property*. The family Cof of all subsets of $\mathbb{N}$ whose complement is finite, is an example of a filter.

(4) A filter $\mathcal{F}$ is called *free* if it contains no finite set.

The filter Cof is free. We shall be especially interested in free filters extending Cof.

(5) A filter $\mathcal{U}$ is called an *ultrafilter* if for all $E \subseteq \mathbb{N}$ we have that $E \in \mathcal{U}$ or $\mathbb{N} - E \in \mathcal{U}$.

From the nontriviality condition, we can prove that for any subset E of $\mathbb{N}$, exactly one of the sets E or $E^c = \mathbb{N} - E$ is in $\mathcal{U}$. Also observe that if a finite union $E_1 \cup \cdots \cup E_n$ is in $\mathcal{U}$ then one of the sets E_i is also in $\mathcal{U}$. The following well-known theorem can be proved using Zorn's lemma or by transfinite induction.

THEOREM A.11 (ULTRAFILTER THEOREM). *There exists free ultrafilters extending the filter* Cof *of cofinite sets.*

Ultrapowers. We can now construct $^*\mathbb{R}$. Let $\mathcal{U}$ be a free ultrafilter and introduce an equivalence relation on sequence in $^{\mathbb{N}}\mathbb{R}$ defined by

$$f \sim_\mathcal{U} g \iff \{n \in \mathbb{N} \mid f_n = g_n\} \in \mathcal{U}.$$

Sets in the ultrafilter are considered as large sets. In fact, we can define a finitely additive measure, μ, on subsets of $\mathbb{N}$ with values zero and one, such that $\mu(A) = 1$ if and only if $A \in \mathcal{U}$. Thus, the equivalence relation we have introduced identifies sequences that agree almost everywhere, with respect to this measure μ.

Since $\mathcal{U}$ is free, two sequence that agree everywhere except on a finite set are equivalent. The set of equivalence classes of $^{\mathbb{N}}\mathbb{R}$ with respect to $\sim_\mathcal{U}$ is $^*\mathbb{R}$, in symbols

$$^*\mathbb{R} = {}^{\mathbb{N}}\mathbb{R}/\mathcal{U}.$$

We identify the equivalence class of the constant sequence $f = \langle r, r, \ldots \rangle$, that is, $f_n = r$, for every $n \in \mathbb{N}$, with the real number r. Thus, we can consider $\mathbb{R} \subseteq {}^*\mathbb{R}$. Also, the sequence $g = \langle 1, 2, 3, \ldots \rangle$, that is, $g_n = n$, for every $n \in \mathbb{N}$,

---

[2] A construction of a nonstandard universe can be found in [78], [52] and [1]. We shall follow mainly the last mentioned reference.

is not equivalent to any constant sequence, thus, it is not in $\mathbb{R}$. Hence, $^*\mathbb{R}$ is a proper extension of $\mathbb{R}$.

We now construct the extension of $V(\mathbb{R})$ that we need. We consider the *bounded* sequences in $^\mathbb{N}V(\mathbb{R})$, i.e., those sequences f such that $f_n \in V_m(\mathbb{R})$, for all n, for a certain m. Similarly as before, we define the equivalence relation

$$f \sim_{\mathcal{U}} g \iff \{n \in \mathbb{N} \mid f_n = g_n\} \in \mathcal{U},$$

but now over the bounded sequences. The set

$$^\mathbb{N}V(\mathbb{R})/\mathcal{U}$$

is the set of equivalence classes. We write $X_{\mathcal{U}}$ for the equivalence class of X. We also define the relation

$$X \in_{\mathcal{U}} Y \iff \{n \in \mathbb{N} \mid X_n \in Y_n\} \in \mathcal{U}.$$

Notice that $^*\mathbb{R}$ can be considered as a subset of $^\mathbb{N}V(\mathbb{R})/\mathcal{U}$.

We interpret the bounded formulas of the language introduced earlier with $\in$ in $^\mathbb{N}V(\mathbb{R})/\mathcal{U}$ by taking $\in$ to be $\in_{\mathcal{U}}$. For any $X \in V(\mathbb{R})$, we define $i(X)$ as the equivalence class of the constant sequence in $^\mathbb{N}V(\mathbb{R})/\mathcal{U}$ with value X. Thus, i is an injection of $V(\mathbb{R})$ into $^\mathbb{N}V(\mathbb{R})/\mathcal{U}$. Notice that if $X \in {^\mathbb{N}V(\mathbb{R})/\mathcal{U}}$, since X is a bounded sequence, there is an $m \in \mathbb{N}$ such that $X \in_{\mathcal{U}} {^\mathbb{N}V_m(\mathbb{R})/\mathcal{U}}$. We have the following theorem:

THEOREM A.12 (LOS' THEOREM). *Let $\varphi(x_1,\ldots,x_n)$ be a bounded formula, $X^1, \ldots, X^n$ be bounded sequences in $^\mathbb{N}V(\mathbb{R})/\mathcal{U}$. Then $X^1, \ldots, X^n$ satisfy φ if and only if the set of $m \in \mathbb{N}$ such that $X^1_m, \ldots, X^n_m$ satisfy φ is in $\mathcal{U}$.*

PROOF. The proof is by induction on the complexity of the formula. For the case that φ is atomic, the conclusion is obtained directly from the definitions of equivalence class and $\in_{\mathcal{U}}$.

Suppose, now, that $\varphi = \neg\psi$ and assume the theorem is true for ψ. Then we have

f satisfies φ in $^\mathbb{N}V(\mathbb{R})/\mathcal{U} \iff f$ does not satisfy ψ in $^\mathbb{N}V(\mathbb{R})/\mathcal{U}$
$\iff \{n \in \mathbb{N} \mid f_n$ satisfies ψ in $V(\mathbb{R})\} \notin \mathcal{U};$

thus, since $\mathcal{U}$ is an ultrafilter

$\iff \{n \in \mathbb{N} \mid f_n$ does not satisfy ψ in $V(\mathbb{R})\} \in \mathcal{U}$
$\iff \{n \in \mathbb{N} \mid f_n$ satisfies φ in $V(\mathbb{R})\} \in \mathcal{U}$

The case when φ is a conjunction is similar, using, now, the finite intersection property.

Consider, finally, φ to be $\exists y \in x_1 \psi(y, x_2, \ldots, x_n)$, and assume the theorem proved for ψ. We have that $X^1, \ldots, X^n$ satisfy φ in $^\mathbb{N}V(\mathbb{R})/\mathcal{U}$ if and only if there is a $Y \in_{\mathcal{U}} X_1$ such that $Y, X^2, \ldots, X^n$ satisfy ψ. This is so, if and only if the set of $m \in \mathbb{N}$ such that $Y_m, X^2_m, \ldots, X^n_m$ satisfy ψ in $V(\mathbb{R})$ is in $\mathcal{U}$.

Suppose first that $X^1, \ldots, X^n$ satisfy φ in $^\mathbb{N}V(\mathbb{R})/\mathcal{U}$. Then, by the assertion in the previous paragraph, the set of m such that there is a $y \in X_m^1$ such that y, $X_m^2, \ldots, X_m^n$ satisfy ψ is in $\mathcal{U}$. Therefore, the set of m such that $X_m^1, \ldots, X_m^n$ satisfy φ is in $\mathcal{U}$.

Suppose now that the set of m such that $X_m^1, \ldots, X_m^n$ satisfy φ is in $\mathcal{U}$. Then, the set of m such that there is a $y \in X_m^1$ such that $y, X_m^2, \ldots, X_m^n$ satisfy ψ is in $\mathcal{U}$. For each m in this set, choose Y_m equal to such a y, and if m is not in this set, choose Y_m arbitrarily in X_m^1. Since X^1 is bounded, the sequence Y is also bounded, and, we have, $Y \in_\mathcal{U} X^1$. It is clear that $Y, X^2, \ldots, X^n$ satisfy ψ. □

As an immediate corollary, we get:

COROLLARY A.13. *Let $\varphi(x_1, \ldots, x_n)$ be a bounded formula and $X_1, \ldots, X_n \in V(\mathbb{R})$. Then $\varphi(X_1, \ldots, X_n)$ is true in $V(\mathbb{R})$ if and only if $\varphi(i(X_1), \ldots, i(X_n))$ is true in $^\mathbb{N}V(\mathbb{R})/\mathcal{U}$.*

We now define an injection j of $^\mathbb{N}V(\mathbb{R})/\mathcal{U}$ into $V(^*\mathbb{R})$, called *Mostowski's collapsing map*. If $X \in {^*\mathbb{R}} = {^\mathbb{N}V_0(\mathbb{R})}/\mathcal{U}$, then $j(X) = X$. Suppose, now, that j is defined for $Y \in {^\mathbb{N}V_m(\mathbb{R})}/\mathcal{U}$ and assume X to be in $^\mathbb{N}V_{m+1}(\mathbb{R})/\mathcal{U}$. Then

$$j(X) = \{j(Y) \mid Y \in_\mathcal{U} X\}.$$

The function j is well defined, since if $Y \in_\mathcal{U} X$, then $Y \in_\mathcal{U} {^\mathbb{N}V_m(\mathbb{R})}/\mathcal{U}$.

It is clear that j is a function from $^\mathbb{N}V(\mathbb{R})/\mathcal{U}$ into $V(^*\mathbb{R})$. For $X \in V(\mathbb{R})$, we define

$$^*X = j(i(X)).$$

We have the following theorem, Mostowski's collapsing theorem:

THEOREM A.14 (MOSTOWSKI). *Let $X^1, \ldots, X^n \in {^\mathbb{N}V_m(\mathbb{R})}/\mathcal{U}$ and let*

$$\varphi(x_1, \ldots, x_n)$$

be a bounded formula. Then $\varphi(X^1, \ldots, X^n)$ is true in $^\mathbb{N}V(\mathbb{R})/\mathcal{U}$ if and only if $\varphi(j(X^1), \ldots, j(X^n))$ is true in $V(^*\mathbb{R})$.

PROOF. The proof is an easy induction on the complexity of the formula φ. The case of atomic formulas is obtained from the definitions. Negations and conjunctions are easily obtained. So assume that φ is $\exists y \in x_1 \psi$. Then $\exists y \in X^1 \psi$ is true in $^\mathbb{N}V(\mathbb{R})/\mathcal{U}$ if and only if there is a $y \in_\mathcal{U} X^1$ such that y satisfies ψ. We have that $y \in_\mathcal{U} X^1$ if and only if $j(y) \in j(X^1)$. So the conclusion follows. □

We are now able to prove the transfer principle:

THEOREM A.15 (TRANSFER PRINCIPLE). *Let $X_1, \ldots, X_n \in V(\mathbb{R})$ and let*

$$\varphi(x_1, \ldots, x_n)$$

be a bounded formula. Then

$$\varphi(X_1, \ldots, X_n)$$

is true in $V(\mathbb{R})$ if and only if
$$\varphi(^*X_1, \ldots, {}^*X_n)$$
is true in $V(^*\mathbb{R})$.

PROOF. By the previous theorems, we have that
$$\varphi(X_1, \ldots, X_n)$$
is true in $V(\mathbb{R})$ if and only if
$$\varphi(i(X_1), \ldots, i(X_n))$$
is true in ${}^{\mathbb{N}}V(\mathbb{R})/\mathcal{U}$, if and only if
$$\varphi(j(i(X_1)), \ldots, j(i(X_n)))$$
is true in $V(^*\mathbb{R})$. □

It is clear that the Extension Principle is true in $V(^*\mathbb{R})$. We now prove Axioms 8, countable saturation. In order to do this, we notice that X is internal if $X \in {}^*Y$ for a certain $Y \in V(\mathbb{R})$. Thus, since $^*Y = j(i(Y))$, X must be in the image of ${}^{\mathbb{N}}V(\mathbb{R})/\mathcal{U}$ by j.

We need two more facts about internal sets. In the first place, since $V_m(\mathbb{R})$ is transitive, i.e., $x \in y \in V_m(\mathbb{R})$ implies $x \in V_m(\mathbb{R})$, by transfer, $^*V_m(\mathbb{R})$ is also transitive. Second, we have that if X is internal, then $X \in {}^*V_m(\mathbb{R})$, for a certain m. This is so, because if X is internal, then $X \in {}^*Y$ for a certain $Y \in V_m(\mathbb{R})$, for some m. Hence, by transfer, $^*Y \in {}^*V_m(\mathbb{R})$, and since $^*V_m(\mathbb{R})$ is transitive, $X \in {}^*V_m(\mathbb{R})$.

THEOREM A.16 (SATURATION). *Let $A_1 \supseteq A_2 \supseteq \cdots \supseteq A_k \supseteq \ldots$ be a countable decreasing sequence of nonempty internal sets, elements of an internal set, in $V(^*\mathbb{R})$. Then*
$$\bigcap_{k=1}^{\infty} A_k \neq \emptyset.$$

PROOF. Each A_k is internal, and thus of the form $j(A'_k)$ where
$$A'_k = \langle A_1^k, A_2^k, \ldots \rangle_{\mathcal{U}}.$$
Since all A_k are elements of an internal set A, and $A \in {}^*V_m(\mathbb{R})$, for a certain m, we have, as $^*V_m(\mathbb{R})$ is transitive, that all $A_i^k \subseteq V_m(\mathbb{R})$, for a certain fixed m. For convenience, we adjoin $A'_0 = \langle V_m(\mathbb{R}), V_m(\mathbb{R}), \ldots \rangle_{\mathcal{U}}$. Define, for $k \geq 0$
$$I_k = \{i \geq k \mid A_i^0 \supseteq A_i^1 \supseteq \ldots \supseteq A_i^k \neq \emptyset\}.$$
We see that: $I_0 = \mathbb{N}$; $I_k \in \mathcal{U}$; and $\bigcap_{k=0}^{\infty} I_k = \emptyset$. For each $i \in \mathbb{N}$, let
$$m(i) = \max\{m \mid i \in I_m\}.$$
Since $I_0 = \mathbb{N}$ and $\bigcap_{m=0}^{\infty} I_m = \emptyset$, $m(i)$ is well defined. Let B_i be some element of $A_i^{m(i)}$. We shall prove that $B = \langle B_1, B_2, \ldots \rangle_{\mathcal{U}} \in A'_k$, for all $k \geq 0$. We have that $B \in_{\mathcal{U}} A'_k$ if and only if $\{i \in \mathbb{N} \mid B_i \in A_i^k\} \in \mathcal{U}$. But we have that

$I_k \subseteq \{i \in \mathbb{N} \mid B_i \in A_i^k\}$, because $i \in I_k$ implies $m(i) \geq k$. Since $I_k \in \mathcal{U}$, we obtain that $\{i \in \mathbb{N} \mid B_i \in A_i^k\} \in \mathcal{U}$.

Since $B \in_\mathcal{U} A'_k$, we have that $j(B) \in j(A'_k) = A_k$, for every $k \geq 0$. Thus, $\bigcap_{k=1}^\infty A_k \neq \emptyset$. □

APPENDIX B

Extensions of integrals and measures

1. Internal abstract theory of integration

In this appendix, I discuss an abstract nonstandard theory of integration derived from the Daniell integral. It is known that in any general theory of integration one should be able to prove some form of the *monotone convergence theorem*, so that the culmination of the appendix will be a proof of a nonstandard version of this theorem.

Although the least upper bound principle is not true in $^*\mathbb{R}$, we need the following approximate version. We say that s is a *near least upper bound (greatest lower bound)* of a set $A \subset {}^*\mathbb{R}$, if

(1) $s \gtrsim x$ ($s \lesssim x$), for every $x \in A$.
(2) If $x \lesssim b$ ($x \gtrsim b$) for every $x \in A$, then $s \lesssim b$ ($s \gtrsim b$).

We say that a set $A \subset {}^*\mathbb{R}$ is *S-bounded from above (below)* if there is a finite $M \in {}^*\mathbb{R}$ such that $M \geq x$ ($M \leq x$) for every $x \in A$.

THEOREM B.1. *Let $A \subset {}^*\mathbb{R}$ be S-bounded from above (below). Then A has a real near least upper bound (greatest lower bound).*

PROOF. We prove the theorem for A S-bounded from above. Let

$$B = \{r \in \mathbb{R} \mid x \gtrsim r, \text{ for some } x \in A\}.$$

Then B is a set of real numbers bounded from above and hence it has a least upper bound $s \in \mathbb{R}$.

Suppose that there is an $x \in A$ such that $x \not\lesssim s$. Then $s \ll x$, and hence there is a real r, $s < r \ll x$. But this contradicts the fact that s is the least upper bound of B.

Assume, now, that $s \not\lesssim b$ for a certain b. Then $b \ll s$ and hence, $b \ll r < s$ for a certain real r. Thus, $b \ll r \lesssim x$, for some $x \in A$. That is, $x \not\lesssim b$ for a certain $x \in A$. □

We now begin the construction of an internal analogue of the definition of the Daniell integral, adapted to the case of probability spaces, which is inspired by the standard version given in [95, Chapter 6].[1]

We start with an algebra of random variables from an internal (in general, hyperfinite) set Ω into $^*\mathbb{R}$. We write $X \approx Y$, if $X(\omega) \approx Y(\omega)$ for every $\omega \in \Omega$. In particular, $X \approx 0$, or X is *infinitesimal*, if $X(\omega)$ is infinitesimal for every $\omega \in \Omega$. We also write $X \lesssim Y$ for $X(\omega) \lesssim Y(\omega)$ for every $\omega \in \Omega$.

An algebra of random variable $\mathcal{F}$ over a probability space $\langle \Omega, \text{pr} \rangle$ is a vector lattice of functions on Ω to $^*\mathbb{R}$, i.e., a nonempty class of functions on Ω to $^*\mathbb{R}$ such that $X + Y$, $X \vee Y$, $X \wedge Y$ and cX are in $\mathcal{F}$, whenever $X, Y \in \mathcal{F}$ and $c \in {^*\mathbb{R}}$. Here, we define for $\omega \in \Omega$

$$(X \vee Y)(\omega) = \max\{X(\omega), Y(\omega)\},$$

$$(X \wedge Y)(\omega) = \min\{X(\omega), Y(\omega)\}.$$

For any X, we define $X^+ = X \vee 0$ and $X^- = -(X \wedge 0)$. Then $X = X^+ - X^-$ and $|X| = X^+ + X^-$. It is easy to see that instead of requiring that $\mathcal{F}$ be closed under the lattice operations, we may ask that $|X| \in \mathcal{F}$, if $X \in \mathcal{F}$.

We assume, as usual, that our vector lattice, the algebra of random variables $\mathcal{F}$, and, hence, every $X \in \mathcal{F}$, are internal. By internal induction, we can show that $\mathcal{F}$ is closed under the hyperfinite versions of addition and the lattice operations. For instance, if $X_1, X_2, \ldots, X_\nu$ are in $\mathcal{F}$ for a $\nu \in {^*\mathbb{N}}$, finite or infinite, $X_1 + X_2 + \cdots + X_\nu \in \mathcal{F}$.

Let $\mathbf{E} : \mathcal{F} \to {^*\mathbb{R}}$ be the expectation defined on the space $\langle \Omega, \text{pr} \rangle$. We have that if $X \approx 0$, then, since Ω is hyperfinite, $|X| \leq \varepsilon \approx 0$, where ε is the maximum of $\Lambda_{|X|}$. Hence

$$|\mathbf{E} X| \leq \mathbf{E}|X|$$
$$= \sum_{\omega \in \Omega} |X(\omega)| \text{pr}(\omega)$$
$$\leq \varepsilon \sum_{\omega \in \Omega} \text{pr}(\omega)$$
$$= \varepsilon$$
$$\approx 0.$$

The internal expectation mapping $\mathbf{E} : \mathcal{F} \to {^*\mathbb{R}}$ is called an *elementary integral*, because $\mathbf{E}$ satisfies:

(1) $\mathbf{E}(X + Y) = \mathbf{E}(X) + \mathbf{E}(Y)$.
(2) $\mathbf{E}(cX) = c\,\mathbf{E}(X)$, if $c \in {^*\mathbb{R}}$.
(3) $\mathbf{E}(X) \geq 0$, if $X \geq 0$.
(4) $X \approx 0 \implies \mathbf{E} X \approx 0$.

---

[1] A more general version appears in [22].

1. INTERNAL ABSTRACT THEORY OF INTEGRATION

We can easily check that the algebra of all random variables over a probability space $\langle \Omega, \mathrm{pr} \rangle$ (i.e., the class of all internal functions from Ω to $^*\mathbb{R}$) is a vector lattice of functions on Ω to $^*\mathbb{R}$, and $\mathbf{E}$ is an elementary integral on this vector lattice.

We have:

THEOREM B.2. *If $X, Y \in \mathcal{F}$, and $X \lesssim Y$, then $\mathbf{E}(X) \lesssim \mathbf{E}(Y)$.*

PROOF. We have, in general

$$X \leq Y + ((X - Y) \vee 0). \tag{1}$$

Also, since $X \lesssim Y$

$$(X - Y) \vee 0 \approx 0. \tag{2}$$

Thus, by (2)

$$\mathbf{E}((X - Y) \vee 0) \approx 0.$$

By (1), $\mathbf{E}(X) \leq \mathbf{E}(Y) + \mathbf{E}((X - Y) \vee 0)$. Hence, we obtain the conclusion of the theorem. □

We now extend the elementary integrals, $\mathbf{E}$, to an integral on a wider class. We define the class $\mathcal{L}$ as the set of all functions (internal or external) $X : \Omega \to {}^*\mathbb{R}$, that satisfy:

(1) The set $B = \{\mathbf{E}(Y) \mid Y \in \mathcal{F}, Y \gtrsim X\}$ is S-bounded from below and the set $C = \{\mathbf{E}(Z) \mid Z \in \mathcal{F}, Z \lesssim X\}$ is S-bounded from above.
(2) If a is a near greatest lower bound of B and b a near least upper bound of C, then $a \approx b$.

The number a is called an *upper near integral*, and b a *lower near integral* of X. If $a \approx b$, then either of these numbers is called a *near integral*, or *near expectation* of X.

The following properties are easy to check.

THEOREM B.3. *Let X be any function on Ω. Then*

(1) *If $B = \{\mathbf{E}(Y) \mid Y \in \mathcal{F}, Y \gtrsim X\}$ is S-bounded from below and the set $C = \{\mathbf{E}(Z) \mid Z \in \mathcal{F}, Z \lesssim X\}$ is S-bounded from above, a is a near g.l.b. of B and b a near l.u.b. of C, then $a \gtrsim b$.*
(2) *If a and b are near expectations of X, then $a \approx b$.*

PROOF. We prove (1). Let $Y \in B$ and $Z \in C$. Then $Y \gtrsim Z$, and, hence, by Theorem B.2, $\mathbf{E}(Y) \gtrsim \mathbf{E}(Z)$.
The proof of 2 is easy. □

We have the following characterizations of functions in $\mathcal{L}$.

THEOREM B.4.

(1) *$X \in \mathcal{L}$ if and only if for every $\varepsilon \gg 0$ there are $Z, Y \in \mathcal{F}$ such that $Z \lesssim X \lesssim Y$, $\mathbf{E}(Z)$ is finite, and $|\mathbf{E}(Y) - \mathbf{E}(Z)| < \varepsilon$.*
(2) *If X is internal, then $X \in \mathcal{L}$ if and only if there are $Z, Y \in \mathcal{F}$ such that $Z \lesssim X \lesssim Y$, $\mathbf{E}(Z)$ is finite, and $\mathbf{E}(Z) \approx \mathbf{E}(Y)$.*

PROOF OF (1). Suppose first that $X \in \mathcal{L}$, a is an upper near integral of X, and b is a lower near integral of X, and let $\varepsilon \gg 0$. Then $a \approx b$. There are Y and $Z \in \mathcal{F}$ such that $Z \lesssim X \lesssim Y$, $|\mathbf{E}(Y) - a| < \varepsilon/3$, and $|\mathbf{E}(Z) - b| < \varepsilon/3$. We have
$$|\mathbf{E}(Y) - \mathbf{E}(Z)| \leq |\mathbf{E}(Y) - a| + |a - b| + |b - \mathbf{E}(Z)| < \varepsilon.$$

Suppose, now, that for every $\varepsilon \gg 0$ there are $Z, Y \in \mathcal{F}$ such that $Z \lesssim X \lesssim Y$, $\mathbf{E}(Z)$ is finite, and $|\mathbf{E}(Y) - \mathbf{E}(Z)| \leq \varepsilon$. Let $\varepsilon \gg 0$ and find Y and Z as in the condition. Then $\mathbf{E}(Y)$ is also finite. The set $B = \{\mathbf{E}(V) \mid V \in \mathcal{F}, V \gtrsim X\}$ is bounded below by $\mathbf{E}(Z) - \varepsilon$. Similarly, the set $C = \{\mathbf{E}(V) \mid V \in \mathcal{F}, V \lesssim X\}$ is bounded above by $\mathbf{E}(Y) + \varepsilon$. Let b be a near l.u.b. of C and a a near g.l.b. of B. We know that $b \lesssim a$. We have that $\mathbf{E}(Y) < \mathbf{E}(Z) + \varepsilon$. Then $a < b + \varepsilon$. Since this is true for every $\varepsilon \gg 0$, $a \lesssim b$. Therefore, $a \approx b$. □

PROOF OF (2). Suppose first that $f \in \mathcal{L}$, a is an upper near integral of X, and b is a lower near integral of X. Then, for each $n \in \mathbb{N}$, there are Z_n and $Y_n \in \mathcal{F}$ such that
$$\mathbf{E}(Y_n) \leq a + \frac{1}{n}, \quad b \leq \mathbf{E}(Z_n) + \frac{1}{n}, \quad Z_n \lesssim X \lesssim Y_n.$$

By denumerable comprehension, extend the sequences Z_n, Y_n. The internal (because X is internal) set A of the $n \in {}^*\mathbb{N}$ that satisfy

(1) $Y_n, Z_n \in \mathcal{F}$,
(2) for all $k \leq n$, $Y_k \geq X - \frac{1}{k}$ and $\mathbf{E}(Y_k) \leq a + \frac{1}{k}$,
(3) for all $k \leq n$, $Z_k \leq f + \frac{1}{k}$, and $b \leq \mathbf{E}(Z_k) + \frac{1}{k}$.

contains all finite natural numbers, and hence it contains an infinite number ν. Then, for this ν, we have

(1) $Z_\nu, Y_\nu \in \mathcal{F}$.
(2) $Z_\nu \leq X + 1/\nu$ and $Y_\nu \geq X - 1/\nu$. Hence, $Z_\nu \lesssim X \lesssim Y_\nu$.
(3) $\mathbf{E}(Z_\nu) \geq b - 1/\nu$ and $\mathbf{E}(Y_\nu) \leq a + 1/\nu$. Thus, $\mathbf{E}(Z_\nu) \approx b$, $\mathbf{E}(Y_\nu) \approx a$, and thus, $\mathbf{E}(Z_\nu)$ is finite. Since $a \approx b$, we get that $\mathbf{E}(Y_\nu) \approx \mathbf{E}(Z_\nu)$.

Thus, Z_ν and Y_ν are the required Z and Y.

Suppose, now, that $Z, Y \in \mathcal{F}$ with $Z \lesssim X \lesssim Y$, $\mathbf{E}(Z)$ finite, and $\mathbf{E}(Y) \approx \mathbf{E}(Z)$. Let a be a near upper integral of X and b a near lower integral. Then $\mathbf{E}(Z) \lesssim b$ and $\mathbf{E}(Y) \gtrsim a$. Hence $a \approx b$, and a and b are finite. □

The equivalence given in (2) of the last theorem is not true for external functions, as the following example shows. Let $\mathcal{F}$ be the set of random variables defined on $\langle \Omega, \mathrm{pr} \rangle$, where Ω is an equally spaced near interval for $[0, 1]$, and pr is the uniform distribution. Consider the external function $X : \Omega \to {}^*\mathbb{R}$ defined by

$$X(\omega) = \begin{cases} 1, & \text{if } \omega \not\approx 1, \\ 0, & \text{if } \omega \approx 1. \end{cases}$$

It is easy to show that, according to the definition, 1 is a near integral of X. On the other hand, if Z is any (internal) function in $\mathcal{F}$ such that $Z \lesssim X$, then there is an $n \in \mathbb{N}$ such that $Z(\omega) \lesssim 0$ for all $\omega \geq 1 - 1/n$, and, hence, $\mathbf{E}(Z) \ll 1$. Therefore, there is no $Z \in \mathcal{F}$, $Z \lesssim X$, such that $\mathbf{E}(Z)$ is a near integral of X.

1. INTERNAL ABSTRACT THEORY OF INTEGRATION

THEOREM B.5.

(1) $\mathcal{L}$ is a real vector lattice.
(2) Let $X, Y \in \mathcal{L}$ and c a finite number. Then, if a is a near integral of X and b is a near integral of Y, then $a + b$ is a near integral of $X + Y$, and ca is a near integral of cX.

The class $\mathcal{L}$ may not be a hiperreal vector lattice, because X may belong to $\mathcal{L}$, while, if $c \approx \infty$, cX may be out of $\mathcal{L}$.

PROOF. The proof that $\mathcal{L}$ is closed under addition and multiplication by a finite number is simple, using B.4. We now prove that $X \in \mathcal{L}$ implies $|f| \in \mathcal{L}$. Let $\varepsilon \gg 0$. Find $Z, Y \in \mathcal{F}$ such that

$$Z \lesssim X \lesssim Y \quad \text{and} \quad |\mathbf{E}(Y) - \mathbf{E}(Z)| < \frac{\varepsilon}{2}.$$

Then

$$Z^+ \lesssim X^+ \lesssim Y^+ \quad \text{and} \quad Y^- \lesssim X^- \lesssim Z^-.$$

By Theorem B.2

$$\mathbf{E}(Z^+) \lesssim \mathbf{E}(Y^+) \quad \text{and} \quad \mathbf{E}(Y^-) \lesssim \mathbf{E}(Z^-). \tag{1}$$

We also have

$$\mathbf{E}(Y^+) - \mathbf{E}(Y^-) = \mathbf{E}(Y) < \mathbf{E}(Z) + \varepsilon = \mathbf{E}(Z^+) - \mathbf{E}(Z^-) + \varepsilon. \tag{2}$$

Then

$$\mathbf{E}(Y^+) + \mathbf{E}(Z^-) < \mathbf{E}(Z^+) + \mathbf{E}(Y^-) + \varepsilon.$$

We now prove by contradiction that $\mathbf{E}(Y^+) < \mathbf{E}(Z^+) + \varepsilon$ and $\mathbf{E}(Z^-) < \mathbf{E}(Y^-) + \varepsilon$. Suppose that $\mathbf{E}(Y^+) \geq \mathbf{E}(Z^+) + \varepsilon$. Then, by (1), $\mathbf{E}(Y^+) + \mathbf{E}(Z^-) \geq \mathbf{E}(Z^+) + \mathbf{E}(Y^-) + \varepsilon$, contradicting (2). The other case is similar.

Thus, we proved that $X^+, X^- \in \mathcal{L}$, proving that $|X| \in \mathcal{L}$.
The proof of 2 is left to the reader. □

We also have the following theorem that is obvious from the definitions.

THEOREM B.6. If $X \in \mathcal{L}$ and $Y \approx X$, then $Y \in \mathcal{L}$.

We now turn to an internal version of the *monotone convergence theorem*. We assume the construction of $\mathcal{L}$, with $\mathcal{F}$ internal as above.

THEOREM B.7 (MONOTONE CONVERGENCE THEOREM). *Suppose that X_1, X_2, ..., X_n, ..., for $n \in \mathbb{N}$, is an increasing sequence of functions in $\mathcal{L}$ which S-converges to a function X on Ω. Suppose, also, that a_n is a near integral of X_n for every $n \in \mathbb{N}$, that $\{a_n \mid n \in \mathbb{N}\}$ is S-bounded, and that a, finite, is a near least upper bound of $\{a_n \mid n \in \mathbb{N}\}$. Then $X \in \mathcal{L}$ and a is a near integral of X.*

PROOF. Suppose that the sequence $\{X_n\}$ is in $\mathcal{L}$ and let $\varepsilon \gg 0$ be given. There are $Z_n, Y_n \in \mathcal{F}$ such that $Z_n \lesssim X_n \lesssim Y_n$ and $|\mathbf{E}(Y_n) - \mathbf{E}(Z_n)| < \varepsilon/2^{n+1}$, for every $n \in \mathbb{N}$. We can extend the sequences Z_n, Y_n to internal sequences such that $Z_n, Y_n \in \mathcal{F}$ for every $n \leq \nu_1 \approx \infty$. Define

$$V_n = Z_1 \vee Z_2 \vee \cdots \vee Z_n$$

$$U_n = Y_1 \vee Y_2 \vee \cdots \vee Y_n$$

for every $n \leq \nu_1$. Then $V_n, U_n \in \mathcal{F}$. We have:

(1) V_n and U_n are increasing sequences.
(2) $V_n \lesssim X_n \leq X$ and $\mathbf{E}(V_n)$ is finite, for every $n \in \mathbb{N}$.
(3) $X_n \lesssim U_n \leq U_\mu$ for every $n \in \mathbb{N}$ and every infinite $\mu \leq \nu_1$. Let $\delta \gg 0$ and $\omega \in \Omega$. Then, there is an $n_0 \in \mathbb{N}$ such that $X(\omega) < X_n(\omega) + \delta$ for every $n \geq n_0$, $n \in \mathbb{N}$. Therefore, $X(\omega) < U_n(\omega) + \delta$. Thus, $X(\omega) < U_\mu(\omega) + \delta$, for every $\delta \gg 0$. Hence, $X \lesssim U_\mu$, for every infinite $\mu \leq \nu_1$.
(4) For a given $\omega \in \Omega$, $U_n(\omega) = Y_i(\omega)$, for some $i \leq n$. For this i and ω, $V_n(\omega) \geq Z_i(\omega)$ and so $U_n(\omega) - V_n(\omega) \leq Y_i(\omega) - Z_i(\omega)$. But $Y_j(\omega) - Z_j(\omega) \gtrsim 0$, for every j and ω, and so we see that

$$0 \lesssim U_n - V_n \lesssim (Y_1 - Z_1) + \cdots + (Y_n - Z_n)$$

for every $n \in \mathbb{N}$. It follows, by Theorem B.2, that

$$0 \lesssim \mathbf{E}(U_n) - \mathbf{E}(V_n) \lesssim \sum_{j=1}^{n} \mathbf{E}(Y_j) - \mathbf{E}(Z_j)$$

for every $n \in \mathbb{N}$. If $\mathbf{E}(U_n) - \mathbf{E}(V_n) < 0$, then $\mathbf{E}(U_n) - \mathbf{E}(V_n) \approx 0$, and hence

$$|\mathbf{E}(U_n) - \mathbf{E}(V_n)| \lesssim |\sum_{j=1}^{n} \mathbf{E}(Y_j) - \mathbf{E}(Z_j)|. \qquad (1)$$

Also, if $\sum_{j=1}^{n} \mathbf{E}(Y_j) - \mathbf{E}(Z_j) < 0$, then $\mathbf{E}(U_n) - \mathbf{E}(V_n) \approx 0$, and hence, we have (1). In the other cases, (1) is clear, so we have in general

$$|\mathbf{E}(U_n) - \mathbf{E}(V_n)| \lesssim \sum_{j=1}^{n} |\mathbf{E}(Y_j) - \mathbf{E}(Z_j)| < \sum_{j=1}^{n} \frac{\varepsilon}{2^{j+1}} < \frac{\varepsilon}{2}$$

for every finite n.

Thus, by overflow, Theorem VIII.7, there is an infinite $\nu_2 \leq \nu_1$ such that

$$|\mathbf{E}(U_\mu) - \mathbf{E}(V_\mu)| < \frac{\varepsilon}{2}$$

for every $\mu \leq \nu_2$.

(5) Since $\mathbf{E}(V_n) \lesssim a_n$, for every $n \in \mathbb{N}$, $\{\mathbf{E}(V_n)\}$ is a bounded increasing sequence and, hence, it S-converges. By Theorem VIII.11, it S-converges to $\mathbf{E}(V_\nu)$, for some infinite $\nu \leq \nu_2$. Thus, there is a finite n such that $|\mathbf{E}(V_\nu) - \mathbf{E}(V_n)| < \varepsilon/2$.

(6) By (4) and (5), we have

$$|\mathbf{E}(U_\nu) - \mathbf{E}(V_n)| \leq |\mathbf{E}(U_\nu) - \mathbf{E}(V_\nu)| + |\mathbf{E}(V_\nu) - \mathbf{E}(V_n)| < \varepsilon.$$

Thus, we have found U_ν, $V_n \in \mathcal{F}$, such that $V_n \lesssim X \lesssim U_\nu$, $\mathbf{E}(V_n)$ is finite, and $|\mathbf{E}(U_\nu) - \mathbf{E}(V_n)| < \varepsilon$. This proves that $X \in \mathcal{L}$. □

2. Null sets

In Definition XI.1, we defined null sets as (internal or external) subsets A of Ω such that for every $\varepsilon \gg 0$ there is an internal set $B \supseteq A$ such that $\Pr B < \varepsilon$. We now give a characterization with the extended measure.

In order to do this, we extend the notion of probability of a set to external sets. We fix the probability space $\langle \Omega, \mathrm{pr} \rangle$. We say that a is a *near probability* of the set $A \subseteq \Omega$ (internal or external), if a is a near integral of the characteristic function of A, χ_A. We have the following characterization of null sets:

THEOREM B.8. *A subset A of Ω is a null set if and only if zero is a near probability of A.*

PROOF. Suppose, first, that A is a null set. Then, for every $\varepsilon \gg 0$, there is an internal B, such that $\chi_B \geq \chi_A \geq 0$ and $\mathbf{E}(\chi_B) < \varepsilon$. Since $\chi_B \in \mathcal{F}$, and the constant function zero is also in $\mathcal{F}$, we have, by Theorem B.4, that $\chi_A \in \mathcal{L}$. Since a near integral a of χ_A satisfies $0 \lesssim a \lesssim \varepsilon$, for every $\varepsilon \gg 0$, we have that $a \approx 0$.

Suppose, now, that 0 is a near probability of A, and let $\varepsilon \gg 0$ be given. By Theorem B.4, there are Y and $Z \in \mathcal{F}$ such that $Z \lesssim \chi_A \lesssim Y$, $\mathbf{E}(Z)$ is finite, and $|\mathbf{E}(Y) - \mathbf{E}(Z)| < \varepsilon/3$. We also have that $\mathbf{E}(Z) \lesssim 0 \lesssim \mathbf{E}(Y)$, and, hence, $\mathbf{E}(Z)$ is infinitesimal or negative, and $\mathbf{E}(Y)$ is infinitesimal or positive. Thus

$$\mathbf{E}(Y) \leq \frac{\varepsilon}{3} + \mathbf{E}(Z) \leq \frac{\varepsilon}{2}.$$

Let B be the internal set of all $\omega \in \Omega$ such that $Y(\omega) + \varepsilon/2 \geq 1$. Then $B \supseteq A$. We have that, if $\omega \in B$

$$1 = \chi_B(\omega) = ((Y(\omega) + \frac{\varepsilon}{2}) \wedge 1) \leq Y(\omega) + \varepsilon/2,$$

and, if $\omega \notin B$, since $Y(\omega)$ is infinitesimal or positive

$$0 = \chi_B(\omega) \leq Y(\omega) + \frac{\varepsilon}{2}.$$

Thus

$$\Pr B = \mathbf{E}(\chi_B) \leq \mathbf{E}(Y) + \frac{\varepsilon}{2} \leq \varepsilon.$$

□

3. Real valued probability measures

Finally, I include a few words about how to obtain a real valued probability measure from our hyperfinite spaces. This construction is similar to Loeb's construction of a measure as it appears in [45]. For $X \in \mathcal{L}$, we define

$$^\circ\mathbf{E}X = \operatorname{st} a$$

where a is a near integral of X. By the monotone convergence theorem, $^\circ\mathbf{E}$ has the properties of an integral.

Let $\mathcal{A}$ be the family of subsets A of Ω such that $\chi_A \in \mathcal{L}$, and, for $A \in \mathcal{A}$, let

$$^\circ\operatorname{Pr} A = {}^\circ\mathbf{E}(\chi_A).$$

Then it is not difficult to check that $\mathcal{A}$ is a σ-algebra of subsets of Ω and $^\circ\operatorname{Pr}$ is a σ-additive probability measure on $\mathcal{A}$. Thus, $\langle \Omega, \mathcal{A}, {}^\circ\operatorname{Pr}\rangle$ is a probability space in the usual, standard, sense of the term, but defined over a nonstandard set Ω and algebra, $\mathcal{A}$.

We shall now define a standard measure over a standard space. Let Ω be an S-bounded set of hyperreal numbers and let $S = \operatorname{st}''\Omega$, the image of Ω under st. We define the algebra $\mathcal{B}$ of subsets of S, by

$$\mathcal{B} = \{\operatorname{st}''A \mid A \in \mathcal{A}\},$$

and the probability measure

$$^{\circ\circ}\operatorname{Pr}(\operatorname{st}''A) = {}^\circ\operatorname{Pr} A,$$

for $A \in \mathcal{A}$ (and, hence, $\operatorname{st}''A \in \mathcal{B}$). Then, it is easy to show, that $\mathcal{B}$ is a σ-algebra of subsets of S, which is a subset of $\mathbb{R}$, and $^{\circ\circ}\operatorname{Pr}$ is a σ-additive probability measure on $\mathcal{B}$. Thus, $\langle S, \mathcal{B}, {}^{\circ\circ}\operatorname{Pr}\rangle$ is a probability space in the usual sense of the term.

The construction indicated at the end of Section 4 in Chapter XVI for Brownian motion can be carried out using the methods of this appendix.

Bibliography

[1] S. Albeverio, J. E. Fenstad, R. Høegh-Krohn, and T. Lindstrøm, *Nonstandard methods in stochastic analysis and mathematical physics*, Academic Press, New York, 1986.

[2] R. M. Anderson, *A nonstandard representation of Brownian motion and Itô integration*, Israel J. Math. **25** (1976), 15–46.

[3] A. Arnauld and P. Nicole, *La logique, ou l'art de penser*, 1643, The fifth edition has been translated into English by P. and J. Dickoff, New York, 1964.

[4] V. Barnett, *Comparative statistical inference*, John Wiley & Sons, Chichester, New York, Brisbane, Toronto, Singapore, second ed., 1982.

[5] J. Bernoulli, *Ars conjectandi*, 1713, There is an English translation of Part IV by Bing Sung, Harvard University Department of Statistics Technical Reports 2, 1966.

[6] A. Birnbaum, *Concepts of statistical evidence*, Philosophy, Science, and Method (Morgenbesser, Suppes, and White, eds.), St. Martin Press, New York, 1969, pp. 112–141.

[7] E. Borel, *Elements of the theory of probability*, Prentice-Hall, Englewood Cliffs, NJ, 1950, English translation by Freund of the French edition of 1950.

[8] M. Bunge, *Treatise on basic philosophy*, vol. III, Ontology, D. Reidel, Dordrecht, 1977.

[9] R. Carnap, *Logical foundations of probability*, The University of Chicago Press, 1950.

[10] _____, *A basic system of inductive logic*, Studies in Inductive Logic and Probability, Volume I (R. Carnap and R. C. Jeffrey, eds.), University of California Press, Berkeley and Los Angeles, 1970.

[11] C. C. Chang and H. J. Keisler, *Model theory*, Studies in logic and the foundations of mathematics, North-Holland Publishing Co., Amsterdam, 1973.

[12] R. Chuaqui, *Nonstandard analysis without transfer*, To appear in the Proc. of the First Iranian Congress of Logic.

[13] _____, *A semantical definition of probability*, Non-Classical Logics, Model Theory and Computability. Proc. of the Third Latin American Symposium on Mathematical Logic (Campinas, Brasil, 1976) (A. I. Arruda, N. C. A. da Costa, and R. Chuaqui, eds.), Volume 89 of Studies in Logic and the Foundations of Mathematics, North-Holland Pub. Co., Amsterdam, 1977, pp. 135–168.

[14] _____, *Simple cardinal algebras and invariant measures*, Notas Matemáticas, Universidad Católica de Chile **6** (1976), 106–131.

[15] _____, *Measures invariant under a group of transformations*, Pacific J. Math. **68** (1977), 313–329.

[16] _____, *Foundations of statistical methods using a semantical definition of probability*, Mathematical Logic in Latin America. Proc. of the Fourth Latin American Symposium on Mathematical Logic (Santiago, Chile, 1978) (A. I. Arruda, R. Chuaqui, and N. C. A. da Costa, eds.), Volume 99 of Studies in Logic and the Foundations of Mathematics, North-Holland Pub. Co., Amsterdam, 1980, pp. 103–119.

[17] _____, *Models for probability*, Analysis, Geometry and Probability. Proc. of the First Chilean Symposium on Mathematics (Valparaíso, Chile, 1981) (R. Chuaqui, ed.), Volume 96 of Lecture Notes in Pure and Applied Mathematics, Marcel Dekker Inc., New York, 1985, pp. 89–120.

[18] _____, *Factual and cognitive probability*, Rev. Colombiana Mat. 19 (1985), 43–57, Proc. of the Fifth Latin American Logic Symposium on Mathematical Logic (X. Caicedo, N. C. A. da Costa, and R. Chuaqui, ed.), (Bogotá, Colombia), 1981.

[19] _____, *How to decide between different methods of statistical inference*, Mathematical Logic and Formal Systems (L. P. de Alcantara, ed.), Volume 94 of Lecture Notes in Pure and Applied Mathematics, Marcel Dekker Inc., New York, 1985, Book in honor of Newton C. A. da Costa, pp. 43–56.

[20] _____, *Sets of relational systems as models for stochastic processes*, Methods and Applications of Mathematical Logic. Proc. of the Eighth Latin American Symposium on Mathematical Logic (Campinas, Brazil, 1985) (L. P. de Alcantara and W. Carnielli, eds.), vol. 69, Contemporary Mathematics, American Mathematical Society, 1988, pp. 117–148.

[21] _____, *Probabilistic models*, Logic Colloquium '88. Proc. of the European Logic Colloquium (Padova, Italy, 1988) (Ferro, Bonotto, Valentini, and Zanardo, eds.), North-Holland Pub. Co., Amsterdam, 1989, pp. 287–317.

[22] R. Chuaqui and N. Bertoglio, *Nonstandard theory of integration*, To appear in Proc. of the Third Chilean Symposium on Mathematics, Notas Matemáticas, Soc. Mat. de Chile.

[23] R. Chuaqui and L. Bertossi, *Approximation to truth and theory of errors*, Methods of Mathematical Logic. Proc. of the Sixth Latin American Symposium on Mathematical Logic (Caracas, Venezuela, 1983) (C. A. Di Prisco, ed.), Volume 1130 of Lecture Notes in Mathematics, Springer-Verlag, Berlin, Heidelberg, New York and Tokyo, 1985, pp. 13–31.

[24] R. Chuaqui and J. I. Malitz, *The geometry of legal principles*, To appear in Theory and Decision.

[25] R. Courant and F. John, *Introduction to calculus and analisis*, vol. II, John Wileyn & Sons, 1974.

[26] N. Cutland, *Nonstandard measure theory and its applications*, Bull. London Math. Soc. 15 (1983), 529–589.

[27] N. C. A. da Costa, *Pragmatic probability*, Erkenntniss 25 (1986), 141–162.

[28] Newton C. A. da Costa and R. Chuaqui, *On Suppes' set theoretical predicates*, Erkenntniss 29 (1988), 95–112.

[29] B. de Finetti, *Foresight: its logical basis, its subjective sources*, Studies in Subjective Probability (H. E. Kyburg, Jr. and H. Smokler, eds.), John Wiley & Sons, New York, London, Sydney, 1964, English translation of an article which appeared in Ann. Inst. H. Poincaré, Volume 7, 1937.

[30] _____, *Theory of probability*, vol. I and II, John Wiley & Sons, Chichester, 1974,1975.

[31] P. J. de Laplace, *A philosophical essay on probabilities*, Dover, New York, 1951, English translation of the French edition of 1820.

[32] H. E. Enderton, *A mathematical introduction to logic*, Academic Press, New York, 1972.

[33] T. L. Fine, *Theories of probability — an examination of foundations*, Academic Press, New York, 1973.

[34] R. A. Fisher, *Statistical methods and scientific inference*, Hafner Press, New York, third ed., 1973, First Edition in 1956.

[35] P. Fletcher, *Nonstandard set theory*, J. Symbolic Logic 54 (1989), 1000–1008.

[36] R. N. Giere, *Objective single-case probabilities and the foundations of statistics*, Proceedings of the Fourth International Congress for Logic, Methodology and Philosophy of Science (P. Suppes, L. Henkin, A. Joja, and G. Moisil, eds.), North-Holland Pub. Co., Amsterdam, 1973, pp. 467–483.

[37] D. A. Gillies, *An objective theory of probability*, Methuen & Co. Ltd., London, 1973.

[38] J. Gordon and E. H. Shortliffe, *The Dempster-Shafer theory of evidence*, Uncertain Reasoning (G. Shafer and J. Pearl, eds.), Morgan Kaufmann Pub. Inc., San Mateo, CA, 1990, pp. 529–539.

[39] I. Hacking, *The logic of statistical inference*, Cambridge University Press, 1965.
[40] _____, *The emergence of probability*, Cambridge University Press, Cambridge, 1975.
[41] C. Hempel, *Deductive nomological vs. statistical explanation*, Scientific Explanation, Space, and Time (Feigl and Maxwell, eds.), University of Minnesota Press, Minneapolis, 1962, pp. 150–151.
[42] D. N. Hoover and H. J. Keisler, *Adapted probability distributions*, Trans. Amer. Math. Soc. **286** (1984), 159–201.
[43] K. Hrbáček, *Axiomatic foundations for nonstandard analysis*, Fund. Math. **98** (1978), 1–19.
[44] _____, *Nonstandard set theory*, Amer. Math. Monthly **86** (1979), 659–677.
[45] A. E. Hurd and P. A. Loeb, *An introduction to nonstandard real analysis*, Academic Press Inc., New York, 1985.
[46] R. C. Jeffrey, *New foundations for bayesian decision theory*, Logic, Methodology and Philosophy of Science (Y. Bar-Hillel, ed.), Studies in logic and the foundations of mathematics, North-Holland Pub. Co, Amsterdam, 1965.
[47] _____, *The logic of decision*, The University of Chicago Press, Chicago, 1983, Second Edition.
[48] H. Jeffreys, *Theory of probability*, Oxford, Clarendon Press, Oxford, 1948, Second Edition.
[49] _____, *The present position in probability theory*, British J. Philos. Sci. V (1955), 275–289.
[50] T. Kawai, *Nonstandard analysis by axiomatic method*, Proceedings of Southeast Asian conference on logic (Singapore, 1981), Volume 111 of Studies in Logic and the Foundations of Mathematics. North-Holland, Amsterdam, 1983, pp. 55–76.
[51] H. J. Keisler, *Elementary calculus*, Prindle, Weber and Schmidt, Boston, 1976.
[52] _____, *Foundations of infinitesimal analysis*, Prindle, Weber and Schmidt, Boston, 1976.
[53] _____, *An infinitesimal approach to stochastic analysis*, Mem. Amer. Math. Soc. **48** (1984), 1–184.
[54] _____, *Probability quantifiers*, Model Theoretic Logics (J. Barwise and S. Feferman, eds.), Springer-Verlag, New York, Berlin, Heidelberg, Tokyo, 1985, pp. 510–556.
[55] J. M. Keynes, *A treatise on probability*, Macmillan, 1957.
[56] A. N. Kolmogorov, *Foundations of the theory of probability*, Chelsea, New York, 1956, Translation of Grudbegriffe der Wahrscheinlichkeitrechnung, 1933.
[57] C. H. Kraft, J. W. Pratt, and A. Seidenberg, *Intuitive probability on finite sets*, Ann. Statist. **30** (1959), 408–419.
[58] H. Kyburg, Jr., *Probability, rationality and a rule of detachment*, Logic, Methodology, and Philosophy of Science (Y. Bar-Hillel, ed.), Studies in Logic and the Foundations of Mathematics, North-Holland Pub. Co., Amsterdam, 1965, pp. 301–310.
[59] _____, *Logical foundations of statistical inference*, Reidel, Dordrecht, 1974.
[60] _____, *Conditionalization*, J. Philos. **77** (1980), 98–114.
[61] _____, *Principle investigation*, J. Philos. **78** (1980), 772–778.
[62] D. Lewis, *A subjectivist's guide to objective chance*, Studies in Inductive Logic and Probability, Volume II (R. C. Jeffrey, ed.), University of California Press, Berkeley and Los Angeles, 1980, pp. 263–293.
[63] J. Lucas, *The concept of probability*, Oxford University Press, Oxford, 1970.
[64] L. B. Lusted, *Introduction to medical decision making*, Charles C. Thomas, Springfield, Il, 1968.
[65] P. Martin-Löf, *The literature on von Mises' Kollectives revisited*, Theoria **35** (1969), 12–37.
[66] B. Mates, *Elementary logic*, Oxford University Press, New York, 1972.
[67] E. J. McShane, *Integration*, Princeton U. Press, Princeton, N. J., 1947.
[68] I. Mikenberg, N. C. A. da Costa, and R. Chuaqui, *Pragmatic truth and approximation to truth*, J. Symbolic Logic **51** (1986), 201–221.
[69] D. Miller, *A paradox of information*, British J. Philos. Sci. **17** (1966), 59–61.

[70] E. Nelson, *Internal set theory: A new approach to nonstandard analysis*, Bull. Amer. Math. Soc. **83** (1977), 1165–1198.
[71] _____, *Radically elementary probability theory*, Annals of Mathematics Studies, Princeton U. Press, Princeton, N. J., 1987.
[72] J. Neyman, *Lectures and conferences on mathematical statistics and probability*, Graduate School of the U.S. Department of Agriculture, Washington, DC, 1952, Second edition, revised and enlarged.
[73] J. Neyman and E. S. Pearson, *On the use and interpretation of certain test criteria for purposes of statistical inference*, Joint Statistical Papers of J. Neyman and E. S. Pearson, Cambridge University Press, Cambridge, 1967, Originally published in 1928, pp. 1–98.
[74] K. Popper, *The logic of scientific discovery*, Hutchinson & Co., London, 1959, English translation of the German original of 1934.
[75] _____, *The propensity interpretation of probability*, British J. Philos. Sci. **10** (1959), 25–42.
[76] _____, *Realism and the aim of science*, Rowman and Littlefield, Totowa, NJ, 1983.
[77] F. P. Ramsey, *Truth and probability*, The Foundations of Mathematics and other Logical Essays (R. B. Braithwaite, ed.), Routledge & Kegan Paul Ltd., London, 1950, Also in *Studies in Subjective Probability*, Kyburg and Smokler, editors, John Wiley & Sons, 1964, pp. 61–92, pp. 156–198.
[78] A. Robinson, *Non-standard analysis*, Studies in Logic and the Foundations of Mathematics, North-Holland Pub. Co., Amsterdam, 1966.
[79] L. J. Savage, *Foundations of statistics*, John Wiley & Sons, New York, 1954.
[80] S. Scheffler, *The rejection of consequentialism*, Oxford University Press, Oxford, 1982.
[81] S. Scheffler, ed., *Consequentialism and its critics*, Oxford University Press, Oxford, 1988.
[82] D. Scott and P. Krauss, *Assigning probabilities to logical formulas*, Aspects of Inductive Logic (P. Suppes and J. Hintikka, eds.), Studies in logic and the foundations of mathematics, North-Holland Pub. Co., Amsterdam, 1966, pp. 219–264.
[83] T. Seidenfeld, *Philosophical problems of statistical inference. Learning from R. A. Fisher*, Reidel, Dordrecht, 1979.
[84] E. H. Shortliffe and B. G. Buchanan, *A model of inexact reasoning in medicine*, Uncertain Reasoning (G. Shafer and J. Pearl, eds.), Morgan Kaufmann Pub. Inc., San Mateo, CA, 1990, pp. 259–273.
[85] R. Sikorski, *Boolean algebras*, Springer-Verlag, Berlin, 1960.
[86] J. J. C. Smart and B. Williams, *Utilitarianism, for and against*, Cambridge University Press, Cambridge, 1973.
[87] P. Suppes, *The probabilistic argument for a nonclassical logic of Quantum Mechanics*, Philos. Sci. **33** (1966), 14–21.
[88] _____, *Set-theoretical structures in science*, Stanford University, 1967.
[89] _____, *A probabilisitic theory of causality*, North-Holland Pub. Co., Amsterdam, 1970, Acta Philosophica Fennica, Fasc. XXIV.
[90] _____, *Propensity representations of probability*, Erkenntnis **26** (1987), 335–358.
[91] P. Szolovits and S. G. Pauker, *Categorical and probabilistic reasoning in medical diagnosis*, Uncertain Reasoning (G. Shafer and J. Pearl, eds.), Morgan Kaufmann Pub. Inc., San Mateo, CA, 1990, pp. 282–297.
[92] A. Tarski, *Der Wahrheitsbegriff in den formaliesierten Sprachen*, Studia Philosophica **1** (1935), 261–405, English translation in *Logic, Semantics and Metamathematics*; 2nd ed. (J. Corcoran, editor), Hackett Publishing, Indiana, 1983. 1st ed. (ed. and translated by J. H. Woodger), Oxford, 1956.
[93] _____, *Cardinal algebras*, Oxford U. Press, New York, 1949.
[94] _____, *Truth and proof*, Sci. Amer. **220** (1969), 63–77.
[95] A. E. Taylor, *General theory of functions and integration*, Blaisdell Pub. Co., Waltham, 1965.
[96] M. van Lambalgen, *Random sequences*, Ph.D. thesis, University of Amsterdam, 1987.
[97] J. Venn, *The logic of chance*, Macmillan, London, 1962, Reprint of the 1888 edition.

[98] R. von Mises, *Probability, statistics, and truth*, Hilda Geiringer, London and New York, 1957, English translation of *Wahrscheinlichkeit, Statistik, und Wahrheit*, J. Springer, Berlin, 1928.
[99] S. Wagon, *The Banach-Tarski paradox*, Encyclopedia of Mathematics, Cambridge U. Press, Cambridge, 1985.

Index

The page (or pages) of the main occurrence of an item appears usually in italics. The symbols are listed at the beginning of the index in the order of their appearance.

∪, 1
∩, 1
⋃, 1
⋂, 1
⊆, 1
⊂, 1
$A - B$, 1
A^c, 1
$\mathcal{P}$, 1
$\mathbb{N}$, 1
$\mathbb{R}$, 1
$A \times B$, 2
$^B A$, 2
$^n A$, 2
$\langle \tau(b) \mid b \in B \rangle$, 2
$f''A$, 2
$f \circ g$, 2
$f^{-1''}A$, 2
↾, 3
$\prod_{i \in I} \tau(i)$, 3
⟹, implies, 4, 444
⟺, if and only if, 4, 444
⊨, 79
¬, negation, 81, 367, 444
∧, conjunction, 81, 367
∨, disjunction, 81, 367
∧, conjunction, 82, 444
∨, disjunction, 82, 444
Fr, 82
Mod, 82, 368, 372
pr, distribution of probabilities, 130
Pr, probability measure, 130
$[\varphi(X)]$, 135
Λ_X, 135
pr_X, 135
Pr_X, 135
$\text{at}(\mathcal{V})$, 142
Λ_ξ, 155
pr_ξ, 155
Pr_ξ, 155
$\sim_G$, 164
$\simeq_G$, 164
$R^\sharp$, 165
$^*\mathbb{N}$, 173
$^*\mathbb{R}$, 173
$x \approx \infty$, 173
$x \approx -\infty$, 173
$x \ll \infty$, 173
$-\infty \ll x$, 173
$x \approx y$, 173
$x \lesssim y$, 173
$x \approx y \ (\varepsilon)$, 174
st, 175
°, 175
$\mathbb{C}$, 175
$^*\mathbb{C}$, 175
$\|\bar{x}\|$, 175
$V_n(X)$, 176
$df(x, dx)$, 189
$\partial_x z$, 190
$F_x(x, y)$, 191
$dF(x, y)$, 191
$DF(x, y)$, 191
$x \asymp y$, 198
$R^\#$, 206
$N(F)$, 207
$I(F)$, 207
$\overline{\sum\sum}_{R^\#, D} F$, 208
$\underline{\sum\sum}_{R^\#, D} F$, 208
$^*\mathcal{P}\Omega$, 227
$^*(^\Omega \mathbb{R})$, 227
$|\mathbf{K}|$, 282
$\mathbf{K}_0$, 282
$\mathbf{K}_1$, 283
$\mathbf{K}_2$, 283
A^f, 284
$G_\mathbf{K}$, 284, 310
$\mathbf{K}_3$, 285
$\sim_\mathbf{K}$, 285
$\langle \mathbf{K}, \sim_\mathbf{K} \rangle$, 286
$\sim$-closed, 286
$\mathcal{F}_b \mathbf{K}$, 287
$\mathcal{F}_\mathbf{K}$, 287
$\langle \mathbf{K}, \mathcal{F}_\mathbf{K}, \text{Pr}_\mathbf{K} \rangle$, 287
R^{-1}, 294
$R''B$, 294
$R''\{t\}$, 294
$(-\infty, t)$, 294
$(-\infty, t]$, 294
$\langle T, \preceq \rangle$, 295
R^∞, 295
$R^{\infty=}$, 295
T_n, 296
$|\lambda|$, 300

INDEX

$\hat{\lambda}$, 300
$\lambda \upharpoonright S$, 300
$g''\lambda$, 301
$|\mathbf{K}|$, 303
$\mathbf{A}_{\lambda_t}$, 304
$\mathbf{A} \upharpoonright S$, 304
$\mathbf{A}(t)$, 304
$\mathbf{A}_t$, 304
$\mathbf{A}_{t]}$, 304
$\mathbf{A}_n$, 304
$\mu \frown \omega$, 304
I_n, 306
$\mathbf{IC}$, 306
$\mathbf{DC}$, 307
L_2, 307
$\mathbf{UB.}$, 307
$\mathbf{D}$, 307
$\mathrm{P}_\mathbf{K}$, 309
$\bar{g}$, 309
g_λ, 309
$\mathbf{A}^g$, 309
$\mathbf{A} \sim_\mathbf{K} \mathbf{B}$, 310
$\mathbf{A}^\circ_n$, 314
$\mathbf{A}^\circ$, 314
λ_t, 320
$\lambda_{t]}$, 320
λ_n, 320
$\mathbf{K} \upharpoonright S$, 321
$\mathbf{K}_t$, 321
$\mathbf{K}_{t]}$, 321
$\mathbf{K}_n$, 321
$\mathbf{K}_\lambda$, 321
$\lambda \frown \omega$, 321
$\mathbf{A}^\circ_n$, 321
$\mathbf{A}^\circ$, 321
$\langle \mathbf{K}, \preceq_T \rangle$, 321
Ho_2, 321
f_t, 323
$f_{t]}$, 323
f_{λ_t}, 323
f_m, 323
$\langle \mathbf{K}, \preceq_T, G \rangle$, 326
$\sim_{t]}$, 327
$\sim_t$, 327
$\sim_{\lambda_t}$, 327
Ho_3, 328
$\mathcal{A} \upharpoonright S$, 330
$(\mathrm{Pr} \upharpoonright S)$, 330
Pr_t, 330
$\mathrm{Pr}_{t]}$, 330
Pr_m, 330
Pr_{λ_t}, 330
$\widetilde{\mathcal{A}}_\mathbf{K}$, 339
$\widetilde{\mathrm{Pr}}_\mathbf{K}$, 339

$\langle \mathbf{K}, \mathcal{A}_\mathbf{K}, \widetilde{\mathrm{Pr}}_\mathbf{K} \rangle$, 339
$\widetilde{\mathrm{Pr}}_{\mathbf{KH}}(\mathbf{A})$, 339
$\widetilde{\mathrm{Pr}}_{f(\lambda_t)}$, 339
$\mathcal{F}_\mathbf{H}$, 340
$\mathrm{Pr}_\mathbf{H}$, 341
$\mathcal{F}_{b\mathbf{H},t}$, 346
$\mathcal{F}_{b\mathbf{H},m}$, 346
$\mathcal{F}_{b\mathbf{H},t]}$, 346
$\mathcal{F}_{b\mathbf{H}}$, 347
$\mathbf{MB}$, 353
$\mathbf{BE}$, 355
$\mathbf{FD}$, 355
$\mathbf{H}_2$, 356
$\mathbf{H}_3$, 357
$\mathbf{Er}$, 358
$\mathbf{BM}$, 359
$[\tau = r]$, 367
$[\tau \in S]$, 368
$\overline{Y}_n$, 388
$\mathbf{Ev}$, 389
$\mathbf{U}$, 434
$\mathbf{S}$, 444
$\forall$, 444
$\exists$, 444
Cof, 455
$X \vee Y$, 462
$X \wedge Y$, 462
X^+, 462
X^-, 462
$\mathcal{L}$, 463
$^\circ \mathrm{Pr}$, 468
$^{\circ\circ} \mathrm{Pr}$, 468

acceptance
   of a chance setup, 86
   of diseases, 123
   of hypotheses, 92
   dialectical process, 380
   of a proposition, 41
      general contingent, 42
   provisional, 379
   rules, 95, 96
      dialectical rule, 106
      ideal rule, 102
a.e., see almost everywhere
algebras
   of events, 68, 131, *146*
      atoms, *146*
      and disjunctive algebras, 160
   disjunctive algebra of events, 68, 74, *159*
   duality between algebras of random variables and of events, 147
   of random variables, *141*

atoms, *142*
    trivial, 142
    generated algebra, *142*
almost everywhere, 254
almost surely, 254
    eventually in a set, 100, 387
Anderson, R. M., 359
Aristotle, 78
a.s., see almost surely
asymptotically near, 198
at least as bad result, 96, 97, *100*, 393, 395
    for diseases, 123
    for significance tests, 390
average measures, 14, 76, *153*, 331, 333, 341, 344
    in disjunctive spaces, *162*
automorphisms, 70
    of chance structures, *310*
    of functional structures, 322
    of simple chance structures, *284*

Barnard, G., 422
Bayes, T., 23
Bayes formula, 7, 88, 120, 134, 364
    in logical languages, 375
Bernoulli, J., 50, 59, 434
Bernoulli random variables, 11, 138, 232
Bertossi, L., 357
Bertrand, 60
Bertrand's mixture paradox, 61, 241
Bertrand's random chord paradox, 62, 116
    degree of support, 374
    solution, 356
beta family of distributions, 247
beta function, 247
binomial random variables, 11, 139, 232
Birnbaum, A., 85, 407, 422, 423
Borel, E., 23
Böse-Einstein statistics, 354
Brownian motion, 359, 468
    standard, 361
    Wiener walks as approximations, 274
Bunge, M., 44, 45, 48
Butler, Bishop, 19

Carnap, R., 33, 35, 36, 37, 42, 43, 44, 48, 63, 64, 89
    Carnap's principle of symmetry, 63, 65
Cartesian product, 2
    generalized, 3
causal relations, see relation, causal
causality, 118
central limit theorem, 274
    de Moivre-Laplace theorem, 246

chance, 19, 34, 45, 48, 49
chance setups, 49, 51, 57
    acceptance of, 86
chance structures, *303*
    automorphisms, 310
    basic blocks, 304
    causal relation, 303
    causal universe, 303
    classical, *312*, 315
        existence, 345
    group of invarariance, 310
    object universe, 303
    isomorphism, 309, 310
    simple, *282*
        automorphisms, 284
        family of basic events, 287
        family of events, 287
        isomorphisms, 284
        measure-determined events, 286
        measurement system of functions, 289
        universe, 282
    strict, *303*
change of variables theorem, 221
Chebyshev's inequality, 12, 141
chi square distributions, 247
classical definition of probability, 23, 343
    representation theorem, 351
coherence, 107
    of the two principles, 109
coin tossing, 83, 101, 103, 111, 293, 385
    unusual experiment, 94
collective, 26
complement, 1
composition of functions, 2
concatenation of functions, 304
conditionalization, 31, 42, 89, 108, 109
conditionality principle, 422
    couterexample, 424
confidence intervals, 407
    for binomial parameter, 407
    dialectical rule for, 409
    for nearly normal distributions, 410
confidence principle, 407
    contradicts the likelihood principle, 422
confidence region estimation, 383, 406, 420
    randomized, 420
conjunction, 81, 367, 444
consequentialism, 431, 432
    local consequentialism, 432
consistency, weak and strong, 404
constant random variables, 11, 138
continuous functions, 81
continuity
    *-continuity, 188

INDEX 477

near continuity, 188
S-continuity, 188
uniform continuity, 188
convergence
    of functions
        *-convergence at infinity, 187
        near convergence at infinity, 187
        S-convergence at infinity, 187
        *-convergence at a point, 187
        near convergence at a point, 187
        S-convergence at a point, 187
    monotone convergence theorem, 465
    of random variables
        near convergence, 254
        near convergence a.s., 254
        near convergence in probability, 255
        S-convergence in probability, 255
    of sequences
        *-convergence of sequences, 185
        near convergence of sequences, 184
        S-convergence of sequences, 184
    of series
        *-convergence, 185
        near convergence, 185
        S-convergence, 185
correlation coefficient, 10, *137*
covariance, 10, *137*
critical region, 393, 414
    for hypotheses tests, 413
critical value, 392
curves, 212
    simple, 212
    closed, 212
    regular, 212

da Costa, N. C. A., 32, 36, 37, 42, 108, 109, 110
d.e., see discriminating experiments
decision theory, 19, 86, 431
    based on evidence, 435
    Bayesian, 431, 432
        Bayes decision rule, 437
        Bayes decision rule in medicine, 439
        in the law, 439, 440, 442
        in medicine, 438
    uniform distribution for decisions, 438
de Finetti, B., 31, 32, 35, 43, 108, 109, 435
degree
    of belief, 30, 34, 35, 86, 107
        assignment of, 107
    of confirmation, 42, 63
    of possibility of truth, 53
    of support, 34, 35, 37, 45, 49, 52, 54, 86

    as a logical notion, 365
    Bertrand's chord paradox, 374
    formal definition for simple languages, 369
    formal definition for compound languages, 373
de Laplace, P. J., 23, 24, 25, 60
de Moivre, 23
de Moivre-Laplace central limit theorem, 246
density function, 235
denumerable comprehension, 183, 451
dependence, 51
    and independence, 69
derivatives, 190
    partial, 191
    total, 191
difference of sets, 1
differentiable, 190
    for several variables, 192
differentials, 189
    partial, 190
    total, 191
discriminating experiments, 95, *101*, 112
    for diseases, 122
    for hypotheses tests, *395*, 408
    for point estimation, *404*
    for probability matrices, 388
    theorems, 388, 396, 404
direct inference, principle of, 34, 36, 45, 49, 53, 54, 86, 88, 89, 97, 107, 109, 119, 120
    for decisions, 431
    extensional form, 54
    widend form, 89
diseases, 117
    acceptance, 123
    at least as bad results, 123
    Bayesian model, 119
    categories, 118, 123
    classical model, 121
    discriminating experiment, 122
    as processes, 122
    provisional rejection, 123
    rejection set, 123, 124
disjunction, 81, 367, 444
distributions
    functions, cumulative, 9
    joint, 156
    of probabilities, 130
    probability distributions, *228*
        Bernoulli, 232
        beta family, 247
        binomial, 232

chi square, 247
   for decisions, 438
   discrete, 229
   F distributions, 249
   gamma family, 247
   nearly continuous, 235
   nearly normal, 243
   Poisson, 233, 417
   t distributions, 250
   uniform distributions, 240
   of random variables, 135
   sampling, 381
   of stochastic processss, 155
distributions of r balls into n cells, 353

efficiency, 406
Einstein, A., 94
enlargements, 443
equiprobability, 18, 23
equiprobability structures, *326*
   classical, 328
      probability space, 329
   homomorphisms, *335*
   measure-determined events, *329*
      weakly, 340
   probability space, 340
      provisional, *339*
   simple, *285*
      simple classical, *287*
   super classical, 328
equivalence relations, 3
   additive, 165
   closed under, 165
   refining, 164, *165*
   strictly positive, 164, *165*
   invariant measures under, *165*
errors
   theory of, 357
   type I, 413, 414
   type II, 413, 414
estimation of parameters, 382
   Bayesian, 429
estimator
   Bayes, 429
   moment, 404
   maximum likelihood, 405
expectation, 10, *136*
   as truth-average, 406
   conditional, *150*
   of discrete distributions, 231
   of nearly continuous distributions, 239
   near expectation, 463
extension principle, 452
   proof, 452

   proof of axioms from extension and transfer principles, 454
external
   induction, principle of, 179
   invariance, principle of, 52, 69, 73, 75, 340, 344
   objects, 178, 449
evaluation function, 156
events, 49, 59, 67, *130*
   algebra of, 131, *146*
   disjunctive algebra of, 68, 74, 159
   family of, 67, 74
      events for equiprobability structures, 340
      basic events for simple structures, 287
      events for simple structures, 287
   in hyperfinite spaces, 227
   impossible, 131
   mutually exclusive, 131
   single, 50
evidential equivalence, 96, 97, *100*, 123, 387, 394

F distributions, 249
Fermat, P., 18, 23, 59
Fermi-Dirac statistics, 355
filters, 455
   free, 455
filtration, 348
Fine, T. L., 24
finite intersection property, *455*
Fisher, R. A., 19, 23, 38, 85, 379, 380, 390, 417, 422
Fisher-Neyman theorem, 405
frequency
   as a discriminating experiment, 101, 103, 113, 391
   functions, 228
      domain sequence, 228
   long run frequencies, 19
   and probabilities, 110
   rules for acceptance as probabilities, 112, 113
   theories of probability, 25
functionals, 264
   continuous, 265
   finite, 264
functional structures, 321
   automorphisms, 322
   full, 321
   homomorphism, 322
   isomorphisms, 322
functions

continuity
    *-continuity, 188
    near continuity, 188
    S-continuity, 188
    uniform continuity, 188
continuous, 81
convergence
    near convergence at infinity, 187
    S-convergence at infinity, 187
    *-convergence at infinity, 187
    near convergence at a point, 187
    S-convergence at a point, 187
    *-convergence at a point, 187
image, 2
inverse image, 2
fundamental theorem of calculus, 200
    approximate, first form, 202
    approximate, second form, 203

games of chance, 18, 52
gamma family of distributions, 247
gamma function, 247
Giere, R. N., 29
Gillies, D. A., 413, 415, 417, 418, 427
    examples, 415
    rule of falsification, 427
        counterexample, 428
groups
    of invariance, 63, 64, 66, 67, 70
        of chance structures, 310
        of simple chance structures, 284
    invariance under, 63
    measures invariant under, 164
    of permutations, 164
    of roulette, 64
    of symmetries, 63, see of invariance
    of transformations, 63 see of invariance

Hacking, I., 19, 20, 29, 35, 37, 426
Hempel, C., 90
homomorphisms, 73, 75
    of equiprobability structuresn, 335
    of functional structures, 322
Hosiasson, J., 71, 73, 321, 328, 334
Hrbáček, K., 443
hyperfinite
    approximation of a rectangle, 206
    probability spaces, 227
        events, 227
        random variables, 227
    sets, 182
    sums, 197
hyperreals, 173

finite, 173
infinite, 173
hypotheses
    alternative, 94, 98, 106, 417
    composite, 380, 383
    estimation, 383
    null, 382
    simple, 380, 383
    tests, 382, 383, 394
        discriminating experiments, 395, 408
        justification, 394
        of nearly normal distributions, 396, 398, 399, 414
        randomized, 419
        working, 382

improper priors, 438
independence, 7
    and dependence, 51, 69
    events, 134
    independent increments condition, 267
    models for independent repetitions, 381
    near independence of random variables, 238
    random variables, 9, 136
indifference, principle of, 60, 438
inequalities
    Jensen's, 12, 140, 152
    Markov's, 12, 140, 152
    Chebyshev's, 12, 141
infinitely close, 173
infinite natural numbers, 173
infinitesimals, 173
    subset of $^{n*}\mathbb{R}$, 207
    vectors, 175
integrable, 203
    for several variables, 208
integral, 204
    Daniell, 461
    elementary, 462
    in higher dimensions, 209
    improper, 222
    near, 463
        lower, 463
        upper, 463
interior, 189, 213
internal
    formulas, 444
    induction, principle of, 179, 450
    invariance, principle of, 52, 59, 64, 66, 70, 73, 344
    objects, 178, 444
        properties, 449
    power set, 227

set theory, 443
intersection, 1
    finite intersection property, *455*
invariance, 51
    group of, 66, 67, 70
    under groups, 63
invariant measures, 53
    extendible, *169*
    for equiprobability structures, *329*
    ∼-invariant measures, 286
    under groups, *164*
inverse inference, 88
    principle of, 97, 109, 380, 382, 391, 407, 431
isomorphisms, 74
    of chance structures, *309*
    of functional structures, *322*
    of outcomes, *301*
    of relational systems, *281*
    of simple chance structures, *284*

Jacobian, 211
Jeffreys, H., 33, 71, 425
Jensen's inequality, 12, 140, 152
Jordan content, 208

Keynes, J. M., 33, 60
Keisler, H. J., 359
Kolmogorov, A., 24, 31, 111
    Kolmogorov's axioms, 131
Kraft, Pratt, and Seidenberg, theorem, 108
Kyburg, Jr., H., 19, 20, 33, 35, 37, 39, 40
    Kyburg's lottery paradox, 90

law
    decision theory in the, 439
    provisional acceptance in a court, 440
    utilities, 441, 442
law of large numbers, 12
    strong, 13, 102, *259*, 375
    weak, *258*
languages, formal, 54, 78
    Bayes formula, 375
    Brownian motion, 374
    compound, 370
        adequate probability structures, 372
        possible probability models, 372
        sentences, 371
        sets of values of terms, 371
        similarity type, 370
        terms, 371
        truth, 371
        value of terms, 371
    laws of large numbers, 375

    and probability models, 81
    random variables, 381
    simple, 365
        sentences, *367*
        similarity type, 366
        terms, *367*
        truth, *367*
        values of terms, *367*
    theory of errors, 374
Leibniz, G., 47, 50
Lewis, D., 32, 36, 37, 43, 108, 110
likelihood principle, 422
    contradicts the confidence principle, 422
Loeb space, 361
logical theories of probability, 33, 35
Łos' theorem, 456

Malitz, J. I., 439, 442
Mates, B., 44
Markov's inequality, 12, 140, 152
Maxwell-Boltzmann statistics, 353
mean, 10, *136*
    as truth-average, 406
measure-determined events, 67, 74
    for equiprobability structures, *329*, 345
    for simple structures, *286*
    weakly, *340*
measurement functions, 58, 59, 79, 289
measurement process, *347*
measurement system of functions, 289
medical diagnosis, 117
    actions in medical example:treatment, 438
    computer-aided, 117, 119, 124, 125
    consequences in medical example, 438
    decisions, 438
        Bayes decision rule, 439
    provisional rejection of diseases, 123
    rejection set for diseases, 123, 124
models, 51, 77, see also each type of model
    for decision, 76, 363
    for independent repetitions, 381
    for inference, 76, 361
    in logic, 77
    in the sciences, 77
monotone convergence theorem, 465
Mostowski's collapsing theorem, 457

natural numbers 1, 173
    properties of *$^*\mathbb{N}$, 179, 450
near intervals, 183
nearly normal distribtuions, 243
    tests 396, 398, 399, 414
    confidence intervals, 410

INDEX

near volume, 208
negation, 81, 367, 444
Nelson, E., 171, 199, 264, 443
Neyman, J., 28, 35, 85, 89, 111, 379, 380, 407, 413, 417, 418
Neyman-Pearson theory, 38, 85, 379, 390, 394, *413–415*, 417
    justification, 413
nonstandard
    analysis, 171
    model, 456
    notions, 175
    numbers, 173
null sets, 13
    external characterization, 467
    in a probability space, *254*
    null subset of $^{n*}\mathbb{R}$, 207

operation, 3
ordered set, 69
    causal orderings, *296*
        causal moments, 70
        of an outcome, 300
        strict, *296*
    linear, see total
    nonreflexive partial, 295
    partial, 4, 295
        finite partial orderings always well-founded, 295
    total ordering, 4, 295
outcomes, 49, 57
    compound, 70, *300*
        as functions, 70, 298, 300
        causal orderings of, 300
        causal universes of, 300
        isomorphisms, *301*, 302
        object universes of, 300
        similarity type of, 300
overflow, 179

parameter space, 381, 385
Pascal, B., 18, 23, 59
partition
    of an interval, 199
    of a rectangle, 206
pdf, 385
Pearson, E. S., 85
point estimation, 383, 403
    discriminating experiment, *404*
Poisson distributions, *233*, 417
Popper, K. R., 24, 25, 27, 28, 29, 35, 71, 92, 93, 111
Port-Royal Logic, 86
possibility, 41, 44

antecedently real epistemic, 46, 48, 92
conceptual, 44, 45
factual, see real
epistemic, 45
logical, 42, 44, 45
physical, 44, 45
real, 42, 44, 45, 46, 48, 49, 50, 92
technological, 44
possible worlds, 47, 48
    logically, 48
    really, 48
    set-theoretic, 48
power set, 1
    internal, 227
power of a test, 414
pragmatic probability, 32
prediction intervals, 420
preference relation, 435
principles of
    coherence of the two principles, 109
    conditionality, 422
        couterexample, 424
    confidence, 407
    direct inference, 34, 36, 45, 49, 53, 54, 86, 88, 89, 97, 107, 109, 119, 120
        extensional form, 54
        widend form, 89
    extension principle, 452
        proof, 452
    external induction, 179
    external invariance, 52, 69, 73, 75, 340, 344
    indifference, 60, 438
    internal induction, 179, 450
    internal invariance, 52, 59, 64, 66, 70, 73, 344
    inverse inference, 97, 109, 380, 382, 391, 407, 431
    likelihood, 422
    sufficiency, 402, 403, 405, 419, 423
    symmetry, 52, 60, 63
        Carnap's, 63, 65
        Chuaqui's, 64
    transfer, 179, 443
        statement, 452
        proof, 452
        proof of axioms from extension and transfer principles, 454
        proof of transfer in the model, 457
probability
    average, 14, 76, *153*, 331, 333, 341, 344
        in disjunctive spaces, *162*
    axioms, 4, *130*, 131
    classical definition, 23, 343

representation theorem, 351
cognitive, 19
conditional, 6, *133*
 as a random variable, 151
epistemic, 19, 41, 47, 49
equiprobability, 18, 23
factual, 19, 49, 50
frequency theories, 25
logical theories, 33, 35
in ordinary language, 17
invariant under equivalence relations, 165
pragmatic, 32
product, 14, 76, *154*, 331, 333, 341, 344
propensity theories, 29, 50
real, see factual
real valued, 468
semantic conception of probability, 48
subjectivistic theories, 30, 42, 89, 107, 108
total probability formula, 88, 133
probability models, 57, 77, 81
probability spaces, 70, 77, *130*
 disjunctive, 160
 of an equiprobability structure, *340*
  of a classical, *329*
  random variables in, 161
  extendible, 163
 fibration, 152
 hyperfinite, 227, see hyperfinite probability spaces
 Kolmogorov, *131*, 148
  weak, 132
 provisional probability space, 339
 of a random variable, 135
 of a simple chance structure, *287*
 weak, *130*
probability structures, 348
 simple, 289
  weak simple, 289
 weak, 348
product algebra of disjunctive algebras, *161*
product measures, 14, 76, *154*, 331, 333, 341, 344
propensity theories of probability, 29, 50
p-value, 387, 391

Ramsey, F. P., 31, 108, 109, 435
Ramsey-de Finetti, theorem, 31, 107, 108
randomness, 66, 284
random processes, *155* see stochastic processes
random variables, 8, 51, 58, *135*
 algebras, *141*
 Bernoulli, 11, 138
 binomial, 11, 139
 conditional probability as, 151
 constant, 11
 correlation coefficient, 10, *137*
 covariance, 10, *137*
 in disjunctive spaces, *161*
 distribution, 135
 equivalence, *136*
 expectation, 10, *136*
  conditional expectation, *150*
  near expectation, 463
 in hyperfinite spaces, 227
 independent, 9, *136*
 in a language, 79, 381
 interpretations, 381
 mean, 10, *136*
 near convergence, *254*
  near convergence a.s., *254*
  near convergence in probability, *255*
 probability space, 135
 range, *135*
 S-convergence in probability, *255*
 standard deviation, 10, *136*
 variance, 10, *136*
rectangle, 206
 closed, 206
 closure, 206
region, 223
 closed, 222
  bounded, 212
 volume, 215
Reinhardt, W. N., 94, 107
rejection of hypotheses, 92
 dialectical, 95, 380
  of H_0 agianst H_1, 395
 provisional, 95
 of diseases, 123
 rules, 92
  ideal rule, 103, 112
  dialectical rule, 103
rejection set, *100*, 105, 387, 395
 for diseases, 123, 124
relations, 3
 antisymmetric, 4
 asymmetric, 295
 automorphisms, 296
 causal, *296*
  causal moments, 296
 equivalence relations, 3
  equivalence class, 3
  refining, 164, *165*
  strictly positive, 164, *165*

INDEX 483

additive, *165*
field, 294
first element, 294
forwardly linear, *296*
image of a set, 294
immediate predecessor under, 294
immediate successor under, 294
incomparable elements, 294
inverse, 294
last element, 294
nonreflexive partial orderings, 295
partial ordering, 4, 295, see orderings
reflexive, 3
strict causal relations, *296*
symmetric, 3
total ordering, 4, 295
transitive, 3
transitive closure, 295
transitive reflexive closure, 295
well-founded, 295
    height, 296
    levels, 296
relational system, 57, 77, 79, *280*
    isomorphisms, 281
    similarity type, 280
relative approximate equality, 174
representation theorem, 349
restriction, 2, *300*
Robinson, A., 171, 181
Robinson's lemma, 181
roulette, circular, 58, 69, 82, 283, 290, 368
    group of symmetry, 64

sampling distributions, 381
sampling evidence, 381
saturation, countable, 446
    proof in the model, 458
Savage, L. J., 31, 35, 43, 108, 435
S-bounded
    sets, 206
    functions of several variables, 207
semantic conception of probability, 48
smooth functions, 194
sequences, 1
    bounded, 456
    *-convergence, 185
    near convergence, 184
    S-convergence, 184
series, 185
    comparison test, 186
    *-convergence, 185
    near convergence, 185
    S-convergence, 185
significance level, 392

significance tests, 390
    at least as bad results, 390
    composite, 393
similarity type
    appropriate for a probability structure, 370
    of compound languages, 370
    of a compound outcome, 300
    of a relational system, 280
    of simple languages, 366
size of a test, 414
standard deviation, 10 *136*
standard
    Brownian motion, 361
    notions, 175, 444
    numbers, 173
    objects, 176
        axioms, 445
        properties, 176, 177
    overflow, 180
    part, 174
    rectangle, 206
    vectors, 175
statistical inference, 19, 379, 382
    basis for, 89
    models for, 361
statistical models of experiments, 384, 385
statistics, 392
    sufficient, 402
step function, 207
Stirling's formula, 244
stochastic processes, *155*
    adapted to probability structures, *348*
    diseases as, 122
    distribution, *155*
    equivalence, *156*
    nearly equivalent, *264*
    probability space *155*
    trajectory, *155*
    very nearly equivalent, *264*
stochastic structure, 348
subjectivistic theories of probability, 30, 42, 89, 107, 108
sufficiency, 404
    sufficiency principle, 402, 403, 405, 419, 423
sufficient statistics, 402
Suppes, P., 32, 68, 70, 118
support of a function, 207
surfaces, 213
    regular, 213
    simple closed, 213
symptoms, 117

tail probabilities, 390

Tarski, A., 42, 44, 54, 78, 79, 367
test statistics, 392
tolerance limits, 420
total evidence, 120, 403
total probability formula, 88, 133
transfer principle, 179, 443
    statement, 452
    proof, 452
    proof of axioms from extension and transfer principles, 454
    proof of transfer in the model, 457
transformations of $^{n*}\mathbb{R}$ into itself, 210
truth, 41, 78, 82
    for compound languages, *371*
    degree of possibility of, 53
    mathematical definition of, 79
    for simple languages, *367*

ultrafilters, 455
ultrapowers, 455
unbiasedness, 406
UMP, 415
UMPU, 415
UMPUI, 426
uniformly most powerful tests, 415
union, 1
unlikely results, 93, 94, 96, 97
untenable system of hypotheses, 107

urn models, 69, 70, 71, 74, 75, 76, 81, 84, 87, 119, 293, 307, 312, 363, 437
utilities, 20, 31, 86, 121
    of consequences, 434
    in law, 441, 442
    preference relation as a basis, 435
    success in the long run as justification for utility rule, 434
    expected
        of actions, 434
        as average goodness, 434
        based on evidence, 437
        maximum expected, 86, 433
        of a penalty, 442

variance, 10, *136*
    of discrete distributions, 231
    of nearly continuous distributions, 239
vector lattice of functions, 462
Venn, J., 25
von Mises, R., 23, 24, 26, 28, 35, 111
Wiener walks, 262
    approximations to Brownian motion, 274
    T-Wiener walks, 266
        near equivalence to, theorem, 268
worse result, 96, see at least as bad results